Computer Controlled
Urban Transportation

Wiley IIASA International Series on Applied Systems Analysis

1. CONFLICTING OBJECTIVES IN DECISIONS
   Edited by David E. Bell, *University of Cambridge,*
   Ralph L. Keeney, *Woodward-Clyde Consultants, San Francisco,*
   and Howard Raiffa, *Harvard University.*

2. MATERIAL ACCOUNTABILITY
   Rudolf Avenhaus, *Nuclear Research Center Karlsruhe,*
   and University of Mannheim.

3. ADAPTIVE ENVIRONMENTAL ASSESSMENT AND MANAGEMENT
   Edited by C. S. Holling, *University of British Colombia.*

4. ORGANIZATION FOR FORECASTING AND PLANNING: EXPERIENCE IN THE SOVIET UNION AND THE UNITED STATES
   Edited by W. R. Dill, *New York University,*
   and G. Kh. Popov, *Moscow State University.*

5. MANAGEMENT OF ENERGY/ENVIRONMENT SYSTEMS
   Edited by Wesley K. Foell,
   Energy Research Center,
   University of Wisconsin–Madison.

6. SYSTEMS ANALYSIS BY MULTILEVEL METHODS
   Yvo M. I. Dirickx, *University of Louvain,*
   and L. Peter Jennergren,
   Odense University.

7. CONNECTIVITY, COMPLEXITY, AND CATASTROPHE IN LARGE-SCALE SYSTEMS
   John Casti, *New York University.*

8. PITFALLS OF ANALYSIS
   Edited by Giandomenico Majone *and* E. S. Quade.
   International Institute for Applied Systems Analysis.

9. CONTROL AND COORDINATION IN HIERARCHICAL SYSTEMS
   W. Findeisen, *Institute of Automatic Control,*
   Technical University, Warsaw,
   F. N. Bailey, *University of Minnesota,*
   M. Brdyś, K. Malinowski, P. Tatjewski,
   and A. Woźniak, all *Institute of Automatic*
   Control, Technical University, Warsaw.

10. COMPUTER CONTROLLED URBAN TRANSPORTATION
    Horst Strobel, *Institute for Transportation*
    and Communication, 'Friedrich List', Dresden

10 International Series on
Applied Systems Analysis

Computer Controlled Urban Transportation

Horst Strobel
*Institute for Transportation and Communication
'Friedrich List', Dresden, GDR*

A Wiley–Interscience Publication
International Institute for Applied Systems Analysis

JOHN WILEY & SONS
Chichester–New York–Brisbane–Toronto–Singapore

Copyright © 1982 International Institute for Applied Systems Analysis

All rights reserved.

No part of this book may be reproduced by any means, nor transmitted, nor translated into a machine language without the written permission of the publisher.

British Library Cataloguing in Publication Data

Strobel, Horst
   Computer controlled urban transportation.
   —(International series on applied systems
   analysis; 10)
   1. Urban transportation 2. Automatic control
   —Data processing
   I. Title II. International Institute for
   Applied Systems Analysis III. Series
   388.4'028   HE193

ISBN 0 471 10036 6

Illustrations prepared by Oxford Illustrators Ltd., Oxford, England
Typeset by Activity, Salisbury, Wilts.,
and printed at The Pitman Press, Bath, Avon

Preface

This book addresses a question of interest to many cities in both developed and developing countries in the East and the West:

What benefits can a city realistically expect to receive from solving its present and future traffic problems by technological innovations, and especially by implementing large-scale computerized traffic and transportation control systems?

This question consists of two major parts:

(1) What basic concepts and methods for control and automation have been proposed, developed, and implemented?
(2) What experience with these concepts and methods has so far been gained in real applications in different cities and nations?

This monograph is the first publication that tries to present a consistent survey of both parts, i.e., of concepts and methods as well as of international experiences, considering all modes of urban transportation in a unified manner.

Part One analyzes the role of automation and computer control within the framework of general urban and transportation development policies, with special attention to the differences and similarities between the control problems occurring in the individual transport modes.

Parts Two to Four present detailed analyses of automobile traffic control (Part Two), control and monitoring of public transport systems (Part Three), and new modes of urban transportation (automated guideway transit and the dual-mode concept) (Part Four). Each part consists of three categories of chapters: (1) methodology (concepts and methods), (2) international experiences, summarized in specially prepared case descriptions, and (3) a summary of major findings.

A reader who is mainly interested in methodological aspects may simply skip the case descriptions (printed in smaller type, in order to limit the size of the book). On the other hand, if international experiences are of interest, he may read the corresponding case descriptions independently of the remaining parts of the book.

This structure and the chosen style of writing fit with the interdisciplinary character of the subject and will be of interest for a broad audience, including

- policy makers, managers, planners, and technical advisers dealing with urban and transportation planning problems (cf. Part One and chapters on basic concepts, international experiences, and findings and summary), as well as
- scientists, engineers, and students of disciplines like transportation planning and engineering, control theory, and engineering and computer science (with special reference to the methodological oriented chapters where the remaining part of the book could serve as useful background information on the practical relevance of methodology)

Acknowledgments

The methodological part of this book has its main roots in research work carried out in the Scientific Department for Automation at the Hochschule für Verkehrswesen "Friedrich List" in Dresden. The integration of the methodology and international experiences, however, was only made possible by cooperation between the International Institute for Applied Systems Analysis (IIASA) and the various National Member Organizations (NMOs) of IIASA, which supported the work by welcoming visits to study their own systems and by recommending authors for the case descriptions. Therefore special thanks are extended to the authors of the case descriptions for their constructive cooperation as well as to the IIASA NMOs. The work that led to the book was encouraged from the very beginning by Academician Professor Helmut Koriolek, an honorary IIASA scholar and the then chairman of the GDR NMO, as well as by its present chairman Professor Dr. Karl Bichtler.

At IIASA, Harry Swain, the then leader of the former Urban Project, and Professor Howard Raiffa, the first director of IIASA, expressed interest in the project and provided the necessary support to make international cooperation possible. Roger Levien, director of IIASA until November 1981, recommended that the book be included in the Wiley IIASA *International Series on Applied Systems Analysis* and gave much encouragement in the subsequent stages of the work.

Moreover, the assistance given by IIASA's staff is appreciated very much. In this respect I am especially grateful for the very active assistance I received from Olivia Carydias.

Paul Makin, managing editor, handled editing, design, and production and supervised all the final details necessary to bring the book to its present form. His efforts and contributions are greatly appreciated.

Finally, I am grateful to my wife Margot and to my family for the understanding they gave me during the preparation of the manuscript.

Authors of Case Descriptions

		Section
FRG	Dr. K. Becker, Abteilungsleiter DEMAG Foerdertechnik, Wetter (Ruhr)	19.2
	O. Böhringer, Hamburger Hochbahn AG, Hamburg	12.2
	K.-H. Suwe, Technische Bundesbahnamtsrat, Bundesbahnzentralamt, München	13.3
	Dr. H. Tappert, Sprecher des Vorstandes der Hamburger Hochbahn AG, Hamburg	12.2
France:	M. Gien, l'Ingenieur des TPE, Centre d'Études Techniques de l'Équipement, Ministère de l'Équipement, Bordeaux	7.3.2
	M. Pouliquen, l'Ingenieur des TPE, Centre d'Études Techniques de l'Équipement, Ministère de l'Équipement, Bordeaux	7.3.2
	E. Sacuto, SRE d'ile-de-France, Unité Exploitation Opérationelle Quest, Boulogne	7.3.1
Ireland:	Dr. B. J. Fitzgerald, Manager, Dublin City Services, Dublin, Republic of Ireland	12.1
Japan:	Dr. Toshiharu Hasegawa, Professor, Department of Applied Mathematics and Physics, Kyoto University, Kyoto	6.3.2
	Dr. Masakazu Iguchi, Professor, Faculty of Engineering, The Univesity of Tokyo, Tokyo	19.1
	Dr. Hiroshi Inose, Professor, Department of Electronic Engineering, Faculty of Engineering, The University of Tokyo, Tokyo	6.3.1
	Dr. Takemochi Ishii, Professor, Department of Mechanical Engineering, The University of Tokyo, Tokyo	19.1
	Dr. Masaki Kochi, Professor, Institute of Industrial Sciences, Tne University of Tokyo, Tokyo	19.1
	Yoshimbumi Miyano, Head, Traffic Facilities Section	6.3.1

	Traffic Division, Tokyo Metropolitan Police Department, Tokyo	
	Hiroyuki Okamoto, Technical Councillor, Traffic Bureau, National Police Agency, Tokyo	6.3.1
Kenya:	L. W. Situma, Traffic Engineer, Nairobi City Council, Nairobi	6.5
USSR:	Dr. V. J. Astrakhan, Department of Automatic and Remote Control, Moscow Institute of Railway Engineers (MIIT), Moscow	13.1
	Dr. L. A. Baranov, Professor, Department of Automatic and Remote Control, Moscow Institute of Railway Engineers (MIIT), Moscow	13.1
	Dr. B. G. Khorovich, Mosgortransproject, Moscow	6.4
	Dr. M. P. Pechersky, Mosgortransproject, Moscow	6.4
	Dr. A. V. Schileiko, Professor, Department of Electronics, Moscow Institute of Railway Engineers (MIIT), Moscow	13.1
UK:	Dr. R. E. Allsop, Professor, Transport Studies Group, University College, London	6.2
	Dr. I. A. Ferguson, Transport Operations Research Group, Department of Civil Engineering, University of Newcastle-upon-Tyne, Newcastle-upon-Tyne	6.2
USA:	Steven A. Barsony, Director, Office of Automated Guideway Transit Applications, Urban Mass Transportation Administration, Department of Transportation, Washington, DC	18.2
	Dr. Dennis M. Elliott, Director of Engineering, Dallas–Fort Worth Airport, Texas	18.1
	Krishna V. Hari, Director of Systems Engineering, Bay Area Rapid Transit District, Oakland, California	13.2
	William R. McCasland, Research Engineer, Freeway Surveillance and Control Department, Texas Transportation Institute, Texas A & M University, Houston, Texas	7.1
	Dr. J. L. Schlaefli, General Manager, Applied Transportation Systems, Inc., Gulf & Western Industries Ltd., Palo Alto, California	6.1
	Dr. Nigel H. M. Wilson, Professor, Department of Civil Engineering, Massachusetts Institute of Technology, Cambridge, Massachusetts	11.1
	Dr. Eldon Ziegler, Program Manager, Bus and Paratransit Division, Urban Mass Transportation Administration, Department of Transportation, Washington, DC	11.1

Contents

Introduction 1

PART ONE TRANSPORTATION, AUTOMATION, AND URBAN DEVELOPMENT: REVIEW OF THE BASIC PROBLEMS AND CONCEPTS

1 URBAN TRAFFIC PROBLEMS 7
  1.1 Mobility 7
  1.2 Traffic Safety 9
  1.3 Environment 12
  1.4 Resources 14
  1.5 Effectiveness and Attractiveness of Public Transit 16

2 TRANSPORT AND URBAN DEVELOPMENT: GENERAL CONCEPTS 19
  2.1 Objectives 19
  2.2 Options 22
  2.3 Constraints 24

3 NEW TECHNOLOGY OPTIONS: TOWARDS COMPUTERIZED TRANSPORT CONTROL 27
  3.1 Short-Term Strategies: Operational Innovations 27
  3.2 Long-Term Strategies: Total Systems Innovations 29
  3.3 Basic Transport Control Concepts 32
  3.4 Feasibility Considerations: Impacts of the Computer Revolution 39

PART TWO   AUTOMOBILE TRAFFIC CONTROL

4   BASIC SYSTEMS CONCEPTS — 53
  4.1   Traffic Signal Control Systems — 56
  4.2   Comprehensive Automobile Control Systems — 60
  4.3   The Automated Highway Concept — 63
  4.4   Concluding Remarks — 63

5   CONCEPTS AND METHODS OF CONTROL — 65
  5.1   The Control Task Hierarchy — 65
  5.2   Route Guidance — 66
  5.3   Urban Street Traffic Flow Control — 78
  5.4   Freeway Traffic Flow Control — 97
  5.5   Vehicle Control — 111

6   INTERNATIONAL AREA TRAFFIC CONTROL SYSTEMS EXPERIENCES — 127
  6.1   USA Experiences: San Jose and UTCS — 127
  6.2   British Experiences: Glasgow and London — 138
  6.3   Japanese Experiences: Tokyo and Osaka — 148
  6.4   USSR Experiences: Moscow and Alma-Ata — 170
  6.5   Experiences of a Developing Country: Nairobi, Kenya — 176

7   INTERNATIONAL FREEWAY TRAFFIC CONTROL SYSTEMS EXPERIENCES — 182
  7.1   The Dallas Freeway Traffic Corridor Control Project — 183
  7.2   The Hanshin Expressway Traffic Control System — 196
  7.3   French Experiences: The Paris Freeway Corridors and the Project *Opération Atlantique* — 205

8   FINDINGS AND SUMMARY — 221
  8.1   Route Guidance Systems — 222
  8.2   Area Traffic Control Systems — 223
  8.3   Freeway Traffic Control Systems — 226
  8.4   Vehicle Control — 229

PART THREE   CONTROL AND MONITORING OF PUBLIC TRANSPORT SYSTEMS

9   BASIC SYSTEMS CONCEPTS   233
   9.1   Train Operation and Traffic Control Systems   235
   9.2   Bus/Tram Operation and Traffic Control Systems   236
   9.3   Passenger Guidance and Information Systems   237
   9.4   The Dial-A-Ride Concept and Computerized Para-Transit Systems   237

10   CONCEPTS AND METHODS OF CONTROL AND SURVEILLANCE   242
   10.1   Route Guidance in Dial-A-Ride Systems   242
   10.2   Bus/Tram Monitoring and Control   249
   10.3   Train Operation and Traffic Control   260
   10.4   Automatic Fare Collection and Passenger Information   274

11   INTERNATIONAL DIAL-A-RIDE SYSTEMS EXPERIENCES   284
   11.1   USA Dial-A-Ride Experiences: Haddonfield and Rochester   284
   11.2   Further Dial-A-Ride Projects   292

12   INTERNATIONAL BUS MONITORING EXPERIENCES   298
   12.1   Irish Experiences: The Dublin System   298
   12.2   FRG Experiences: The Hamburg System   307
   12.3   Further Bus Monitoring Projects   310

13   INTERNATIONAL RAPID RAIL TRANSIT CONTROL SYSTEMS EXPERIENCES   315
   13.1   USSR Experiences: The Moscow Metro and the ASU-PP System   315
   13.2   USA Experiences: The San Francisco BART System   322
   13.3   FRG Experiences: The New Munich S-Bahn   332
   13.4   Further Computerized Rapid Rail Transit Systems   343

14   FINDINGS AND SUMMARY   347
   14.1   Dial-A-Ride (Para-Transit) Systems   347
   14.2   Bus and Tram Transit Systems   348
   14.3   Rapid Rail Transit Systems   350
   14.4   Passenger Guidance and Service Systems   351

PART FOUR NEW MODES OF URBAN TRANSPORT:
AUTOMATED GUIDEWAY TRANSIT AND
THE DUAL-MODE CONCEPT

15	**BASIC SYSTEMS CONCEPTS**	355
	15.1 AGT (Automated Guideway Transit) Systems	357
	15.2 Combined AGT and Dial-A-Ride Systems	362
	15.3 The Dual-Mode Concept	364
	15.4 Concluding Remarks	367
16	**CONCEPTS AND METHODS OF CONTROL AND AUTOMATION**	369
	16.1 The Control Tasks Hierarchy	369
	16.2 Automatic Lateral Vehicle Guidance	373
	16.3 Automatic Longitudinal Vehicle Guidance: Speed and Position Control	375
	16.4 Automatic Longitudinal Vehicle Guidance: Headway Regulation	381
	16.5 Traffic Flow Control	393
	16.6 Route Guidance and Empty-Car Disposition	402
17	**INTERNATIONAL SLT SYSTEMS EXPERIENCES**	412
	17.1 Airport Ground Transportation Systems	412
	17.2 SLT Systems in Urban Cities	415
	17.3 Further SLT Installations	416
18	**INTERNATIONAL GRT SYSTEMS EXPERIENCES**	418
	18.1 AIRTRANS at Dallas–Fort Worth Airport	418
	18.2 The Morgantown Group Rapid Transit (GRT) System	430
	18.3 Further GRT Systems	443
19	**INTERNATIONAL PRT SYSTEMS EXPERIENCES**	449
	19.1 Computer Controlled Vehicle System (CVS): Personal Rapid Transit in Japan	449
	19.2 Cabinentaxi—A New Concept of Urban Transport Developed in the FRG	461
	19.3 Other PRT Proejcts	470

20 FINDINGS AND SUMMARY	474
20.1 SLT Systems	475
20.2 GRT Systems	476
20.3 PRT Systems	477
Indexes	483

Introduction

MOTIVATION

Many cities face serious urban passenger transport problems caused by increasing use of motorcars. The environmental, economic, and social impacts of these problems are well known:

- The average risk of being killed or injured as a result of an automobile accident is far larger than that for natural catastrophes, fire, explosions, etc.: worldwide, about 150,000–200,000 human beings, corresponding to the population of a medium-sized city, are killed in automobile accidents every year.
- Traffic congestion causes time delays, which represent large economic losses.
- The urban environment is endangered by increasing air and noise pollution levels.
- Limited resources, in terms of energy and land, are used ineffectively.
- The attractiveness and effectiveness of public transport systems are decreasing.

Urban traffic problems of this type are occurring more or less in all industrialized nations of both the East and the West as well as in an increasing number of developing countries. They are exerting a considerable influence on the quality of urban life.

What solutions can be offered? Roughly, the known proposals are of three kinds: (1) change the rules of the game, that is, change peak demands by staggering working hours, limit fuel consumption, create pedestrian zones, and so forth; (2) improve existing transportation systems; and (3) provide new technological options.

This monograph aims to analyze the role of new technologies in the last two approaches.

In the past, new technologies such as the wheel, the sail, the steam engine, the electric motor, the internal combustion engine, the jet engine, and others (cf. Figure 1) have created breakthroughs to entirely new modes of transportation, resulting in significant changes in the structure of cities and in the way of urban living. Where formerly cities grew up along waterways, railroads, and streetcar lines, they now grow up along highways or around airports.

It seems reasonable to ask whether the fundamental new technology of our age—modern automation and computer technology—could contribute to a new breakthrough or, at least, to basic improvements in urban transportation (Figure 1). But what contribution can one realistically expect from extensive application of advanced automation and computer technology?

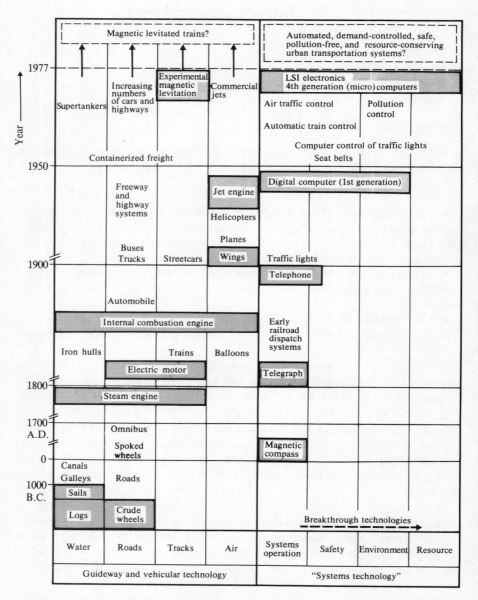

FIGURE 1 New technologies that have or probably will create breakthroughs to new modes of transportation, resulting in essential changes in urban structure and the quality of urban living (cf. Cannon 1973).

This new and fast developing technology seems to offer for the first time the possibility of changing from *extensive* to *intensive* development of transportation systems. What does this mean? Most of the earlier developments in transportation systems were achieved by brute force—more and bigger traffic areas using more concrete, stronger engines, and more vehicles—at higher cost. In principle, it is becoming impossible in more and more cities to continue in this manner, and the digital computer and related automation technology provide a promising alternative. The extensive use of automation in urban transportation systems is supposed to lead to a new level of transportation service, to an increase in capacity, to a decrease in operating costs (including a decreased impact on energy reserves and environment), and to new standards of safety.

It is reasonable to assume (cf. Cannon 1973) that this new *systems technology* will give an impulse to improving urban transportation similar to that given by the magnetic compass to extending sea transport from the local to the global arena, or by the telegraph and telephone to developing nationwide railway dispatching systems (cf. Figure 1).

SUBJECT AND PURPOSE

This monograph addresses the following two questions:

1. What benefits may a city expect from implementing computerized control systems for existing modes of urban transportation?

2. During the next ten years or so, will it be possible to create entirely new and highly automated urban transportation systems characterized by their potential for a demand-oriented, safe, pollution-free, and resource-conserving operation (cf. Figure 1)?

In dealing with such questions one is faced with truly interdisciplinary problems ranging from rather general to very specific topics:

- general urban transport development strategies
- planning and operating principles used in the individual transport system
- concepts and methods of control, monitoring, and automation

Therefore, it is believed that a comprehensive analysis of these two questions is of interest for scientists, engineers, and practitioners coming from various disciplines:

- urban and transportation planning
- traffic engineering
- control engineering
- computer science

This audience will be dealing with corresponding problems in urban or national government offices, planning and operation offices of transport companies (highway departments, railways, and other public transit authorities), traffic and urban research institutes, different branches of industry (e.g., in computer and automation departments), and universities.

With this prospective audience in mind, the following framework was chosen for dealing with the above-mentioned questions.

Part I of the monograph presents a review of basic problems and concepts by

- summarizing the various impacts of urban traffic problems (Chapter 1)
- discussing the resulting general urban and transport development concepts (Chapter 2)
- analyzing the specific contributions of advanced computer and automation technology (Chapter 3)

Parts II-IV present an analysis of computerized transport and traffic control systems used for

- freeway and area traffic control and guidance (Part II)
- controlling and monitoring public transport systems (Part III)
- automation and control of completely new modes of urban transport (Part IV)

Parts II-IV are each structured in the same way:

1. A survey is presented first of the *basic system concepts* dealing with fundamental features of the different automation and control principles proposed and used for the individual transport modes.
2. The *methodology* developed for control and automation is then discussed. In order to limit the size of the book, only the most important methods are presented.
3. The general considerations are supplemented by presenting a survey of the *international experience* gained so far in real applications. For this purpose, case descriptions on advanced or interesting projects have been prepared by 30 scientists from eight countries: USA, USSR, Japan, FRG, France, UK, Eire, and Kenya.

The case descriptions were selected by considering geographical aspects (covering experience gained in North America, western Europe, eastern Europe, Japan, and a developing country), and the different philosophies used in designing, implementing, and operating the various systems.

4. Finally, each part concludes with a survey of major *findings* and a *summary*.

REFERENCE

Cannon, R. H. 1973. Transportation, automation, and societal structure. Proceedings IEEE, 61 (5): 518–525.

Part One

Transportation, Automation, and Urban Development: Review of the Basic Problems and Concepts

Part 1 of this monograph analyzes the contribution of advances in computer control technology to the solution of urban passenger transport problems within the framework of general strategies for coordinated development of urban transportation systems and the urban area as a whole. Chapters 1 and 2 present a survey of the various impacts of urban traffic problems and the resulting general transport and urban development concepts. Chapter 3 analyzes the specific contribution of computerized control systems with respect to both short- and long-term strategies and summarizes basic transport control concepts. Finally, the basic technical and economical feasibility of these short- and long-term strategies is discussed by reviewing the various impacts of the revolutionary developments in computer technology.

1 Urban Traffic Problems

The serious social, economic, environmental, and other effects of urban traffic problems are well known. A brief discussion of these problems is, however, useful, for two reasons: (1) to illustrate the size and character of the problems facing many cities all over the world, and (2) to provide the necessary background information for analyzing the potential of advanced computer and automation technology within general transport- and urban-development concepts. For this purpose, the following five categories of problems will be discussed: (1) mobility, (2) traffic safety, (3) environment, (4) resources, and (5) effectiveness and attractiveness of public transit.

1.1 MOBILITY

Traffic congestion represents the best-known problem caused by the increasing use of motorcars in cities. It is estimated that the time losses caused every day by congestion in Paris are approximately equal to the daily working time of a city with 100,000 inhabitants. The Road Research Laboratory has found (Holroyd and Robertson 1973) that in Britain the loss to the community from delays in a city with about 100 intersections is of the order of £4 million per year. For Tokyo it has been estimated that the annual overall losses caused by inefficient traffic flow through the main 268 intersections amount to 57 billion yen, i.e., about $200 million (cf. Toyota 1974b). These values do not take into account the fact that traffic congestion also results in a remarkable increase in air pollution.

Traffic congestion causes a significant decrease in both traffic throughput and travel speed. This is illustrated for freeway-type traffic by the so-called fundamental diagram of traffic flow shown in Figure 1.1, which describes the relations among the three basic macroscopic traffic-flow variables under stationary flow conditions (Gazis 1974):

- traffic volume x_C (in number of cars per hour per lane)
- traffic-flow speed x_S (in kilometers per hour)
- traffic density x_D (in number of cars per kilometer per lane)

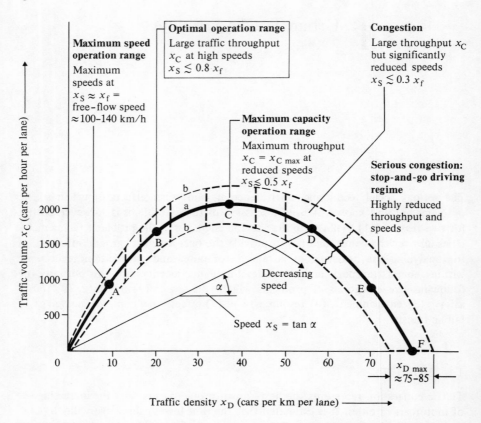

FIGURE 1.1 Fundamental diagram of traffic flow (curve a) and corresponding tolerances (curves b) for various free-flow speeds x_f (100–140 km/h) and jam concentrations $x_{D\ max}$ (75–85 cars/km/lane). (cf. Gazis 1974.)

Typical traffic-flow states, depending on the traffic density x_D and indicated by points A–F in Figure 1.1, can be distinguished:

- state A, *free traffic flow*, with no interaction between the individual vehicles
- state B, operation at *slightly reduced flow speeds* caused by interactions between vehicles, and at traffic volumes near the lane-capacity limit
- state C, operation at the *lane-capacity limit*
- state D, *congestion*, i.e., significantly reduced flow speeds, reduced traffic throughput (compared with state C), and increased sensitivity to small disturbances like speed changes of a leading car
- state E, *serious congestion*, i.e., stop-and-go driving regime characterized by frequent alternations between acceleration and deceleration maneuvers, including short-time stops

- state F, *complete traffic breakdown*, i.e., traffic speed and throughput are approaching zero, since the density reaches the critical jam concentration x_{Dmax}

The impact of congestion on travel time is illustrated by the following example for vehicles in the stop-and-go driving regime surrounding point E in Figure 1.1.

If a motor-car is accelerated and decelerated between 7 km/h and 88 km/h, one thousand times in a cyclic manner, then an average time loss of 6 h results. In the case of a truck, this can amount to 21 h (Neuberger 1971, Curry and Anderson 1972). Moreover, the traffic throughput decreases by a factor of about two or more (cf. Figure 1.1). Therefore, it is important to find a means of keeping the traffic flow near the optimal state, i.e., in a range near the points B and C in Figure 1.1, otherwise a large part of the physical capacity of the limited traffic areas will be lost. This holds true for freeway-type traffic, and also in a modified form for urban street traffic.

Increased congestion also causes a dispersion of travel time, thus leading to uncertainties in estimating travel time, and to irregularities in the operation of public transport means (buses, streetcars), which are not able to keep to prescribed timetables.

1.2 TRAFFIC SAFETY

In many countries the average risk of being killed or injured as a result of an automobile accident is far larger than that for natural catastrophes, fire, explosions, etc. The size of this safety problem is illustrated by the number of human beings killed every year: 46,000–56,000 in the USA, 14,000–19,000 in the FRG, 13,000–17,000 in France, 2,000–3,000 in Austria, 14,000–17,000 in Japan, 2,000–2,500 in the GDR, 3,000–4,200 in Poland, about 2,000 in the CSSR, 1,400–1,800 in Hungary, 3,600–4,500 in Yugoslavia, and about 90,000 in the whole of Europe (ECE 1974a,b and 1976).

In the heavily motorized countries of North America and western Europe, the annual fatality rate, expressed as the number of persons killed per 10,000 inhabitants, reached values within the range 2–3.6 during the period 1962–1974. In Japan, this parameter lies between 1.4 and 1.6 and for most of the eastern European countries is in the range 0.6–2.0. This is illustrated in Figure 1.2, which shows the relations between the following three aggregated variables for selected countries:*

- number of fatalities per 10,000 inhabitants per year
- number of fatalities per 100 kilometers of the road network per year
- the relative road network length in kilometers per 100 inhabitants

---

* For reasons of data availability, in Figures 1.2 and 1.3 the length of all roads of a particular nation is used instead of the length of the urban streets alone. The conclusions drawn from Figures 1.2 and 1.3 also hold true, however, for urban traffic situations.

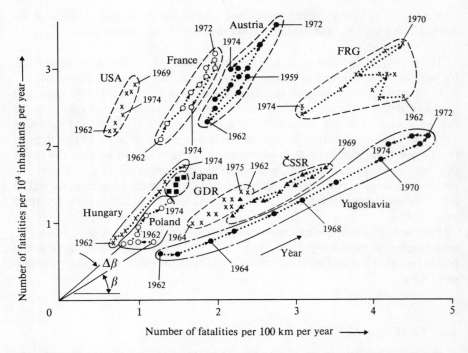

FIGURE 1.2 The annual fatality rate from 1962 to 1974 as a function of the length of the road network for selected countries (cf. ECE 1976, IRF 1976, Fischer and Barthel 1974); tan β = relative road network length (km per 100 inhabitants); tan $\Delta\beta$ = change of relative road network length.

The safety problem, as characterized by the number of fatalities per inhabitant, is currently less serious in the eastern European countries than in the USA and in western European countries. However, the rapidly increasing number of registered automobiles in East Europe (Figure 1.3a) may eventually cause similar difficulties, if future developments are not analyzed and controlled carefully (CMEA 1975). This is made obvious by considering the traffic space available per vehicle, which may be measured by the aggregated variable "length of all auto roads divided by the total number of automobiles" (Figure 1.3b). In the early sixties this parameter was 5-20 times larger in Japan and in most of the eastern European countries than in the countries of the West. However, in the middle of the seventies this situation changed fundamentally: in 1974, for example, the road-network length that was on average available for a motorcar reached the same order of magnitude (50-80 m per automobile) for many eastern and western countries like Austria, CSSR, France, GDR, Japan, USA, and Yugoslavia (cf. Figure 1.3b).

The aggregated variable for traffic space introduced above does not distinguish between urban and rural roads or one-lane and multi-lane roads. Thus, it character-

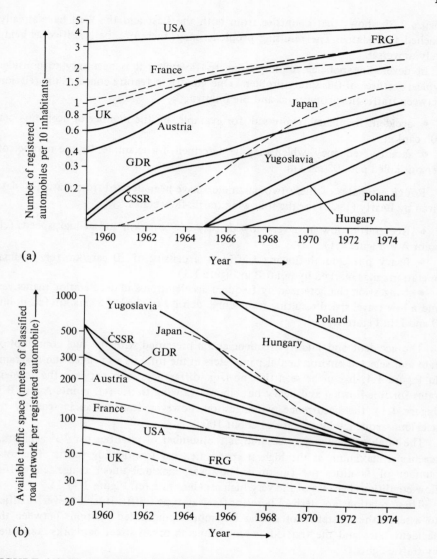

FIGURE 1.3 Time-dependent development of (a) the number of registered automobiles (cf. ECE 1976), and (b) the traffic space available on average per registered automobile (the length of all roads according to the statistics in IRF (1976)).

izes the available traffic space only roughly on a nationwide scale. Nevertheless, it serves as an indicator, i.e., if the relative length of the road network decreases to values of the order of 100 m per automobile, then the occurrence of traffic problems—among them safety problems—on a nationwide scale becomes very likely.

Figure 1.3b shows that countries from both the East and the West have already reached that status. The resulting problems may, therefore, be classified as being truly universal.

In developing policies for improving traffic safety it is necessary to consider typical features of this safety problem. One important feature concerns the relation between traffic-flow conditions and the following:

- accident frequency, expressed, for example, by the number of accidents per 10^6 car-km
- fatality frequency, which may be described, for example, by the number of persons killed per 10^8 car-km

Rough estimates of these two parameters have been derived from statistical data given by Beatty (1972) for the following traffic-flow states:

- free traffic flow at a very low density of 7 cars/km and at high speeds (cf. point A in Figure 1.1)
- heavy but smooth-flowing traffic at a density of 20 cars/km (cf. optimal operation range marked by point B in Figure 1.1)
- congestion characterized by frequent acceleration and deceleration maneuvers and a low travel speed resulting from a high density of about 55 cars/km (cf. points D and E in Figure 1.1)

The accident and fatality frequencies are presented with the fuel consumption rate and selected environmental parameters in the form of a so-called star diagram* in Figure 1.4. It can be seen that no large differences exist between the accident rates for free-flowing and heavy but smooth-flowing traffic (cf. points A and B in Figure 1.1); the occurrence of congestion, however, leads to an increase of the accident frequency by a factor of about 100.

The fatality frequency, measured as the number of fatalities per 10^2 accidents, reaches a maximum at the highest speed, i.e., for free-flowing traffic. The total number of fatalities per car-km is, however, maximal under congested traffic-flow conditions because of the very high accident rate (cf. Figure 1.4).

These relations are derived here for freeway-type traffic. It is, however, justified to assume the existence of similar but more complicated relations between the accident rates and the traffic-flow conditions in urban street networks controlled by traffic lights.

1.3 ENVIRONMENT

Increasing levels of air pollution and noise, vibration of buildings, and visual intrusion and severance of the urban area by more and bigger freeways and arterial

---

* The scales of this star diagram have been chosen in such a way that values characterizing undesirable states are located near the centre, i.e., the most favorable state of the system is given by the curve that covers the largest area.

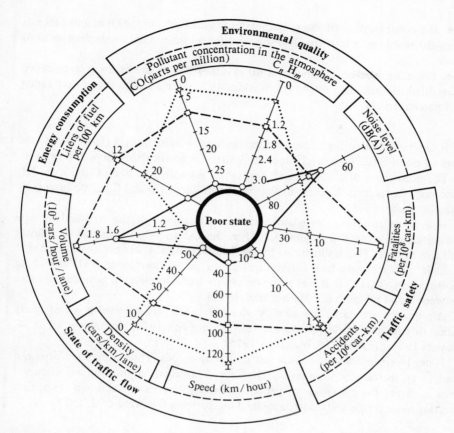

FIGURE 1.4 Relation between traffic-flow conditions and traffic safety (cf. Beatty 1972), environmental quality (cf. Marcus 1974, Ullrich 1973, Ullrich and de Veer 1974, Baerwald 1976, Dare 1976), and energy consumption (cf. Baerwald 1976): ○——○, congestion; □- - -□, optimum operation point; △· · · ·△, free flow.

streets form the third negative factor caused by the increased use of automobiles (cf. Curry and Anderson 1972, Chreswick 1973, Watkins 1973, Horowitz 1974, Toyota 1974, Afanasyev 1975, Hirao 1975, Windolph 1975).

The air pollution problem concerns the increase in the atmosphere of toxic substances like carbon monoxide CO, hydrocarbons C_nH_m, nitrogen oxides NO_x, aldehydes, and lead, as well as the abstraction of more oxygen from the air. It is important to consider the existing relation between traffic-flow conditions, on the one hand, and parameters characterizing the environmental quality, on the other hand. This relation is illustrated in Figure 1.4 for the same traffic-flow states considered above (cf. points A, B, D, E in Figure 1.1). Rough estimates of the following parameters have been estimated using results published by Baerwald (1976), Koshi and Okura (1974), Ullrich (1973), Ullrich and de Veer (1974), Marcus (1973, 1974), and Weiss (1970):

- the concentration of the pollutants* CO and C_nH_m in the air at a distance of 20 m either side of a four-lane highway and for a longitudinal wind component of 5 m/s
- the noise level at a distance of 40 m either side of the highway in so-called A-weighted decibels, i.e., dB(A) (cf. Watkins 1973), for a traffic stream containing 17.5 percent trucks

It can be seen that the increase of traffic density from point A (free flow) to the optimal point B (cf. Figure 1.1) does not lead to a dramatic rise in air pollution levels.

The occurrence of medium-sized congestion according to point D in Figure 1.1, however, results in an increase in the concentration of CO and C_nH_m pollutants by a factor of about 3-6.

If even more severe congestion arises, for example, of the stop-and-go type illustrated by point E in Figure 1.1, then the CO and C_nH_m air pollution levels will rise further. This is illustrated by the following data: it has been estimated that a single motorcar sets free, on average, 27 kg carbon monoxide (CO) and 11 kg hydrocarbon (C_nH_m) if it has to be decelerated and accelerated a thousand times between 0 and 88 km/h (cf. Masher et al. 1975).

Air quality standards are already violated in densely traveled areas of many cities all over the world (cf. Watkins 1973, Horowitz 1974, Sawaragi et al. 1974, Toyota 1974a, Hirao 1975, Windolph 1975).

Noise levels increase with increases in driving speeds. Thus congestion does not lead to an increase in noise levels. The air and noise pollution problems discussed here motivate the migration of people from city centers to suburban areas, thus creating more traffic and undesirable land-use patterns (cf. Ward et al. 1977).

1.4 RESOURCES

The fourth main problem concerns ineffective consumption of resources, i.e., energy and land (cf. Chreswick 1973).

Energy consumption: It is well known that a remarkably large part of all energy consumed in heavily motorized countries is devoted to the highway transportation system. This is especially true for liquid resources in the form of petroleum. In the USA, for example, about 38 percent of all petroleum is used by highway vehicles, 27 percent by cars, and 11 percent by buses and trucks (cf. Ward et al. 1977, Chreswick 1973, French 1974, Pierce 1974). In other countries less energy is used for motor vehicles, although with the increase in the rate of motorization shown in Figure 1.3a, an ever increasing part of the available energy resources has to be assigned to the highway transportation system.

*The author would like to thank Dr. H. B. Kuntze for computing these parameters by means of the renewal model introduced by Marcus (1973).

It is important to mention that energy consumption is highly dependent on the operational state of the highway transportation system. There is a significant increase in the mean rate of fuel consumption with increase in driving speed. However, travel under congested conditions can raise fuel consumption rates even more because of ineffective acceleration and deceleration maneuvers. This is illustrated in Figure 1.4. Driving under congested flow conditions according to points D and E in Figure 1.1 can result in fuel consumption rates that are twice as high as those for heavy, smooth-flowing traffic given by point B. Two other figures taken from Claffey (1971) characterize the size of the problem: the fuel consumption of a motorcar can increase by about 60 liters if it has to accelerate and decelerate 1,000 times between 7 km/h and 88 km/h; a truck needs an additional 144 liters. Similar results were obtained by Haslböck and Huttmann (1977) for downtown traffic conditions in ten large cities of the FRG. Using a VW Minibus equipped with a special fuel-consumption measuring device, they made trips through the individual city centers covering distances of 8–15 km during both the off-peak and the rush-hour periods.

Figure 1.5 shows that the total consumption rates for off-peak trips were 1.4–2.1 times higher than the nominal consumption rate; for rush-hour traffic these factors increased to 1.9–3.3, and reached maximal values of 3.5–8.8 in short, heavily congested sections of the driving route. According to the estimates made by Haslböck and Huttman ineffective traffic flow caused by congestion results in an additional fuel consumption of about 2.5 million liters per day for the ten cities considered.

Land consumption: About 28 percent of the area of US cities is on average devoted to highway vehicles. In the city of Atlanta (Georgia), 54 percent of the downtown area is reserved for parking and driving; this is often still insufficient during rush hours (cf. Volpe 1969, French 1974, Hirten 1974). Experience gained during the last 25 years shows that the construction of more and bigger traffic areas, like urban freeways, only provides a reduction in traffic problems for a limited time period. This is illustrated in Figure 1.2 for the relation between the number of fatalities and the increase in the length of the road network (cf. angle β). The reason is that the number of cars and the resulting traffic demand is growing faster than the potential of a city to provide the necessary traffic space. In Japan, for example, the number of automobiles increased from 2.2 million to 12.5 million, i.e., by a factor of 5.7, during the period 1965–1972 (cf. Figure 1.3a). During this seven-year period it was obviously not feasible to create the space needed for the operation of the increased numbers of vehicles, although remarkable highway construction programs were started, leading among other things to urban freeways in Tokyo and Osaka (Toyota 1974a). Congestion occurred more frequently and to a larger spatial extent. Between 1969 and 1970 the Tokyo Metropolitan (Shuto) Expressway experienced about six complete traffic breakdowns every day (cf. Spring 1974). These phenomena were first observed in the USA at the end of the fifties (cf. Figure 1.3a). In the mid-sixties the same development made progress in western Europe (FRG, France, UK, Italy) (cf. Carter et al. 1968, ECE 1974).

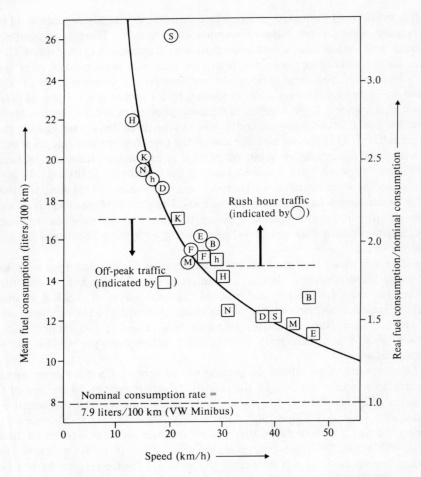

FIGURE 1.5 Relation between fuel consumption and travel speed from measurements by Haslböck and Huttmann (1977) for 10 large cities in the FRG: E, Essen; M, Munich; B, Bremen; S, Stuttgart; D, Düsseldorf; N, Nuremberg; H, Hamburg; h, Hannover; F, Frankfurt; K, Cologne.

Most East European countries are presently faced with the beginning of a similar development, as discussed earlier in Section 1.2 with reference to Figure 1.3b.

1.5 EFFECTIVENESS AND ATTRACTIVENESS OF PUBLIC TRANSIT

The automobile not only changed the extent to which the city area is used for parking and driving, it also changed the land-use patterns in many countries, i.e., the locations of residential and industrial zones. Since World War II, in the USA and in many other countries, suburban areas have been growing faster than central cities (cf. Ward 1977). Jobs and activities have followed people to these suburbs, decentralizing the functions once largely confined to a single central business district. However, conventional public transport systems like fixed-route bus, trolley, streetcar lines, or urban railway systems were designed to serve city structures typical of the pre-war period, when more than two-thirds of the urban population lived in the high-density central city. Thus, the changing land-use patterns caused by the automobile produce a situation where conventional fixed-route transit is more and more incompatible with the spatial distribution of activity and population in many cities with large, multinucleated suburbs and diffuse trip patterns.

Changing land-use patterns and increased use of the private car caused a decrease in the ridership of public transportation systems in many countries. This often resulted in less frequent services, which again supported the decision of another group of public transit users to prefer private cars for their daily trip to work. The loss of attractiveness of public transit, however, is not only produced by decreases in service frequency: since buses and streetcar lines are using the same traffic areas as motorcars, they will experience the same time losses connected with irregularities with respect to timetables. Moreover, most public transport companies are faced with increases in personnel costs and/or labour shortages.

What solutions can be offered for the problems summarized in this chapter? This question represents the subject of Chapter 2.

REFERENCES

Afanasyev, L. L. 1975. The tendency towards fuel economy and decreasing air pollution in automobile transport of the USSR. Transportation Research, 9 (2/3): 107–110.
Baerwald, J. E. (ed.). 1976. Transportation and Traffic Engineering Handbook. Englewood Cliffs, New Jersey: Prentice-Hall.
Beatty, R. L. 1972. Speed analysis of accidents. Public Roads, 37 (3): 89–102.
Carter, A. A., E. A. Hess, E. A. Hodkins, and I. Raus. 1968. Highway traffic surveillance and control research. Proceedings IEEE, 56 (4): 566–576.
Chreswick, F. A. 1973. Energy, environment, and ground transportation. Research Outlook (Battelle), 5 (1): 28–32.
Claffey, P. J. 1971. Running costs of motor vehicles as affected by road design and traffic. Washington, D.C.: Research Report NCHR-111.
CMEA. 1975. Problems of Street Traffic Safety. Council of Mutual Economic Assistance, First Scientific Conference, Alma Ata, USSR, Moscow (in Russian).
Curry, D. A. and D. G. Anderson. 1972. Procedures for estimating highway users costs, air pollution, and noise effects. Washington, D.C.: NCHRP-Report 133.
Dare C. E. 1976. Transportation, land use, and air quality programs. Transportation Engineering Journal, Proc. ASCE, 102 (TE2): 411–426.
ECE Annual Bulletin of Transport Statistics for Europe, 1973. 1974a. New York: United Nations, Economic Commission for Europe.
ECE Statistics of Road Traffic Accidents in Europe, 1973. 1974b. New York: United Nations, Economic Commission for Europe.

ECE Annual Bulletin on Statistics of Road Traffic Accidents in Europe, 1975. 1976. New York: United Nations, Economic Commission for Europe.

Fischer, P. and S. Barthel. 1974. Statistical considerations on the development of personal public traffic in CMEA countries, 1950–1970. Wiss Z. d. Hochschule für Verkehrswesen "Friedrich List", 21 (2): 341–356; 21 (4): 685–708 (in German).

French, A. 1974. Highway transportation and the energy crisis.: US Department of Transportation, Federal Highway Administration.

Gazis, D. C. 1974. Traffic Science. New York: John Wiley and Sons.

Hirao, O. 1975. Automobiles and air pollution. Technocrat, 8 (8): 15–20.

Hirten, J. E. 1974. Transportation and community development—new trends and prospects. Paper presented at the International Urban Seminar, Vienna, Austria.

Holroyd, J. and D. J. Robertson. 1973. Strategies for area traffic control systems: present and future. Crowthorne, Berkshire: Transport and Road Research Laboratory, Report LR 569.

Horowitz, J. L. 1974. Transportation controls are really needed in the air-cleanup fight. Environmental Science and Technology, 8 (9): 800–805.

Haslböck and Huttmann. 1977. Quick, No. 52, December 15–21: 28–32.

IRF. 1976. World Road Statistics 1971–1975. Geneva/Washington D.C.: International Road Federation (IRF).

Koshi, M. and I. Okura. 1974. Motor vehicle emissions related to the vehicle operating conditions. Report from the Institute of Industrial Science, University of Tokyo, Japan.

Marcus, A. H. 1973. A stochastic model of microscale air pollution from highway traffic. Technometrics, 15: 353–363.

Marcus, A. H. 1974. Effect of stochastic traffic flow on perceived noise. Paper presented at the 6th International Symposium on Transportation and Traffic-Flow Theory, Sydney, 1974. Amsterdam: Elsevier Publ. Co.

Masher, D. P. et al. 1975. Guidelines for design and operation of ramp control systems. Research Report from the Stanford Research Institute, California, Contract NCHRP3-22, SRI Project 3340.

Neuberger, H. 1971. The economics of Heavily Congested Roads. Transportation Science, 5: 283–293.

Pierce, J. R. 1974/75. CALTECH Seminar Series on Energy Consumption in Private Transportation. Technical Report DOT-OS-30119, from the California Institute of Technology, Pasadena, California.

Sawaragi, T. et al. 1974. Environment pollution control. Research Abstracts, Special Research Program, Ministry of Education, Japan.

Spring, A. 1974. The automation of traffic on urban freeways. Paper presented at the Road Research Group T9 Workshop in Tokyo, 1973. Straße und Verkehr, No. 1: 25–31 (in German).

Toyota. 1974a. Transport countermeasures. The Wheel Extended, A Toyota Quality Review, 4 (1): 14–32.

Toyota. 1974b. Transport in Tokyo. The Wheel Extended, A Toyota Quality Review, 4 (1): 4–13.

Ullrich, S. 1973. Speed limitations—a means for reducing noise from freeways and expressways? Straßenverkehrstechnik, 17 (1): 9–13 (in German).

Ullrich, S. and H. de Veer. 1974. On the problem of reducing noise by means of speed limitation on expressways. Straßenverkehrstechnik, 18 (3): 69–71 (in German).

Volpe, J. A. 1969. Urban transportation tomorrow. The American City: 59–62.

Ward, J. D. et al. 1977. Toward 2000: opportunities in transportation evolution. Report No. DOT-TST-77-19 from the US Department of Transportation, Office of the Secretary, Office of Research and Development Policy, Washington, District of Columbia.

Watkins, L. H. et al. 1973. Effects of traffic and roads on the environment in urban areas. Paris: Organization for Economic Cooperation and Development (OECD).

Weiss, G. H. 1970. On the noise generated by a stream of vehicles. Transportation Research, 4: 229–233.

Windolph, J. 1975. Automobile traffic and environment. DDR-Verkehr, 8 (11): 462–467 (in German).

2 Transport and Urban Development: General Concepts

It is obvious that the complexity of urban traffic problems requires the application of a set of approaches in the sense of a general *combined transport and urban development strategy*. Elements of such a strategy are discussed here with reference to (1) objectives, (2) options, and (3) constraints.

2.1 OBJECTIVES

The transport development task represents by nature a multiobjective decision-making problem. Three groups of conflicting objectives have to be considered (cf. Figure 2.1):

- the objectives of the public transit travellers and private-car drivers
- the objectives of the inhabitants of the city, i.e., of the city as a whole
- the objectives of public transit companies

Public transit users and private-car drivers. "The success of the auto should be a lesson" (Kieffer 1972) in dealing with the development of concepts for the reduction of urban traffic problems. This success indicates that relatively small gasoline-powered vehicles fulfill to a great extent the expectations and requirements of its users. Therefore, every car owner compares the service delivered by the available public transit systems with the potential provided by his car. From sociological studies it is known (cf. Demag/MBB 1974) that the decision to use a private car or public transit, respectively, depends mainly on the following criteria:

- *independence* with respect to departure time and destination (no need for using more than one transport mode for a trip, i.e., no vehicle changing, availability around the clock)

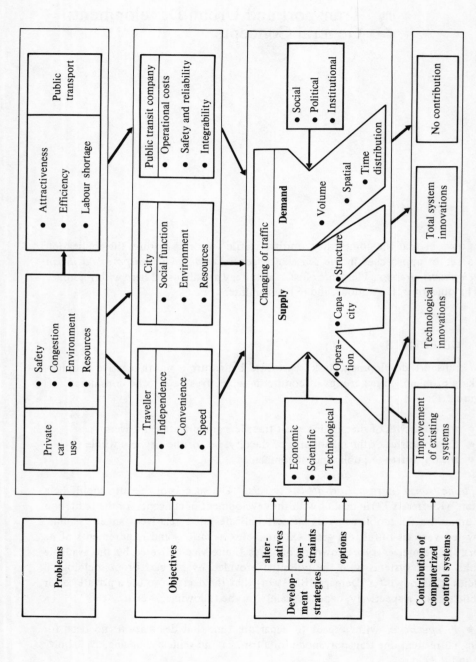

FIGURE 2.1 The contribution of computerized traffic control systems within the framework of general urban city and possible transport development strategies.

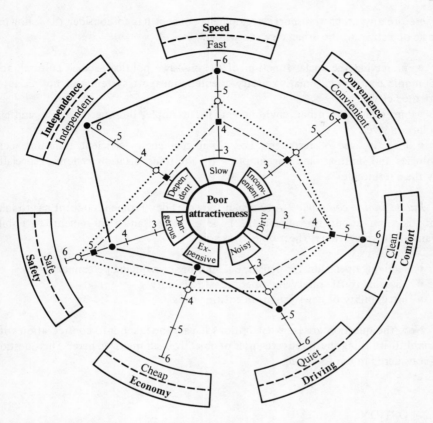

FIGURE 2.2 The attractiveness of private-car travel versus public transit judged by automobile owners using a nondimensional scale ranging from 1 (poor attractiveness) to 7 (high attractiveness) (derived from results of an inquiry carried out by INFAS in Hamburg, Bremen, and Dortmund in the FRG in 1968; cf. Demag/MBB 1974): ——, private car; - - - -, bus, tram; · · · · ·, subway, bus, tram.

- *convenience* (comfortable seats, air-conditioning, car radio)
- *speed*

Figure 2.2 illustrates that most car owners consider private-car travel to be much more attractive, especially with respect to these criteria. Therefore, most of them prefer using their own automobile, even though it is more expensive and less safe than public transit (cf. Figure 2.2).

The city. From the viewpoint of the inhabitants of the city or the city government, fulfilling the objectives of the private-car users can only be accepted if basic social, economic, environmental, and other criteria of the whole population of the city are not violated. This leads to unsolvable conflicts, as illustrated in Chapter 1.

Therefore any urban transport development concept has to consider the following basic objectives of the urban city as a whole:

- preserving the social function of the city as a political, social, cultural, and economic centre, i.e., creating a city-oriented transport system and not an auto-oriented city
- protecting the urban environment, for example, limiting the noise- and air-pollution levels
- effective use of resources, i.e., supporting energy-efficient transport technologies and limiting land consumption by the transportation systems, especially by the automobile

Public transit companies. Any urban transport development concept has to take into account public transit and the basic objectives and requirements of public transport organizations. These objectives are:

- reducing operational expenses, i.e., limiting or reducing personnel
- ensuring traffic safety and operational reliability
- integrability of new systems in existing ones

Now the question arises: What options have to be taken into consideration with regard to fulfilling the three categories of objectives summarized here? This question is considered in the following.

2.2 OPTIONS

The gap between automobile transportation demand and supply can only be reduced by means of complex policies to control (cf. Figure 2.3 and MITI 1975):

- the transportation demand, which changes in time and space
- the transportation supply, characterized by the capacity of roads, parking lots, and other transport facilities

In developing such control policies, the time span between their formulation and the benefits realized from them must be considered carefully, i.e., a distinction is necessary between short-term, medium-term, and long-term policies.

2.2.1 CONTROL OF DEMAND (FIGURE 2.3)

Long-term strategies. Automobile transportation demands arise and are concentrated as a result of the configuration of urban land use and activities. The first level of policy to control demand, therefore, is to implement controls on the

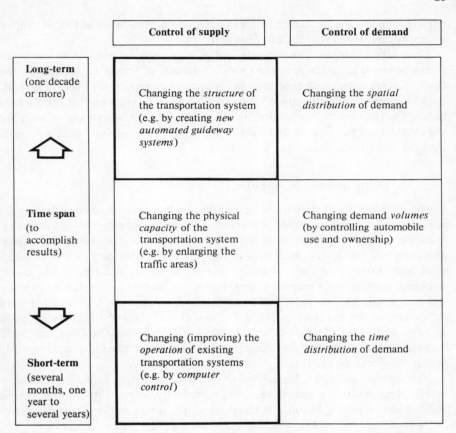

FIGURE 2.3 Long- and short-term transportation supply and demand policies (cf. MITI 1975).

spatial distribution of origin-destination demands generated in the city. These include measures such as urban redevelopment, construction of new towns, factory and market relocation, and reorganization of other urban facilities. These measures are long-range and require sustained expenditures over long periods of time to implement.

Medium-term strategies. The second level of policy to control demand is to implement controls on demand volumes. Given the existing configuration of urban land use and activities, these controls aim to reduce the use and ownership of the automobile. Such controls could take the form of increased automobile taxes, stricter requirements for automobile ownership, curbs on driving into the central business district, increased gasoline taxes, etc. In addition to these steps to discourage driving, measures could be taken to encourage people to use public transportation or to use the telephone instead of travelling. Unfortunately, existing public transportation systems cannot, in many cases, provide an attractive alternative to private cars, as has already been discussed.

Short-term strategies. The third level of policy to control demand is to institute measures aimed at controlling the time distribution of transport demands in the city — for example, those generated by commuting to work and school. In general, traffic congestion is caused merely by excessive concentration of demand. By controlling the time distribution of demand, therefore, a better balance between supply and demand can be achieved. Included in this category are measures such as staggering work and school hours.

2.2.2   CONTROL OF SUPPLY (FIGURE 2.3)

Long-term strategies. The first and most basic level of policy to control the supply of transportation concerns the structure of the system, e.g., road networks, subway lines. By constructing new roads, bypasses, overpasses, and parking lots, and by abolishing old roads, it is possible to change the overall structure of the road network and thereby the flow of traffic over it. However, because of the reasons discussed in Section 1.4 increased construction of more and bigger highways alone will not present an acceptable solution. Moreover, as mentioned above (cf. Figure 2.2), changing the structure of existing modes of public transportation will not in all cases lead to a reduction of automobile traffic demand. Therefore, one basic long-term policy consists of developing entirely new (automated, demand-oriented) public transportation systems.

Medium-term strategies. The second level of policy to control the supply of automobile transportation concerns the physical capacity of the system to link origin and destination points. Included in this category are measures that alter the capacity of the road network, such as widening roads and enlarging parking facilities.

Short-term strategies. The third level of policy to control supply concerns the operation of the various parts of the transportation system as well as of the whole system. Included in this category are measures that restrict or limit the use of a link or its connection with other links, e.g., establishing one-way roads and speed limits. Moreover, the optimization of system operation by automation and computer control belongs to this level. Policies of this kind focus on short time units of weeks, months or even a few years.

2.3   CONSTRAINTS

It is quite obvious that the feasibility of the supply and demand control policies summarized here depends on constraints that differ from country to country. For example, measures of the first (long-term) levels mentioned above aim at fundamentally restructuring transportation supply and demand by changing the configuration of urban land use and activities, and thus will require many years to formulate, implement and realize. It is equally difficult to restructure transportation demand, since demand is closely related to the pattern of land use, the location of urban functions, and social customs and practices.

The medium-term measures for controlling the demand volumes of automobile traffic provide the advantage that they can be implemented with very low investments. This is especially true for such measures as increasing automobile and gasoline taxes. It is obvious, however, that the feasibility of these measures depends on the socioeconomic system in the specific country, i.e., on special institutional, social, and political constraints (cf. Figure 2.1).

The problems already discussed, related to the enlargement of the urban road network, i.e., long- and medium-term measures for changing the structure and capacity of the urban highway transportation system (cf. Figure 2.3), have led to a so-called "no-action philosophy" in several countries (cf. Loder 1974). This school of thought believes that any enlargement of the street network in the downtown area attracts more traffic, creating traffic congestion after a certain period of improvement. Therefore, this no-action school of thought recommends that nothing be changed with respect to the structure and capacity of the urban street network in the downtown area; the congestion will stimulate many drivers not to use their cars to go to the city center. This could lead to an equilibrium between traffic demand and available traffic supply. It is an open question whether an equilibrium based on congestion can be accepted from the point of view of the danger to the urban environment, the increase in energy consumption, and the other objectives summarized in Section 2.1. However, it is very uncertain, for the reasons discussed above (cf. Figure 2.2), whether long- and medium-term measures for increasing the capacity of existing public transportation systems (cf. Figure 2.3) will provide a feasible alternative to this no-action philosophy.

Because of this situation, more and more urban traffic experts have come to the conclusion that a simple extensive further development of the existing urban highway and public transport systems will in the long run lead to a dead end (cf. Bahke 1973, BMFT 1974, 1975, Brand 1976, Carroll 1976, Day and Shields 1973, DHUD 1968, DOT 1974, Filion 1975, Hamilton and Wetherbee 1973, Hamilton and Nance 1969, Loder 1974, Kieffer 1972, Ross 1969, Volpe 1969).

Qualitative changes are needed with regard to the structure of the urban transport system, as well as concerning the operation of the existing systems. Now the question arises: How can these qualitative changes be achieved? The answer is: Mainly by extensive use of advanced systems technology, i.e., that of computers and automation (cf. Figure 1, Introduction).

Chapter 3 analyzes these new technology options in more detail.

REFERENCES

Bahke, E. 1973. Transportation Systems Today and Tomorrow. In the book series Conveying and Elevating. Mainz, FRG: Krauss-Kopf Verlag (in German).

Brand, D. 1976. Bringing logic to urban transportation innovation. Technology Review, January: 39–45.
BMFT. 1974. Program for Rapid Transit Research. Bonn, FRG: Bundesministerium für Forschung und Technologie (in German).
BMFT. 1975. Rapid Transit Research '75. Status Seminar II. Bonn-Bad Godesberg, FRG: Bundesministerium für Forschung und Technologie (in German).
Carroll, J. 1976. Next steps in urban transit. Datamation, 22 (2): 86–94.
Day, J. B. and C. B. Shields. 1973. Prospects for change in urban transportation. Research Outlook (Battelle), 5 (1): 16–21.
Demag/MBB. 1974. The personal rapid transit system "Cabinentaxi" – a description of the system. A Report from DEMAG, MBB, FRG (in German).
DHUD. 1968. Tomorrow's transportation – new systems for the urban future. Washington, D.C.: USA Department of Housing and Urban Development.
DOT. 1974. Innovation in public transportation: a directory of research, development and demonstration projects. Washington, D.C.: US Department of Transportation, Urban Mass Transportation Administration.
Filion, A. 1975. Urban Transport – An Innovation Policy. Paris: Editions Eyrolles (in French).
Hamilton, C. W. and J. K. Wetherbee. 1978. Ground transportation in the US. Research Outlook (Battelle), 5 (1): 6–15.
Hamilton, W. F. and D. K. Nance. 1969. Systems analysis of urban transportation – computer models of cities suggest that in certain circumstances installing novel "personal transit" systems may already be more economic than building conventional systems such as subways. Scientific American, 221 (1): 19–27.
Kieffer, I. A. 1972. The success of the auto should be a lesson to us. In J. E. Anderson et al. (eds.) Personal Rapid Transit – A Selection of Papers on a Promising New Mode of Public Transportation. Minnesota: University of Minnesota, Institute of Technology.
Loder. J. L. 1974. Personal automated transportation – a PAT solution for Australian cities. Transportation Planning and Technology, 2: 221–262.
MITI. 1975. Comprehensive Automobile Control Systems. Preliminary Report by the Japanese Ministry of International Trade and Industry, Tokyo, Japan.
Ross, H. R. 1969/70. Future development in personal transit. Proceedings of the Institution of Mechanical Engineers, 184 (3S): 38–53.
Volpe, J. A. 1969. Urban transportation tomorrow. The American City: 59–62.

3 New Technology Options: Towards Computerized Transport Control

The possible contribution of new technologies is restricted to the following two levels of the general urban and transport development hierarchy according to Figures 2.1 and 2.2:

- introduction of operational innovations into existing transport systems (short-term supply strategies with a time horizon of several months or years)
- creation of completely new public transportation systems that meet the three categories of objectives summarized in Section 2.1 in the sense of an optimal compromise (long-term supply strategies with a time horizon of one to several decades).

3.1 SHORT-TERM STRATEGIES: OPERATIONAL INNOVATIONS

Chapter 1 illustrated that congestion, especially in the form of the so-called stop-and-go driving regime (cf. point E in Figure 1.1), can lead to a dramatic increase in accident frequency, as well as in the levels of fuel consumption, air pollution (cf. Figure 1.4), and time losses. Thus, the question arises whether the flow conditions can be improved for given automobile-traffic-demand patterns and volumes in such a way that congestion will not occur or will occur less frequently and less severely.

One promising way of reaching this target is to operate the urban highway system so that the capacity of the road network is automatically adapted, as far as possible, to the automobile transportation demand changing in time and space. This would required a solution to the following three tasks (cf. Figure 3.1):

- measuring the state of traffic flow in the various parts of the network by means of automatic traffic detectors

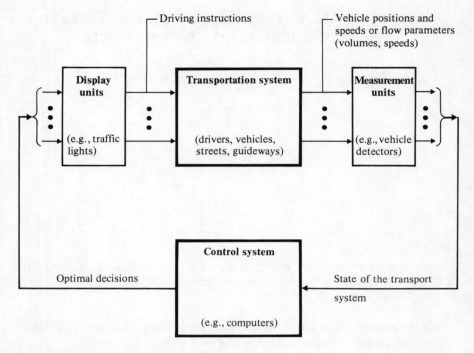

FIGURE 3.1 Simplified diagram of a traffic control system.

- transmitting this information to a control center where it is analyzed and used to determine an optimal control strategy that will allow the most effective use of the available traffic areas
- displaying the optimal control instructions to the drivers by means of traffic lights, speed signals, route signals, etc.

All three tasks have to be solved for a large area within seconds, i.e., in a real-time operation mode. High speed control computers, with the capability of handling a large amount of data in a short time are obviously needed for this purpose. These computers became available during the sixties and led to the first traffic-responsive computerized traffic-flow control systems.

The creation of such systems is one of the major efforts within the framework of new technology based short-term supply strategies, which aim to optimize the operation of the urban highway transportation system (cf. Figure 2.3). This application of modern computer control technology represents the subject of Part 2 of this monograph.

A second category of short-term measures concerns the following improvements in the operation of public transport systems:

- in the case of buses and trams, to reduce time losses and deviations from timetables caused by the automobile traffic
- to increase the efficiency and attractiveness of public transport

Two possibilities have to be taken into account in reaching the first target:

- giving priority to public transport at intersections controlled by traffic lights
- supervising the operation of the whole rail, bus or streetcar system by a dispatching center that should have the capacity to identify the positions of all transport means, detect irregularities, and take measures for reducing or avoiding such irregularities.

Both concepts need the automated measurement of a large amount of data and the determination of control actions in a real-time operation mode. Thus, their feasibility also depends on the availability of powerful computer surveillance and control systems.

The second target mentioned above concerns improvements in the operation of public transport systems that could lead to a reduction of operational costs, including personnel levels and energy consumption, as well as to an increase in line and station capacities and in service frequency. One approach for reaching these objectives is to apply advanced automation and computer technology extensively, especially to urban railway systems.

Another way of increasing the attractiveness of public transport is to introduce operational innovations. Transport service in terms of bus routes and schedules can be adapted to stochastically changing trip demands, i.e., trip origins, destinations, and starting times. Such demand-responsive bus systems, known under the names "demand-bus", "dial-a-bus", or "dial-a-ride", require the implementation of a dispatching center with the potential to solve the following tasks within a short period of time: (1) collect trip requests sent to the dispatching center by telephone calls or other means; (2) prepare optimal routes and schedules for the available fleet of vehicles (buses, taxis); (3) transmit these routes and schedules to the vehicles.

If the number of buses exceeds a certain limit, then it is necessary to use advanced communication and computer technology. This is analyzed along with other topics in Part 3, which is devoted to computer control and automation problems occurring in connection with urban public transport systems such as streetcar lines, bus systems, and urban railways.

3.2 LONG-TERM STRATEGIES: TOTAL SYSTEMS INNOVATIONS

What role will or can advanced computer and automation technology play in the framework of long-term supply strategies in accordance with Figure 2.3?

The expected contribution of this technology is concerned with the creation of completely automated, computer-controlled public transport systems, which are supposed to motivate many car drivers not to use their own cars, especially for their daily work trips, and which will have the potential of a demand-responsive, pollution-free, resource-conserving, and safe operation (cf. Figure 2.3). The reason why automation and computer contol are of vital importance for the creation of these new modes of urban transport becomes obvious if one tries to identify the dominant features of a hypothetical public transportation system, which should be able to fulfill the three categories of objectives described in Section 2.1. Figure 3.2 illustrates that the objectives independence, convenience, and travel speed of the public transit user lead to:

- small and comfortable vehicles that have a comparable size with the auto and can be used individually
- a guideway separated from the ordinary street network
- a dense traffic network with short walking distances to the stations
- a demand-responsive operation with nonstop origin–destination travel and small waiting times at stations, as well as operation around the clock

The objectives of the city, to preserve its social function, protect the environment, and conserve resources, require:

- pollution-free, quiet and energy-efficient vehicles
- narrow guideways
- a network that can be adapted to an existing city structure
- energy-saving operation

These objectives of the traveller and the city can only be met by a system that does not violate the basic criteria of the public transport company, namely, the required number of employees, the operational costs, safety and reliability, and integrability into existing transport systems. This leads to the following dominating features for the essential components of the new system (cf. Figure 3.2):

1. *Vehicles:* A large number of small vehicles cannot be operated economically by any public transport company if one employee has to be assigned to each vehicle. It follows that implementing the desired new system is only feasible if driverless, i.e., completely automated, operation of these vehicles is possible.

2. *Network:* A dense network with a large number of stations leads to a large number of employees dealing with various tasks such as ticket selling, watching vehicle operation and pedestrian traffic. Therefore, automated passenger information and guiding systems are needed to limit or reduce the number of employees.

3. *Operation:* Demand-responsive, nonstop, origin–destination travel obviously requires the coordinated control of the operation of all vehicles and all stations, i.e., of the whole system. Such a complicated control task cannot be solved by human operators. Therefore, the implementation of an automated computer control

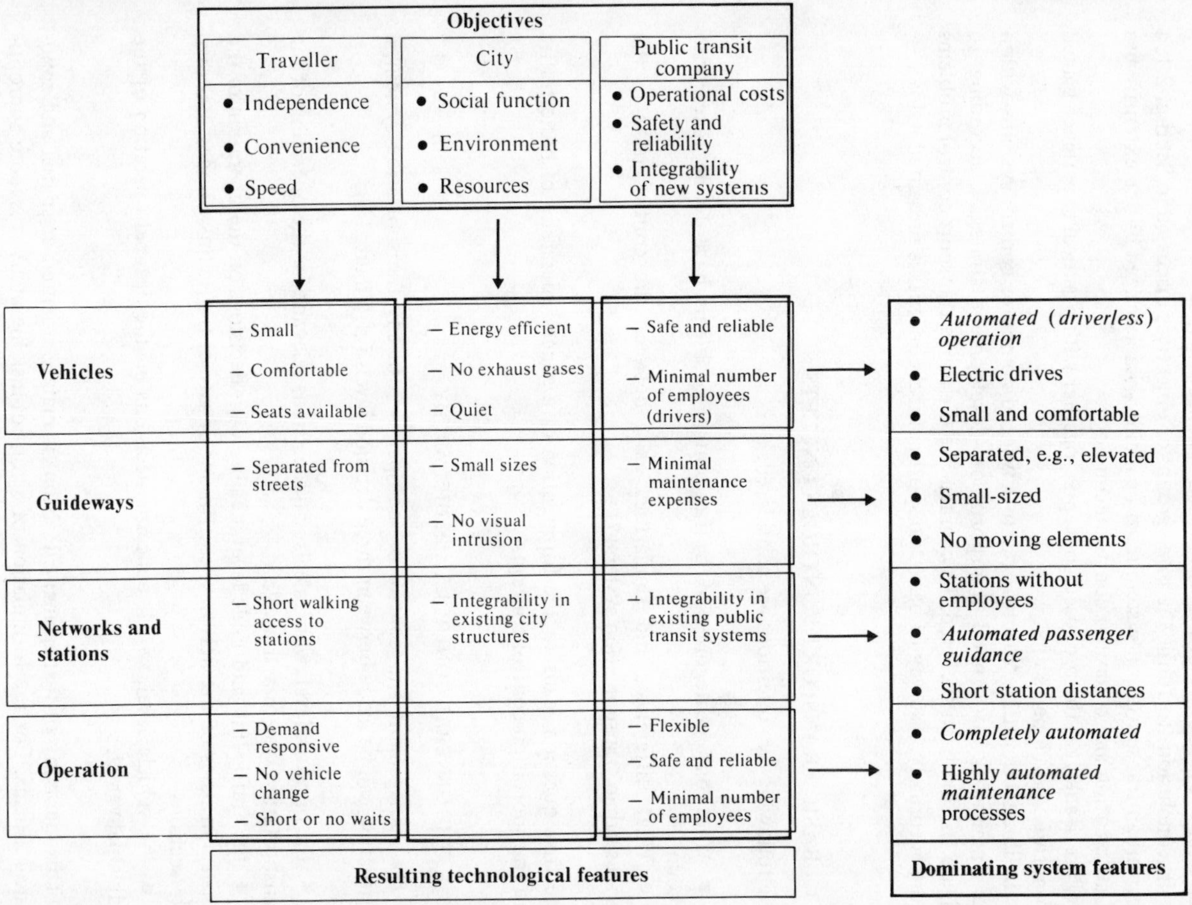

	Objectives		
	Traveller • Independence • Convenience • Speed	City • Social function • Environment • Resources	Public transit company • Operational costs • Safety and reliability • Integrability of new systems
Vehicles	– Small – Comfortable – Seats available	– Energy efficient – No exhaust gases – Quiet	– Safe and reliable – Minimal number of employees (drivers)
Guideways	– Separated from streets	– Small sizes – No visual intrusion	– Minimal maintenance expenses
Networks and stations	– Short walking access to stations	– Integrability in existing city structures	– Integrability in existing public transit systems
Operation	– Demand responsive – No vehicle change – Short or no waits		– Flexible – Safe and reliable – Minimal number of employees

Resulting technological features

↓

Dominating system features

- *Automated (driverless) operation*
- Electric drives
- Small and comfortable

- Separated, e.g., elevated
- Small-sized
- No moving elements

- Stations without employees
- *Automated passenger guidance*
- Short station distances

- *Completely automated*
- Highly automated maintenance processes

FIGURE 3.2 Dominating system features of new urban transportation systems resulting from an evaluation of the objectives of the traveller, the city, and the public transit company.

system that supervises and guides the vehicles through the network is of vital importance for the feasibility of the whole system concept.

In conclusion, to fulfill the three groups of objectives discussed in Section 2.1, a new urban transportation system must be mainly characterized by the extensive use of automation and computer control technology.

Part 4 analyzes the proposed concepts, together with practical experience gained from those developed so far.

In this section the expected role of advanced computer control systems within the framework of general transportation development strategies according to Figure 2.3 has been sketched. To characterize the urban transport control problems more accurately, a survey of basic transport control concepts is presented below.

3.3 BASIC TRANSPORT CONTROL CONCEPTS

The following two questions are considered:

- What basic categories of traffic control problems have to be taken into account?
- What differences and similarities exist between the control tasks in the various urban transportation systems?

The first question leads to the formulation of a general hierarchy of basic traffic and transportation control concepts.

3.3.1 THE GENERAL CONTROL TASK HIERARCHY

In the management and control of the operation of any transportation system, a distinction can be made between three decision levels (cf. Figure 3.3):

- the network level, which deals with decisions relevant for the operation of the transportation system as a whole
- the traffic-link and node level, dealing with decisions on the operation of the vehicles in parts of the network, e.g., in one or several traffic links, nodes, stations, intersections
- the vehicle level, which concerns decisions on the optimal operation of the individual vehicles

This hierarchy of management decisions is mainly oriented at spatial and topological aspects. For the formulation of a control-task hierarchy it is more appropriate to focus on functional aspects, which refer to basic control tasks that dominate the operation of the system at the different levels. This consideration leads to the following general three-level control hierarchy (cf. Figure 3.3):

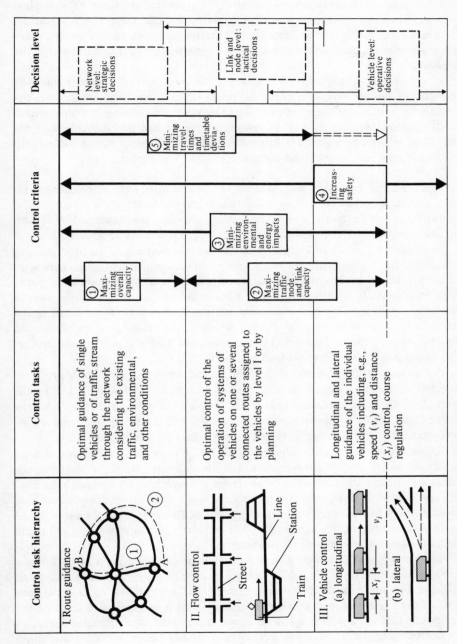

FIGURE 3.3 The hierarchy of traffic control tasks.

I. *Optimal route guidance and control,* i.e., optimal guidance of individual vehicles, of groups of vehicles, or of whole traffic streams, respectively, through a given network, taking into account the real traffic situations and other conditions like maintenance operations or environmental problems in the different parts of the network.

II. *Optimal flow control,* i.e., optimizing the operation of a larger number of vehicles on one or several connected routes, which have been assigned to the vehicles by operational transport planning or by the route-guidance system of level I.

III. *Optimal vehicle control,* i.e. ensuring optimal and safe movement of the individual vehicles taking both lateral vehicle guidance and longitudinal control into account, e.g., speed, position and distance regulation (cf. Figure 3.3).

This hierarchy of control tasks is closely related to a hierarchy of objective functions that have certain connections with the general objectives of car drivers, public transit users, public transit companies, and of the city, as summarized in Section 2.1. These control criteria can be classified in 1-5, as follows (cf. Figure 3.3).

1. *Maximizing the overall capacity of the network by equable use of all parts of the network.* This objective function aims to make optimal use of the network's capacity reserves, by optimal distribution of traffic loads over the available traffic links and nodes. It is obvious that only the control task at the top of the control hierarchy of Figure 3.3 can contribute to such a criterion. Therefore, optimal route-guidance systems may be considered an aid to creating — within certain limits — an equilibrium between traffic demand and traffic supply, thus contributing to the stability of the whole traffic system.

2. *Maximizing the capacity of critical links and nodes playing the role of bottlenecks.* This criterion can be fulfilled mainly by optimal flow control, i.e., by the level II of the control hierarchy. However, one should consider that for a special public transport mode, namely, new automated public transit, optimal vehicle control plays a key role in maximizing link capacity (cf. Part 4 and Figure 3.3).

3. *Minimizing the negative impacts of traffic on the environment (air and noise pollution) and on energy consumption.* The discussion in Chapter 1 illustrates the influence of congestion on air-pollution levels and energy consumption. Since control levels I and II in Figure 3.3 aim to reduce congestion, they will also contribute to minimizing fuel consumption and air pollution. However, control level III could also play a significant role in this respect, if one introduces energy-saving and pollution-reducing controls on the motorcar engine.

4. *Increasing traffic safety.* This objective can be directly influenced by the control task in level III, i.e., by vehicle control, and especially by means of automated collision prevention and other safety systems. In contrast, the relation between traffic-flow conditions and accident frequency shown in Figure 1.4 leads to the conclusion that route guidance and flow control may also improve traffic safety.

5. *Minimizing travel times and deviations from timetables.* Fulfilling this objective is the main subject of the flow control task at level II of the hierarchy. More-

over, route guidance systems should help to reduce travel times. In addition to the control task in level II, longitudinal control of vehicle movement can play an important role in minimizing deviations from timetables of public transport systems. This concerns the headway regulation problem in a string of moving vehicles such as bus or subway trains.

The control task hierarchy and the corresponding objectives formulated here and illustrated by Figure 3.3 are a useful "scheme of thinking" or framework for dealing with computer control problems occurring in the various modes of urban transport. It is obvious for automobile traffic, public transport systems, and new modes of (automated) urban transport that the different levels of the general control hierarchy and the corresponding control criteria will be modified and characterized by different levels of significance. The following consideration illustrates the relation between the general control hierarchy of Figure 3.3 and the specific control tasks of the various urban transport systems (cf. Figure 3.4 and Barwell 1974, Drew 1968, Gazis 1971, Dörrscheidt 1974, IFAC 1970, 1974, 1976, Strobel 1971, 1975, 1976, Strobel et al. 1977, Tabak 1973).

3.3.2 FREEWAY AND URBAN STREET TRAFFIC (cf. PART 2)

For the urban highway transportation system, the general control task hierarchy of Figure 3.3 can only partly be implemented in the form of a real hierarchically structured traffic control system. This concerns levels I and II:

- guidance of the main vehicular traffic streams through a network of freeways and surface streets
- optimal traffic-flow control at critical nodes and links, e.g., intersections, freeway ramps, tunnels, long bridges, (cf. Figure 3.4)

Clearly there is a close connection between these two categories of control tasks.

In systems currently in operation, the control tasks at level II play the dominant role in the form of well-known traffic light control and coordination systems for urban street networks, and ramp metering, speed control and merging control systems for freeways. A detailed analysis of the methodological aspects of these control concepts is presented in Chapter 5; Chapters 6 and 7 present experience obtained with real applications.

Level III, i.e., vehicle control, is currently of minor importance, since the drivers fulfill the tasks of vehicle controllers. However, the creation of microcomputers (cf. Section 3.4) seems to be opening the way for installing computer control systems onboard vehicles; these will not replace the driver but will assist him in obtaining better economy and safety (Adlerstein 1976).

The following control concepts are currently the subject of fundamental and applied research:

Control task hierarchy	Automobile traffic (cf. Part 2): Freeways and arterial roads (cf. Chapters 5 and 7)	Automobile traffic (cf. Part 2): Urban street networks (cf. Chapters 5 and 6)	Public transport (cf. Part 3): Bus systems (cf. Chapters 10, 11 and 12)	Public transport (cf. Part 3): Tram systems (cf. Chapters 10, 11 and 12)	Public transport (cf. Part 3): Rail transit (cf. Chaps. 10 and 13)	New (automated) transit systems (cf. Part 4, Chapters 16–19)
I. Route guidance 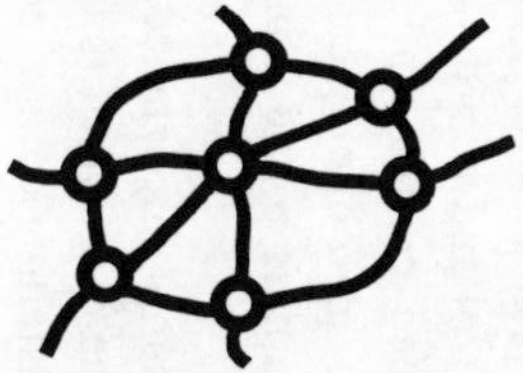	• Regional Traffic stream guidance	• Urban Traffic stream guidance	• Demand bus systems	Not applicable (except emergency control)	Not applicable (except emergency control)	• Network control (route guidance and empty car distribution)
II. Flow control 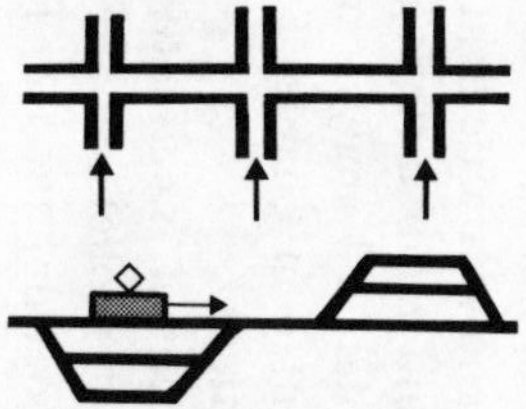	• Ramp metering • Speed control • Merging control	Traffic light coordination and control	• Bus and tram traffic-light priority control systems • Bus and tram monitoring and control systems	• Bus and tram traffic-light priority control systems • Bus and tram monitoring and control systems	• Automated train operation surveillance and control	• Flow control (automated merging, and diverging)
III. Vehicle control (a) longitudinal 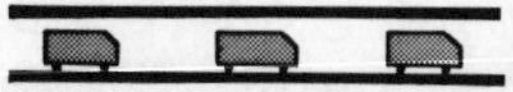	• Distance regulation • Anti-blocking control • Engine control	• Distance regulation • Anti-blocking control • Engine control	• Headway regulation	• Headway regulation	• Headway regulation • Speed control	• Automated speed, position and headway control
(b) lateral 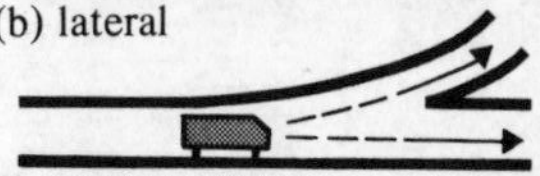	Not applicable	Not applicable	Not applicable	Not applicable	• Automated interlocking systems	• Automated lateral vehicle guidance

FIGURE 3.4 The applicability of various control concepts.

- collision prevention systems, i.e., distance regulation in a string of vehicles using onboard radar measuring devices
  - anti-locking wheel brakes (cf. Adlerstein 1976)
  - energy-saving and pollution-reducing controls on engines (cf. Figure 3.4)

The limits and potentials of these control concepts are analyzed in Chapter 5.

3.3.3 PUBLIC TRANSPORT SYSTEMS (cf. PART 3)

In conventional public transport systems, e.g., urban railways, streetcars, and bus systems, the selection of routes in a network is not a subject of control but of operational planning. Therefore, level I of the general control task hierarchy according to Figures 3.3 and 3.4 does not usually exist. Operational rerouting becomes important only in emergency situations, i.e., in the case of destroyed streets and rails, or if a tram is out of order.

However, there is a concept for a public transport system that is based on operational route guidance and control; this concerns the demand-responsive bus operation systems known as dial-a-ride, dial-a-bus, or demand-bus mentioned in Section 3.1. Chapters 10 and 11 show that real-time routing and scheduling are the main features of this public transport system, which obviously can no longer be considered a normal bus system (cf. Figure 3.4).

Level II of the general control task hierarchy of Figures 3.3 and 3.4 is concerned with the different control problems of various public transport systems:

1. One basic flow-control concept is to integrate buses and streetcars using the same traffic areas as the automobile into the traffic light control system. Here, the particular aim is to implement priority control schemes, which ensure that a bus or a tram occupied with 50 or 100 passengers experiences less delay at an intersection than an automobile used by only one to four persons (cf. Chapter 10).

2. Automated monitoring and computer-aided control of the operation of the whole bus or tram system is the subject of a second control concept. The main objective of such computer-assisted dispatching systems is to discover irregularities as quickly as possible, and to take proper measures against them without significant delay. Chapters 10 and 12 discuss the potential of this concept in detail.

3. For urban railway systems the control problem in level II is mainly concerned with computer-assisted surveillance and coordinated control of the operation of trains in large stations and densely occupied lines. The control objective is similar to that mentioned earlier for the bus and tram dispatching system, i.e., reducing delays and other irregularities. Moreover, an increase in station and line capacity is expected from these systems (cf. Figures 3.3 and 3.4, and Chapters 10 and 13).

Control level III of the general hierarchy is of minor importance for bus systems and streetcar lines. An exception is the regulation of time-distances, i.e., of the headways between buses.

For railway systems, the longitudinal and lateral control of the movement of the individual trains represents — as is well known — a major area for the application of automation and control technology. Typical control tasks are:

- regulation of the headways between the individual trains of a subway line
- speed control and target-stopping control
- energy optimal control, i.e., minimizing traction energy
- automated switching and signaling

The present status and current trends concerning the use of computers for these control tasks is analyzed in Chapters 10 and 13.

3.3.4 NEW (AUTOMATED) TRANSIT SYSTEM (cf. PART 4)

To produce a completely automated transit system in which the basic features and objectives of the new urban transport concept outlined in Section 3.2 (cf. Figure 3.2) are not lost, it is necessary to implement a hierarchically structured control system containing all three levels of the general tasks hierarchy of Figures 3.3 and 3.4. Thus, a sophisticated longitudinal and lateral vehicle control system is needed to make safe driverless operation of the vehicles feasible. If very small vehicles, e.g., motorcars containing from four to six people, are to be used, then a powerful headway control system is of vital importance in reaching a sufficiently large link capacity (cf. Chapter 16).

The control task in level II, i.e., flow control, concerns the operation of a system of driverless vehicles at conflicting points of the network as, for example, at intersections, merging points, stations (cf. Figure 3.4 and Section 16.5).

Finally, the route guidance task at the top of the hierarchy consists of two parts (cf. Section 17.6):

- automated guidance of the individual driverless vehicles from a certain origin to a certain destination, taking into account the traffic status in the different parts of the network
- guiding empty cars to those stations where they are needed

It is obvious that there are close connections between the three levels of control hierarchy. Therefore, the complete implementation of the general control tasks hierarchy of Figures 3.3 and 3.4 is indispensable. This is discussed in Chapter 16, and Chapters 17-19 analyze the practical experience gained so far with new public transport systems, especially with respect to the concepts and methods used for computerized control and guidance.

The implementation of the control concepts discussed here, and the creation of the corresponding complex and complicated computer control systems, will only be technically and economically feasible if high level computer technology is available at acceptable costs. However, will the computer technology available now or in the near future permit the implementation of the highly sophisticated transport control systems that are expected to make a significant contribution to the solution of present and future urban traffic problems?

The technical and economical feasibility of the short- and long-term strategies discussed in Sections 3.1 and 3.2 of this chapter are considered in the following. For this purpose, a survey of the impact of developments in computer technology on the creation of transport control systems is presented.

3.4 FEASIBILITY CONSIDERATIONS: IMPACTS OF THE COMPUTER REVOLUTION

The contribution of advanced computerized traffic control systems to urban traffic problems cannot be understood completely without considering that computer technology has been up till now, and still is, in a process of revolutionary development.

Since the creation of the first usable digital counter during the forties, four computer generations have been developed, used, and partly phased out (cf. Figure 3.5):

- first generation computers using electromagnetic relays and electronic vacuum tubes (until about 1960-1965)
- second generation computers using transistors and other discrete semiconductor components (about 1959-1972)
- third generation computers characterized by the use of integrated circuits (IC) performing the functions of a large number of discrete components, e.g., transistors, by means of a single semiconductor chip (about 1964-1980)
- fourth generation computers using large scale integrated (LSI) circuits, e.g., so-called microprocessors, containing all the functions of a complete central processing unit (CPU) on a single semiconductor chip

The next generation of computers is already available. This new generation, using very large scale integrated (VLSI) circuits, brings an operational computer consisting of a CPU, a medium-sized memory and an input/output control unit on a single semiconductor chip (cf. Faggin 1977).

What are the consequences of this development with respect to the application of computers to traffic control purposes?

3.4.1 THE BEGINNING: LARGE-SIZED AND EXPENSIVE PROCESS COMPUTERS

Figure 3.5 illustrates that the first digital computers that could be used for control tasks in an online operation mode according to Figure 3.1 appeared on the market around 1960 in the form of second generation machines. They were characterized by the following general features (cf. Williams 1969, Anke et al. 1970):

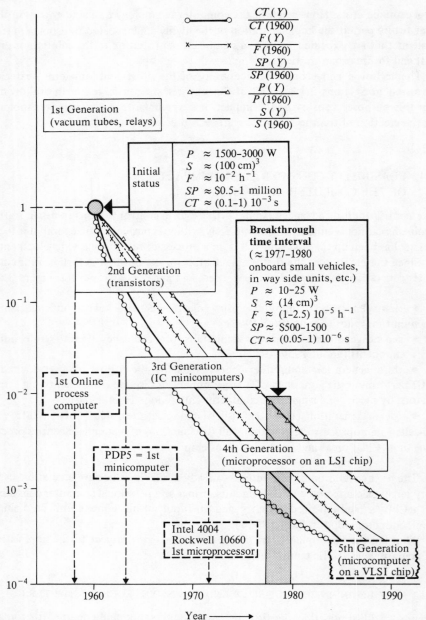

FIGURE 3.5 The various computer generations and the time-dependent development of selected parameters (rough estimates for a minimal (standard) configuration including a 4 K words memory, input/output control unit etc.): P, power consumption; S, size; F, failure rate; SP, selling price; CT, mean computing time per addition.

- high costs of the order of about $0.5-1 million
- low reliability given by a mean time between failures (MTBF) of approximately 100 hours, i.e., a breakdown of the control system could be expected on average every 100 hours
- relatively low computing speed of about 1,000-10,000 additions per second
- a large size, i.e., the process computer covered the space of several huge racks
- large power consumption of the order of one to several kilowatts and requiring special facilities like a motor generator set
- strict requirements concerning the working environment, i.e., regarding the admissable ranges of temperature, humidity, vibrations, etc., which necessitated the use of air-conditioned computer rooms
- availability of a relatively large number of well-skilled personnel for operation and maintenance

Because of these initial problems the number of process computer applications remained low until the next generation became available.

3.4.2 MINI- AND LOW-COST COMPUTERS: BROAD APPLICATION OF CONTROL COMPUTERS

The breakthrough to the broad application of process computers occurred in connection with the introduction of the second computer generation, i.e., that of the so-called mini- or low-cost computers, around the year 1965. During the first half of the seventies, more than 150 different types of these minicomputers, characterized by the extensive use of IC technology, were developed and produced by more than 70 companies. In 1972, the number of installed minicomputers was estimated to be in the order of about 10,000 (cf. Hollingworth 1973, Jürgen 1970, Köhler 1973, Kull 1972, Münchrath 1973, Theis and Hobbs 1971). This success is understandable if one considers the dramatic change of the five selected parameters price, computing time, reliability, power consumption, and size as a function of time (cf. Figure 3.5). Compared with the first process computer of the year 1960 these parameters were improved by factors of the order of about 20-200 within ten years.

The progress in performance-to-cost ratio led to the extensive use of minicomputers for implementing computerized traffic control systems. However, in most cases their application was — and still is — restricted to an operation in special central control rooms. The use of minicontrol computers onboard relatively small vehicles, such as automobiles, trucks, buses, locomotives, and other transportation means, as well as in roadside units, and ticket-selling machines, was in general not feasible for technical and economical reasons.* Figure 3.5 illustrates that the breakthrough to this qualitatively new area of application was made feasible by the development of the fourth computer generation, i.e., that of microprocessors and microcomputers.

---

* There are, of course, several exceptions especially with respect to spaceflight and military applications (cf. Frost 1970).

3.4.3 THE MICROCOMPUTER: A BREAKTHROUGH TO QUALITATIVELY NEW APPLICATION AREAS

A microprocessor is the CPU of a microcomputer on a single semiconductor chip.* The term chip denotes a very thin slice of semiconductor material in the form of a square or a rectangle with a surface area from $(4 \text{ mm})^2$ to $(7 \text{ mm})^2$. The microprocessor has the potential to perform logic and arithmetic operations, i.e., decode and execute instructions, as well as to handle input/output operations and perform synchronization and control functions.

A microprocessor has to be complemented by additional units to become an operational computer:

- random access memory (RAM)
- read only memory (ROM)
- input/output devices (I/O)
- interface components

One basic feature of the RAMs and ROMs is that they are also produced as LSI chips. RAM chips are primarily used for variable data and ROM chips for storing instruction sequences and invariant, important numbers. The data in the ROMs must be stored at the time they are created, so there is a production delay associated with them, as well as a certain programming cost. Thus, so-called programmable read only memories (PROMs) have been developed, which can be erased by ultraviolet light and reprogrammed, and which are used in place of ROMs when only small quantities are required. The introduction of ROMs and PROMs represents great progress from the reliability point of view, since their memory content cannot be changed or lost by operator mistakes, electric disturbances, breakdown of electric power supply, etc.

The fabrication technologies for LSI circuits are still in a process of rapid development (cf. McGlynn 1976, ERA 1973, Wolf 1974). A distinction can already be made between four microprocessor generations (McGlynn 1976).

- The first generation of microprocessors, based on the PMOS technology, leads to LSI chips that contain about 9,000 transistor functions on an area of about 40 mm^2. The first representative of this generation appeared in the form of the 4-bit processor Intel 4004 in 1971 (cf. Figure 3.5). Microprocessors of this generation are characterized by "calculator"-type chips with a limited instruction set and a relatively low speed (Cushman 1974b).
- In 1973/74 the use of NMOS technology permitted the creation of the second microprocessor generation characterized by a higher chip density of about 11,000–19,000 transistor functions per chip, a larger instruction set (up to 80 instructions) and higher speed; the 8-bit processor Intel 8080 is a typical example

* See Table 3.1 for an explanation of these and other terms from microelectronics.

TABLE 3.1 Selected Terms from Microelectronics (cf. Theis 1974)

Term	Definition
1 *Computers, processors, memories*	
MC	Microcomputer is a computer that uses a central processing unit (CPU) in the form of a microprocessor, and LSI memories in the form of RAMs, ROMs, PROMs etc.
MP or μP	Microprocessor is a CPU on a single semiconductor chip
RAM	Random access memory is a memory on an LSI chip with both read and write capabilities
ROM	Read-only memory is a memory on an LSI chip that cannot be rewritten; ROM requires a masking operation during production to record program or data patterns permanently in it
PROM	Programmable read-only memory is a memory type that is not recorded during its fabrication; some PROMs can be erased by means of ultraviolet light and reprogrammed
2 *Integrated circuits, fabrication technologies*	
IC	Integrated circuit is a complex electronic circuit fabricated on a single chip
Chip	A small piece of silicon impregnated with impurities in a pattern to form transistors, diodes, and resistors. Electrical paths are formed on it by depositing thin layers of aluminum or gold
LSI	Large scale integration refers to a component density of more than 100 transistors per chip
Bipolar	The most popular fundamental kind of IC, formed from layers of silicon with different electrical characteristics
T^2L or TTL	Transistor-transistor logic, a kind of bipolar logic
I^2L	Integrated injection logic, an advanced kind of bipolar logic characterized by high computing speed, low power dissipation and high component density
MOS	Metal oxide semiconductor, a term referring to the second basic (unipolar) process for fabricating ICs. MOS circuits achieve the highest component densities
PMOS	P-channel MOS refers to the oldest type of MOS circuit, where the electrical current is a flow of positive charges
NMOS	N-channel MOS circuits use currents made up of negative charges resulting in processors twice as fast as PMOS devices
CMOS	Complementary MOS refers to a combination of PMOS and NMOS technology that results in processors as fast as those based on NMOS, but consuming less power

(Cushman 1974a, Shima and Faggin 1974). The use of CMOS technology led to a further increase in computing speed and to a further decrease in power dissipation.

• The third microprocessor generation, which appeared around 1976, made use of the faster Schottky bipolar technology and a more sophisticated microprocessor architecture. The resulting processors are characterized by a large instruction set, and by a computing speed that is more than ten times faster than that of the first generation. The Intel 3001/2 is typical of this generation (McGlynn 1976).

- The fourth generation, which first appeared in 1976 in the form of the Texas Instruments SPB 0400, is based on the most advanced bipolar technology, i.e., integrated injection logic (I^2L). The basic features of this latest generation are highly sophisticated microprogramming structure, maximal chip density of the order of about 20,000-30,000 transistor functions per chip, and minimal power consumption at high speed.

The cost of the last two generations based on bipolar technologies is in general from two to ten times higher than that of MOS processors. However, the rapid price decrease resulting from the continuous improvements in fabrication technologies has to be taken into account: between 1970 and 1975 price decreases by factors from six to ten and from three to six occurred for the bipolar and MOS technologies respectively (McGlynn 1976).

3.4.3.1 The impact of microelectronic technologies. The extensive use of LSI technology enabled the construction of control microcomputers that are characterized by the following features (cf. Figure 3.5):

1. A control microcomputer will become available at a price as low as $200-1,500 or even less ($50-200 for minimal configurations) during the next five years; the price for the microprocessor chip itself will approach a level of about $5-50.
2. The computing speed and the memory size will be large enough to handle even complicated control tasks in a real-time operation mode.
3. The reliability, expressed by the MTBF is supposed to reach about 10^4-10^5 hours, i.e., using sufficiently large redundancy in the form of doubled or tripled microcomputers it will be possible to fulfill safety standards comparable with those required in railway signalling systems.
4. The power required drops to values of about 10-30 watts, which can easily be provided, e.g., by an auto battery.
5. Increased robustness is another significant feature; the admissable temperature range will approach standards valid for military applications.
6. The whole operational control microcomputer can be put into a box small enough to be installed in even small machines, devices, vehicles, etc.

To judge the impact of the microcomputer on the development of transportation systems it is useful to make a comparison with the consequences of the invention of the electric motor or internal combustion engine (cf. Figure 1 in the Introduction).

The electric motor and the internal combustion engine provided individual vehicles like the automobile, the streetcar, the motorcycle, as well as production and other (small) machines with their own driving unit. This produced the well-known changes to the structure of transport systems (cf. Figure 1 in the Introduction) and to the way of industrial production.

The control microcomputer will enable even small vehicles, machines, and devices to have their own computer intelligence (cf. Figure 3.5), permitting their optimal control in the sense of more effective and safe operation. Thus, the present state of microcomputer technology has already created a qualitative breakthrough to new areas of application of computer control. Control microcomputers have already been installed onboard small vehicles like automobiles (McGlynn 1976, Temple and Devlin 1974), trucks, buses, streetcars, locomotives, as well as in roadside units as, for example, in traffic light control cabinets at an intersection, and in small devices like sophisticated ticket selling and passenger information equipment (cf. Figure 3.5).

It is clear that the new computer generation will influence the whole philosophical approach to implementing computerized traffic control systems. Some of these consequences are discussed in the following, with regard to the problem of structuring a complex computer control system.

3.4.4 COMPUTER CONTROL CONCEPTS: CENTRALIZED OR DISTRIBUTED SYSTEMS

One basic feature of a transportation system is its large spatial extension and large number of elements. Therefore, the optimal management of such a large-scale system requires the creation of a dispatching center that has to be provided with information on the state of the system and that should have the potential to send control instructions to the vehicles and to stationary units.

The first generation of central dispatching systems was created more than twenty years ago in the form of conventional remote-control systems using hardwired electronics (cf. Figure 3.6). The main disadvantage of this central control system was that it did not possess the capabilities of evaluating automatically a large amount of measured data in a real-time operation mode. The capability to determine an optimal operation regime for a transportation system was first provided by the second central control system generation, i.e., centralized computer control (cf. Figure 3.6, and Williams 1969, Anke et al. 1970).

The high costs of the first process computers (cf. Figure 3.5) permitted the installation of only one computer in a control center, as a so-called stand-alone system. However, reliability considerations motivated the use of two or three computers, if basic cost-benefit considerations were not violated. Figure 3.6 (cf. column "cost aspects") illustrates that in the early days of computer control the use of control computers was only more economical than the installation of conventional hardwired logic for complex control tasks, e.g., controlling traffic lights in an urban street network. The Toronto area traffic control system, which was put into operation at the beginning of the sixties, represents such an example (cf. Chapter 6).

The emergence of relatively small and cheap minicomputers between 1965 and 1975 (cf. Figure 3.5) revealed two potential impacts for computerized traffic control systems (Hollingworth 1973, Jürgen 1970, Theis and Hobbs 1971):

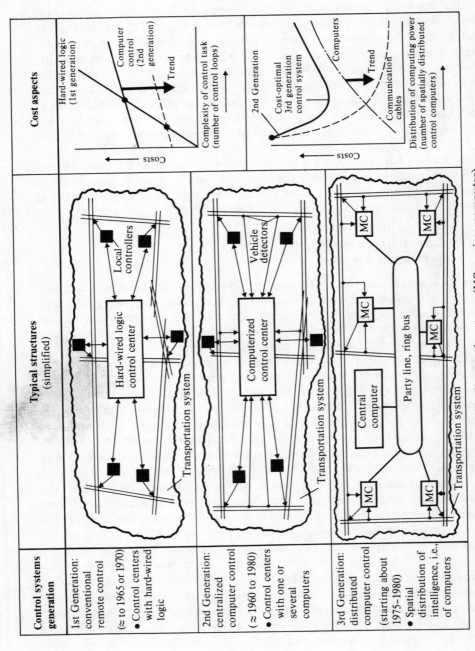

FIGURE 3.6 The three generations of centralized control systems (MC = micro-computer).

1. The installation of a computer control center became feasible from an economical viewpoint, even for relatively small transportation processes like the control of a single intersection.

2. The minicomputer allowed the functional decentralization of complex control tasks, i.e., the distribution of the various control functions to a set of smaller and more reliable computers located in the control center.

These two features of the application of minicomputers lead to the following results:

- even very complex transportation systems can be safely and effectively controlled by one computer control center; significant examples are the area traffic control centers of Tokyo and Osaka, which cover several thousands of intersections (cf. Section 6.3)
- the number of computer controlled transportation systems of medium or low complexity has increased rapidly

In conclusion it can be stated that the second generation of centralized control systems using third generation minicomputers (cf. Figure 3.5) represents a fairly well developed and proven technology, which is used with remarkable success in many countries for the various modes of urban transportation (cf. Chapters 6, 7, 11-13, 17-19).

In spite of this success, several problems connected with the use of centralized computing power remain unsolved:

- a breakdown of the control center may cause serious problems concerning the operation of the whole transportation system
- centralizing the intelligence requires the installation of a large number of information-transmission lines that often represent the most expensive part of the whole control system
- transmitting a large amount of data to and from the control center can raise serious reliability problems
- the necessary restrictions on the data transmission capacity mean that the control center can only be partly informed about the real state of the transportation system; this limits the capacity of the control systems to react in an optimal manner to changing traffic conditions

To overcome these difficulties it is necessary to spatially decentralize the intelligence needed for determining control decisions by means of measured data. The feasibility of this so-called distributed computer control concept (cf. Baily 1974, Färker 1973, Levine 1974, Neubert 1974, Reyling 1974), which leads to the third generation of centralized control systems (cf. Figure 3.6), depends on the availability of small, cheap, reliable, and robust computers that can be operated in a rough environment, e.g., in roadside units or onboard vehicles.

Figure 3.5 illustrates that a breakthrough to this new area of computer control has been created during recent years by introducing LSI microprocessors and microcomputers. The impact of the new computer generation on implementing transportation control systems is summarized as follows:

1. Even for control tasks of low complexity, the microcomputer provides a more economical solution than conventional hardwired logic (cf. Figure 3.6, upper diagram); thus, microcomputers are expected to replace a large part of conventional electronic control systems in various applications, e.g., vehicle guidance and control by means of microcomputers onboard locomotives, buses, trucks, and automobiles.

2. The significant increase in the reliability (cf. Figure 3.5) of LSI technology permits computer applications to control tasks with safety requirements, e.g., in railway signalling and train control systems.

3. The creation of the third generation of centralized control systems with spatially distributed microcomputers (cf. Figure 3.6) will result in:

- a reduction of the overall costs of the whole control system by decreasing the expenses for the communication cables (cf. the second diagram in the column "cost aspects" in Figure 3.6)
- an improvement of the total systems reliability by making it insensitive to failures of a single decentralized computer
- low installation costs for the system and the capacity for incremental expansion
- a simplification of software development
- improvements concerning traffic-responsive control by increased local data-processing capabilities

REFERENCES

Adlerstein, S. 1976. Transportation: computers are everywhere − in cars, trucks, trains, and even taxi meters. Electronic Design, 24 (22): 66−71.
Anke, K., H. Kaltenecker, and R. Oetker. 1970. Process Computers. Munich, FRG: R. Oldenbourg Verlag; Berlin, GDR: Akademie Verlag. (In German).
Baily, S. J. 1974. Microprocessor: candidate for distributed computing control. Control Engineering, 21 (3): 40−44.
Barwell, F. T. 1974. Automation and Control in Transportation. Oxford: Pergamon Press.
Cushman, R. H. 1974a. Intel 8080: the first of the second-generation microprocessors. EDN, 19 (9): 30−36.
Cushman, R. H. 1974b. Don't overlook the 4-bit microprocessors: they're here and they're cheap. EDN, 19 (4): 44−50.
Dörrscheidt, F. 1974. Applications of control engineering in ground transportation. VDI-Zeitschrift, 116 (3): 217−224 (in German).
Drew, D. R. 1968. Traffic Flow Theory and Control, New York: McGraw-Hill.
ERA. 1973. Microprocessors: an ERA assessment of LSI computer components. Report by the Electrical Research Association, Computers and Automation Division. London: Orum.
Faggin, F. 1977. The future of microelectronics and microcomputers. Paper No. 10 presented in Section O of the World Congress of Electrical Engineering, Moscow, USSR.

Färker, G. 1973. Decentralized data-processing concepts by means of microprocessors. On-Line, 11 (11): 784–789 (in German).
Frost, C. R. 1970 Military CPU's. Datamation, 16 (12): 87–103.
Gazis, D. C. 1971. Traffic control from hand signals to computers. Proceedings IEEE, 59 (7): 1090–1099.
Hollingworth, D. 1973. Minicomputers: a review of current technology, systems, and applications. Report R-1279, Santa Monica, CA: The Rand Corporation,.
IFAC. 1970. Traffic Control and Transportation Systems. Proceedings of the First International IFAC/IFIP/IFORS Symposium, Versailles, France.
IFAC. 1974. Traffic Control and Transportation Systems. Proceedings of the Second IFAC/IFIP/IFORS World Symposium, Monte Carlo, Monaco. Amsterdam: North Holland Publ. Co.
IFAC. 1976. Control in Transportation Systems. Proceedings of the Third International IFAC/IFIP/IFORS Symposium, Columbus, Ohio. Published by International Federation of Automatic Control, and distributed by the Instrument Society of America, Pittsburgh, Pennsylvania.
Jürgen, R. K. 1970. Minicomputer applications in the seventies. IEEE Spectrum: 37–52.
Köhler, R. 1973. Computer of the future: situation and development trends. On-Line, 11 (5): 342–350 (in German).
Kull, W. F. E. 1972. Compact computers, structure, model survey, comparable criteria, applications, Elektronik, 21: 233–240, 271–274, 309–314 (in German).
Levine, S. T. 1974. Decentralized networks allowing the computer to be moved to the job. Electronic Design, 22 (9): 66–71.
McGlynn, D. R. 1976. Microprocessors: Technology, Architecture, and Applications, New York: John Wiley and Sons.
Münchrath, R. 1973. Tendencies of minicomputer development. Elektronik, No. 10: 345–352 (in German).
Neubert, H. 1974. Systems technique using microcomputers. Elektronik, 23 (10): 319–395 (in German).
Reyling, G. Jr. 1974. Performance and control of multiple microprocessor systems. Computer Design, 13 (3): 81–86.
Shima, M. and F. Faggin. 1974. In switching to n-MOS microprocessor one gets a 2-microseconds cycle time. Electronics, 47 (8): 95–100.
Strobel, H. 1971. State-of-the-art and development tendencies for process computer applications in transportation. Wiss Z. d. Hochschule für Verkehrswesen "Friedrich List", Dresden, 18 (4): 759–787 (in German).
Strobel, H. 1975a. Computerized urban traffic control and guidance systems. In IIASA Project on Urban and Regional Systems: A Status Report. Laxenburg, Austria: International Institute for Applied Systems Analysis, pp. 48–54.
Strobel, H. 1975b. Transportation, automation and the quality of urban living. IIASA Research Report RR-75-34. Laxenburg, Austria: International Institute for Applied Systems Analysis.
Strobel, H. 1976a. Transportation, automation and urban development. Paper presented at the IIASA Conference '76, Laxenburg, Austria. Laxenburg, Austria: International Institute for Applied Systems Analysis, Vol. 2 pp. 85–91.
Strobel, H. 1976b. Computerized urban traffic control systems. Plenary paper, in Proceedings of the IFAC Symposium, Udine, Italy.
Strobel, H., R. Genser, and M. M. Etschmaier. 1977. Optimization applied to transportation. Proceedings of an IFAC/IIASA Workshop held in Vienna, Austria, February 1976. IIASA Collaborative Paper CP-77-7, (IIASA), Laxenburg, Austria: International Institute for Applied Systems Analysis.
Tabak, D. 1973. Application of modern control and optimization techniques to transportation systems. Advances in control theory, 10: 345–435.
Temple, R. H. and S. S. Devlin. 1974. Proceedings of the second IFAC/IFIP/IFORS World Symposium, Monte Carlo, Monaco. Amsterdam: North Holland Publ. Co. Use of microprocessors as automobile onboard controllers. IEEE International Convention Technical Papers, Paper No. 24/3.
Theis, D. J. and L. C. Hobbs. 1971. The minicomputers revisited. Datamation, May: 24–34.
Theis, D. I. 1974. Microprocessor and microcomputer survey. Datamation. 20 (12) 90–91, 96–100.

Williams, T. J. 1969. Interface requirements, transducers, and computers for on-line systems. Survey paper presented at the Fourth IFAC World Congress, Warsaw, Poland.

Wolf, H. 1974. Microcomputers – A New Revolution. EE/Systems Engineering Today, 33 (1): 80–83.

Part Two

Automobile Traffic Control

Part 2 analyzes the role of computerized control and surveillance systems in improving traffic flow conditions and in reducing the negative effects of increased use of automobiles. Chapter 4 presents a brief survey of basic systems concepts, and Chapter 5 reviews concepts and methods of control under the headings: route guidance; urban street traffic flow control; freeway traffic flow control; vehicle control. Experience gained with computerized area and freeway traffic control systems is presented in Chapters 6 and 7 in the form of specially prepared case studies.

4 Basic Systems Concepts

Many of the traffic problems summarized in Chapter 1 are closely related to two types of information problem:

- the non-availability of the traffic information required for the optimal response of the car drivers, e.g., with respect to selection of the optimal driving route
- the limited information processing capabilities of a human driver, especially with regard to the speed, accuracy, and reliability

Every car driver is faced with three levels of information processing and control (cf. Figure 4.1):

Level I   Processing of information generated and/or displayed inside the vehicle
Level II Processing of information generated and/or displayed in the immediate (visible) environment
Level III Processing of information on the traffic situation in the peripheral (not yet visible) environment

The main objective of any automobile traffic control system is to improve the information exchange processes at the three levels of the complex system illustrated in Figure 4.1, i.e., the man-machine-environment system.

In uncontrolled traffic systems, level III of the information processing system, i.e., a data link between the peripheral environment and the driver, does not exist. Therefore one key problem of automobile traffic control is to provide a communication link between the not yet visible environment and the driver.

This is illustrated in more detail in Figure 4.2, which shows the relation between the general objectives of traffic control, according to Chapters 2 and 3 (cf. Figure 3.3), and the functions required of traffic control technology to achieve them. The most important functions are:

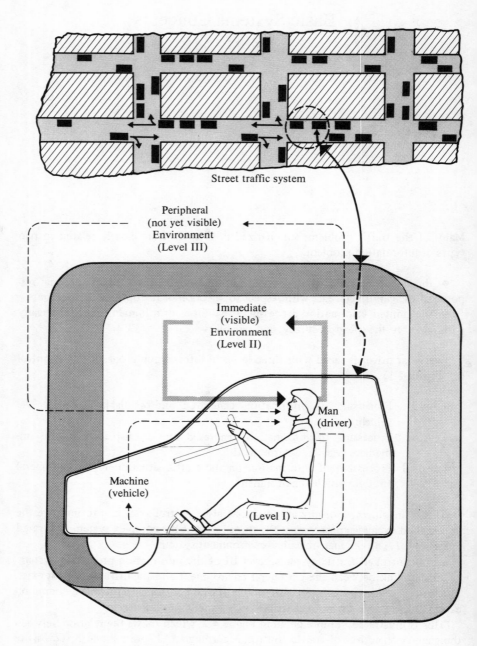

FIGURE 4.1 The man–machine–environment system.

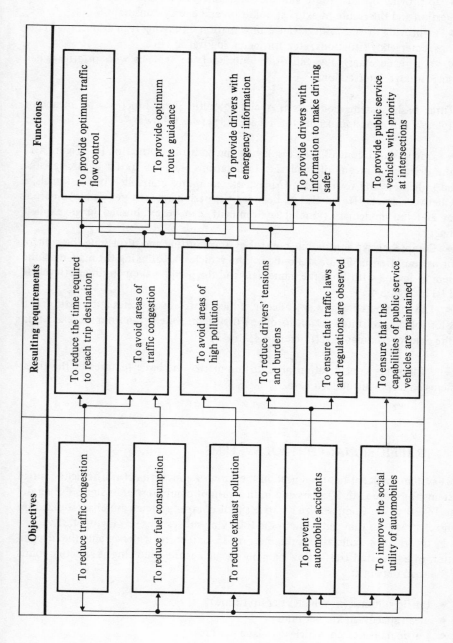

FIGURE 4.2 Objectives and resulting functions of traffic control.

- to provide optimal route guidance and traffic flow control in order to reduce congestion and the resultant exhaust pollution and energy consumption
- to provide drivers with advance information on driving regulations, road conditions, emergency situations, etc., for increasing traffic safety
- to provide emergency and other public-service vehicles with priority when passing through intersections

Three basic systems concepts that aim to fulfill these functions by means of completely different approaches (cf. Figure 4.3) are as follows:

- Traffic signal control, i.e., displaying the control information (cf. Figure 4.1) to the drivers by means of road side installations, such as traffic lights, variable speed indicators and route signs. Using these computer controlled traffic signals, a link between levels III and II, i.e., between the peripheral (not yet visible) environment and the immediate (visible) environment, can be established in an explicit manner
- Comprehensive automobile control by transmitting and displaying the control information given in Figure 4.3 onboard the vehicles, i.e., linking the man-machine system at level I with both the immediate and the peripheral environment (levels II and III in Figure 4.1)
- The automated highway concept, which not only assumes automatic transmission of control information to onboard systems, but also automatic execution of the control commands (cf. Figure 4.3, row 3)

The following consideration aims to summarize the basic features of the three systems concepts introduced here.

4.1 TRAFFIC SIGNAL CONTROL SYSTEMS

This systems concept has been used in the form of conventional traffic light control systems for more than 50 years and in the form of computerized systems during the last 15-20 years. Computerized traffic signal control systems are characterized by control structures that can contain the following subsystems (cf. Figure 4.3).

1. Information collection units, i.e., traffic detectors, which enable the identification of aggregated traffic flow parameters such as the following at selected points of the road network:

- traffic volumes (in vehicles per unit time)
- mean speeds (in km per hour)
- traffic densities (in vehicles per lane per km)
- headways between cars (in seconds)

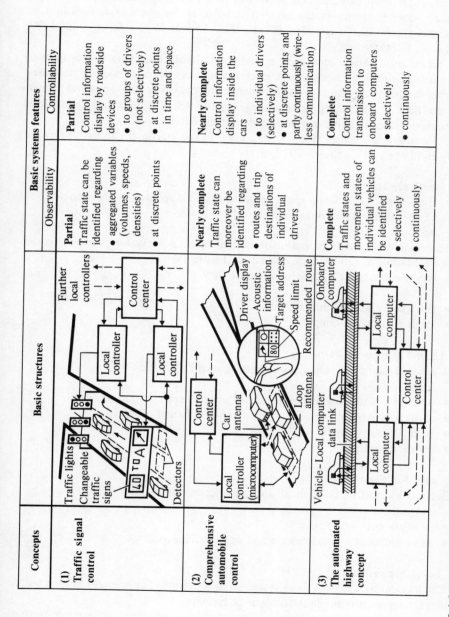

FIGURE 4.3 Basic systems concepts.

The inductive loop detector, which consists of a wire loop embedded in the road surface and an electronic evaluation unit installed at the roadside, is used in most applications (cf. Baerwald 1976, Baker 1970, Everall 1972, Hillier et al. 1972, Krell et al. 1971).

2. Local information processing systems, i.e., local controllers in the form of conventional electromechanical and electronic devices or as microcomputer-based systems that evaluate detector measurements, and control traffic lights and changeable traffic signs according to a strategy prescribed by the central information processing system, i.e., the computerized control center.

3. Information display systems, i.e., controllable traffic signals (cf. Baerwald 1976, Byrd et al. 1973, Everall 1972) in the form of

- traffic lights
- changeable route signs, speed indicators, and displays for emergency information (accidents, congestion) or climatic conditions

According to the spatial extension of the controlled traffic system, three categories of control problems can be distinguished (cf. Figure 4.4):

1. Junction control, i.e., control of traffic flow at

- single, e.g., isolated intersections
- single freeway on-ramps

(cf. Level 4 in Figure 4.4).

2. Arterial road control, i.e., coordinated control of traffic flow

- over arterial urban streets with several interconnected intersections
- along long freeway sections with or without on- and off-ramps
- through long tunnels
- over long bridges

(cf. Level 3 in Figure 4.4).

3. Network control, i.e., control and guidance of traffic flow

- in urban street networks
- in networks of urban freeways and neighbouring roads

(cf. Level 2 in Figure 4.4).

The integration of these three systems categories into one complex system leads to two classes of automobile traffic control systems:

1. Area traffic control systems for supervising, controlling, and guiding traffic in urban streets and parking lots (cf. Hillier et al. 1972).

2. Freeway traffic surveillance and control systems for supervising, controlling,

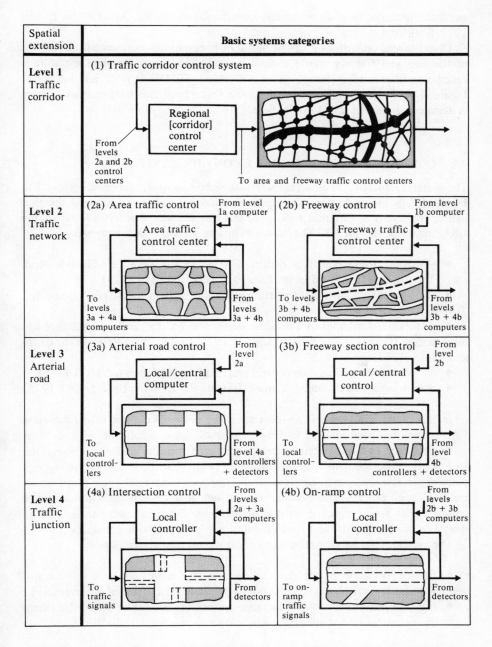

FIGURE 4.4 Survey of basic concepts of computerized traffic signal control systems.

and guiding traffic flow on urban and suburban freeways (cf. Athol 1974, Everall 1972, Krell et al. 1971).

Under special conditions it can be useful or necessary to coordinate the operation of the area and freeway traffic control centers. The resulting system structure is called a traffic corridor control system (cf. May 1969, OECD 1975, and Level 1 of Figure 4.4). Figure 4.4 illustrates the basic features of the traffic signal control systems categories described briefly here.

4.2 COMPREHENSIVE AUTOMOBILE CONTROL SYSTEMS

The systems concept of comprehensive automobile control, according to row 2 in Figure 4.3, is relatively new. During the second half of the sixties and the first half of the seventies, different versions of this concept were developed in Japan, the USA, the FRG, and the UK. They came to be known under such names as:

- electronic route guidance system (ERGS) in the USA (cf. Covault et al. 1967, Hanysz et al. 1967, Rosen et al. 1970, French and Lang 1973)
- multifunctional automobile communication system (MAC) in Japan (cf. Toyota 1973, 1974)
- Autofahrer-Lenkungs und Informationssystem (ALI = car driver guidance and information system) in the FRG (cf. Baier 1975, Friedl 1975, Bolle 1976, Groth and Pilsack 1974)
- comprehensive automobile control system (CAC) in Japan (cf. MITI 1975)
- road information transmitted aurally (system RITA) in the UK (cf. RRL 1975)

In this book, the name of the most advanced system, i.e., that of the Japanese CAC system, has been adopted to denote the whole systems concept.

The creation of CAC systems cannot and does not aim to replace the computerized traffic signal control systems used worldwide. The CAC system merely supplements traffic light control systems, thus increasing the effectiveness of the whole traffic control process. This becomes clear if one analyzes the basic limitations of computerized traffic signal control systems. These limitations are concerned with questions of observability and controllability as well as with other basic ideas.

The systems concept of traffic signal control described above allows

- the identification of the state of the traffic system merely in the form of aggregated traffic variables at discrete points of the network (partial observability)
- the transmission of control information only to groups of drivers by means of traffic signals located at specific roadside points (partial controllability)

(cf. Figure 4.3, row 1). The CAC concept tries to improve these observability and controllability conditions in the following ways:

1. The onboard systems transmit the desired trip destinations to the local controllers and — if necessary — to the central computer (improvement of traffic observability).
2. The stationary control system transmits the following to the individual drivers via onboard audio and/or visual displays:

- information on recommended driving routes
- driving information such as advance information concerning speed limits, stop signs, and other regulations
- emergency information regarding accidents, fires, earthquakes etc.

(improvement of the overall controllability of the automobile traffic system).

A CAC system consists of three major components: vehicle units, roadside units, control center.

1. *Vehicle units.* The onboard system contains the following units (cf. Figure 4.3):

- an operating keyboard that receives data input by the driver (e.g., code number for desired destination)
- a two-way digital communication unit that receives from and transmits data to roadside communication units
- a display panel that provides a visual display of data received from roadside communication units
- an automatic radio activator that automatically activates the car radio to receive information transmitted from roadside or central broadcasting units

2. *Roadside units.* Two types of roadside units may be distinguished:

- roadside communication units that memorize data (i.e., route guidance tables) received from the central processing unit, and conduct two-way communication with individual vehicles via a loop antenna embedded in the road surface and vehicle antennas
- roadside broadcasting units that are linked directly to the control center and conduct one-way communication (roadside-to-vehicle) with passing vehicles via leak cable antennas

Moreover, traffic detectors and changeable route signs that are used in the concept described in Section 4.1 can also be coupled to the roadside units of the CAC system.

3. *Control center.* The control center is equipped in a manner similar to that in the traffic signal control concept.

In the Japanese CAC project (cf. MITI 1975) it is planned to coordinate the CAC system with the area traffic signal control system and the freeway traffic surveillance and control system, as well as with further urban management and information systems like the air pollution surveillance system and the vehicle fleet

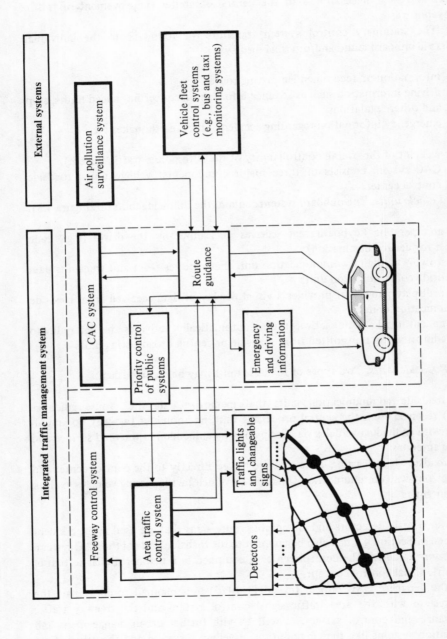

FIGURE 4.5 Anticipated coordination between different automobile traffic control systems and further urban management systems (cf. MITI 1975).

control system (particularly bus, taxi, and other public-service companies). Figure 4.5 illustrates the anticipated relationship between these information, management, and control systems (cf. MITI 1975).

4.3 THE AUTOMATED HIGHWAY CONCEPT

The third basic systems concept is that of automated highways. Here it is assumed that not only the information processing and exchange processes are automated, but also the operation and guidance of the individual vehicles. This would result in further improvements in the observability and controllability conditions of the whole automobile traffic system, as illustrated in row 3 of Figure 4.3.

The development of such systems cannot be considered as a measure for improving the operation of existing transport systems, but rather as a long range strategy for creating completely new modes of urban transport (cf. Figure 2.3). Therefore the automated highway concept is analyzed in detail in Part 4 of this monograph (cf. Section 15.3, The Dual-mode Concept), which is devoted to new modes of urban transport.

4.4 CONCLUDING REMARKS

The three basic systems concepts discussed in this chapter are of quite different significance for practical implementation of automobile traffic control systems. Computerized traffic signal control systems represent a well developed technology used world-wide. The comprehensive automobile control concept has reached the status of large-scale demonstration projects and its introduction into practical use is expected during the next decade or so. The concept of automated highways, however, is still a subject of fundamental research; its future is at present uncertain.

For this reason Chapter 5 is restricted to an analysis of the *concepts and methods of control* proposed and used for implementing the first two basic system concepts.

REFERENCES

Athol, P. 1974. Urban freeway surveillance and control: the state of the art. Research Report from The Technology Service Corporation, Bethesda, Md, USA.
Baerwald, J. E. (ed.). 1976. Transportation and Traffic Engineering Handbook. Englewood Cliffs, NJ. Prentice-Hall.
Baker, J. L. 1970. Radar, acoustic, and magnetic vehicle detectors. IEEE Transactions on Vehicular Technology, VT-19 (1): 30–43.
Baier, W. 1975. "ALI", a Guide for Drivers. Internationale Elektronik Rundschau (in German).
Bolle, G. 1976. System ALI, traffic guidance by computer prediction. Bild der Wissenschaft, No. 4: 76–82 (in German).

Byrd, L. G., et al. 1973. Changeable message signs: a state-of-the-art report. Highway Research Circular, No. 147.

Covault, D. C., T. Derrish, and A. C. Kanen. 1967. A study of the feasibility of using roadside communications for traffic control and driver information. Highway Research Record, No. 202: 32–66.

Everall, P. F. 1972. Urban freeway surveillance and control: the state-of-the-art. Research Report FCP 22 E 1613, from the Department of Transportation, Federal Highway Administration, Washington, D.C., USA.

French, R. L. and G. M. Lang. 1973. Automatic route control system. IEEE Transactions on Vehicular Technology, VT-22 (2): 36–42.

Friedl, H. 1975. Guidance of motorcar traffic streams — ALI. Radio Elektronik Schau, No. 5: 266–269 (in German).

Groth, G. and O. Pilsack. 1974. An electronic system for individual and active route guidance on motorways. In Traffic Control and Transportation Systems, Proceedings of the Second International IFAC/IFIP/IFORS Symposium in Monte Carlo, September 1974. Amsterdam: North-Holland Publishing Co.

Haynsz, E. A., et al. 1964. DAIR — a new concept in highway communications for added safety and driving convenience. IEEE Transactions on Vehicular Technology, VT-16: 33–45.

Hillier, J. A. et al. 1972. Area traffic control systems. Report of the OECD Road Research Group, Paris.

Krell, K. et al. 1971. Electronic aids for freeway operation, Report of the OECD Road Research Group, Paris.

May, A. D. 1969. Bay area freeway operations study: summary on the comprehensive phase freeway corridor operation studies. Report of the University of California, Institute of Transportation and Traffic Engineering, Berkeley.

MITI. 1975. Comprehensive automobile control system. Preliminary Report from the Japanese Ministry of International Trade and Industry — MITI, Tokyo.

OECD. 1975. International corridor experiments. Report of the OECD Road Research Group, Paris.

Rosen, D. A., F. J. Mammano, and R. Favaut. 1970. An electronic route guidance system for highway vehicles. IEEE Transactions on Vehicular Technology, VT-19 (1): 143–152.

RRL. 1975. RITA: road information transmitted aurally. Report RG 11 GAU, LF 117, from the Transport and Road Research Laboratory, Department of the Environment, Crowthorne, Berkshire.

Toyota. 1973. Research and development of a multifunctional automobile communication system (MAC). Report from the Advanced Group for Transportation, Toyota Motor Sales Co. Ltd., Tokyo, Japan.

Toyota. 1974. For a better tomorrow: Toyota MAC system. Report from the Advanced Group for Transportation, Toyota Motor Sales Co. Ltd., Tokyo, Japan.

5 Concepts and Methods of Control

This chapter is concerned with the following two questions:

1. What basic concepts of automobile traffic control are known and in use?
2. What methods of control have been proposed, developed, and implemented?

A large number of publications on traffic control methodology has appeared during the last 10-15 years, among them survey papers and monographs by FHWA (1974), Gazis (1971, 1974), Holroyd and Robertson (1973), Inose and Hamada (1974), Kuntze (1977, 1978), May (1964, 1974, 1975), Masher et al. (1975), Pitzinger and Sulzer (1968), Rach et al. (1974, 1975, 1976), Ross (1972), Schlaefli (1972), Schnabel (1975), Stockfisch (1972), Strobel and Ullmann (1973), Strobel (1977), Tabak (1973), and others (cf. references Everall 1972, Hillier 1972, Krell 1971, to Chapter 4). It is thus neither feasible nor useful to present here a detailed description of the individual traffic control algorithms and the related design methods. Considering the general purpose of the monograph, as outlined in the Introduction, it is more appropriate to focus on the following two topics:

- presenting a survey of basic control concepts and the most important fundamentals of control methodology
- evaluating control concepts and methods with respect to their present and future significance, especially regarding their possible contribution to the reduction of the traffic problems summarized in Chapter 1

5.1 THE CONTROL TASK HIERARCHY

The known control concepts may be assigned to a three-level control hierarchy as already discussed in a general framework in paragraph 3.3.2 (cf. Figure 3.4).

The three basic control concepts

1. Route guidance
2. Traffic flow control
3. Vehicle control

are illustrated in more detail in Figure 5.1. In general a route guidance system may cover both urban/suburban freeways and main urban streets. The problems of vehicle control are basically the same for freeway and urban street traffic. For these reasons no distinction is made between freeway and urban street traffic in dealing with the first and the third concepts, i.e., route guidance and vehicle control (cf. Figure 5.1, levels 1 and 3).

However, the traffic flow control problems at level 2 of the hierarchy differ significantly for freeway and urban street traffic. Therefore in the following a distinction is made between freeway traffic flow control, and urban street traffic flow control.

5.2 ROUTE GUIDANCE

Two basic approaches for the creation of route guidance systems may be distinguished:

1. Explicit route guidance, i.e., the traffic control system recommends certain routes to the drivers or forces the drivers to use alternative routes, respectively (cf. Figure 5.2).
2. Implicit route control by means of special measures, e.g., by closing a freeway on-ramp or by switching on information signs such as "congestion ahead" drivers are forced or motivated to use alternative routes where it is their decision which route they really choose.

Implicit route control is obviously closely related to traffic flow control problems. Therefore, the following consideration is restricted to explicit route guidance concepts and methods.

5.2.1 CONCEPTS

Two basic route guidance concepts are known:

- route guidance by means of changeable route signs (cf. Figure 5.2, and Mahoney et al. 1973, Cleveland 1972, Heathington et al. 1971, Zajkowski 1972, Everts et al. 1972, Jasper and Wienand 1975, Kuntze 1977, 1978, Ziegler and Baron 1973, Knoll and Ullrich 1972, and reference Everall 1972, to Chapter 4)
- route guidance by means of onboard driver displays (cf. Figures 5.3 and Section 4.2)

Levels	Control concepts			Control methods	
	Freeway traffic	Urban street traffic		Freeway traffic	Urban street traffic
(1) Route guidance	(freeway/urban streets origin to destination network)			• Route planning, static traffic assignment by mathematical programming • Time-of-day and traffic-responsive route pattern selection • Traffic-responsive route computation	
(2) Traffic flow control	(2.1a) Ramp metering 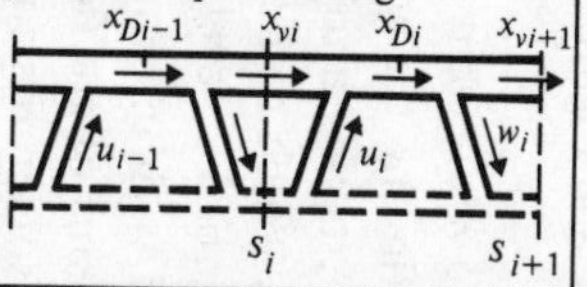	(2.1b) Area traffic control 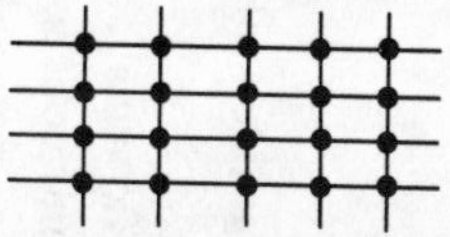		• Static optimization (open-loop ramp metering) • Dynamic optimization (feedback ramp metering)	• Time-of-day and frequency-responsive selection of precomputed signal plans • Signal-plan modification • Signal-plan generation
	(2.2a) Speed and lane-changing control 	(2.2b) Arterial street control (progressive green system) 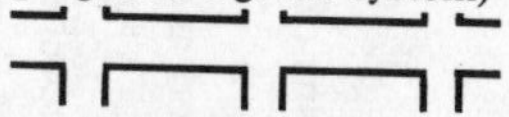		• Speed control • Lane use control • Information systems • Incident detection	
	(2.3a) Merging control 	(2.3b) Intersection control 		• Open-loop control of on-ramp vehicles (green-band system) • Closed-loop control (pacer systems)	• Fixed-time control • Vehicle-actuated control • Selfoptimizing control
(3) Vehicle control	(3.1) Headway control 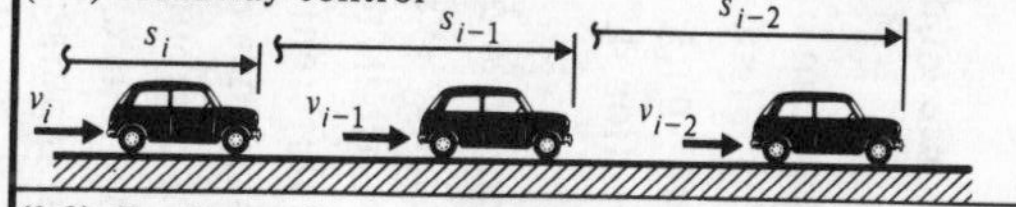			• Headway control by head-up display • Radar distance-warning systems • Radar headway control systems	
	(3.2) Engine and braking system control 			• Engine control (fuel minimizing and emission limitation by spark timing, air-fuel ratio, EGR control) • Anti-skid braking, etc.	

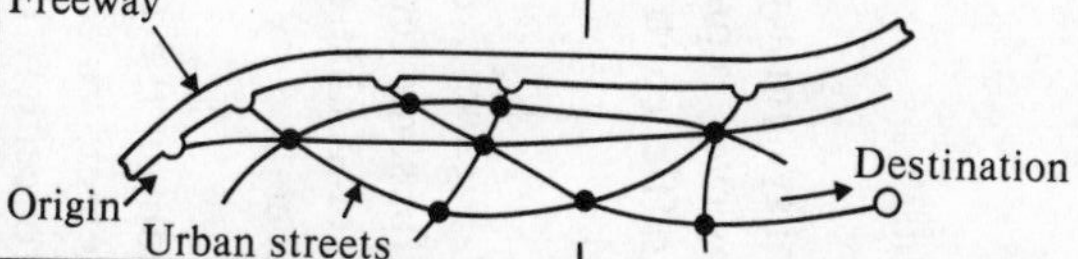

FIGURE 5.1 Concepts and methods of automobile traffic control.

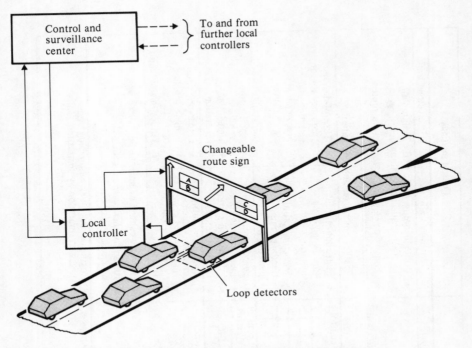

FIGURE 5.2 Route guidance by means of changeable route signs.

The first concept can be implemented as a traffic signal control system according to Section 4.1 and Figure 4.3, row 1. The second concept, however, represents the most important part of the comprehensive automobile traffic control system described in Section 4.2 (cf. Figure 4.3, row 2, and Figure 5.3).

1. *Route guidance by means of changeable route signs (cf. Figures 5.1 and 5.2).* The route guidance system assists drivers in finding the best (in some senses) route from a certain origin to a desired destination, taking into account changing traffic conditions in different parts of the network, which could be caused by accidents, weather, maintenance operations, etc. (cf. Figure 5.1, level 1).

The control computing system is provided with information on the traffic situations in the various alternative routes by means of traffic detector measurements and/or manual inputs made by traffic supervisors in the control center. The control computer has to use these data to predict the development of the traffic situation over a time horizon of about 5-15 minutes. Depending on the prediction, a decision has to be made on whether or not to change the displayed routes. If a route change appears to be necessary, then the corresponding control command is transmitted to the local controllers, which change the route displays as required. At present, these local controllers are in most cases designed as conventional electronic devices. It is

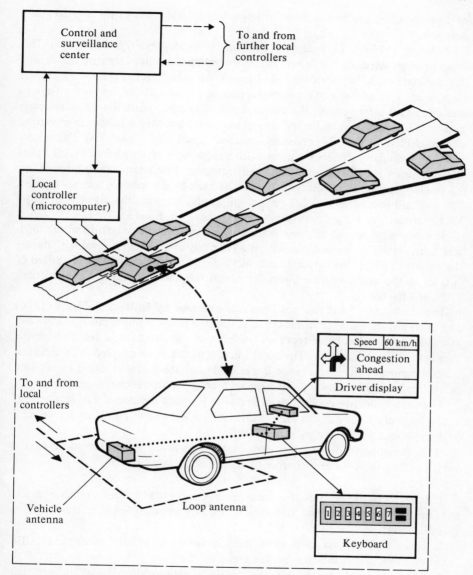

FIGURE 5.3 Route guidance by means of onboard driver displays.

expected that in future systems they will contain microcomputers, which will provide the capacity to store different route guidance tables and to modify them in accordance with the time of day and/or the local traffic conditions. Local controllers and the corresponding changeable route signs are in general located in front of freeway off-ramps and in front of important intersections of the arterial street network.

2. *Route guidance by means of onboard driver displays (cf. Figure 5.3).* This concept assumes that the driver puts a code number characterizing his desired destination into a keyboard installed inside the vehicle (cf. Figure 5.3). As the driver's car approaches certain decision points, in the form of freeway off-ramps and important intersections, the onboard microcomputer transmits the stored code number via a vehicle antenna and a loop antenna embedded into the street surface to a roadside unit that also contains a microcomputer (cf. Figure 5.3). This microcomputer compares the code number with a memorized route guidance table, takes the recommended driving direction, and transmits the corresponding information via the loop antenna and the vehicle antenna back to the onboard microcomputer. The onboard computer switches on a visual display showing the driver whether he has to turn left, turn right or proceed straight ahead (cf. Figure 5.3).

The whole process of information exchange between the onboard and the roadside units has to be carried out within a relatively short time interval, i.e., during the period in which the car antenna is moving over the loop antenna. At a speed of 200 km/h the available time interval is about 0.07 seconds (cf. reference Friedl 1975, to Chapter 4).

The feasibility of this task has been demonstrated by Groth and Pilsack (1974) and others. If the car approaches the next decision point then the same information exchange process has to be repeated. In this way the driver is guided through the network to his destination. The local microcomputers are coupled with a central control computer that can change the individual route guidance tables memorized in the local roadside units in accordance with changing traffic conditions.

The route guidance concept described here provides the essential advantage that it can be applied not only to private automobiles but also to public service vehicles such as police cars, emergency vehicles, etc.

It is possible to distinguish between three systems levels, according to the spatial extension of the area covered by the route guidance system.

1. *Local level*: application of route guidance concepts to a small number of intersections in order to improve traffic flow, e.g., by avoiding left turn traffic under special conditions.

2. *Urban and regional level*: application of the route guidance concept to traffic networks covering urban and suburban areas.

3. *National level:* route guidance within a nationwide network of national expressways and rural roads.

The CAC concept has been designed for both the national and the urban regional level. The ALI system, for example, is proposed as a nationwide route guidance system covering the whole national freeway network of the FRG (cf. reference Friedl 1975, to Chapter 4). Any trip destination is described and coded by four characters — the first defines one of sixteen zones in the FRG, the second defines one of sixteen areas within each zone, and the remaining two define an area of about 7.3 square kilometers. The resolution is sufficient to direct the driver to 65,000 destinations, each with a size of only 2.7 × 2.7 kilometers.

5.2.2 METHODOLOGY

Each method of route control may be characterized by two features:

- the control criterion used
- the control algorithm resulting from a given control criterion, a set of constraints and the available detector measurements characterizing the traffic state

5.2.2.1 Control Criteria. Two kinds of control criteria should be taken into consideration:

1. Control criteria of the systems users, i.e., of the drivers. The main objective of the drivers is to *minimize* the overall triptime needed to reach a desired destination from a given origin. Moreover, the limitation of fuel consumption and other costs (e.g., of tolls) can play a certain role (cf. Belenfant and Vesval 1976).
2. Control criteria of systems operators, i.e., of the city or public authorities. Here a whole set of criteria exists. The most important objectives are:

- optimal, i.e., equable distribution of a given traffic demand over available network resources, thus maximizing the overall network capacity
- minimizing environmental impacts, especially regarding air pollution
- minimizing overall fuel consumption

Which control criterion or combination of criteria is suitable depends heavily on the specific features of the traffic network. In most cases the minimum travel time criterion, i.e., the objective function of the drivers is preferred; a small number of Japanese authors tried to introduce environmental criteria (cf. Sasaki and Inouye 1974, Iida 1974).

5.2.2.2 Control Algorithms. There are three basic approaches to developing route control algorithms:

- route planning using historical traffic data and time-of-day related selection of route patterns from a set of precomputed routes
- traffic-responsive selection of suitable route patterns by means of traffic detector measurements
- traffic-responsive generation of suitable (optimal) route patterns

The first task corresponds to the well-known traffic-assignment problem for which a fairly well developed methodology based on various techniques of mathematical programming is available (cf. Nguyen 1974). Using these methods it is obviously feasible to precompute sets of alternative routes for different traffic conditions and to store them in the memories of the central control computer and the local microcomputers according to Figures 5.2 and 5.3.

The simplest version of a route control algorithm can then be implemented in the form of a time-of-day open-loop control system in which the route signs are changed during the course of the day, e.g., different routes are recommended for the morning peak period and for the afternoon rush hour. It is obvious that the effectiveness of such a system can be improved if measured traffic data are used for selecting suitable routes from the set of precomputed routes. However, such a traffic-responsive open-loop control system can only adapt the route guidance systems to a limited (small) number of traffic situations.

Therefore the third algorithm mentioned above, traffic-responsive route generation, which is supposed to react in an effective way even under complicated and unforseeable traffic situations (e.g., in those caused by accidents), is a very promising control principle. However, the implementation of this algorithm in the form of a so-called closed-loop control system is connected with several difficulties related to the network structure, i.e., the 'origin–destination" relations, as well as to other factors.

Four types of such origin–destination relations have to be taken into consideration:

- one-to-one systems containing one major origin zone and one major destination area connected via two or more alternative routes
- many-to-one systems characterized by many origins and one destination
- one-to-many systems with one main origin and many destinations
- many-to-many systems with many origins and many destinations

5.2.2.3 One-to-One Route Control Systems. In the case of a one-to-one situation the essential differences between the second and the third principles, i.e., between traffic-responsive route selection and route generation, may vanish. This is illustrated by Figure 5.4, which shows an example for a closed-loop route controller proposed and implemented by Berger and Shaw (1975).

By means of this example, some general and basic features of closed-loop route control algorithms can be discussed. Figure 5.4 illustrates that a route controller has to solve two tasks:

- trip-time estimation, i.e., determination of the mean travel times τ_1 and τ_2 along nominal route 1 and alternative route 2 using detector measurements x_{a1}, ..., x_{ar} on traffic volumes and speeds occurring at discrete points $TD_1, \ldots, TD_N$ on routes 1 and 2
- decision making, i.e., determining the optimal route using the estimation results

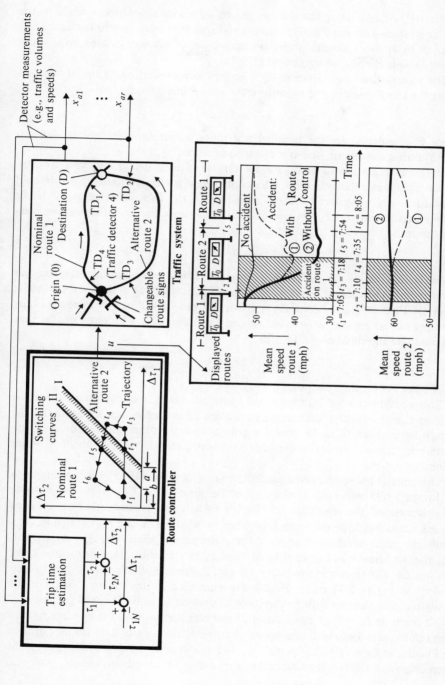

FIGURE 5.4 Closed-loop route control algorithm applied to a one-to-one situation, and the corresponding simulation results obtained by Berger and Shaw (1975).

The task of estimating travel times can be solved by simulation models, which provide a relation between traffic volumes and speeds at selected points and the trip times, or by applying special estimation techniques to so-called macroscopic input-output models introduced by Strobel (1977).

The second task, i.e., determining a route recommendation, is solved by comparing the travel times τ_1 and τ_2, and/or the travel time delays

$$\Delta\tau_1 = \tau_1 - \tau_{1N} \quad \text{and} \quad \Delta\tau_2 = \tau_2 - \tau_{2N}$$

where τ_{1N} and τ_{2N} are the nominal trip times occurring under normal (uncongested) traffic conditions, i.e., at low traffic densities.

In the example shown in Figure 5.4 the travel time delays $\Delta\tau_1$ and $\Delta\tau_2$ are chosen as the essential variables. The control law is characterized by a threshold constant b and a hysteresis factor a leading to the following operation: let us assume that at time $t = t_1$ the nominal route 1 is recommended. This route recommendation is changed to route 2 if according to

$$\Delta\tau_1 - \Delta\tau_2 \leqslant b$$

the difference $\Delta\tau_1 - \Delta\tau_2$ between the two travel time delays $\Delta\tau_1$ and $\Delta\tau_2$ exceeds the threshold constant b (cf. time $t = t_2$ and switching curve I in the route controller block of Figure 5.4). In contrast, a return to route 1 will be recommended only for the cases where the difference between travel time delays becomes smaller than the modified threshold $b - a$, i.e., for

$$\Delta\tau_1 - \Delta\tau_2 \leqslant b - a$$

(cf. Figure 5.4 for the time point $t = t_3$).

This combination of threshold and hysteresis features of the route controller ensures that drivers will meet traffic conditions on the recommended route that are clearly better than those he would experience on the other route. Moreover, the hysteresis factor a avoids undesirable frequent changes between the two route recommendations.

The simulation results obtained by Berger and Shaw (1975) for the Interstate Highways I 695 and I 95 of the city of Baltimore (USA) illustrate the possible effectiveness of the algorithm (cf. Figure 5.4). It was assumed that an accident caused a lane blockage on route 1 between $t_1 = 7{:}05$ a.m. and $t_4 = 7{:}35$ a.m. As a result, the mean travel speed on route 1 starts decreasing, as shown in the corresponding diagram shown in Figure 5.4. At time $t_2 = 7{:}10$ a.m., the travel time delay $\Delta\tau_1$ reaches the threshold level $\Delta\tau_2 + b$ (cf. diagram inside the block "route controller" in Figure 5.4) and route 2 is recommended to the drivers. It takes about 8 minutes more before definite results of this measure can be observed (cf. curves 1 and 2 for time $t_3 = 7{:}18$ a.m.). After 25 minutes the speed on the nominal route 1 starts to increase again while the speed on recommended route 2 decreases slightly.

Finally, at time $t_5 = 7{:}54$ a.m., i.e., 44 minutes after the alternative route 2 has been displayed, the trip time delay $\Delta\tau_1$ approaches the threshold level $\Delta\tau_2 + (b - a)$

and the route sign is reset to the original state. The curves 2 given in Figure 5.4 demonstrate that without the application of the route control algorithm the mean travel speed along route 1 remains significantly lower for more than 1 h.

5.2.2.4 Many-to-Many Route Control Systems. In a "many-to-many" situation it can be very complicated to develop powerful real-time and closed-loop route control algorithms, and to implement them by means of a central control computer. One way of reducing the requirements on the central computing system is to distribute the necessary computation work to a number of local (micro-) computers installed at the individual route diversion points, i.e., by implementing a spatially distributed computer control system. The basic idea for the corresponding control algorithm can be derived from procedures already used for the routing problem that occurs in data networks (cf. Butrimenko 1970, Butrimenko and Strobel 1974, Meditch 1978).

The control principle is illustrated in a simplified manner in Figure 5.5. It is assumed that there are four possible destinations: the main intersections N (north), W (west), C (center) and E (east). The main traffic origins should be intersections S (south) and S'. It is further assumed that each of the main intersections N, W, C, E, S and S' shown in Figure 5.5 is equipped with

- changeable route signs $CRS_N, \ldots, CRS_{S'}$
- traffic detectors $TD_N, \ldots, TD_{S'}$ and
- route-guidance microcomputers $MC_N, \ldots, MC_{S'}$

Each of the local microcomputers MC_i can make decisions on the driving directions that are optimal for the individual intersections, if these computers are provided with information on the various travel times. This can be achieved by:

- estimating the trip times between neighboring intersections from traffic detector data
- exchanging this information between neighboring computers

This process will be described in more detail for the microcomputer MC_S, which is installed at intersection S and has to control the changeable route sign denoted CRS_S in Figure 5.5. In this computer, the travel-time matrix in eq. (5.1) has to be stored in the memory; it contains the minimal trip times T_{IJ} needed for reaching the destination J from the neighbouring intersection I:

$$(S) = \begin{bmatrix} & | & | & \\ \tau_W & | & \tau_C & | & \tau_E \\ & | & | & \end{bmatrix} = \begin{bmatrix} T_{WN} & T_{CN} & T_{EN} \\ T_{WW} & T_{CW} & T_{EW} \\ T_{WC} & T_{CC} & T_{EC} \\ T_{WE} & T_{CE} & T_{EE} \end{bmatrix} \quad (5.1)$$

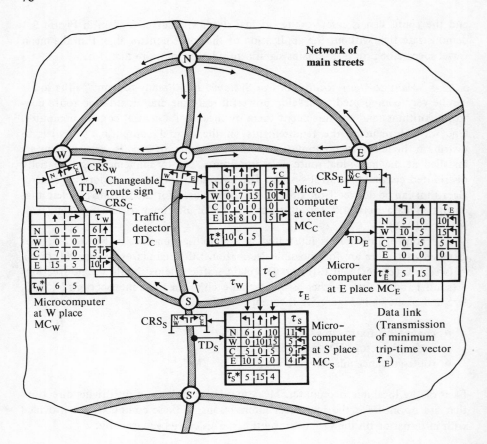

FIGURE 5.5 Decentralized closed-loop algorithm designed for a many-to-many route guidance problem (cf. Butrimenko and Strobel 1974).

To the first row of matrix **(S)** are assigned the minimal trip times T_{IN} that are required to reach the destination from the route signs CRS_W, CRS_C, and CRS_E located in front of the intersections W, C, and E, respectively. It is obvious that the condition $T_{II} = 0$ is valid.

The row vector τ_S^* shown under the matrix **(S)** in Figure 5.5 is defined by

$$\tau_S^* = (T_{SW}\ T_{SC}\ T_{SE}) \tag{5.2}$$

where T_{SW}, T_{SC}, and T_{SE} denote the trip times needed for driving from the CRS_S route sign in front of intersection S to the corresponding signs CRS_W, CRS_C, and CRS_E in front of W, C, and E, respectively. These travel times have to be estimated from traffic data measured by the traffic detectors TD_S, TD_W, TD_C, and TD_E (cf. Strobel 1977).

Using the matrix **(S)** and the vector τ_S^* as given by eqs. (5.1) and (5.2), respectively, gives the vector

$$\tau_S = \begin{bmatrix} T_{SN} \\ T_{SW} \\ T_{SC} \\ T_{SE} \end{bmatrix} \qquad (5.3)$$

of minimal overall trip times T_{SJ} and the corresponding optimal driving directions, where the T_{SJ} can be obtained from

$$T_{SJ} = \min \{T_{WJ} + T_{SW}; T_{CJ} + T_{SC}; T_{EJ} + T_{SE}\} \qquad (5.4)$$

for J = N, W, C, E.

Using the simple numerical example given in Figure 5.5, the vector τ_S takes the form

$$\tau_S = \begin{bmatrix} T_{SN} \\ T_{SW} \\ T_{SC} \\ T_{SE} \end{bmatrix} = \begin{bmatrix} \min\{\; \underline{6+5};\;\; 6+15; 10+4\} \\ \min\{\; \underline{0+5};\; 10+15; 15+4\} \\ \min\{\; 5+5;\;\; 0+15;\; \underline{5+4}\} \\ \min\{10+5;\;\; 5+15;\; \underline{0+4}\} \end{bmatrix} = \begin{bmatrix} 11 \\ 5 \\ 9 \\ 4 \end{bmatrix} \qquad (5.5)$$

The recommended time-optimal driving directions may be given as follows:

- to N, turn left
- to W, turn left
- to C, turn right
- to E, turn right

(cf. Figure 5.5.)

In summarizing, the local microcomputer MC_S has to solve two tasks in determining the optimal driving directions:

- to identify the elements of vector τ_S^* according to eq. (5.2) using traffic detector measurements
- to compute the minimal overall travel-time vector τ_S using matrix **(S)** and vector τ_S^* according to eqs. (5.1)–(5.4)

But, how can the matrix **(S)** be obtained? Figure 5.5 illustrates that the columns of **(S)** are identical with the minimal overall trip-time vectors τ_W, τ_C, and τ_E, which are valid for neighboring intersections W, C, and E, respectively. They have to be computed in the way just described for τ_S. This task has to be solved by the corresponding microcomputers MC_W, MC_C, and MC_E. The determined vectors $_W, \tau_C$, and τ_E are then transmitted to the microcomputer MC_S, which forms the required matrix **(S)** from the transmitted data.

The same procedure has to be used to set up the travel-time matrices **(W)**, **(C)**, and **(E)** shown in Figure 5.5. If the various vectors and matrices are updated by means of detector measurements, then it seems feasible to determine and display in a real-time operation mode those routes that are optimal with regard to travel time for specific traffic conditions.

The route guidance algorithm described here is illustrated in Figure 5.5 for simplified conditions only. In a real many-to-many situation one has to consider as many travel-time matrices and vectors τ and τ^* per intersection as are necessary for the changeable route signs, i.e., the maximal number of changeable route signs, travel-time matrices and vectors τ and τ^* per intersection is equal to the number of streets carrying traffic in the direction of the corresponding intersection.

5.2.3 APPLICATION OF ROUTE GUIDANCE METHODOLOGY: PRESENT STATUS AND CURRENT TRENDS

Route guidance systems based on changeable road signs have been installed with remarkable success in France for freeway-type traffic (cf. case descriptions presented in Chapter 7). Moreover, experiments have been carried out in the USA (Berger 1975), the FRG (Jasper and Wienand 1975), the GDR (Bennewitz 1977), and a few other countries. Most of the experimental and operational systems are of the one-to-one structure discussed in paragraph 5.2.2.3. In some applications the decision to change route signs is made by a traffic supervisor using the computer as an advisor.

So far, it has not been possible to implement many-to-many route guidance algorithms, in accordance with paragraph 5.2.2.4, for either freeway traffic or urban street traffic. However, the availability of cheap, small and robust microcomputers could open the way to create such systems at reasonable costs. The route guidance algorithms of the many-to-many type are of vital importance for the creation of the CAC systems described in Section 4.2, i.e., for providing route guidance by means of onboard driver displays (cf. Figure 5.3).

A research and development program for the creation of a CAC system, which contains route guidance as its main feature, was started in Japan in 1966. A prototype of the new system was studied in a large-scale demonstration project in a 25 km^2 area in the south-western section of Tokyo during the years 1977/1978 (cf. Rossberg 1977). For this purpose, more than 1000 motorcars were equipped with the onboard devices illustrated in Figure 5.3 (cf. reference MITI 1975 to Chapter 4).

5.3 URBAN STREET TRAFFIC FLOW CONTROL

5.3.1 INTRODUCTION

The most widely used traffic flow control concept is traffic light control and coordination.

The age of traffic signals began on December 10th, 1868, with the installation of mechanical street crossing signals in front of the Houses of Parliament in London (cf. von Stein 1969). In 1914 the first traffic light was installed in Cleveland (USA) (cf. Gazis 1971). The first notable innovation in traffic light control was the synchronization of a string of traffic lights to provide uninterrupted driving through them, preferably in both directions. In 1918 the first so-called progression system was installed in Salt Lake City (USA) (cf. Gazis 1971, Marsh 1927). These progression systems were designed as "multi-dial" systems, possessing three options for synchronization: one for morning rush hours, one for afternoon peak periods, and one for average conditions. The transition from one synchronization scheme to another was carried out at fixed times during the day. One of the first and best-known systems that permits a traffic-responsive transition from one signal coordination scheme to another was installed in Baltimore (USA) in the early 1950s (cf. Gazis 1971). This system used traffic detector measurements at some key points in the arterial network.

A new age of urban street traffic control was heralded in 1959 when a pilot study on computer control of urban traffic lights was initiated in Toronto (Canada) using an IBM 650 computer, one of the earliest general purpose digital computers (cf. Cass and Casciato 1960). In 1964, Metropolitan Toronto became the first municipality in the world to install a digital computer system consisting of two Univac computers (1107 and 418) for controlling traffic lights (cf. Hewton 1967). Further cities like San Jose, California, in the USA (cf. Section 6.1), and West London and Glasgow (cf. Section 6.2) in the UK followed the Toronto example. At present, more than one hundred cities around the world are operating centralized computer systems for controlling traffic lights.

However, another new age of traffic light control has started. This new age is characterized by a change from centralized to spatially distributed computer control of traffic lights, i.e., by the use of microcomputers as the hearts of local traffic light controllers installed at the individual intersections (cf. Brinkman et al. 1975, Weiss 1974, Bång and Nilsson 1976, Hahn and Eustis 1977, Krayenbrink and Vlaanderen 1975, Siemens 1977, Sealbury 1974, Cutlar 1976).

The long history of traffic light control, which has been sketched here very roughly, has been associated with extensive research projects on the development and implementation of traffic light control concepts and methods. The research efforts were increased significantly during the last twenty years, especially after the creation of the Toronto computerized system in the early sixties. These activities led to a very large number of publications dealing with methodological topics of traffic light control and coordination. It is obviously neither possible nor useful to discuss details of traffic light control methodology in this section. It is more appropriate to restrict the following consideration to:

- basic concepts of traffic light control
- basic control systems structures
- basic features of the existing methodology for single intersection and co-ordinated control, and current trends in its further development

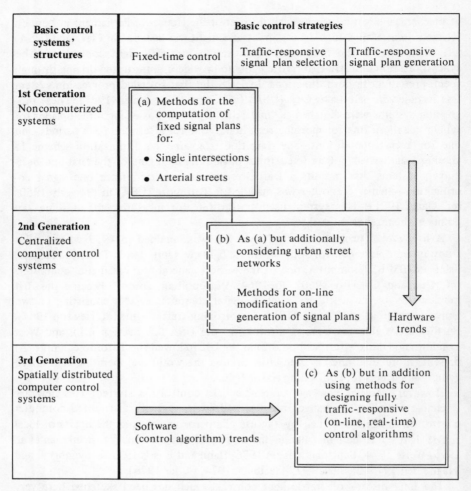

FIGURE 5.6 Survey of basic traffic light control concepts.

5.3.2 BASIC TRAFFIC LIGHT CONTROL CONCEPTS

Three basic control strategies can be distinguished (cf. Figure 5.6).

1. *Fixed-time control.* For typical situations, such as morning rush hours, afternoon peak periods, several signal plans are precomputed by means of historical traffic data and are stored in local and/or central controllers. A transition from one program to another is carried out manually by the policeman in charge, or automatically, depending on the time of the day.

2. *Traffic-responsive signal plan selection and modification.* The control system selects, from previously calculated fixed-time plans, the one that best suits a specific traffic situation identified by traffic detector measurements. Within the framework of the selected signal program, certain small modifications regarding intersection control and synchronization between intersections are permitted.

3. *Traffic-responsive signal plan generation.* Coordination and local adaptation are discarded, and the most appropriate traffic signal settings are obtained by on-line calculations based on detector measurements.

Most of the computerized traffic light control systems currently in existence make use of the second control concept, i.e., traffic responsive signal program selection. Several studies have shown that the concept of traffic-responsive signal plan generation does not lead to improvements in control efficiency (Holroyd and Robertson 1973). However, it should be noted that there is a close interaction between the basic control system structure and the usefulness and feasibility of certain control strategies.

This relation is discussed in the following with reference to the general computer control concepts that were dealt with in paragraph 3.4.4, and illustrated in Figure 3.6.

5.3.3 BASIC TRAFFIC LIGHT CONTROL SYSTEMS STRUCTURES

According to the general classification given in Figure 3.6, three basic categories of traffic light control systems have to be taken into account (cf. Figures 5.6 and 5.7).

1. *First generation: non-computerized systems.* Here the control functions are performed by specially designed hard-wired logic in the form of an electromechanical device, or by electronic logic. The very limited data processing capabilities of these hard-wired controllers do not permit the evaluation of many traffic detector measurements. Therefore, in almost all cases a restriction to the first control strategy mentioned above, i.e., to fixed-time control and time-dependent signal program selection, is required, as illustrated by Figure 5.6.

2. *Second generation: centralized computer control.* Here a control center containing one or several digital computers is established. The individual control tasks can be carried out by a single computer for a relatively small number of intersections. Under complicated conditions, e.g., if several hundreds or even several thousands of intersections have to be coupled to the control center, then the installation of a complete computer hierarchy system may become necessary.

The availability of powerful computers allowed the use of more sophisticated control strategies, such as traffic-responsive signal plan selection and generation. But, as already mentioned, traffic-responsive signal program selection is preferred in most applications, for the following reasons.

- Complicated software and necessary computation speed. The on-line and real-time determination of signal programs requires powerful high-speed computers.

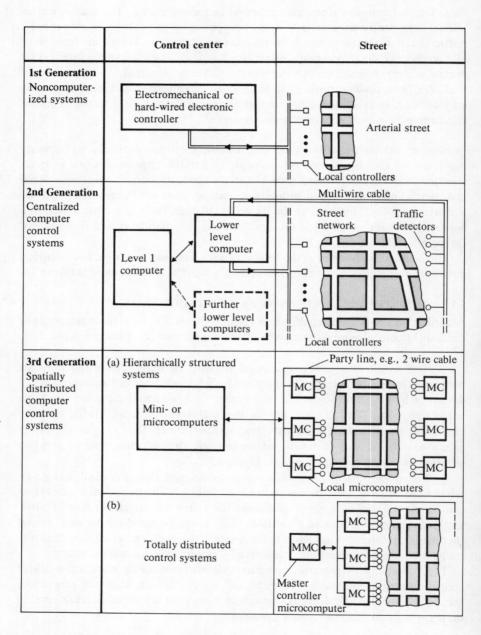

FIGURE 5.7 Basic traffic light control systems structures.

These computers were not available for control purposes during the sixties, but are available now.
- Traffic prediction capabilities. One key problem in the implementation of traffic-responsive signal program generation methods is to solve the traffic prediction problem. Using traffic detector measurements, the state of the traffic flow has to be forecast over short periods of time. The accuracy with which the state can be predicted, and with this the effectiveness of signal program generation, depends on the availability of a sufficiently large number of traffic detectors.
- Detector and communication costs in a traffic control system. The main part of the overall systems expenses is the costs of the data links (cables, modems, etc). For a centralized control system these communication costs are nearly proportional to the average number of detectors per intersection, since all detector information has to be transmitted to the central computing system. According to estimates given by Holroyd and Robertson (1973), the total systems cost can increase by a factor of 1.4 on changing from a fixed-time to a traffic-responsive system with about four detectors per intersection (cf. Figure 5.8).

It is unlikely that a signal program generation strategy will provide significant benefits in comparison with signal program selection unless there is an average of at least four detectors per intersection. However, mainly for reasons of cost, in most of the existing centralized traffic light control systems a restriction is made to

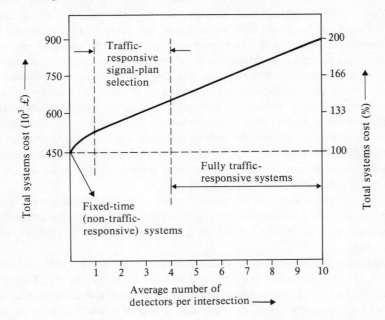

FIGURE 5.8 Total systems cost of a centralized computer control system covering 100 intersections (derived from data published by Holroyd and Robertson 1973).

one or two detectors per intersection. Consequently the implementation of a control strategy based on traffic-responsive signal program generation is not justified (cf. Figure 5.6).

3. *Third generation: distributed computer control.* The availability of small, robust, and cheap microcomputers allows the individual intersections to be provided with their own digital data processing unit. Thus it will no longer be necessary to transmit all detector measurements over long distances to the control center; these measurements can now be evaluated by means of the local microcomputer installed at the intersection. This will lead to a dramatic reduction in the amount of data that will have to be exchanged between the local microcomputer-based controllers and the central computer, as well as in the required data transmission speeds. An increase in the number of detectors per intersection will not increase the number of communication channels needed for linking together the control center with the local controllers.

For this reason, it can be concluded that the barrier discussed above, which exists with respect to the implementation of signal program generation strategies for centralized control systems, can be broken by introducing the concept of distributed control (cf. Figure 5.7). Moreover the local microcomputer can be used to implement effective algorithms for the priority control of public transport.

Two versions of distributed traffic light control systems may be distinguished:

1. *Totally distributed traffic control systems.* Here it is assumed that no control center exists. Each local microcomputer solves the control tasks occurring at its own intersection. To enable synchronization of traffic lights, adjacent microcomputers have to exchange relevant data where one of the local microcomputers, installed at a critical intersection, could play the role of a master controller. This master controller ensures coordinated operation of the individual microcomputers. It determines optimal coordination schemes by evaluating detector data obtained at its intersection as well as from data transmitted from the other microcomputers. The data transmission between the individual microcomputers can be facilitated by a so-called party-line structure, e.g., by a two-wire telephone cable. The reliability of data transmission can be increased by using microcomputer software for such tasks as coding, decoding, error checking, etc.

The totally distributed traffic control system discussed here is very likely a suitable concept for relatively simple traffic control systems, as, for example, the traffic-responsive coordinated control of several (up to about ten) interconnected intersections.

2. *Hierarchically structured distributed control systems.* In more complicated situations, it is necessary to have a control and surveillance center (cf. Figure 5.7). Because of the local microcomputers, the tasks of the central computing system can be reduced significantly. Thus, in many cases, one or a small number of microcomputers have the potential to handle the tasks of the control center (cf. Brinkman et al. 1975). For large urban areas, a structure with three or more levels can be used. Such levels could be:

- central computing system
- microcomputers installed in spatially distributed, unmanned subcenters
- local microcomputers installed at the intersections

Distributed traffic light control systems of the forms described here are currently in the process of development; some installations are already operational or will become operational in the near future (cf. Anderson 1977, Barker et al. 1977, Brinkman et al. 1975, Bång and Nilsson 1976, Cutlar 1976, Hahn and Eustis 1977, Krayenbrink and Vlaanderen 1975, Scott 1975, Sealbury 1974, Siemens 1977, Stanford and Parker 1977).

The benefits expected from the introduction of this new systems generation can be summarized as follows:

- reduction of the total systems costs, mainly as a result of a decrease in the costs of data transmission (cables, modems, etc.)
- increase of the reliability and availability of the system by reduction of the sensitivity of the whole control system to breakdowns of a local microcomputer, a detector or a single data link
- increase of the flexibility of the systems structure, i.e., the distributed control system can be installed and put into operation by starting with one traffic-responsive controlled intersection, which can then be linked with the next one, etc.
- simplification of computer software by dividing a huge software package into a number of smaller ones that can more easily be implemented by means of the local microcomputers
- increase of the efficiency of traffic control by improvement of the adaptability to changing traffic conditions, and by provision of priority control to public means of transport

5.3.4 SINGLE INTERSECTION CONTROL METHODS: BASIC FEATURES AND CURRENT TRENDS

Three categories of intersection control methods have been developed and used (cf. Figure 5.9):

- fixed-time control
- vehicle-actuated control
- self-optimizing control

1. *Fixed-time control.* Here, all control parameters are precomputed and kept constant. These parameters are:

- cycle lengths T, i.e., the total duration of the green, yellow, and red periods
- split g, i.e., the proportion of the sum of red and yellow time to cycle length T
- offset, i.e., the time difference between successive green times

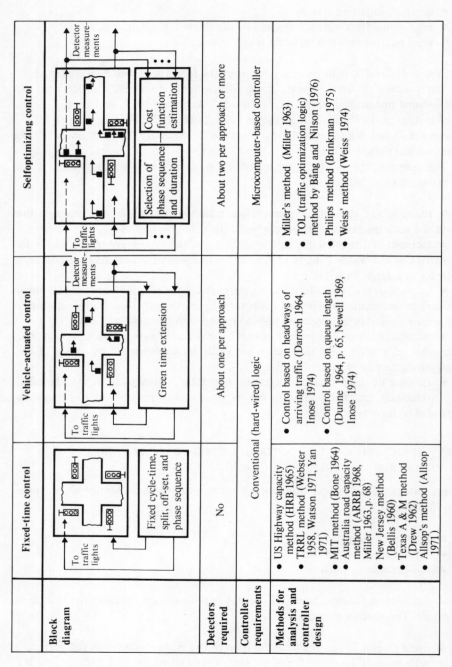

FIGURE 5.9 Survey of single intersection control methods.

- phase sequence, i.e., the succession in which the individual driving directions will become green

Methods for determining these parameters by means of historical traffic data were developed by Webster (1958), Miller (1963), Blunden (1964), Allsop (1968, 1971), and many others (cf. Figure 5.9). Formulas have been derived to calculate capacity, delay, probability of stop and queue length for most distributions of arrivals and service-times (e.g., Inose and Hamada 1974, May 1974, Schnabel 1975). The theoretical results were in many cases consistent with field data. An excellent survey of important methods for designing fixed-time traffic light control programs has been presented by May (1974).

2. *Vehicle-actuated control.* This control principle, which has been known since the 1930s (cf. Inose and Hamada 1974), aims to adapt the length of the green time to real traffic conditions. Two variants have to be distinguished: (1) control based on headways of arriving vehicles; (2) control based on queue length (cf. Figure 5.9).

The first, i.e., the headway method works as follows: the green signal is unconditionally "on" during a given minimum green time; after this time has expired, a small unit green time increment is added if a traffic detector, located a certain distance in front of the stop line, detects the passage of a vehicle. If the detector identifies more vehicles passing during the extension period, the green time is extended successively until a maximum green time is reached. If a vehicle is not detected during one of the extension periods, then the green time is switched to (fixed-time) yellow and then to red. An intersection that uses this control method in all approaches is called a fully actuated signal control system.

The second varient, i.e., the queue length method, modifies the method sketched here in the following manner: In order to use the green time with high efficiency, the green signal is switched to red immediately after the queue vanishes, because saturation flow rate is then always maintained in the green time (cf. Inose and Hamada 1974).

Methods of analyzing and designing vehicle-actuated control systems were developed by Dunne and Potts (1964), Grafton and Newell (1967), Newell (1969), and others (cf. Figure 5.9).

3. *Selfoptimizing control.* This control principle aims to maintain optimal traffic flow control even under heavily changing traffic conditions. Thus the individual controller characteristics, such as cycle time, split, phase duration, and phase sequence, are adapted automatically to changing traffic conditions, while taking into account certain constraints necessary for coordination. Depending on the dynamics of the adaptation processes, one may distinguish between two types of control: macro control and micro control.

Macro control concerns the selection of the cycle time and of the above-mentioned constraints, such as determining suitable off-set times in order to ensure synchronization in coordinated systems. These control parameters, especially the cycle time, cannot be changed frequently; they depend on mean traffic conditions.

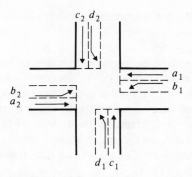

FIGURE 5.10 Phase generating principle: an example intersection.

Micro control, in contrast, deals with the cycle-by-cycle modification of the phase sequences and the durations of the individual phases. Phase sequences and durations are adapted in an optimal manner to short-time stochastic traffic change (cf. Brinkman et al. 1975, Bång and Nilsson 1976).

The principle of phase sequence modification is illustrated by means of a simplified example in Figures 5.10-5.12. The traffic movements indicated by a_1 and a_2 conflict with those denoted by b_2 and b_1, respectively. As soon as the green demand for direction a_1 (or a_2) becomes zero, the corresponding alternative direction of movement, i.e., b_2 (or b_1), becomes green. Using the terminology of control states or phases, this means that the phase $A = (a_1, a_2)$ is transfered to the state $B_1 = (a_1, b_1)$ or $B_2 = (b_2, a_2)$, respectively, depending on which traffic movement ceases first. A direct transition from state $A = (a_1, a_2)$ to $B = (b_1, b_2)$ is chosen if the green demands from both movements a_1 and a_2 cease at the same time or the maximum admitted green is exceeded. A complete cycle of phase transitions is given in Figures 5.11 and 5.12 for the special green demand conditions

$$a_2 > a_1, b_2 > b_1, c_1 > c_2, \quad \text{and} \quad d_1 > d_2$$

In this special case one gets the phase sequence

$$A - B_2 - B - C - D_1 - D - A - \ldots$$

(cf. Figure 5.11).

It is now necessary to determine how the green time phase sequence can be optimally determined in a real-time operation mode. This problem concerns the second part of the micro control task, i.e., determining optimal phase durations. Selfoptimizing phase timing strategies were suggested by Miller (1963), Weinberg (1966), Ross et al. (1973), Brinkman et al. (1975), and Bång and Nilsson (1976).

The basic idea proposed by Bång and Nilsson (1976), which is probably the most advanced method, is discussed here briefly. This method, called TOL (traffic optimization logic), compares the predicted benefits resulting from enlarging a particular phase with the correponding losses occurring in the alternative phases

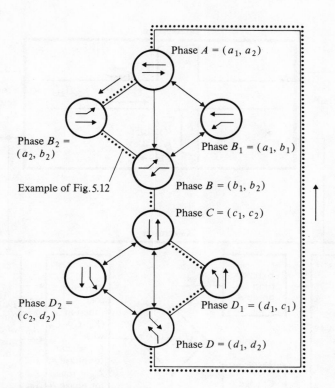

FIGURE 5.11 Phase transition diagram.

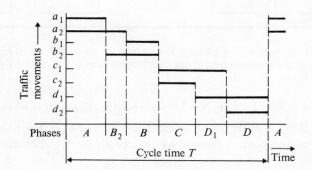

FIGURE 5.12 Example of state transitions for $a_2 > a_1, b_2 > b_1, c_1 > c_2$ and $d_1 > d_2$.

having red lights. Figure 5.13 illustrates this procedure for simplified conditions, i.e., at a crossing of two one-way streets. It is assumed that phase A has green light for the moment. The decision to extend the green by the time increment Δt is based on an evaluation of the objective or cost function Q_A defined as follows:

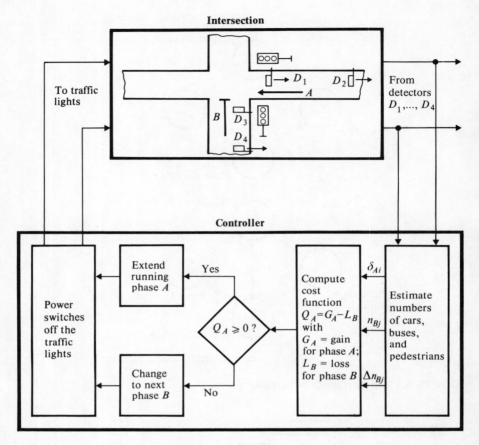

FIGURE 5.13 Selfoptimizing phase timing strategy proposed by Bång and Nilsson (1976), illustrated by means of a simplified intersection (cf. eqs. (5.6)–(5.9)).

$$Q_A = G_A(\delta_{Av}, \delta_{Ab}, \delta_{Ap}) - L_B(n_{Bv}, n_{Bb}, n_{Bp}, \Delta n_{Bv}, \Delta n_{Bb}) \qquad (5.6)$$

where G_A is the gain obtained for phase A and L_B is the corresponding loss resulting from an enlargement of phase A by Δt seconds for the alternative phase, i.e., for B.

The parameters δ_{Ai} describe the number of additional cars (δ_{Av}), buses (δ_{Ab}) or other public transport means, and pedestrians (δ_{Ap}) that can pass the intersection if the green is extended by Δt seconds. The coefficients n_{Bj} denote the number of queueing cars (n_{Bv}), buses (n_{Bb}), and pedestrians (n_{Bp}) in approach B that will suffer an increased delay of Δt seconds if the prevailing green is extended. The parameters Δn_{Bj} are the numbers of extra vehicles (Δn_{Bv}), and buses or trams (Δn_{Bb}) that will be forced to a full stop if the prevailing green is extended by Δt

seconds. Bång and Nilsson (1976) defined the gain G_A and the loss L_B in the following manner:

$$G_A = r_A [a_v \delta_{Av} + a_b \delta_{Ab} + a_p \delta_{Ap}] + b_v \delta_{Av} + b_b \delta_{Ab} \tag{5.7}$$

$$L_B = \Delta t [a_v n_{Bv} + a_b n_{Bb} + a_p n_{Bp}] + b_v \Delta n_{Bv} + b_b \Delta n_{Bb} \tag{5.8}$$

where r_A is the time interval (red and intergreen) for phase A to get a green light again if it is terminated immediately; a_i are the costs of delay per second for cars (a_v), buses (a_b), and pedestrians (a_p); and b_j are the vehicle operating costs to bring a car (b_v) or a bus (b_b) to a complete stop and to resume normal speed.

A suitable choice of parameters a_i and b_j provides the potential basis for giving priority to public transport means. To achieve this, the a_i and b_j have to be set up in such a way that the conditions $a_b > a_v$ and $b_b > b_v$ are valid.

The decision to extend phase A is made by the following rule, considering restrictions of maximum green time:

$$Q_A = \begin{cases} \geq 0, \text{ extension of the running phase } A \\ < 0, \text{ change to the next phase } B \end{cases} \tag{5.9}$$

The implementation of the phase timing and phase sequence generation algorithms requires that all approaches to the intersection are continuously surveyed to obtain estimates of the parameters δ, n, and Δn given above.

Two methods can be used:

- derivation of δ, n, and Δn from passage detectors $D_1 - D_4$ installed at certain distances in front of the stop lines (cf. Figure 5.13)
- direct measurement of the parameters δ, n, and Δn by means of so-called analog long-loop detectors encircling each lane in the approaches (cf. Bång and Nilsson 1976 for more details)

4. *Comparison of fixed-time (FT), vehicle-actuated (VA), and selfoptimizing control.* The basic features of the three control principles are illustrated in Figure 5.9. The main differences may be summarized as follows:

- *Detector expenses*: FT systems do not require detectors, while VA systems demand at least one detector per approach. The implementation of selfoptimizing control principles requires significantly higher installation costs: at least two normal loop detectors or a so-called long-loop detector are needed for the individual approaches.
- *Controllers:* FT and VA traffic light systems require modest local data processing capabilities, i.e., the corresponding local controllers can be designed in the conventional way as hard-wired logic. A selfoptimizing control algorithm is characterized by a large number of logic and arithmetic operations, which have to be performed within a second: hence the local controller has to be designed as a microcomputer based system.

● *Control efficiency*: The efficiency of the individual control principles depends on (1) the topology of the specific intersection, and (2) the traffic conditions. For simple intersections and average traffic volumes, fixed-time control may be an optimal compromise between installation costs and control efficiency. However, under complicated conditions the implementation of a selfoptimizing control strategy can become useful and necessary.

Table 5.1 illustrates this fact by means of results obtained in a comparative study carried out by Bång and Nilsson (1976). They found that their selfoptimizing procedure called TOL (traffic optimization logic) can give substantial reductions in average delay and the proportion of stopped vehicles as compared with conventional FT and VA control. By increasing the bus weighting factors a_b and b_b in eqs. (5.7) and (5.8) further improvements can be gained for buses. These improvements are often cost-effective even if only the reduction of vehicle operating costs (largely energy consumption) is considered (cf. Bång and Nilsson 1976 for more details).

5.3.5 COORDINATED CONTROL METHODS: BASIC FEATURES AND CURRENT TRENDS

The implementation of the three basic control concepts discussed in paragraph 5.3.2, requires methods for signal plan precomputation, modification, and on-line generation. The basic features of the corresponding methodology are summarized in the following with regard to coordinated control of networks and arterial streets (cf. Figure 5.14).

TABLE 5.1 Results of Field Tests obtained by Bång and Nilsson (1976) for Three Control Modes[a]

Control mode	Traffic flow (vehicles/hour)		Average delay (seconds/vehicle)		Proportion stopped (%)	
	Cars	Buses	Cars	Buses	Cars	Buses
Fixed time control (FT)	2600	18	21.5 ± 0.9	20.4 ± 2.0	68	63
Vehicle actuated control (VA)	2740 (105)	12 (67)	20.2 ± 3.0 (94)	20.0 ± 3.0 (98)	67 (99)	62 (98)
Selfoptimizing control by TOL method	2620 (101)	20 (111)	15.5 ± 0.4 (72)	12.2 ± 1.2 (60)	59 (87)	55 (87)

[a]Figures in parentheses give the values as a percentage of the value obtained using FT.

Basic principle	Fixed-time control	Signal plan modification	Signal plan generation
	Precomputation of cycle times, splits and offsets from historical traffic data	Local modification of fixed-time programs using detector measurements	On-line and real-time computation of signal programs using detector data
Methods	*Arterial Streets* • Bandwith method of Little et al. (1964) • Method of Bleyl (1967) • Delay-difference of offset method of Whiting (cf. Hillier 1965, given in Section 6.2) • Method of Brooks (1965) • Translation method of Schnabel (1975) *Streets and Networks* • Transyt (Robertson 1969) • Sigop (1966) • Combination method (Hillier 1965, given in Section 6.2) • Mixed-integer linear programming method of Gartner and Little (1975) • UTCS-1 (Bruggeman 1971, cf. Section 6.1) • Sigrid (Rach 1974, p.75)	*Streets and Networks* • Flexiprog (RRL 1965) • Equisat (Holroyd and Hillier 1969, given in Section 6.2)	*Streets and Networks* • Dynamic plan generation (Valdes 1970, Fuehrer 1970) • Plident (Holroyd and Hillier 1971, given in Section 6.2) • Ascot (Ross 1972, p.73, cf. Section 6.1) • RTOP (Rach 1976) • UTCS-2 and 3 (Bruggeman 1971, cf. Section 6.1) • Tokyo multi-criterion method (Inose 1974, Nakahara 1970, cf. Section 6.3)

FIGURE 5.14 Methods for designing coordinated traffic light control systems.

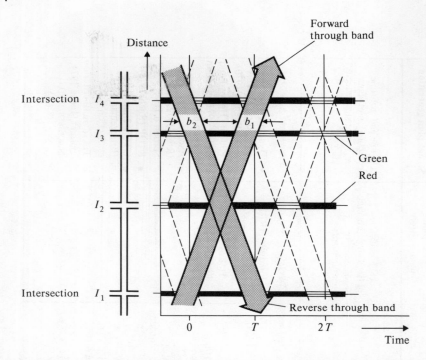

FIGURE 5.15. The bandwidth maximization method (cf. Little et al. 1964, Bleyl 1967, Brooks 1965, May 1974, Inose 1974, Schnabel 1975).

1. *Precomputation of fixed-time control algorithms.* The key problem in developing fixed-time control systems is precomputing signal plans from historical traffic data. These signal plans are described by (1) a common cycle time and (2) a set of offsets, and splits. The required synchronization of traffic lights mainly depends on optimal selection of the offsets.

Two basic, systematic, methods for determining the offsets of coordinated signals are known:

- maximization of green bandwidth
- minimization of delay using delay-offset functions

The concept of bandwidth maximization has been used fairly extensively on single routes. Figure 5.15 illustrates the basic idea: there is a particular time period during which a vehicle is able to continue without stopping at any intersection if it maintains a fixed speed. This time period is called the through band, and its ratio to the cycle length is called the bandwidth. If a sufficient bandwidth can be provided, i.e., the traffic volume is so small that it can be included entirely in the through band, then almost all vehicles can go through all the intersections without stopping. Thus, this is the optimal synchronization scheme, resulting in a maximal bandwidth. Efficient methods for designing maximal bandwidth signal plans have been developed by Little et al. (1964), Bleyl (1967), Brooks (1965) and others (cf. May 1974 and Figure 5.14).

The second basic method, i.e., minimization of delay using delay-offset functions, has been developed more recently and can be used not only for arterial streets but also for complete networks. The method uses an objective function to measure the deviations from an ideal set of offsets. The objective function is minimized by mathematical optimization techniques, such as "hill climbing" methods. Probably the best known and most widely used method of this type is the TRANSYT (traffic network study tool) method of Robertson (1969) (cf. Section 6.2). In essence TRANSYT is an optimization technique for computing signal offsets and splits for minimum delay and stops in a network. The procedure has two main elements : (1) the simulation model, which is used to calculate the performance index of the network for a given set of signal timings; and (2) a hill climbing optimization process, which leads toward optimal signal phasing and offsets. As a performance index the following expression is used:

$$Q = \sum_{i=1}^{n} (d_i + kc_i) \tag{5.10}$$

where d_i is the average delay in passenger-car-units-hours per hour on the ith link of the network; c_i is the average number of passenger-car-units-stops per second on the ith link; k is a weighting factor. (cf. Robertson 1969, and Section 6.2 for more details.)

A third method, which models traffic behaviour and optimizes the fixed-time settings of the traffic signals in an urban road network, uses the SIGOP (traffic signal optimization program) model developed in the USA (cf. SIGOP 1966, May 1974). SIGOP also determines offsets by a hill climbing technique, and calculates splits independently of the offset optimization. All signals are assumed to work on the same cycle, but several cycle times are evaluated in one computer run.

Other important methods are:

• the combination method developed as part of the work for the Glasgow area traffic control system (cf. Hillier and Rothery 1967, Robertson 1967, Allsop 1968, and Section 6.2)
• the mixed-integer linear programming method of Gartner and Little (cf. Gartner et al. 1975)
• the UTCS-1 (urban traffic control system) model (cf. Bruggeman et al. 1971, and Section 6.1)
• the Canadian SIGRID method developed for the Toronto system (cf. Rach et al. 1974, 1975)

2. *Traffic-responsive modification of fixed-time plans.* The methods in this category use fixed-time plans obtained by one of the above-mentioned techniques,

and modify them by means of local detector measurements. Two methods should be mentioned here (cf. Section 6.2 for more details):

- the vehicle-actuated flexible progressive system (FLEXIPROG)
- the EQUISAT method

3. *Fully traffic-responsive signal plan generation.* The methods in this group allow the signal settings to be calculated from traffic detector data in a real-time and on-line operation mode. The following techniques should be mentioned here (cf. Holroyd and Robertson 1973, Rach et al. 1976):

- The dynamic plan generation method, which has been developed for the Madrid area traffic control system (cf. Valdes and de la Ricci 1970, Fuehrer 1970), calculates cycle times and splits according to the principles given by Webster and Cobbe (1966 in Section 6.2). Offsets are determined to minimize delays or stops using measured or estimated speeds and a model of traffic behavior down the link.
- The PLIDENT (platoon identification) method (cf. Section 6.2) identifies the movement of platoons of traffic on the network and, by predicting arrival times, tries to operate the signals so as to pass the platoon unimpeded on the higher priority routes.
- The ASCOT (adaptive signal control optimization technique) developed by Stanford Research Institute (cf. Ross et al. 1973) and the above-mentioned UTCS software package try to achieve very flexible modes of control (cf. case description in Section 6.1).
- The RTOP (real-time optimization program) developed for the Toronto area traffic control system (cf. Rach et al. 1976) computes the signal timing parameters (cycle length, offsets and splits) for a finite control period of 5–30 minutes.
- The Tokyo multi-criterion control method uses not one but a whole hierarchy of control criteria and corresponding control modes (cf. Inose and Hamada 1974, Nakahara et al. 1970, and paragraph 6.2.1).

4. *Comparison of fixed-time and traffic-responsive control methods.* Evaluating the methods discussed above with regard to control efficiency, implementation costs, etc., represents a very complicated and complex task. Considerable research effort has been made during the last 5–10 years to solve this task by means of both simulation studies and full-scale tests.

In a comprehensive Canadian study made in Toronto the methods SIGOP, TRANSIT, Combination, and RTOP were evaluated under real traffic conditions (cf. Rach 1974, 1975, 1976).

The case descriptions included in Chapter 6 of this monograph present further experiences with the methods described above in the USA, UK, and Japan. From these experiences and results published elsewhere, it can be concluded that there is no optimal or best method. The various methods are characterized by both advantages and disadvantages. Nevertheless, in many cases TRANSYT was delivering the most efficient fixed-time control algorithms, and very often the fully traffic-responsive methods did not fulfill their expectations. However, other applications, as for example in the Tokyo system, have demonstrated high efficiency for real-time and on-line methods (cf. case descriptions in Chapter 6 for more details).

It is expected that the introduction of a new hardware generation, i.e., that of spatially distributed computer control systems, will provide a new driving force for further development of the existing control methodology, leading to a new generation of methods and software packages. This new software generation will doubtless be characterized by higher traffic adaptability and lower hardware and software costs.

5.4 FREEWAY TRAFFIC FLOW CONTROL

If the variables volume, density and mean speed, which characterize the traffic flow state on freeways, violate certain limits, then the danger of congestion and accidents increases rapidly, and the installation of freeway traffic flow control and surveillance systems becomes desirable or even necessary. These critical values are as follows (cf. reference Krell 1971, to Chapter 4): volume $\geqslant$ 3,000 vehicles per hour per two lanes; density $\geqslant$ 50 vehicles per km per two lanes; mean speed $\leqslant$ 60 km per hour.

Three basic control concepts have been developed for improving traffic flow in these cases (cf. Figure 5.1):

- ramp metering, i.e., controlling the inflow rates of traffic entering the freeway via the various entrance ramps
- speed and lane-changing control, i.e., traffic-responsive regulation of the maximum driving speed and use of the individual lanes of a freeway, a tunnel, or a bridge, respectively
- merging control, i.e., controlling the process of merging in a dense traffic stream

The main features of these control concepts are discussed in the following.

5.4.1 RAMP METERING

5.4.1.1 Control Concepts. The capacity of a traffic lane decreases if the traffic density becomes larger than an optimal value. This well-known phenomenon, illustrated by the fundamental diagram of traffic flow given in Figure 1.1, explains the occurrence of natural congestion on freeways and in tunnels when too many cars enter traffic links. The aim of a ramp metering system is therefore to maintain traffic demand along all parts of the freeway below the critical levels by means of traffic lights at the entrance ramps (cf. Figures 5.1, 5.16 and 5.17). If such a traffic

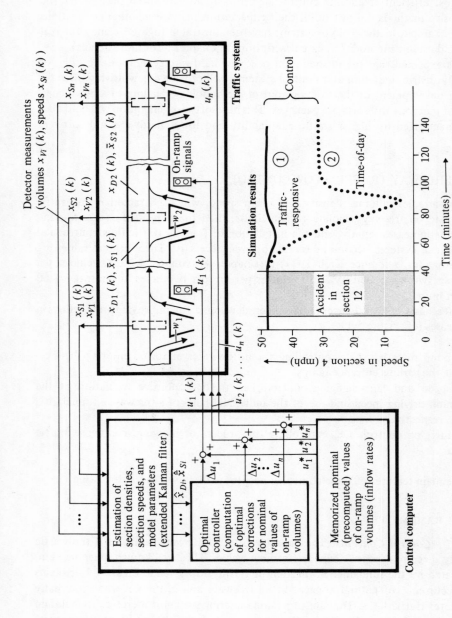

FIGURE 5.16 Concept of a traffic-responsive ramp metering system, and simulation results obtained by Isaksen and Payne (1973) for a freeway in Los Angeles (USA).

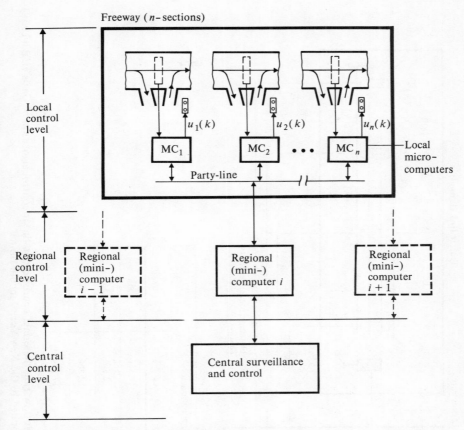

FIGURE 5.17 Concept of distributed freeway inflow control (cf. Donner 1975).

light is switched to green for a period of about two seconds, then one car is permitted to enter the traffic link. By changing the frequency of green, a suitable inflow rate, i.e., an optimal number of cars entering the freeway per unit time, can be selected. This control concept may be applied in a slightly modified version to tunnel traffic flow control, as illustrated by Figure 5.18; here inflow traffic metering is established by means of a traffic light installed in front of the tunnel entrance.

5.4.1.2. Methodology. Two types of optimization problems have to be solved in developing optimal ramp metering algorithms:

1. *Static optimization and open-loop control* (cf. Gershwin 1975, Isaksen and Payne 1972, 1973, Payne and Thompson 1974). Using demand patterns obtained from historical traffic data, nominal values for the inflow rates u_i have to be determined in such a way that the overall traffic throughput is maximized. This gives sets

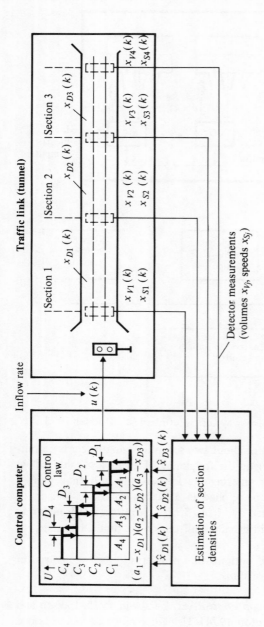

FIGURE 5.18 Tunnel traffic flow control by inflow rate metering; heuristic non-linear control algorithm of Gazis et al. (1968) and Gazis and Footh (1969), which has been applied to the New York Lincoln tunnel.

of nominal control variables u_i^*, and nominal traffic state variables, i.e., freeway section speeds $\bar{x}_{Si}$ and densities x_{Di} (cf. Figure 5.16). These variables are, of course, no longer optimal if disturbances, e.g., accidents, occur. For such situations (2) should be used.

2. *Dynamic optimization and feedback control* (cf. Gazis and Footh 1969, Houpt and Athans 1975, Isaksen and Payne 1972, 1973, Looze et al. 1975, Payne et al. 1973, Tabak 1973). The task of this control system is to minimize deviations between the nominal, precomputed state variables, i.e., section speeds $\bar{x}_{Si}$ and section densities x_{Di}, and their actual values by real-time computation of corrections to the nominal values of the control variables u_i^* (cf. Figure 5.16).

Simulation studies carried out by Isaksen and Payne (1973) suggest that a feedback control system can lead to promising results in avoiding congestion caused by certain incidents, e.g., an accident. In a study for a Los Angeles freeway they assumed that three of the four lanes in section 12 became blocked for about 30 minutes, causing congestion in the previous (upstream) sections after a certain delay. This is illustrated in Figure 5.16, which shows the time dependence of the mean traffic speed in the upstream section 4. In the case of fixed-time (open loop) control the speed decreases to about 5 mph 50 minutes after the incident happened; after 60 minutes it remained for some time at 20 mph below the initial speed (cf. curve 2 in Figure 5.16).

However, the feedback control system can obviously avoid such serious disturbances by optimal limitation of the inflow rates at the on-ramps lying upstream from the section where the accident happened (cf. curve 1 in Figure 5.16).

In spite of these promising results, the following statements should be considered:

- Static optimization and open loop control lead in many cases to an essential improvement in the stationary traffic flow conditions (cf. Chapter 7).
- Dynamic optimization and feedback control can become important under non-stationary flow conditions, e.g., in the case of accidents; therefore feedback control may primarily be considered as a tool of emergency control.

Moreover, one has to take into account that the implementation of complex feedback ramp metering systems is still hindered by some serious methodological difficulties associated with the following:

- It is not possible to take direct measurements of the state variables section densities x_{Di}, and section speeds $\bar{x}_{Si}$ needed for the feedback control algorithm. Traffic detectors permit the measurement of traffic volumes x_{Vi} and mean speeds x_{Si} only at fixed points, i.e., at the section boundaries (cf. Figure 5.16).
- The parameters of the mathematical traffic flow models, which are required for solving the above-mentioned dynamic optimization problem, change with

conditions, traffic incidents, etc. They have to be determined in a real- and on-line operation mode.

For these two reasons, the implementation of optimal feedback ramp metering systems requires the solution of a combined state and parameter estimation problem, by applying the extended Kalman filter (cf. Orlhac et al. 1975, Payne et al. 1975). This estimation task is still the subject of fundamental research (cf. Chang and Gazis 1975, Gazis and Knapp 1971, Nahi and Trivedi 1973, Szeto and Gazis 1972, Orlhac et al. 1975, Payne et al. 1975). Therefore in most practical applications feedback ramp metering algorithms are implemented in a simplified manner using heuristic control methods. The simplest version of these heuristic algorithms is a bang-bang control principle, e.g., the ramp is closed for a certain period of time if congestion occurs in the flow direction.

A more sophisticated heuristic algorithm developed for metering tunnel traffic inflow is illustrated in Figure 5.18. This algorithm determines the admissible values of the control variable "inflow rate" u by means of the product $(a_1 - x_{D1})(a_2 - x_{D2})(a_3 - x_{D3})$, which contains the differences $a_i - x_{Di}$ between nominal section densities a_i and real, i.e., estimated, section densities x_{Di}. Different threshold constants $C_1, \ldots, C_4$ and hysteresis factors $D_1, \ldots, D_4$ avoid frequent changing of inflow rates and ensure the stability of the control process.

A main direction of current research and development work is the creation of spatially distributed ramp metering systems (cf. Figure 5.17). Here it is assumed that at each on-ramp a microcomputer is installed to evaluate detector measurements, and control on-ramp traffic lights. These local microcomputers can memorize the nominal control variables and change them during the course of the day. Thus the tasks of the higher level computers shown in Figure 5.17 may be restricted to coordinating the operation of the local microcomputers and solving the dynamic optimization tasks affecting many or all freeway sections. This control structure leads to the important advantage that the data exchange between the local microcomputers and the higher level computers can be achieved by means of two-wire telephone cables (cf. Donner et al. 1975). Thus it is expected that replacing the centralized ramp metering system according to Figure 5.16 by the distributed control system shown in Figure 5.17 will lead to significant savings in installation costs.

5.4.2 SPEED AND LANE USE CONTROL

This control principle aims to improve traffic flow and safety by means of the following measures:

- limiting the speed if traffic densities exceed critical values or if weather conditions are bad
- forbidding overtaking maneuvers or prescribing the use of certain lanes
- providing drivers with information on traffic conditions and emergency situations, e.g., accidents in the direction of traffic flow

(cf. Le Pera and Nenzi 1974, and references Everall 1972, and Krell 1971, to Chapter 4).

Figures 5.19 and 5.20 illustrate two traffic situations in which speed control or combined speed and lane changing control are suitable tools for increasing traffic safety and reducing the negative effects of congestion and accidents. If shock waves, e.g., caused by accidents, occur (cf. Figure 5.19), then driving speeds in the upstream sections have to be reduced as early as possible in order to reduce the probability of rear-end accidents and the duration of lane blockages. If only one lane is blocked by an incident then a combination of lane changing and speed control is required, as illustrated by Figure 5.20.

The implementation of the speed and lane control systems shown in Figures 5.19 and 5.20 requires the corresponding part of the freeway to be divided into certain sections, and changeable road signs and traffic detectors to be installed at the section boundaries. Because of the large spatial extension of such control systems, local microcomputers are used in order to reduce the costs of the necessary data links.

Lane use control not only represents a tool for emergency control but also a measure for increasing traffic link capacity and for giving priority to public transport means and car pools. The following control principles are used (cf. reference Krell 1971, to Chapter 4).

1. *Reversible lanes.* Three possibilities may be distinguished:

- tidal flow at fixed locations using permanent equipment at fixed times
- tidal flow at fixed locations using permanent equipment at times selected according to traffic conditions
- reversible lanes working at bottlenecks arising from major road works requiring temporary equipment

2. *Reserved lanes.* The corresponding control problem is to assign special lanes to special classes of vehicles (buses, car pools, etc.) at certain periods of time. This assignment problem can be handled as a planning problem in which special lanes are permanently reserved for special vehicles, or as a traffic-responsive control task where the lane reservation is made according to the traffic conditions.

Another method of speed and lane use control does not use changeable speed and lane use signs but variable message signs presenting driving recommendations in words.

By means of computer controlled bulb matrix message signs, the following types of information can be displayed in sections located upstream of an incident (cf. Grover 1973, Green 1973):

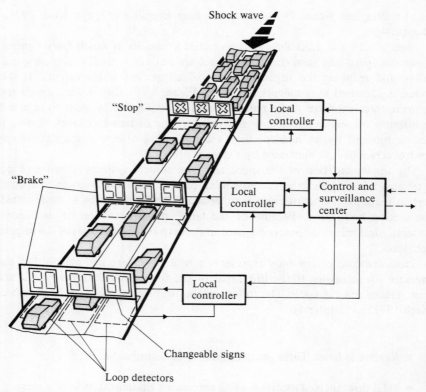

FIGURE 5.19 Concept of speed control by means of changeable, computer-controlled speed signs.

- CONGESTION
  ... MILES AHEAD
- ACCIDENT AHEAD
  USE CAUTION
- ACCIDENT
  ... MILES AHEAD
- CONGESTION
  NEXT ... MILES
- RIGHT LANE
  BLOCKED AHEAD
- CONGESTION ENDS
  ... MILES AHEAD
- 15 MINUTES
  TO ...

The primary benefits from providing motorists with information are expected to be reductions in:

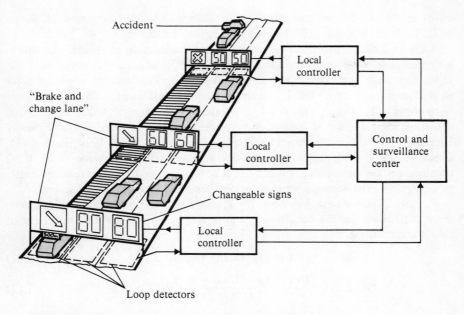

FIGURE 5.20 Concept of combined speed and lane use control.

- motorist aggravation
- secondary accidents
- delays, by providing diversions to alternate routes, or by reducing demand to permit rapid recovery after an incident

The creation of these motorist information systems involves high installation costs. Thus, they are only likely to be justified for more heavily traveled freeways such as the Los Angeles urban freeways (cf. Green 1973).

Most of the speed and lane control systems of the type shown in Figures 5.19 and 5.20, as well as those of the motorist information systems, are operated in a semi-automatic way, i.e., the decision to change the speed and lane use signs or the variable message signs is made by operators using information sources such as

- detector measurements
- information provided by local TV cameras, helicopter pilots and ground patrols

A key problem in trying to improve the efficiency of the systems sketched above is to reduce the time required to discover an incident, e.g., an accident. Therefore the creation of reliable automatic incident detection systems represents one major effort in the development of freeway surveillance and control systems (cf. Green 1973, Foth 1974, Kahn 1974, Payne 1976, Payne et al. 1975).

Such an incident detection system identifies the location of an incident within a minimum of time and with a low probability of a false alarm. Several heuristic algorithms have been developed and implemented. The basic principle of these algorithms is to compare traffic features — simple functions of traffic detector data — to thresholds. This is illustrated by means of the so-called California incident detection logic developed for and implemented in the Los Angeles Area Freeway Surveillance and Control Project (cf. Figure 5.21, and Kahn 1974, Foth 1974, Green 1973). This algorithm uses the one minute average occupancy OCC (i, t) obtained for the detectors $i = 1, 2, \ldots, n$ at time interval t as the relevant traffic variable, where the occupancy measurements are updated every $B = 20$ seconds.

For two adjacent detectors i and $i + 1$, with i counted in the direction of travel, three conditions for the occurrence of an incident can be identified by means of heuristic considerations:

$$\text{OCCDF} = \text{OCC}(i, t) - \text{OCC}(i + 1, t) \geqslant K_1 \tag{5.11}$$

$$\text{OCCRDF} = \frac{\text{OCC}(i, t) - \text{OCC}(i + 1, t)}{\text{OCC}(i, t)} \, 100 \geqslant K_2 \tag{5.12}$$

$$\text{DOCRDF} = \frac{\text{OCC}(i+1, t - B) - \text{OCC}(i + 1, t)}{\text{OCC}(i + 1, t - B)} \, 100 \geqslant K_3 \tag{5.13}$$

If all three conditions are satisfied then the existence of an incident is assumed and an alarm lamp is switched on in the control center (cf. Figure 5.21).

The threshold constants K_1, K_2, and K_3 are functions of both location and time of day. The performance of an incident detection algorithm, in terms of the detection and false alarm rates, depends directly on the choice of these thresholds (cf. Payne 1976). The incident detection algorithm sketched here has been extensively tested in the Los Angeles Area Freeway Surveillance and Control Project mentioned above, with the following results: Over 80 percent of all verified incidents were detected in under four minutes. A detection rate of over 90 percent was achieved for congestion that caused incidents, with a false alarm rate of 32 percent (cf. Kahn 1974, Green 1973).

5.4.3 MERGING CONTROL

Modern freeways are approximately twice as safe to travel on as conventional highways. However, the entrance ramp, where the motorist makes the transition from the conventional highway to the freeway, very often experiences high accident rates. This is especially true when a queue of vehicles builds up on the ramp, waiting to enter a crowded freeway. About 20 percent of urban freeway accidents in the USA at the beginning of the seventies occurred on entrance ramps and acceleration lanes (cf. FHWA 1971). This motivated the development of merging control systems, with the following aims:

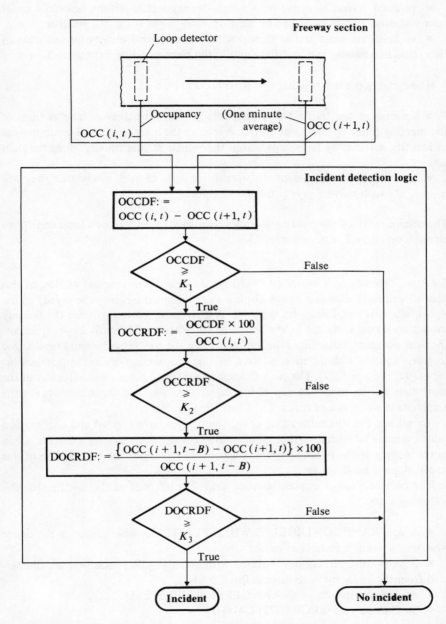

FIGURE 5.21 Concept of incident detection, illustrated by the so-called "California logic" (cf. Kahn 1974, Payne 1976, Green 1973).

- to assist drivers to merge with a high density traffic stream, especially under poor visibility conditions caused by ramp-freeway geometry or the weather
- to keep the main traffic stream within a stable and undersaturated state of flow, thus maximizing traffic throughput in the corresponding freeway section

The merging control task has two parts (cf. Figure 5.22):

- to analyze the traffic flow conditions in the right freeway lane in front of the merging point by means of special detectors the merging control computer has to identify sufficiently large gaps within the vehicle stream moving along the right lane
- to control the movement of on-ramp vehicles in such a way that they will meet safely with moving gaps at the merging point

The second part of the control task can be solved by either open-loop control or closed-loop control of on-ramp vehicles.

5.4.3.1 Open-Loop Control of On-Ramp Vehicles. The simplest version of this control principle uses as a driver display a conventional (green–yellow–red) traffic signal, which is installed at the rampside. A driver who intends to enter the freeway must stop his car if the red light is switched on. When a sufficiently large gap occurs then the computer determines the moment when the traffic light should be changed to green and the waiting driver allowed to start the merging process by accelerating his car (cf. Figure 5.22). The car will reach the merging point on the freeway at the same time as the predicted gap if the driver uses assumed average values for the acceleration and speed of his car (cf. Drew 1968).

To reduce the difficulties that could result from wrong speed and acceleration values, a more advanced version of open-loop control systems uses a band of green lamps moving at the required speed. Figure 5.22 illustrates the functioning of this so-called green band system.

The complete driver display installed along the left side of the on-ramp has the following units:

- a sign RAMP CONTROLLED WHEN FLASHING, which informs the driver whether the ramp is controlled or not
- a green band display unit covered with green acrylic panels that are illuminated from the inside by incandescent flood lights
- an advisory speed sign showing GREEN BAND SPEED . . .
- an illuminated MERGE WITH CAUTION sign

Such an on-ramp installation permits three states of operation: (1) moving; (2) stopped gap–acceptance; (3) fixed-time metering (cf. FHWA 1971).

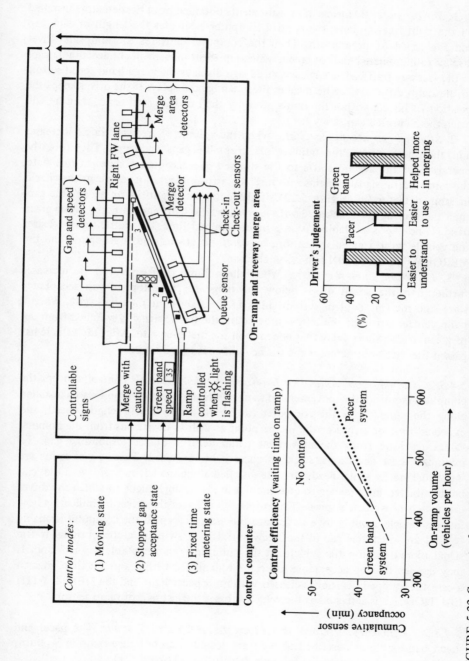

FIGURE 5.22 Concept of merging control (further entrance ramp displays are: 1, static sign, DRIVE BESIDE GREEN BAND, 2, static sign, STOP HERE ON RED, 3, moving green band display) (cf. FHWA 1971, True 1973).

1. *Moving state*. By means of measurements obtained from the detectors installed on the right freeway lane, the control computer calculates the length of each gap and the speed of progression. Then the computer activates the rampside driver display to illuminate bands of green, which represent movements of acceptable gaps in the freeway traffic. The advisory speed sign displays the green band speed setting to the ramp driver before he reaches the green band display. If the driver keeps the position of his car within the limits given by the moving green band, then he will reach the identified gap safely.

2. *Stopped gap–acceptance state*. When the volume of freeway traffic increases and the speed drops, the number of acceptable gaps decreases. Therefore the operation of the system shifts to the stopped gap-acceptance state. In this state, the ramp traffic signal remains red until the arrival of a ramp vehicle at the check-in ramp detector (cf. Figure 5.22). If the computer identifies an acceptable gap, then the waiting car is released with a green traffic light and an accelerating green band. If no gap is available within a predetermined time, the traffic signal releases the vehicle with a green indication; however, no green band is generated and the MERGE WITH CAUTION sign is switched on.

3. *Fixed-time metering state*. When the volume of freeway traffic increases further and the speed drops below the range for the stopped gap-acceptance state, the merging control system shifts to the fixed-time metering state. When a ramp vehicle arrives at the check-in detector, the traffic signal is switched from red to green after a short delay that depends on the metering rate. After the vehicle has passed the check-out detector the traffic light goes to red again.

5.4.3.2 Closed-Loop Control of On-Ramp Vehicles. In this control concept the computer monitors the movement of ramp vehicles by pairs of loop sensors installed along the ramp. Thus the computer can identify the vehicle length, speed, and expected time of arrival at the merge point as well as deviations from the nominal values resulting from the movement parameters of the corresponding gap. To minimize these deviations, so-called pacer lights are used. These pacer lights are individual traffic signal heads that are installed along the left rampside and that can be illuminated in sequence to give the illusion of a single light to which the driver can match his vehicle's speed. This display thus paces the driver by indicating how fast his vehicle should move to fit into the assigned gap. If the computer is unable to find an acceptable gap in the freeway traffic, it switches a conventional traffic signal to red to stop the vehicle on the ramp until an acceptable gap is found. It may happen that no acceptable gap can be identified within a preset time. In such a case the stopped vehicle is released without a pacer light and the MERGE WITH CAUTION sign is activated as the vehicle crosses the last pair of ramp sensors.

5.4.3.3 Efficiency of Open- and Closed-Loop Control Systems. The pacer and green band systems described above were tested at an entrance ramp in Woburn, Massachusetts (USA) with the following results (cf. FHWA 1971):

- The control efficiencies of the two systems are approximately equal. This is illustrated by the relation between the cumulative sensor occupancy, corresponding to the on-ramp waiting time, and the on-ramp volume. The decidedly steeper slope of the "no control" curve at high on-ramp volumes illustrates the capability of merging control to reduce delays and to increase on-ramp capacity.
- Of the drivers who were using both the pacer and the green band system there was a general preference for the green band system, as illustrated in the corresponding diagram given in Figure 5.22.
- No merging accident occurred during the test phase, which covered several months.

In spite of the benefits summarized here, merging control systems have not yet found widespread application. This is mainly because of high installation costs. In general, the much simpler fixed-time metering system, which does not need green band and pacer lights, is preferred (cf. Cook et al. 1970, Drew 1968, Fowler 1967, Gervais 1964, Masher et al. 1975, McDermott 1968, Messer 1969, Newman et al. 1970, Pinnell et al. 1967, True and Rosen 1973, Wattleworth 1971, Wiener et al. 1970, Yagoda 1970).

5.5 VEHICLE CONTROL

This section presents a survey of the control tasks occurring at level 3 of the hierarchy shown in Figure 5.1. Three categories of problems can be distinguished:

- headway control
- engine/power train control
- antiskid braking control, and further control and surveillance tasks

5.5.1 HEADWAY CONTROL

The objective of this control task is to assist individual drivers in choosing adequate headways when driving in a string of vehicles.

The main control criteria concern safety, i.e., reducing the danger of rear-end collisions. Moreover, vehicle headways are closely related to traffic throughput, i.e., a second control criterion is preserving the road capacity. The safety and the capacity problems may be characterized as follows.

1. *Safety problem.* Car driving in a string of vehicles is a complicated control process. The dynamics of this man–machine system can be described using the results of the so-called microscopic traffic flow theory (cf. Gazis 1974). In the microscopic modeling approach it is assumed that every driver who finds himself

in a single-lane traffic situation reacts to a stimulus from his immediate environment according to the relation

Reaction of driver i at time $t = \lambda_i$ {Stimulus at time $t - \tau_i$}

The reaction of the driver may be expressed by the acceleration, $\ddot{s}_i(t)$, of his car. The parameter λ_i describes the sensitivity of the driver's reaction to a given stimulus, and τ_i is a reaction time lag.

It has been shown that the main stimulus is caused by the speed difference $v_{i-1} - v_i$ resulting in the well-known linear car-following model

$$\dot{s}_i(t) = v_i(t)$$
$$\dot{v}_i(t) = \lambda_i \{v_{i-1}(t - \tau_i) - v_i(t - \tau_i)\} \tag{5.14}$$

and leading to the speed-transfer function

$$G_i(p) = \frac{\mathcal{L}\{v_i(t)\}}{\mathcal{L}\{v_{i-1}(t)\}} = \frac{\lambda_i \exp(-p\tau_i)}{p + \lambda_i \exp(-p\tau_i)} \tag{5.15}$$

which describes the dynamics of a two-car system (cf. Figure 5.1, level 3.1, and Gazis 1974, Strobel 1977, Ullmann 1978). It is well-known from the fundamentals of control theory that a system described by eq. (5.15) will become unstable for $\lambda_i \tau_i > \pi/2$, resulting in collisions of the two-car system. The dynamics of a long queue of vehicles can be determined by means of eq. (5.15) using certain values of λ_i and τ_i, and assuming special driving maneuvers of the leading car. Figure 5.23 illustrates such an example for $\pi/2 > \lambda_i \tau_i > 0.5$. Although the two-car system remains stable and collision free, collisions may occur at the end of the queue (cf. positions of cars numbers 7 and 8 in Figure 5.23). Instability and collisions occur if a driver reacts too slowly (large τ_i) or too sharply (large λ_i) to speed changes of the leading car: small speed changes of the leading car are then amplified, resulting, for long strings of vehicles, in collisions between the cars at the end of the queue.

If one assumes the same model for all drivers, i.e., $\lambda_i = \lambda$ and $\tau_i = \tau$, then this result occurs as soon as $\lambda\tau > 0.5$ — a value of the same order of magnitude as the experimentally determined values given by Herman and his co-workers (cf. Gazis 1974, p. 89 ff).

2. *Capacity problem.* The danger of a rear-end collision can obviously be reduced by choosing large headways. However, large headways lead to a reduction of traffic throughput. According to the well-known equation

$$C = \frac{v}{L + H_{\min}} = \frac{v}{L + \tau v + cv^2} \tag{5.16}$$

the lane capacity depends on speed v, average car length L, and the minimal headway $H_{\min} = \tau v + cv^2$. Here τ represents the time lag mentioned above of the driver and c is inversely proportional to the maximum rate of deceleration.

Figure 5.24 illustrates that the road capacity reaches a maximum at a certain speed. This maximum can be increased by reducing the driver's time lag τ.

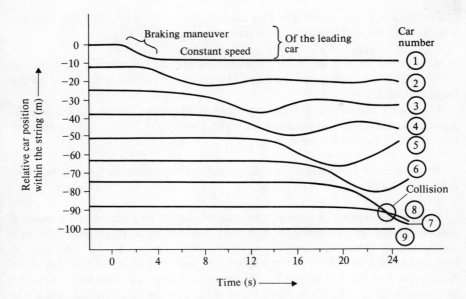

FIGURE 5.23 Simulation example illustrating the phenomenon of queue instability (cf. Gazis 1972, and eq. (5.15) for $\lambda\tau > 0.5$).

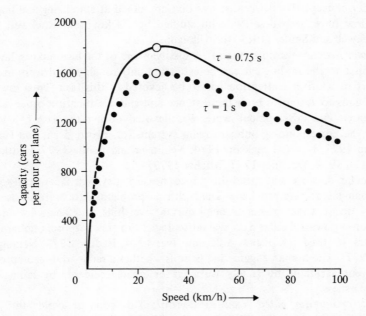

FIGURE 5.24 Capacity of a single lane as a function of speed v and the driver's time lag τ ($L = 5$ meters, $c = 0.6$ hour^2 per km).

Summarizing, one may state that fulfilling both the safety and the capacity criterion requires two types of measures:

- reducing the time lag τ and increasing the accuracy in estimating the state of movement of the leading car, i.e., improving the state observer functions of the human controller
- assisting the drivers in selecting a proper reaction to given stimuli, i.e., in selecting suitable controller parameters λ and τ with $\lambda\tau < 0.5$ (improving the control algorithm used by human drivers)

Three concepts have been developed to achieve the improvements mentioned here (cf. Figure 5.25):

1. *Headway control by means of head-up displays*: This concept uses a so-called head-up display in the form of two vertical lines, which are projected at the car's windscreen by means of a specially designed optical device. An electronic control unit changes the horizontal distance d between these two lines as a function of the driving speed. If the windscreen projection of the leading car remains between the two vertical lines then the distance between his own car and that immediately in front is sufficiently large. A violation of the head-up display lines by the windscreen projection of the leading car warns the driver that the headway is too short and that a speed reduction is required.

Further concepts for driver displays that are aimed at stabilizing and improving traffic flow have been described and studied by Rackoff and Rockwell (1973), and Rockwell and Snider (1969) (cf. Ullmann 1978).

2. *Radar distance-warning system*: A disadvantage of the head-up display is that the distance to the leading car can be estimated only roughly by drivers and a very large part of a driver's attention has to be devoted to this task. Thus the second concept aims to automate both the distance and speed estimation process. Extensive research and development work was devoted during the last 5–8 years to creating the corresponding onboard radar systems (cf. Figure 5.25, and Hahlganss and Hahn 1977, Ives and Jackson 1974, Neininger and Pählig 1977, Radtke 1977, Shefer et al. 1974, Wüchner 1977, Wocher 1977).

The radar distance and speed difference measuring system faces both technical and human-engineering problems. The technical problem is to supress false alarms that may occur when driving through curves, over hills, etc. The solution to this task requires powerful radar data evaluation logic. For this purpose, onboard microcomputers are used (cf. Figure 5.25, and Ives 1974, Radtke 1977, Neininger and Pählig 1977). The human-engineering issue is whether a radar-equipped automobile will prevent accidents by timely warnings or create accidents by lulling drivers into carelessness.

3. *Microcomputer aided headway control*: This concept avoids the human-engineering problem in the following way. The onboard system not only warns

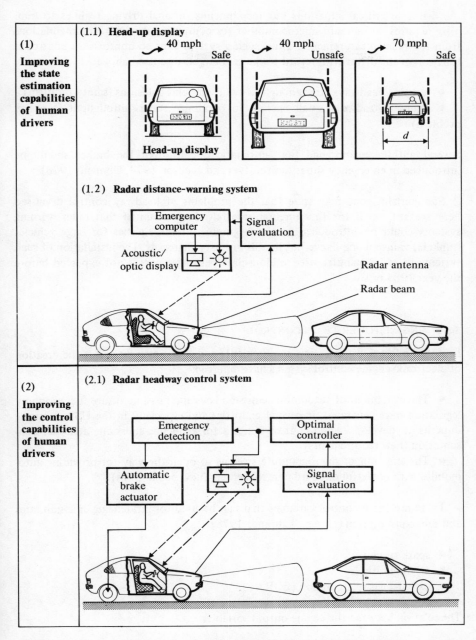

FIGURE 5.25 Survey of concepts for improving the state estimation and controller performances of human drivers (cf. Ullmann 1978).

the driver in critical situations but recommends optimal driving regimes to him. This requires an onboard microcomputer for computing driving recommendations from the radar measurements. Thus, such a system aims to improve the controller function of the driver with regard to the two targets discussed above:

- keeping headways large enough to avoid rear-end collisions (safety problem)
- keeping headways small enough to maximize traffic throughput (capacity problem)

Several authors recommend that automatic operation of the brakes should be introduced in emergency situations (cf. Ives and Jackson 1974, Ullmann 1978).

Summarizing, one may state that the problems of headway control discussed here are still open for fundamental research. It is assumed that radar warning systems could be introduced at the beginning of the eighties for large vehicles (tankers) transporting dangerous liquids or heavy materials. The installation of such systems in series-manufactured middle-class passenger cars is not expected before the year 1985 or so.

5.5.2 ENGINE/POWER TRAIN CONTROL

During the last 5-8 years, new requirements and new possibilities for the creation of electronic engine control systems have emerged.

- The enactment of automobile emission laws and possible future fuel economy legislation have generated, in several countries and especially in the USA, a strong impetus to develop engine control systems for reducing emissions and fuel consumption (new requirements)
- The creation of microcomputers seems to open the way for providing automobiles with powerful onboard control centers (new possibilities)

There are three control variables that can be used for influencing emission rates and fuel consumption (cf. e.g., Laurance 1978):

- spark timing
- air-fuel ratio
- exhaust gas recirculation (EGR)

The relation between the engine output variables

- fuel consumption rate
- carbon monoxide (CO) emission rate

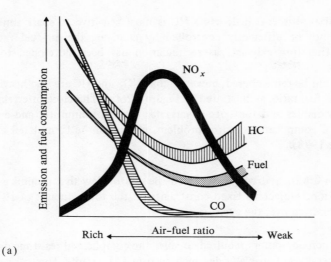

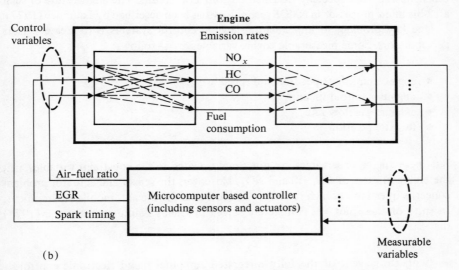

FIGURE 5.26(a) Relation between fuel consumption, pollutants emission, and air-fuel ratio at constant speed/load, constant EGR, with the spark advance maintained optimum (cf. Laurance 1978). (b) Simplified block diagram of the engine control system.

- emission rate of oxides of nitrogen (NO_x)
- emission rate of hydrocarbons (HC)

and the control variable air-fuel ratio is illustrated in Figure 5.26. It can be observed that CO is most sensitive to the air-fuel ratio. The emission of CO increases rapidly if the mixture is enriched, while HC is most sensitive to spark timing. However, NO_x cannot be efficiently controlled by changing the air-fuel ratio and spark timing. Therefore exhaust gas recirculation has been developed to control this pollutant.

It should be remembered, however, that NO_x and HC are also heavily dependent on the air-fuel ratio, as illustrated by Figure 5.26(a). This close interrelation between control variables and the output variables (fuel consumption and emission rates) leads to a complicated control problem, which can be formulated as follows (cf. Laurance 1978):

> For an arbitrary driving cycle, find the control law that minimizes fuel consumption, subject to fixed constraints on the total amount of HC, CO, and NO_x emitted over the cycle.

The microcomputers required to solve the complicated real-time data-processing tasks generally use specially designed custom LSI circuits. The architecture of such a 12-bit microprocessor in NMOS technology has been described by Laurance (1977).

One key problem in implementing engine control systems is to create sensors and actuators. Some measurable engine variables are as follows:

- engine crankshaft position and speed
- engine manifold vacuum
- exhaust gas flow rate
- throttle position

But these engine observables are not simply related to the constraint variables, i.e., the emission rates of CO, HC and NO_x. Moreover, the sensor and actuator problem concerns questions of high reliability and low cost for the automobile environment. In spite of these not yet completely solved problems Laurance has stated (1978) that

> "with the advent of the fully integrated computer based electronic control system for automobile engines, a new era in the approach to engine control has begun"

(cf. Special Report on Automotive Electronics 1973, Temple and Devlin 1974, Oswald et al. 1975, Moyer and Mangrulkar 1975, Prabhaker et al. 1975).

It is expected that the general introduction of microcomputer based engine control systems during the next ten years or so will make a major contribution to the solution of the air pollution problems discussed in Chapter 1. Moreover, a reduction of fuel consumption rates of the order of 10-20 percent is considered to be feasible.

5.5.3 FURTHER CONTROL AND SURVEILLANCE TASKS

The availability of microcomputers onboard cars raises the question: What further control and surveillance tasks can be solved by an onboard central computer? Four categories of control tasks may be distinguished, which include the problems of headway and engine control already discussed above (cf. Figure 5.27).

1. *Safety related tasks*:
- headway control
- automatic antiskid braking
- driver's fitness, prestart checking
- air bag control
- seat belt supervision

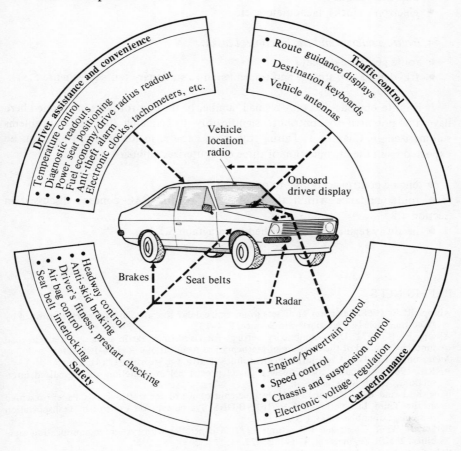

FIGURE 5.27 Survey of onboard control problems.

2. *Car performance-increasing tasks*:
- engine/power train contol
- speed control
- chassis and suspension control
- electronic voltage regulation

3. *Driver assistance and convenience tasks*:
- temperature control
- diagnostic readouts
- power seat positioning
- fuel economy/drive radius readout
- anti-theft alarm
- electronic clocks, tachometers, etc.

4. *Route guidance and traffic control tasks*:
- route guidance displays
- trip destination storage and transmission to stationary units (cf. Figure 5.3)

It is quite obvious that only a small number of the control tasks mentioned here have the potential to contribute significantly to reducing the traffic problems summarized in Chapter 1. During the next decade, major contributions can be expected from the introduction of three computerized onboard systems:

- engine control according to paragraph 5.5.2
- route guidance within the framework of the CAC concept discussed in Section 4.2
- headway regulation as described in paragraph 5.5.1

REFERENCES

Allsop, R. E. 1968. Selection of offsets to minimize delay to traffic in a network controlled by fixed-time signals. Transportation Science, 2: 1–13.
Allsop, R. E. 1971. Delay-minimizing settings for fixed-time traffic signals at a single road junction. Journal of the Institute of Mathematical Applications, 8: 164–185.
Anderson, P. E. 1977. A revolution in electronics. Traffic Engineering, 47 (4): 17–18.
ARRB. 1968. Australian road capacity guide: provisional introduction and signalized intersections. ARRB Bulletin No. 4.
Bång, K. L and L. E. Nilsson. 1976. Optimal control of isolated traffic signals. Paper presented at the Third International IFAC/IFIP/IFORS Symposium on Control in Transportation Systems, Columbus, Ohio, USA.
Barker, J., W. R. Goetz and J. F. Hunt. 1977. A controller with system communication capability. Traffic Engineering, 47 (4): 23–27.
Belenfant, G. and Y. Vesval. 1976. Closed toll system. Paper presented at the Third International IFAC/IFIP/IFORS Symposium on Control in Transportation Systems, Columbus, Ohio, USA.

Bellis, W. R. 1960. Capacity of traffic signals and traffic signal timing. Highway Research Board Bulletin 271: 45-67.
Bennewitz, W. 1977. Hardware and Traffic Engineering Analysis of a Street Traffic Control System. Dissertation, Hochschule für Verkehrswesen, Friedrich List, Dresden, GDR (in German).
Berger, C. R. and L. Shaw. 1975. Diversion control of freeway traffic. Paper presented at the Sixth IFAC World Congress, Boston/Cambridge, Massachusetts, USA.
Bleyl, R. 1967. A practical computer program for designing traffic signal system timing plans. Highway Research Board Record 211: 19-33.
Blunden, W. R. 1964. On the computer simulation of a single channel queueing facility for a wide range of arrival and departure distributions. Proceedings of Australian Road Research Board.
Bolle, G. 1976. System ALI – traffic guidance by computer prediction. Bild der Wissenschaft, No. 4: 76-82 (in German).
Bone, A. J., B. V. Martin, and T. N. Harvey. 1964. The selection of a cycle length for fixed-time traffic signals, Research Report R 64-09, from the Department of Civil Engineering MIT, Cambridge, Massachusetts, USA.
Brinkman, A., J. Nijmeyer, and H. van Peeteren. 1975. Application of the all-purpose traffic processor. Philips telecommunication review, 33 (1): 16-33.
Brooks, W. D. 1965. Vehicular traffic control: designing arterial progressions using a digital computer. IBM Corporation.
Bruggeman, J. M., et al. 1971. Network Flow Simulation for Urban Traffic Control Systems. Report PB 207268, from the National Technical Information Service.
Butrimenko, A. 1970. Adaptive routing technique and simulation of communication networks. Paper presented at the Fourth International Transport Conference, Munich.
Butrimenko, A. and H. Strobel. 1974. On an algorithm for the adaptive automatic control of the main vehicular traffic streams in an urban area. IIASA Working Paper, WP-74-14. Laxenburg, Austria: International Institute for Applied Systems Analysis.
Cass, S. and L. Casciato. 1960. Centralized traffic signal control by general purpose computer. Proceedings of the Institute of Traffic Engineers: 203-211.
Chang, M. F. and D. C. Gazis. 1975. Traffic density estimation with consideration of lane-changing. Report from the IBM Thomas J. Watson Research Center, Yorktown Heights, New York, NY.
Cleveland, D. E. 1972. Some experiments in urban freeway information and control system. Proceedings of the Intertraffic 72, International Congress on Traffic Engineering RA I, Amsterdam.
Cook, A. R. et al. 1970. Evaluation of the effectiveness of ramp metering operations. Research Report TSP-2 from the Highway Safety Research Institute, University of Michigan, Ann Arbor, USA.
Cutlar, S. E. 1976. Microcomputer networks in automobile traffic control Paper presented at the tenth IEEE Computer Society Meeting on Computer Technology to Reach the People, San Francisco, USA.
Darroch, J. N., G. F. Newell, and R. W. J. Morris. 1964. Queues for a vehicle actuated traffic light. Operations Research, 12: 882-895.
Donner, R. L., L. E. Welsh, and B. C. Fong. 1975. Traffic responsive ramp control through the use of microprocessors. State Division of Highways, Sacramento, California, US Department of Transportation, Federal Highway Administration HRB No. 082699.
Drew, D. R. and C. Pinnell. 1962. A study of peaking characteristics of signalized urban intersections as related to capacity and design. Highway Research Board Bulletin 352: 1-54.
Drew, D. R. 1968. Traffic flow theory and control. New York: McGraw-Hill.
Dudek, C. L. 1971. A study of freeway operational controls. PhD Dissertation, Texas A & M University, Austin, Texas, USA.
Dunne, M. C, and R. B. Potts. 1964. Algorithms for traffic control. Operations Research, 12: 870-881.
Everts, K., G. Spies, and K. B. Thomas. 1972. Changeable route signs in freeway subnetworks. Straβe und Autobahn, 23 (12): 636-641 (in German).
FHWA. 1971. Merging control systems. Report from the US Department of Transportation, Federal Highway Administration, Washington, DC.

FHWA. 1974. Traffic control devices handbook — An operating guide. Report by the National Advisory Committee on Uniform Traffic Control Devices, US Department of Transportation, Federal Highway Administration, Washington, DC.
Foth, J. R. 1974. The California incident detection logic. Information Print from State of California, Business and Transportation Agency, Freeway Operation Branch, District 07, Los Angeles, USA.
Fowler, P. 1967. Hollywood freeway ramp control project. Research Report from the Western Section, Institute of Traffic Engineers, Seattle, Washington, DC.
Fuehrer, H. H. 1970. Area traffic control. Madrid. Paper presented at the First IFAC/IFIP Symposium on Traffic Control, Versailles, France.
Gazis, D. C., B. T. Bennett, R. S. Footh, and L. C. Edie. 1968. Control of the Lincoln Tunnel traffic by an on-line digital computer. Paper presented at the Fourth International Traffic Flow Theory Symposium, Karlsruhe, FRG.
Gazis, D. C. and R. S. Footh. 1969. Surveillance and control of tunnel traffic by an on-line digital computer. Transportation Science, 3: 255–275.
Gazis, D. C. 1971. Traffic control from hand signals to computers. Proceedings of the IEEE. 59 (7): 1070–1099.
Gazis, D. C. and C. H. Knapp. 1971. On-line estimation of traffic densities from time-series of flow and speed data. Transportation Science, 5: 283–302.
Gazis, D. C. 1972. Traffic flow and control: theory and applications. American Scientist, 60 (4): 414–424.
Gazis, D. C., editor. 1974. Traffic Science. New York: John Wiley & Sons.
Gartner, N. H., J. D. C. Little, and H. Gabbay. 1975. Optimization of traffic signal settings by mixed-integer linear programming. Transportation Science, 9 (4): 308–344.
Gershwin, S. B. 1975. Dynamic stochastic control of freeway corridor systems; Vol. II: Steady state optimal traffic assignment using the accelerated gradient projection model. Research report ESL-R-609, Contract DOT-TSC-849, MIT, Cambridge, Massachusetts, USA.
Gervais, E. F. 1964. Optimization of freeway traffic by ramp control. Highway Research Record, No. 59.
Grafton, R. B. and G. F. Newell. 1967. Optimal policies for an undersaturated intersection. Paper presented at the Third International Symposium on Traffic Flow, Amsterdam: Elsevier Publ. Co.
Green, R. H. 1973. The 42-mile freeway surveillance loop, Research report from the California Division of Highways.
Groth, G. and D. Pilsack. 1974. An electronic system for individual and active route guidance on motorways. Paper presented at the Second International IFAC/IFIP/IFORS Symposium on Traffic Control and Transportation Systems, Monte Carlo, Monaco.
Grover, A. L. 1973. Bulb matrix changeable message signs operational characteristics. Research report AASHO from the Subcommittee on Communication and Electronic Applications for Highways, Los Angeles, California.
Hahlganss, G. and L. Hahn. 1977. Development of a distance warning device for motorcars based on the impulse radar principle. In Entwicklungslinien der Kraftfahrzeugtechnik, Status seminar of the Federal Ministry of Research and Technology, FRG. Rheinland, Köln: Verlag TÜV, pp. 695–700 (in German).
Hahn, J. F. and G. F. Eustis. 1977. The microprocessor approach to local controller construction. Traffic Engineering, 47 (4): 19–22.
Heathington, K. W., R. D. Worrall, and G. C. Hoff. 1971. Attitudes and behaviour of drivers regarding route diversion. Highway Research Record, No. 363: 18–26.
Hewton, J. T. 1967. The Metropolitan Toronto signal system. Paper presented at the Symposium on Area Control of Road Traffic, Institution of Civil Engineers.
Hillier, J. A. and R. Rothery. 1967. The synchronization of traffic signals for minimum delay. Transportation Science: 81–94.
Holroyd, J. and D. J. Robertson. 1973. Strategies for area traffic control systems: present and future, Report LR 569, from the Transport and Road Research Laboratory, Crowthorne, UK.
Houpt, P. K. and M. Athans. 1975. Dynamic stochastic control of freeway corridor systems, Vol. I: Summary. Research Report ESL-R-608, Contract DOT-TSC 849, MIT, Cambridge, Massachusetts, USA.

HRB. 1965. Highway capacity manual. Special Report 87 from the Highway Research Board, Washington, DC.
Iida, Y. 1974. Road network operation system considering environmental conditions. Report from the Department of Civil Engineering, Kanazawa University, Kanazawa, Japan.
Inose, H. and T. Hamada. 1974. Road traffic control. Report by the Division of Engineering and Applied Science, California Institute of Technology, Pasadena, California, and the Department of Transportation, Contract No. DOT-OS-30169 (translated from Japanese).
Isaksen, L. and H. J. Payne. 1972. Regulation of freeway traffic. Proceedings of the Joint Automatic Control Conference, Paper No. 4–5.
Isaksen, L. and H. J. Payne. 1973. Freeway traffic surveillance and control. Proceedings of the IEEE, 16 (5): 526–536.
Ives, A. P. and P. M. Jackson. 1974. A vehicle headway control system using Q-band primary radar. SAE Publication No. 740097.
Jasper, L. and K. Wienand. 1975. The control center of the changeable route sign system in the freeway sub-network Rhein–Main. Siemens Informationen zur Straßenverkehrstechnik "Grünlicht", No. 2, Munich, FRG (in German).
Kahn, R. 1974. Incident detection logics for the Los Angeles freeway surveillance and control project. Revised Research Report, State of California Business and Transportation Agency, Department of Transportation, District 07, Los Angeles.
Knoll, E. and J. Ullrich. 1972. Traffic control by changeable route signs. Report from the Hessisches Ministerium für Wirtschaft und Technologie, Wiesbaden, FRG (in German).
Krayenbrink, C. J. and A. Vlaanderen. 1975. The Philips all-purpose traffic processor system. Philips Telecommunication Review, 33 (1): 8–15.
Kuntze, H. B. 1977. Process computer applications for freeway and long-distance road traffic. Research Report FB-77-1, from the Hochschule für Verkehrswesen, Friedrich List, Wissenschaftsbereich Automatisierungstechnik, Dresden, GDR (in German).
Kuntze, H. B. 1978. Algorithms for computer control and surveillance of freeway traffic. Research Report FB-78-1 from the Hochschule für Verkehrswesen, Friedrich List, Wissenschaftsbereich Automatisierungstechnik, Dresden, GDR (in German).
Laurance, N. 1977. A microprocessor architecture for engine control. IEEE, Computer Conference, Washington, DC.
Laurance, N. 1978. The development of an automobile engine control system. Paper presented at the Seventh Triennal IFAC World Congress, Helsinki, June, 1978. Oxford: Pergamon Press, Vol. 1, pp. 377–384.
Le Pera, R. and R. Nenzi. 1974. The Naples tollway computer system for highway surveillance and control. Paper presented at the Second International IFAC/IFIP/IFORS Symposium on Traffic Control and Transportation Systems, Monte Carlo, Monaco.
Little, J. D. C., B. V. Martin, and J. T. Morgan. 1964. Synchronizing traffic signals for maximal bandwidth. Report R 64-08, from the Department of Civil Engineering, MIT, Cambridge, Massachusetts, USA, p. 54. (Also published in Operations Research, 12: 846–912, and as Highway Research Board Record 118: 21–47 (1966)).
Little, J. D. C. 1966. The synchronization of traffic signals by mixed-integer linear programming. Operations Research, 14: 568–594.
Looze, D. P., P. K. Houpt, and M. Athans. 1975. Dynamic stochastic control of freeway corridor systems, Vol. III: Dynamic centralized and decentralized control strategies. Research Report ESL-R-610, Contract DOT-TSC 849, MIT, Cambridge, Massachusetts, USA.
Mahoney, M. A., H. A. L. Lindberg, and G. W. Cleven. 1973. A solution to intercity traffic corridor problem. Public Roads, 37: 173–184.
Marsh, B. W. 1927. Traffic control. Ann Amer. Soc. Political Social Sci., 133: 90–113.
Masher, D. P., et al. 1975. Guidelines for design and operation of ramp control systems. Research Report, Contract NCHRP 3-22, SRI Project 3340, Stanford Research Institute, California, USA. December, 1975.
May, A. D. 1964. Experimentation with manual and automatic ramp control. Highway Research Record, No. 59: 9–38.
May, A. D. 1974. Some fixed-time signal control computer programs. Paper presented at the Second International IFAC/IFIP/IFORS Symposium on Traffic Control and Transportation Systems, Monte Carlo, Monaco.

May, A. D. 1975. Urban freeway corridor research at ITTE – A survey. Institute of Transportation and Traffic Engineering, University of California, Berkeley, USA.
McDermott, J. M. 1968. Automatic evaluation of urban freeway operations. Traffic Engineering, January.
Meditch, J. S. 1978. Multivariable control of data networks. Paper presented at the Seventh Triennal IFAC World Congress, Helsinki, June, 1978. Oxford: Pergamon Press, pp. 1421-1428.
Messer, G. J. 1969. A design and synthesis of a multilevel freeway control system and a study of its associated operational control plan. Dissertation, Texas A & M University, Austin, Texas, USA.
Miller, A. J. 1963. Settings for fixed-cycle traffic signals. Operational Research Quarterly, 14 (4): 373-386.
Miller, A. J. 1968. The capacity of signalized intersections in Australia. ARRB Bulletin, No. 3, March.
Morgan, J. T. and D. C. Little. 1964. Synchronizing traffic signals for maximal bandwidth. Operations Research, 12: 896-912.
Moyer, D. F. and S. M. Mangrulkar. 1975. Engine control by an onboard computer, SAE Paper 750433.
Nahi, N. E. and A. N. Trivedi. 1973. Recursive estimation of traffic variables: section density and average speed. Transportation Science, 7 (3): 269-286.
Nakahara, T., N. Yumoto, and A. Tanaka. 1970. Multi-criterion area traffic control system with feedback features. Paper presented at the First International IFAC/IFIP Symposium on Traffic Control, Versailles, France.
Neininger, G. and K. Pählig. 1977. Advances in the development of a self-sufficient distance warning device with regard to reducing false alarms. In Entwicklungslinien der Kraftfahrzeugtechnik, Status seminar of the Federal Ministry of Research and Technology, FRG. Rheinland, Köln: Verlag TÜV, pp. 679-687 (in German).
Newell, G. F. 1969. Properties of vehicle actuated signals – I. Oneway streets. Transportation Science, 3: 30-52.
Newman, L., A. Dunnet, and G. J. Meir. 1970. An evaluation of ramp control on the harbour freeway in Los Angeles. HRB-Record, 303: 44-55.
Nguyen, S. 1974. An algorithm for the traffic assignment problem. Transportation Science, 8: 209-216.
Orlhac, D., et al. 1975. Dynamic stochastic control of freeway corridor systems, Vol. IV: Estimation of traffic variables via extended Kalman filter methods. Research Report ESL-R-611, Contract DOT-TSC 849, MIT, Cambridge, Massachusetts, USA.
Oswald, R. S., N. L. Laurance, and S. S. Devlin. 1975. Design considerations for an onboard computer system. SAE Paper 750434.
Payne, H. J., W. A. Thompson, and L. Isaksen. 1973. Design of a traffic-responsive control system for a Los Angeles freeway. IEEE Transactions on System, Man, and Cybernetics, SMC-3 (3): 213-225.
Payne, H. J. and W. A. Thompson. 1974. Allocation of freeway ramp metering volumes to optimize corridor performance. IEEE Transactions on Automatic Control, AC-19, (3).
Payne, H. J., D. N. Goodwine, and M. D. Teener. 1975. Evaluation of existing incident detection algorithms. Research Report FHWA-RD-75-39, Technology Service Corporation, Santa Monica, California, USA.
Payne, H. J. 1976. Calibration of freeway incident detection algorithms. Paper presented at the Third International IFAC/IFIP/IFORS Symposium on Control in Transportation Systems, Columbus, Ohio, USA.
Pinnell, C., D. R. Drew, W. R. McCasland, and J. A. Wattleworth. 1967. Evaluation of entrance ramp control on a six mile freeway section. Highway Research Record, No. 157: 22-76.
Pinnell, C., et al. 1974. Traffic control systems handbook, DOT-FHWA Contract, DOT-FH-11-8123 m, Draft Report.
Pitzinger, P. and R. E. Sulzer. 1968. Traffic Light Systems for Road Traffic. Wiesbaden, Berlin: Bauverlag (in German).
Prabhakar, R., S. J. Citron, and R. E. Goodson. 1975. Optimization of automotive engine fuel economy and emissions. ASME Paper 75-WA/Aut.
Rach, L., et al. Improved Operations of Urban Transportation Systems. Report of the Metropolitan Toronto Roads and Traffic Department; Volume 1, Traffic signal control strategies

– state of the art, 1974; Volume 2, The evaluation of off-line area traffic control strategies, 1975; Volume 3, The development and evaluation of a real-time computerized traffic control Strategy, 1976.

Rackoff, N. J. and T. H. Rockwell. 1973. Use of driver displays to stabilize and improve traffic flow. Final Research Report, Project EES 328 B, Ohio Department of Highways, Driving Research Laboratory, Department of Industry and Systems Engineering, The Ohio State University, Columbus, Ohio.

Radtke, Th. 1977. Self-sufficient distance warning device for increasing road traffic safety. In Entwicklungslinien der Kraftfahrzeugtechnik, Status Seminar of the Federal Ministry of Research and Technology, FRG. Rheinland, Köln: Verlag TÜV, pp. 666–672 (in German).

Robertson, D. I. 1967. An improvement to the combination method of reducing delays in traffic networks. RRL Report LR 80.

Robertson, D. I. 1969. TRANSYT: A traffic network study tool. Report LR 253 from the Road Research Laboratory, Crowthorne, UK.

Rockwell, T. H. and J. N. Snider. 1969. Investigations of driver sensory ability and its effect on the driving task. RF 2091 Final Report, Research Foundation, The Ohio State University, Columbus, Ohio.

Ross, D. W. 1972. Traffic control and highway networks. Networks Journal, 1: June–July.

Ross, D. W., et al. 1973. Improved control logic for use with computer controlled traffic. NCHRP-Report 3-18, Stanford Research Institute, Menlo Park, California, USA.

Rossberg, R. R. 1977. Driving without congestion and stress. Der Stern, Hamburg, FRG, No. 49: 224–229.

RRL. 1965. Research on Road Traffic. Report from the Road Research Laboratory, Crowthorne, UK. London: HMSO.

Sasaki, T. and H. Inouye. 1974. Approach to emergency traffic control. Report from the Department of Transportation Engineering, Kyoto University, Japan.

Schnabel, W. 1975. Traffic Light Controlled Street Traffic. Berlin: Transpress, VEB Verlag für Verkehrswesen (in German).

Schlaefli, J. L. 1972. Computerized traffic signal systems: a future. Traffic Engineering, 42 (9).

Scott, E. 1975. Microcomputer networks in automobile traffic control. Proceedings of the COMPCON 75 Spring Conference on Computer Technology Reaches People, New York, NY, pp. 263–266.

Sealbury, T., Micro-Processors Signal in On-Traffic Control, EE/Systems Engineering Today, January, 1974, pp. 76–78.

Shefer, J., et al. 1974. A new kind of radar for collision avoidance. Society of Automotive Engineers, Automotive Engineering Congress, Detroit, Michigan, February 25–March 1.

Siemens. 1977. A step towards the future: a microcomputer at every signal controlled intersection – intelligent control devices of the M-family. Straßenverkehrstechnik, 21 (3) (in German).

SIGOP. 1966. Traffic signal optimization program: a computer program to calculate optimum coordination in a grid network of synchronized traffic signals. Report by the US Department of Commerce, Washington, DC.

Special Report on Automotive Electronics. 1973. Electronics: 94–108.

Stanford, M. R. and H. Parker. 1977. The South Bay traffic signal control system. Traffic Engineering, 47 (4): 28–35.

von Stein, W. 1969. 100 years of traffic signals. Straßenverkehrstechnik, 13 (H. 1): 32–83 (in German).

Stockfisch, C. R. 1972. Selecting digital computer signal systems. Report FHWA-RD-72-20 to US Department of Transportation.

Strobel, H. and W. Ullmann. 1973. Control computer applications in road traffic. Technische Informationen, GRW 11 (H. 3): 135–141 (in German).

Strobel, H. 1977. Traffic control systems analysis by means of dynamic state and input–output models. IIASA Research Report RR-77-12, Laxenburg, Austria: International Institute for Applied Systems Analysis.

Szeto, M. W. and D. C. Gazis. 1972. Application of Kalman filtering to surveillance and control of traffic systems. Transportation Science, 6: 419–439.

Tabak, D. 1973. Application of modern control theory and optimization technique to transportation systems. In Advances in Control Theory, Vol. 10. New York: Academic Press, pp. 345–434.

Temple, R. H. and S. S. Devlin. 1974. The use of microprocessors as automobile onboard controllers. Computer, August: 33–36.

True, J. and D. Rosen. 1973. Moving merge – A new concept in ramp control. Public Roads. 37 (7).

Ullmann, W. 1978. A contribution to the control-theoretic analysis of dynamic processes in vehicle strings. Dissertation, Hochschule für Verkehrswesen, Friedrich List, Dresden, GDR (in German).

Valdes, A. and S. de la Rica. 1970. Area traffic control by computer in Madrid. Traffic Engineering and Control, No. 12: 132–134.

Watson, B. K. 1971. A complete program to calculate signal timings and delays. Paper No. 2718, Institute of Transportation and Traffic Engineering, University of California, Berkeley.

Wattleworth, J. A. 1967. Peak period analysis and control of a freeway system. Highway Research Record, 157: 1–10.

Wattleworth, J. S. 1971. Accomplishments in freeway operations in the United States Highway Research Board Record No. 368, USA Department of Transportation.

Webster, F. V. 1958. Traffic signal settings. Technical Report No. 39, Road Research Laboratory. London: Her Majesty's Stationary Office.

Weinberg, M. F. 1966. Digital computer controlled traffic signal system for a small city. NCHRP Report, USA Department of Transportation, HRB.

Weiss, C. D. 1974. Basic microcomputer software: algorithm for a traffic light controller illustrates principles that can be used with any MOS/LSI microprocessor. Electronic Design, 9: 142–146.

Wiener, R., L. J. Pignatoro, and H. N. Yagoda. 1970. A discrete Markov renewal model of a gap-acceptance entrance ramp controller for expressways. Transportation Research, 4 (2): 151–161.

Wocher, B. 1977. Description of the evaluation logic of a distance warning device. In Entwicklungslinien der Kraftfahrzeugtechnik. Status seminar of the Federal Ministry of Research and Technology, FRG. Rheinland, Köln: Verlag TÜV, pp. 673–678 (in German).

Wüchner, E. 1977. Results of driving experiments with self-sufficient radar distance warning devices and their consequences for further development. In: Entwicklungslinien der Kraftfahrzeugtechnik, Status seminar of the Federal Ministry of Research and Technology, FRG. Rheinland, Köln: Verlag TÜV, pp. 688–694 (in German).

Yagoda, H. N. 1970. The dynamic control of automotive traffic at a freeway entrance ramp. Automatica, 6: 385–393.

Yan, G. and W. Sue. 1971. A computer program for delay calculations and graphical solution to optimal signal setting. Paper Nos. 2910 and 3200, Institute of Transportation and Traffic Engineering, University of California, Berkeley.

Yardeni, L. A. 1965. Algorithms for traffic-signal control. IBM Systems Journal, 4 (2).

Zajkowski, M. M. 1972. Arrow aiming in traffic guide signs: a laboratory investigation. Research Report from the Wayne State University, College of Engineering for Michigan State Highway Commission, Detroit, USA.

Ziegler, M. and P. Baron. 1973. System analysis of changeable route signs. Straßenverkehrstechnik, 17 (1): 17–22 (in German).

6 International Area Traffic Control Systems Experiences

The first computerized area traffic control system was put into operation in Toronto, Canada at the beginning of the sixties. Since then more than one hundred cities all over the world have successfully installed such systems. It is neither possible nor useful to present in this monograph a detailed survey of all area traffic control systems currently in operation. The experience gained so far will merely be sketched by a selected number of specially prepared reports describing the main features of computerized traffic control systems that are in use in the USA, USSR, UK, Japan, and in one developing country – Kenya. Each report covers the following three topics:

- case histories dealing with the initial situation and the decision processes that led to the installation of the individual systems
- the main topics related to the implementation and operation of the systems
- an evaluation of the operational experience regarding the contribution achieved by the control systems to solving the traffic problems summarized in Chapter 1

6.1 USA EXPERIENCES: SAN JOSE AND UTCS*

6.1.1. BACKGROUND

In 1964 the City of San Jose, California, became the first US city to initiate a project involving digital computer control of a major traffic system. It seemed clear at that time that the digital computer could overcome many of the shortcomings of the earlier (analog) devices and open new vistas in traffic control/management capabilities. At present, there are more than thirty-

---

*Based on a case description specially prepared by J. L. Schlaefli, General Manager, Applied Transportation Systems, Inc., Gulf + Western Industries, Inc., Palo Alto, California, USA.

five major systems operating and over seventy-five additional systems in either the planning or implementation stages. The US Federal government recognized early the potential of digital computers in traffic control, and devoted significant resources to the development of its Urban Traffic Control System (UTCS) project using Washington, DC as a demonstration site.

The San Jose and UTCS experiences have been chosen for the case descriptions to be presented here. They are unique, but are representative of the "lessons learned" in US area traffic control. Brief system chronologies are presented in Table 6.1, followed in paragraph 6.1.2 by system details. Operational experiences and lessons learned in area traffic control using digital computers in San Jose and UTCS are then summarized in paragraph 6.1.3.

TABLE 6.1   Systems Chronologies

San Jose Area Traffic Control System		Urban Traffic Control System (UTCS)	
June, 1964	Joint City/IBM system feasibility study initiated		
June, 1965	Initiation of signal control by computer		
December, 1966	Feasibility study and initial control experiments completed. IBM support removed, including the IBM 1710 computer		
February, 1967	Decision to install IBM 1800 computer		
Novemner, 1967	IBM 1800 computer fully operational		
1967–1971	Minor research on surveillance	July, 1968	Initial steps taken to develop UTCS
		February, 1970	System design specifications completed
		July, 1970	System installed
		June, 1971	UTCS system expanded to include bus priority (BPS) and new detectors tested and added to the system
		July, 1971 to ca. 1978	Advanced software development and implementation
August, 1971	Major project to improve control logic initiated. Cooperative effort between Stanford Research Institute (SRI) and San Jose		
June, 1972	Evaluation of new fixed-time signal plans completed. TRANSYT settings installed	November, 1972	Initial UTCS System (111 intersections) operational.

TABLE 6.1 (*Continued*)

San Jose Area Traffic Control System		Urban Traffic Control System (UTCS)	
			First generation software developed.
		August, 1973	Initial second generation software developed
		September, 1973	First generation software reprogrammed in Fortran IV
June, 1974	SRI software (ASCOT) completed, tested and implemented in San Jose	January, 1975	Evaluation of the impact of data error sensitivity in UTCS completed
		February, 1975	Extensive evaluation of first generation control strategy completed
May, 1975 to ca. 1978	Continuing work on ASCOT background control and sensitivities		

6.1.2 SYSTEM DESCRIPTIONS

6.1.2.1 San Jose. The City of San Jose is located in Santa Clara County just south of San Francisco, California. Santa Clara County has a population of over one million people, of which San Jose accounts for almost one-half. Transportation in the area takes place principally on roads, and there are over 750,000 registered vehicles in the County. On a typical 12 hour weekday period, approximately one million vehicle counts are measured by the detectors of the San Jose computer traffic control network. The computer traffic control system operates on a 63 intersection network as depicted in Figure 6.1.

Most of these intersections (i.e., about 50) are located in the downtown grid area, with the remainder located on an arterial street (San Carlos) that serves as a primary access route to the central area. While the efficiency of traffic operations is extremely important to the growth and welfare of San Jose, some observers have stated its traffic problems are minor compared with those found in most major urban areas. Nonetheless, it was estimated early in the San Jose project that motorists using the network were absorbing a loss of nearly $20,000 in vehicle operations cost and lost work time during a typical twelve-hour period.

As San Jose had the first area traffic control system developed in the United States, using a general purpose digital computer, the system hardware used has now been superseded. Nonetheless, the experience that has been gained to date is still worth considering by other potential users of area traffic control systems. The original system configuration is shown in the block diagram presented in Figure 6.2.

The traffic signals in San Jose are currently controlled by an IBM 1800 computer having a 32,768 word (16 bit) 2 microsecond core memory, an IBM disk cartridge data storage unit with three removable disks, a card reader/punch, an IBM line printer, a logging typer, and an input/output printer with keyboard. The IBM 1800 computer has been marketed extensively for pro-

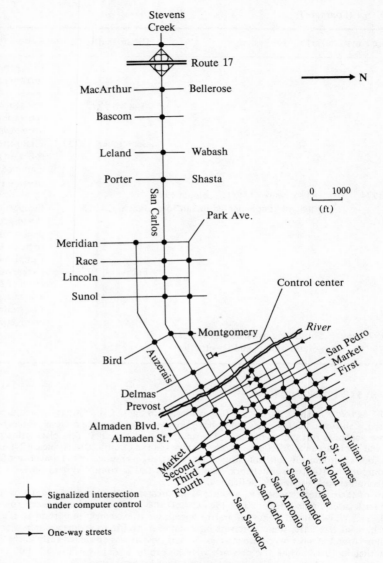

FIGURE 6.1 San Jose intersections under computer control.

cess control applications, but is no longer in production owing to the introduction of new technology systems. The computer itself continues to be reliable and adequate for the San Jose traffic control application. A simple display map showing intersection status and green phases is also included in the system.

A simple relay-type hardwired communications system is used in San Jose. The intersection controllers are of a modified electro-mechanical type commonly found in the United States. Most of the controllers were part of a three-dial interconnected system already in existence,

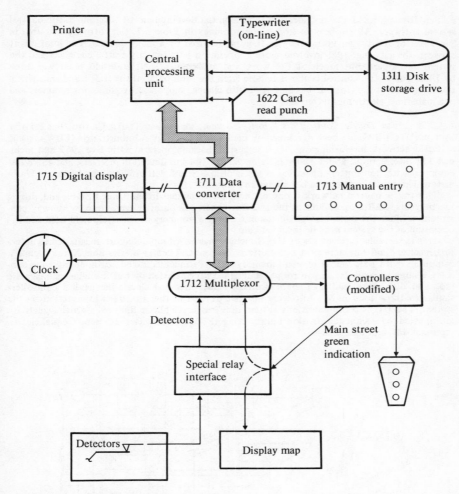

FIGURE 6.2 Block diagram of San Jose system.

which was adapted for computer control by installing three relays in each controller. The hold-on-line relay is used to disconnect the dial advance and release contracts from the drum motor, and in effect switch from local to computer control. The advance relay is then used to step the controller drum through a signal interval, and finally a monitor relay is used to determine when the signal is in "main street green" (see Figure 6.2). There are several multiple phase controllers along San Carlos Street. In these cases, a fourth relay was added to provide for an all-red condition during clearance of vehicles on a particular phase. This relay was also used later to implement phase skipping.

As mentioned above, a dedicated wire communications system with no multiplexing is used. It should be noted that, in the above design, at least three wires are required between each controlled intersection and the computer. In addition, one wire is required for each detector at the intersection, and one wire is used as a common ground. At present, the system uses about 420 detectors for traffic surveillance, which makes San Jose one of the most highly instrumented traffic control systems in the world.

San Jose provided the test location for the initial development of the IBM traffic control system software. As can be seen from the chronology in Table 6.1, this software was used in San Jose for nearly ten years and was then modernized by a major project. The joint effort between the city and IBM involved close cooperation between city traffic engineers and the system developers, but much of the work and management of the development was done by IBM. The traffic control center is remote from the focal point of the traffic and transportation activities in San Jose. It has an engineer in charge, supported by three programmers and one, part-time, electronics technician.

6.1.2.2 Urban Traffic Control/Bus Priority System. The Urban Traffic Control/Bus Priority System (UTCS/BPS) has been developed by the Federal Highway Administration (FHWA) using a traffic network in Washington, DC. The system became operational in late 1972 and included 111 intersections, about 500 vehicle detectors, 144 bus detection receivers, and 450 buses equipped with transmitters operating in the central area and within Wisconsin Avenue in Washington. This initial control area is depicted in Figure 6.3.

This traffic control network is one of the most congested in the United States and, during recent years, traffic operations in the area have been complicated even more by subway construction. Since the system is well documented, only a summary of the equipment used and the operation of the system need be included here.

Traffic surveillance elements of UTCS include magnetic loop detectors, magnetic bus detectors/receivers and bus detection transmitters. The control center houses dual Xerox Sigma V computers with 65K words of main memory each. Peripherals include cathode-ray tube consoles, a large map display, a control panel designed for operation by traffic engineers, magnetic tape and disk units, and high speed printers. New, three-dial electro-mechanical controllers similar to those used in San Jose were installed at each of the controlled intersections in the system. The UTCS communications system uses leased telephone lines over which signals are multiplexed by frequency division multiplexing (FDM), which uses different frequencies to represent different data bits.

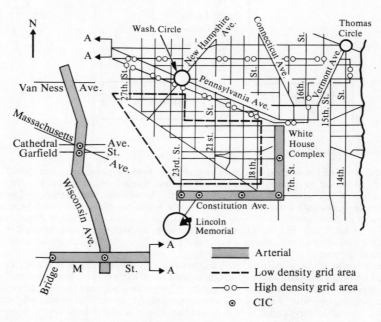

FIGURE 6.3 The UTCS control area in Washington, DC.

A systems contractor was employed who had responsibility for all procurement, installation, and system integration for UTCS. While most of the equipment used was "off-the-shelf", the requirement to incorporate bus priority into UTCS called for development of a special transceiver for detecting buses in the traffic stream. A functional block diagram of the system is presented in Figure 6.4.

During development, FHWA project staff and the systems contractor worked closely with the Washington, DC, traffic engineering staff to coordinate this major undertaking. Once the system was completely installed and operational, the major efforts of the project were devoted to testing system effectiveness and to developing new software that could be applied to other installations throughout the United States. This work included developing and using simulation techniques to evaluate traffic control strategies of standard and new signal control strategies, and in turn evaluating them through a comprehensive manual and computerized analysis of traffic operations data. Throughout the development and testing, UTCS was under the operational control of FHWA. Management of day-to-day operation of the system was contracted to a support organization. Some of the best records on equipment and operating experience have resulted from the UTCS project. While no major equipment problems developed, it was found that the electronic controller interface units and vehicle detectors were the most unreliable items in the system, with a mean time between failures (MTBF) of the order of 150,000 hours (all the other equipment had MTBFs at least twice as large). However, frequent operational failures did occur, and were accounted for, in general, by telephone disconnections, which happended on the average almost once a week. A summary of UTCS component reliability experience is given in Figure 6.5.

6.1.3 OPERATIONAL EXPERIENCES AND LESSONS LEARNED

The two systems described briefly above represent unique experiences. The first system, in San Jose, is a typical example of the pioneering work that has to be done to implement a new program of traffic control in a medium sized city. The second experience reflects the applied research activities of a government organization at Federal level.

6.1.3.1 San Jose Experience. One of the most important lessons learned is that recently it was possible to implement a very sophisticated traffic control software system (ASCOT) that is capable of controlling the San Jose traffic network using equipment that was designed at least 15-20 years ago. It is important because it illustrates the flexibility that can be achieved in traffic control using general purpose digital computers somewhat independently of the total system hardware. It was not necessarily the original design of the equipment in San Jose that allowed advanced techniques to be accommodated; the inclusion of an extensive surveillance network (i.e., one detector per lane per block initially) has contributed in a substantial way. The old electro-mechanical controllers were used effectively, and continue to be used in the advanced system.

Another important lesson from the development in San Jose is that major improvements and evaluations of the system took place only when outside parties were involved. First of all, the project began with a joint effort between San Jose and IBM. Further installation of optimized fixed-time traffic signal settings was accomplished in association with a joint US/UK project whose purpose was to evaluate two offline traffic signal timing optimization programs (i.e., SIGOP and TRANSYT). Finally, as part of a National Cooperative Highway Research Program project (NCHRP) the ASCOT software package was implemented and evaluated in San Jose.

When evaluating the San Jose system in terms of its ability to improve traffic control, it is necessary to go back to some of the initial traffic engineering activities of the project. Experiments with a number of control procedures were initiated upon system installation and continued for about two years. Fixed-time programs of traffic signal timing were developed, including a single program system, a three program system set by time of day, and a three program system that could change as a result of measured changes in traffic demand. In addition, a fixed-time program was developed by IBM using a simulation model. These fixed-time programs as well as an on-line progression program and individual intersection control (micro-loop) techniques were all tried. During this initial experimental period, evaluations of the control

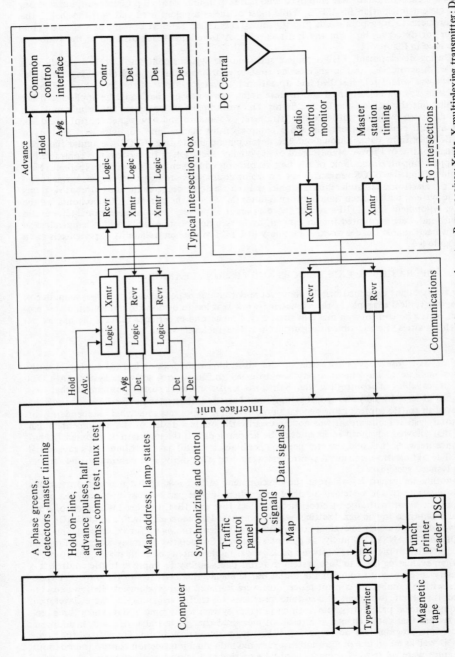

FIGURE 6.4 Block diagram of the UTCS system: DC Central, dispatching control center; Rcvr, receiver; Xmtr, X-multiplexing transmitter; Det, detector; Contr, controller; Aϕg, A phase greens.

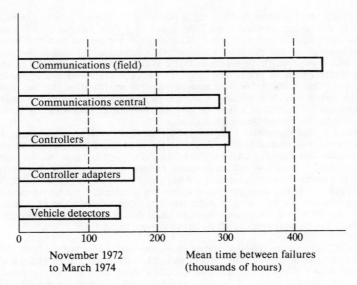

FIGURE 6.5 Reliability of UTCS equipment.

techniques were based mainly on computer generated surveillance data. The result was that the three program fixed-time technique was slightly better than the one program technique, and signal timings based on the IBM simulation were superior to the other fixed-time programs. The micro-loop techniques were only tried on two intersections, where improvements in operation were found. The fixed-time program developed by simulation was given credit for achieving a decrease in average vehicle delay of almost 15 percent in the San Jose network. However, the 15 percent seems to be an overestimate of the contribution to the computer system, since physical interconnection of a number of the intersections not previously coordinated was carried out as part of the project.

In 1971, the development of ASCOT began, which was ultimately to result in one of the major improvements to the operation of the San Jose system. This was a major research project involving development and documentation of a completely new software system and a demonstration of its effectiveness using San Jose as a test network. Comprehensive development of the three level control strategy was demonstrated. This included: (1) on-line background control (optimized fixed-time plans); (2) cycle-to-cycle updates of green time split; (3) phase skipping; and (4) minimum delay, critical intersection control. One of the most comprehensive field studies of the impacts of these control strategies was carried out as part of the project, using automated data collection and floating car techniques. The results of the project were significant in that they showed that modern software could provide functions not previously incorporated in an existing traffic control system. All techniques except the on-line background control seemed to produce good results. However, the results of the background control were no better than those that had been generated by off-line optimization programs, and in some cases they were worse. Later it was discovered that certain software changes were required in order to implement the system properly.

Considering the experience in traffic control system development in San Jose, it may seem strange that the project did not move ahead more quickly. Events that affected the development present some important lessons. At the end of 1966, when IBM removed its hardware and software support from the project, there was a delay of almost one year before the city finally purchased, installed and commissioned its own IBM computer, and provided appropriate programming and traffic engineering support. During this time, the computer center was moved, which caused considerable cost and delay to the project. Communication lines to all of the controllers and detectors had to be reconnected to the new location. Shortly after the new computer became operational, a program of street repair was initiated. The use of heat to level the asphalt pavement in the central business district destroyed over fifty detectors. At the same time, a number of reconstruction projects were taking place in the central area that involved installation and redesign of intersections. Much of the communications wire in the traffic control system in San Jose passes through an area that was completely reconstructed in 1969. During the construction period of nearly two years, 11 or 12 intersections and approximately 119 detectors were disconnected from the system. Another modification that had significant impact on the system was the reversal of a number of one-way streets. Fortunately, most of the traffic detectors are located in mid-block, and the reversing of these one-way streets involved more of a conceptual change than anything else.

In summary, this experience with reconstruction is not unique to San Jose. If one is to operate a comprehensive traffic control system, the importance of good communications and planning between the various agencies in the municipality must be taken into account.

Hardware and software should be designed for redundancy, and so that changes can be readily implemented. Research programs are not always the best way to develop traffic control system capabilities. The turn-on/turn-off situation in a municipal environment can create a political situation where the benefits of a computer traffic control system can be superseded, in the public eye, by the minor disruptions that take place. San Jose vividly demonstrates the pitfalls and credits of the development of traffic control systems in the US.

6.1.3.2 UTCS Experience. The operational experience and lessons learned from UTCS are of a completely different nature to those of San Jose. The major objective of UTCS is the development of advanced traffic signal control strategies. In essence, the UTCS project has evolved into a software development project. Experiences with the first phases of this work have been documented and provide some interesting results. Before discussing these experiences it is important to point out that the US FHWA has learned some important lessons during the development and implementation of UTCS. In traffic control systems, there always seem to be problems with the communications subsystem development, and UTCS was no exception. A delay of six months in the project was principally caused by the inability of the Federal government to negotiate with the local telephone company to provide appropriate lines for system communication. Also, it was very difficult to obtain accurate data on existing conduit locations. Even with substantial financial and political resources behind a traffic control system project, implementation does not always proceed as smoothly as one would expect. The primary results and lessons learned from the UTCS project fall into the area of traffic control strategy. Extensive efforts were spent on the development of software, but an unfortunate problem with UTCS was that a large scale computer was used that is no longer manufactured. In any event, the basic software concepts developed by UTCS have begun to find applications in other US systems. The UTCS program involves the development of three generations of control strategies:

• The first generation uses pre-stored traffic signal timing developed off-line and based on previously collected traffic data. Timing plans can be selected on the basis of time-of-day, operator selection, or automatic matching of the timing program best suited to the existing traffic demand conditions.

• The second generation includes an on-line optimization routine to develop the timing program in real-time on the basis of current traffic conditions. The process is repeated at intervals of 5–10 minutes, whenever changing traffic conditions require a new set of signal times.

• The third generation considers individual intersections on a cycle-to-cycle basis using area-wide optimization criteria.

The first generation has been fully implemented and tested. Over recent years, an extensive field study and evaluation of first generation UTCS traffic control strategies has been carried out. One of the most significant operational experiences and lessons learned in area traffic control in the US has resulted from this evaluation. After a complete evaluation, it was concluded that the first generation control patterns developed by off-line optimization programs were about as good or a little better than patterns developed by traffic engineers using graphical techniques. Using these fixed-time patterns in the time-of-day mode is almost as good as using them in a traffic responsive mode. A critical intersection control concept for cycle-to-cycle adjustment of splits was also evaluated but did not result in improved traffic operations. The use of TRANSYT or a similar method for developing optimal fixed-time patterns seems to be very effective and strongly justified in that it requires fewer man-hours to develop timing plans using this technique than using manual procedures. Another important result from the extensive UTCS field evaluation was that most of the evaluations of new control strategies can be carried out using data available directly from the surveillance subsystem of the traffic control system itself. Extensive field evaluations are very time consuming and costly, and it has been concluded that assessing advanced software control strategy developments can be based on computer generated evaluations.

6.1.3.3 Summary of Lessons Learned. The traffic control experiences in San Jose and UTCS are extensive and it is impossible to fully document all aspects of these important developments here. It is appropriate to summarize in general functional terms some of the most important lessons learned, on the basis of this and other United States system developments. This is outlined in Table 6.2.

TABLE 6.2 Summary of Lessons Learned

Effectiveness of fixed-time plans
To date, fixed-time signal plans developed by off-line optimization techniques (i.e., TRANSYT) have resulted in the best improvements in traffic operations.

Network changes
Street repair and urban development are dynamic factors in any major urban traffic network. A computer traffic control system must be designed to accommodate street network changes.

Communications system
The communications system between controllers/detectors and the central computer is often the most expensive and troublesome system element.

System evaluation
Given a good surveillance system design, the results of new control techniques can be evaluated effectively without extensive engineering field studies.

System development cycle
In the case of both the San Jose and UTCS systems, implementation took much longer than initially planned. Development may take place in spurts. The system implementation should plan for an extensive, long-term effort.

Unique applications
The traffic networks, the political situation and resources available are different at each location. A system design that is good for one location may be far from optimum for another.

System equipment
The absolute latest state-of-the-art equipment is not necessarily required for an effective computer traffic control system. Good improvements can be made using electro-mechanical controllers. Modern general purpose computers should be used to support and to promote new control improvements.

Traffic surveillance
Any feedback control system is only as good as its data inputs. Experience in the United States indicates that having a high level of surveillance resulted in promoting development of traffic control system capabilities.

BIBLIOGRAPHY

SAN JOSE

San Jose traffic control project final report. IBM Document No. 320-0959-0, 1965.
SIGOP/TRANSYT evaluation — San Jose, California. FHWA, July, 1972.
Improved control logic for use with computer controlled traffic. Final Report, Stanford Research Institute, NCHRP 3-18(1), November, 1973.
Improved control logic for use with computer controlled traffic. Follow on Quarterly Progress Report, Stanford Research Institute, September, 1975.

URBAN TRAFFIC CONTROL SYSTEM (UTCS)

Advanced control technology in urban traffic control systems, FHWA, October, 1969.
UTCS/BPS design and installation. Final Report, November, 1972.
UTCS/BPS maintenance manual. Final Report, November, 1972.
UTCS/BPS operator's manual. Final Report, November 1972.
UTCS/BPS software manual — Volumes I and II. Final Report, FHWA, December, 1972.
UTCS third generation control policies. FHWA, 1973.
Evaluation of first generation UTCS/BPS strategy. Final Report, FHWA, February, 1975.

6.2 BRITISH EXPERIENCES: GLASGOW AND LONDON*

Steady growth in the ownership and use of private cars was resumed in Britain soon after 1950 and has been maintained for over 20 years. Economic growth has brought corresponding increases in lorry traffic, and buses have continued to be an important means of travel in towns. In the same period, investment in major urban road construction and improvement has been relatively low, and there has been a strong liking for and financial commitment to existing forms of development in city centers.

As a result, urban traffic congestion became a severe problem in most British towns and cities by 1960, and the operation of urban roads has been the subject of an increasing program of research by the Road Research Laboratory (RRL) (more recently Transport and Road Research Laboratory (TRRL)), the Ministry of Transport (MOT) (more recently Department of the Environment (DOE)), equipment manufacturers, universities and consulting engineers.

One major aspect of this research has been the control of traffic by signals. When signal controlled junctions are not close to each other they can work quite efficiently, dealing with traffic as it arrives at the intersections, irrespective of neighboring conditions. When they are close together, as in the city centers, the operation of one junction has an effect on the others in that area and it is necessary to have coordination, or linking, between the signals at the different junctions. Such coordination over a network of streets is called area traffic control. The development of area traffic control in Britain has stemmed mainly from two experiments, one in Glasgow and the other in London. This case description is therefore largely devoted to progress in these two cities, with occasional references to other relevant British work. The results of the two experiments have led to guidelines (DOE 1970, 1975) for the introduction of area traffic control in other British cities, and about 12 cities have introduced or placed orders for control systems.

6.2.1 HISTORY OF AREA TRAFFIC CONTROL IN GLASGOW AND LONDON

A panel set up in 1960 by the RRL recommended that research in coordinated traffic signals should take the form of a full-scale experiment in an existing city. Glasgow was chosen because

---

*Based on a case description specially prepared by R. E. Allsop, Transport Studies Group, University College, London, and I. A. Ferguson, Transport Operations Research Group, University of Newcastle-upon-Tyne, UK.

the City Corporation there was among the first in Britain to express an interest in area traffic control, and the city center contains enough signals to justify the use of central coordination but is compact enough for the experimental system to cover the whole of the central area. The center of Glasgow is, however, somewhat atypical of British city centers in having a grid road pattern with short block lengths. The experiment has been concerned with assessing and comparing the benefits of a range of wholly automatic techniques of traffic control by computer, with little provision for manual intervention from the control center in the traffic control at individual junctions. The experiment as originally conceived has been described by Hillier (1965/66).

Concurrently, a complementary experiment was undertaken in West London by the Traffic Control Development Division of the MOT, as described by Cobbe (1967) and Mitchell (1967). The experimental system there covered a similar number of signals to that in Glasgow, but in a less compact and geometrically less regular road network forming the south-western approaches to Central London. The aims of this experiment were the development of the computer controlled system and associated equipment, and the testing of its use in conjunction, where appropriate, with remote manual operation of important junctions assisted by closed circuit television.

When these two experiments began, the cities in Europe and North America that had installed systems for the central control of traffic had very little information available on the comparison between the effects of different types of control on the traffic, the savings that may accrue, and the cost of equipment and maintenance necessary to perform each type of control. This information is required if traffic engineers wish to make objective decisions as to the type of control to install in the area for which they are responsible. The aim of the Glasgow and West London experiments was to provide such information. The average journey time of all traffic using the areas concerned was the criterion used in both experiments to assess the effects of control upon traffic. There was also some monitoring of effects on safety.

The Glasgow experiment has compared and established the effectiveness of three different techniques for calculating signal timing plans from historical traffic data, and a development of one of these techniques to give some degree of priority to buses. In addition, three types of traffic-responsive control have been tested, and a fourth such technique is currently being developed.

The West London experiment clearly demonstrated not only the effectiveness of signal timing plans calculated from historical data, but also a substantial reduction in accidents in the area of the experiment. It also contributed to the development of equipment. The experimental system was taken over by the Greater London Council in 1970, and the Council has since introduced area traffic control on broadly similar principals throughout Central London.

6.2.2 CONTROL SYSTEMS

The Glasgow experiment covers an area of about 2.5 square kilometers including about 80 signal-controlled junctions in the city center, as shown diagramatically in Figure 6.6. Over the ten years duration of the experiment, a few extra signals have been added and one or two streets have been pedestrianized, but the network has remained essentially the same. The pattern of traffic has, however, been affected substantially over recent years by the opening of various parts of an inner ring motorway, which now partially encloses the controlled area.

The equipment used has been described by Hillier (1967). Before the experiment, each junction had been equipped for vehicle-actuated operation by means of a local controller and pneumatic tube detectors. For purposes of the experiment, the existing controllers and detectors were retained and an interposing unit was installed next to each local controller. The function of this unit is to communicate information between the control center and the local controller. The information from between 70 and 80 junctions is transmitted along cables laid mainly in disused tramway ducts, and the remaining junctions use Post Office telephone lines with frequency-division multiplexing. It would not have been difficult to install cables to these remaining junctions, but it was decided to use the telephone lines to gain experience of this method. To facilitate checking of external equipment, provision is made for telephone communication between each junction and the control center.

There are 250 detectors and each is interrogated every fortieth of a second. Once per second, the signals are scanned, and if changes are required by the program the requests to

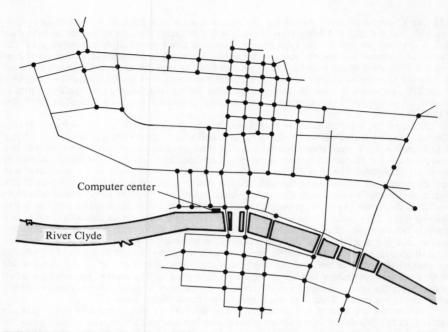

FIGURE 6.6 The network covered by the Glasgow experiment (reproduced from Almond and Lott 1967).

change are sent by the computer to the local controllers. When traffic signals are under the control of the computer, safety requirements built-in to the local controllers must be met before a change in the state of the lights can take place. When the signals are not under control, they are able to revert to local operation by virtue of the local controllers. More recently, loop detectors have replaced a number of the pneumatic tube detectors.

Incoming information is presented in the center on a bank of 1152 contacts giving information, e.g., about the phase showing the green, the state of the pneumatic tube detector and, if the intersection is being controlled, by the computer. Outgoing commands from the computer are presented on 576 pairs of contacts.

The necessary computing facilities could in principle have been avoided either by using part of the capacity of a large general purpose installation, as was done in Toronto (Hewton 1967), or by installing a smaller machine dedicated to traffic control. The latter course was adopted in Glasgow and allowed the implementation of control techniques and the related research to proceed unhindered by competing demands upon the computing facilities.

The main requirement of the computer specification was that the equipment should be able to carry out in 40 percent of the available time the tasks required by the most complicated control scheme considered at the time of the installation. A Marconi Myriad with 16K words of quick access store was chosen. More recently a 32K word store has been installed. For long-term storage a drum with a capacity of 80K 24-bit words is provided. A console typewriter is used to receive monitoring information from the computer and to input instructions to it. A graph plotter can also be controlled by the computer. A map display indicates the green signals on the approaches to each junction and whether the intersection is under computer control or not. This display is not used for control purposes but it enables visitors to appreciate better the actions of the computer, and it allows some checking of the overall timing patterns.

The control programs were developed by the team working on the experiment, and were written in a user code that has a correspondence with the basic computer code. This was done partly to make most effective use of the store and partly because of the lack of a suitable higher-level language when the experiment was being planned (Woolcock 1969).

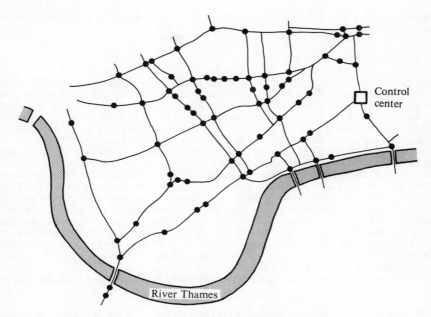

FIGURE 6.7 The network covered by the West London experiment.

The West London experiment covers an area of about 15 square kilometers, and includes about 70 signal controlled junctions, as shown diagramatically in Figure 6.7, and 30 signal controlled pedestrian crossings. As in Glasgow, each junction had previously been equipped for vehicle-actuated operation, and this equipment was retained, computer control being via interposers. There was, however, no counterpart in West London for the disused tramway ducts, which had made it economic in Glasgow to lay special-purpose cables for data transmission. For this reason, Post Office telephone lines were used for all data transmission. After comparison of the costs of associated equipment, time-division multiplexing was chosen in preference to frequency-division (Wheele 1967). Consideration was given to replacing the existing pneumatic detectors by inductive loop detectors in the interests of reliability and resistance to wear and tear. This course was rejected on grounds of cost, but inductive loop detectors were provided at 20 sites to provide vehicle counts of improved accuracy (Green and Ham 1967). The development of such detectors to provide accurate counts on multilane roads was a substantial aspect of the West London experiment (Ham 1969).

Detectors and signals were each scanned once per second; at each scan 24 bits of information were received from each local controller, and 12 bits of command transmitted to it. Controller-specific coding enabled these bits to be used in different ways for different junctions and pedestrian crossings.

As in Glasgow, a dedicated computing installation was used, consisting of two Plessey XL9 general purpose computers each with 16K words of quick access store. One machine served as a central processor; the other was used normally as a data handler, but also acted as a backup to provide some basic facilities in the event of failure of the central processor. For long-term storage a magnetic drum with 100K 24-bit words was provided. Fuller details of the equipment have been given by Halton et al. (1967).

Unlike the Glasgow system, the system in West London was designed to provide for appreciable manual intervention in the control of traffic, and therefore included a substantial control and display subsystem. This consisted of a map display flanked by 24 television monitors, visible from three control desks. Each desk was equipped with a further television monitor and facilities for showing, for any chosen junction, a schematic diagram including current incoming information from the junction. Television cameras were installed at six critical junctions (Jennings 1967).

6.2.3 CONTROL TECHNIQUES

Prior to the experiments, over half of the signals in the center of Glasgow and signals at nine junctions on the busiest roads in West London were linked by master controllers using the flexible-progressive system. Timing plans for such linking in Glasgow had been drawn up manually by trial and error with the aim of obtaining a plan that would normally work, rather than one that would optimize some objective function. For the nine junctions in London, however, timing plans had been calculated by the Combination method (see below).

During the experiment in Glasgow, comparisons were made of the performance of signal timing plans calculated, for morning peak, evening peak, and offpeak periods separately, by each of three methods: the Combination method (Hillier 1965/66), TRANSYT (Robertson 1969), and SIGOP (Traffic Research Corporation 1966, Peat Marwick Livingston & Co. 1968). In West London, the performance of timings calculated by the Combination method was assessed; this method was later improved by Huddart and Turner (1969), and subsequent developments in London have been based mainly on this technique. In addition, in Glasgow three traffic responsive techniques, FLEXIBPROG, EQUISAT (Holroyd and Hillier 1969), and PLIDENT (Holroyd and Hillier 1971) have been assessed. A fourth, known as SCOOT (see below) has been the subject of recent trials. An adaptation of TRANSYT to give priority to buses has also been assessed (Robertson and Vincent 1975).

The three methods of calculating signal timing plans from historical data each consist of two parts:

- a mathematical model that estimates the delay per unit time and the number of stops per unit time as functions of the signal timings
- an optimization technique to find signal timings that minimize a performance index, which is a suitable linear combination of the delay and stops per unit time

The mathematical models have a number of features in common. Each assumes a completely cyclic pattern of traffic throughout the network, with period equal to the common cycle time of all the signals (except those at lightly loaded junctions, which in TRANSYT, and more recently in the other two methods, could operate at half that cycle time). Each requires the common cycle time and the amounts of traffic per hour going straight ahead, turning left and turning right at each stopline to be specified, together with the corresponding saturation flow (i.e., the rate at which traffic crosses the stopline when the signal is green and there is a queue). Each requires the sequence of stages at each junction and the intergreen times between stages to be specified, and each regards as variable the offsets between the times at which the cycles at different junctions begin.

TRANSYT differs from the other two methods in regarding the allocation of green time between stages at each junction as variable also. The other two methods require these allocations to be decided in advance (cf. Webster and Cobbe 1966). Essentially for this reason, TRANSYT also takes some account of random fluctuations in traffic from cycle to cycle, whereas the other two methods assume that the overall effect of such fluctuations will be independent of the offsets.

The TRANSYT traffic model has two more important advantages over the others. The first is that it represents the dispersion of platoons as they move along links — a feature that has since been incorporated in the Combination method by Huddart and Turner (1969). The second is that it allows the estimated delay and number of stops on any link of the network to depend on the timings of all signals through which traffic on that link has passed, whereas the other two methods assume that they depend only on the timings of the signals at the two ends of the link. In its most recent version, TRANSYT is also able to model the movement of buses along each link separately from that of other traffic, in a way that allows not only for differences in speed and the time spent by buses at stops, but also for the interaction of buses and other traffic at the junctions themselves.

From the point of view of optimization, the main characteristics of the resulting models are as follows. In the Combination model the performance index for each link is an arbitrary numerical function of the difference between the offsets of the signals at the ends of the link. In SIGOP it is a piecewise quadratic function of the same variable, and the performance index for the whole network is therefore a piecewise quadratic function of all the offsets. In TRANSYT the latter performance index is an arbitrary numerical function of offsets and allocations of green time at all junctions in the network.

The corresponding optimization techniques are quite distinct. The Combination method uses a dynamic programming technique, which is based on combining pairs of links in parallel and in series, and is exact for networks that can be reduced to a single link in this way. For more complicated networks it is necessary to omit suitably chosen links. SIGOP exploits the piecewise quadratic form of the performance index and the fact that having found the minimum for one piece it is easy to find the minima for adjacent pieces, and so finds a good local optimum. TRANSYT also finds a good local optimum by systematically testing the effects of changes of various sizes in each offset and green time in turn, retaining just those changes that decrease the perfomance index.

By each of these three methods, timing plans were calculated for the Glasgow network for the morning peak, evening peak, and offpeak periods, using in all cases observations of traffic flows and turning movements made shortly before the plans concerned were to be implemented and assessed.

Plans calculated by the Combination method formed the basis of the first two traffic-responsive systems. The first of these, FLEXIPROG, was an adaptation of the flexible-progressive system previously used for the linking of signals by master controller. It enables each junction to operate under isolated vehicle-actuated control for limited periods in each cycle, subject to the requirement that green must always be available on demand for any traffic arriving in accordance with the underlying fixed timings. In the second system, EQUISAT, the midpoints of the green times were as determined by the Combination method settings, but the allocation of green time between stages at each junction was varied from cycle to cycle to equalize the degrees of saturation on critical approaches, as estimated from arrival rates and saturation flows measured over the previous eight cycles (subject to precautions against spurious values).

In the third traffic-responsive scheme, PLIDENT, the aim was to identify platoons of vehicles moving or intending to move along routes designated as priority routes, and wherever possible to allow the unimpeded passage of these platoons through the signalized network. About 40 percent of the links in the area were treated as priority links. Within very wide limits, the changing of signals was based on second-by-second estimates of the arrival times of these platoons at the signals concerned and the length of time that the platoons would take to pass.

The most recent technique tried in Glasgow, SCOOT, allows the offsets and allocations of green time at individual junctions and the cycle time for areas or subareas to be altered rapidly in response to changing traffic conditions. Any changes are based on estimates of advantages that are likely to accrue to the traffic using information on cyclic traffic behaviour from loop detectors on road links carrying significant volumes of traffic. These estimates are used by an optimizing routine to determine whether small changes should be made in the current signal settings. In addition, special procedures can be implemented when congestion is detected. Research and development of this method is continuing with the intention of making it available for general use.

6.2.4 ASSESSMENT AND EVALUATION

In both experiments, the criterion for assessing the effects of various control techniques on the traffic was total travel time for all traffic using the networks. Estimates of this quantity were obtained by observers driving over carefully chosen routes through the network according to a statistically designed scheme (Almond and Lott 1967, Williams 1969). To facilitate observations, the cars were fitted with equipment developed at the RRL (Vincent 1967). Because travel time per vehicle in a road network is known to increase as the flow of traffic increases, it was necessary to measure traffic flows concurrently with travel times. In Glasgow, this was done at first by roadside observers, and later by the computer directly from detector outputs. In London the flows were measured by roadside observers (Williams 1969).

Comparisons between different control techniques were made by plotting vehicle-hours per

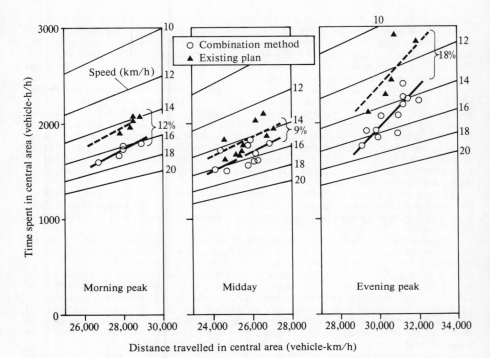

FIGURE 6.8 Comparison of previously existing control and combination method timings in Glasgow (reproduced from Holroyd and Hillier 1971).

hour against vehicle-distance per hour for each type of control and fitting parallel lines to the two sets of points. In Glasgow this was done separately for the morning peak, midday, and evening peak periods. Examples are shown in Figures 6.8 and 6.9. This method of analysis allows a linear hypothesis for the relationship between travel time and traffic flow. The difference between the effectiveness of the two control techniques is indicated by the vertical separation of the two lines. The statistical significance can be assessed by analysis of covariance.

The results of the main comparisons in Glasgow are summarized in Table 6.3, and it is appropriate to make the following brief comments on the individual results at this point. The previously existing control comprised interim settings for the linked vehicle-actuated systems during transition between two types of flexible progressive operation; for this reason the advantage of computer control of all types over the previously existing control may have been slightly exaggerated. The comparisons of SIGOP and TRANSYT were part of a joint Anglo-American experiment (Whiting 1972). The single-cycle TRANSYT timings were assessed to determine whether the slight advantage of TRANSYT over SIGOP could be attributed mainly to the double-cycling facility, which SIGOP did not then possess. The conclusion was that it could, and this suggests that an appreciable part of the advantage shown by TRANSYT over the Combination method may also have been due to double-cycling, because the Combination method as tested did not have this facility. Because FLEXIPROG and EQUISAT were strongly based on Combination method timings, it is hardly surprising that they did not show a large improvement over these timings, but it is at first sight somewhat surprising that they showed no improvement at all. A possible reason is that any advantage gained by traffic at one junction is easily lost at the next if the progression provided by the fixed timings has been disturbed. Further comparisons under light and very low flow conditions also showed no advantage in

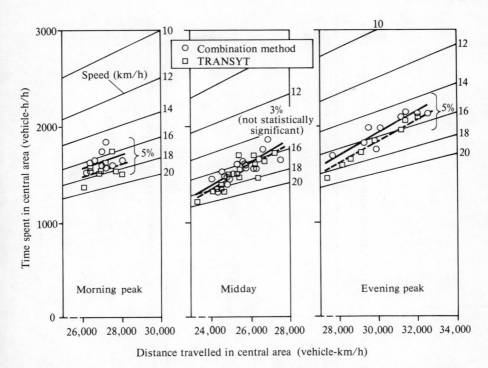

FIGURE 6.9 Comparison of TRANSYT and Combination method timings in Glasgow (reproduced from Robertson 1969).

vehicle-actuated operation (Owens and Holroyd 1973). PLIDENT was a radical technique, but the attraction of the intuitive ideas on which it was based was not borne out by experience.

Comparison of Combination method timings with previously existing control in West London showed a reduction of about 9 percent in travel time for a given flow (Williams 1969). In comparing this with the result in Glasgow, it should be noted that the previously existing equipment in London had been deliberately brought up to a high standard before the experiment, and the area already had the benefit of Combination method timings on one major route.

In comparing conditions under computer control with those under previously existing control, it is important to remember that in addition to changing the signals in different ways according to the various control techniques, the computer also provides very effective monitoring of faults in equipment on the streets, especially detectors and local controllers. Even without any change in control technique, this monitoring, if accompanied by prompt rectification of the fault indicated, would in itself be expected to benefit the traffic. The size of this benefit was investigated in West London (Cobbe 1971) by using the computer first to give a situation that was equivalent to fault-free equipment, and then to artificially apply faults in the street equipment to represent as far as practical the number and types of faults found in equipment surveys. The results of this work indicated that the difference between the fault-free operation and the normal situation that tends to prevail without a monitoring system, can be equivalent to a 5-10 percent reduction in journey times.

Assessment of timings calculated by the version of TRANSYT that gives priority to buses showed an increase of about 8 percent in the mean journey speed of buses with a reduction of about 1 percent (which could well have arisen by chance) in the average speed of other traffic (Robertson and Vincent 1975).

TABLE 6.3 Summary of Main Comparisons of Control Techniques made in the Glasgow Experiment (estimated percentage reductions in travel time for given levels of flow taken from TRRL publications)

	Morning peak	Daytime offpeak	Evening peak
Combination method timings compared with previously existing control	12	9	18
TRANSYT timings compared with Combination method timings	5	3	5
SIGOP timings compared with TRANSYT timings	0	−4	−5
Single-cycle[a] TRANSYT timings compared with TRANSYT timings	0	−5	−4
SIGOP timings compared with single-cycle TRANSYT timings	0	0	0
FLEXIPROG compared with Combination method timings	0	0	0
EQUISAT compared with Combination method timings	0	0	0
PLIDENT compared with Combination method timings	−24	−16	−50

[a]Single-cycle TRANSYT suppresses the facility for allowing sufficiently lightly loaded junctions to operate at half the cycle time of the remainder.

An assessment of control by SCOOT in Glasgow has been carried out for approximately 40 of the junctions in the central area using TRANSYT timings for comparison. Although no significant differences were measured for the morning peak and off-peak periods the average journey times obtained with SCOOT during the evening peak were found to be about 8 percent lower than those obtained using TRANSYT timings.

Economic evaluation of the results of the experiments depends strongly on the values to be attached to small time-savings by various categories of road user. It also depends upon the extent to which the community takes the benefit of improved control in time-savings to existing travellers and the extent to which the benefit is taken in extra travel through the network (thus reducing the time savings to the original travellers). Using conventional methods of evaluation, however, both experiments were found to save substantially more than their capital cost within one year (Hillier and Holroyd 1968, Williams 1969).

REFERENCES

Almond, J. and R. S. Lott. 1967. The Glasgow experiment: implementation and assessment. Ministry of Transport RRL Report LR 142, Crowthorne, UK.

Cobbe, B. M. 1967. The approach to control in West London. Area Control of Road Traffic, Institution of Civil Engineers, London, pp. 1-5.

Cobbe, B. M. 1971. Computer control of traffic. British Symposium of Advanced Technology, BNEC Export Council for Europe and the London Chamber of Commerce, Zurich.

Department of the Environment. 1970. Roads Circular 13/70.

Department of the Environment. 1975. Roads Circular 26/75.

Green, F. B. and R. Ham. 1967. Vehicle detection. Area Control of Traffic, Institution of Civil Engineers, London, pp. 167-179.

Halton, D., D. G. Hornby, and C. H. Wolff. 1967. The design of wide area equipment. Area Control of Road Traffic, Institution of Civil Engineers, London, pp. 57-66.

Ham, R. 1969. Area traffic control in West London, 2. Vehicle counting detectors. Traffic Engineering and Control, 11 (4): 172-176.

Hewton, J. T. 1967. The Metropolitan Toronto signal system. Area Control of Road Traffic, Institution of Civil Engineers, London, pp. 17-45.

Hillier, J. A. 1965/66. Glasgow's experiment in area traffic control. Traffic Engineering and Control, 7 (8): 502-509, and 7 (9): 569-571.

Hillier, J. A. 1967. The Glasgow experiment: schemes and equipment. Ministry of Transport RRL Report LR 95, Crowthorne, UK.

Hillier, J. A. and J. Holroyd. 1968. The Glasgow experiment in area traffic control. Traffic Engineering, 39 (1): 14-18.

Holroyd, J. and J. A. Hillier. 1969. Area traffic control in Glasgow: a summary of results from four control schemes. Traffic Engineering and Control, 11 (5): 220-223.

Holroyd, J. and J. A. Hillier. 1971. The Glasgow experiment: PLIDENT and after. Department of the Environment TRRL Report LR 354, Crowthorne, UK.

Huddart, K. W. and E. D. Turner. 1969. Traffic signal progressions – GLC Combination Method. Traffic Engineering and Control, 11 (7): 320-322 and 327.

Jennings, S. W. 1967. Closed circuit television for traffic surveillance. Area Control of Road Traffic, Institution of Civil Engineers, London, pp. 161-165.

Mitchell, G. 1967. Control concepts of the West London experiment. Area Control of Road Traffic, Institution of Civil Engineers, London, pp. 121-126.

Owens, D. and J. Holroyd. 1973. The Glasgow experiment: assessment under light and very low flow conditions. Department of the Environment TRRL Report LR 522, Crowthorne, UK.

Peat Marwick Livingston & Co. 1968. SIGOP – traffic signal optimization program: user's manual, New York, NY.

Robertson, D. I. 1969. TRANSYT: a traffic network study tool. Ministry of Transport RRL Report LR 253, Crowthorne, UK.

Robertson, D. I. and R. A. Vincent. 1975. Bus priority in a network of fixed-time signals. Department of the Environment TRRL Report LR 666, Crowthorne, UK.

Traffic Research Corporation. 1966. SIGOP – Traffic signal optimization program, New York, NY.

Vincent, R. A. 1967. Traffic survey equipment for measuring journey time and stopped time. Ministry of Transport RRL Report LR 65, Crowthorne, UK.

Webster, F. V. and B. M. Cobbe. 1966. Traffic signals. Road Research Technical Paper No. 56, London: Her Majesty's Stationery Office.

Wheele, D. W. E. 1967. Data transmission systems. Area Control of Road Traffic. Institution of Civil Engineers, London, pp. 67–75.

Whiting, P. D. 1972. The Glasgow traffic control experiment: interim report on SIGOP and TRANSYT. Department of the Environment RRL Report LR 430, Crowthorne, UK.

Williams, D. A. B. 1969. Area traffic control in West London, 1. Assessment of first experiment. Traffic Engineering and Control, 11 (3): 125–129 and 134.

Woolcock, M. 1969. Traffic signal control in Glasgow by computer. Ministry of Transport RRL Report LR 262, Crowthorne, UK.

6.3   JAPANESE EXPERIENCES: TOKYO AND OSAKA

About 30 Japanese cities had implemented computerized traffic control systems prior to March 1976 (cf. Table 6.4). In the fiscal year 1976 the "Second Five Year Plan for Facilities Construction for Traffic Safety Installations" was started, and in this plan 28 additional cities are to come under the facilities planning.

The experience obtained with the two largest and most advanced area traffic control systems, i.e., those in operation in Tokyo and in Osaka, will be described in the following.

6.3.1.   THE TOKYO SYSTEM*

6.3.1.1   Case History. Activities in the Tokyo Metropolitan Area, with 10 million inhabitants, depend heavily upon road traffic, which has been increasing at a rate of 10–14 percent per year.

Tokyo has also suffered an increase in traffic jams and traffic accidents, and the necessity for systematic traffic control has risen sharply since around 1960. In 1963, the Tokyo Metropolitan Police Department deployed the Semi-Automatic Traffic Information and Control System (SATIC). This sytem gathers information on the length of traffic queues, by manually entering data into the traffic congestion reporting equipment installed at major traffic points, according to the observations made by police officers based upon predetermined standards. On the basis of this information gathering network, instructions for traffic control are sent to front line police officers, and as is done by the Traffic Control Center to be described later, directions are given in an organized manner to vehicles by dissemination of traffic control information.

Also, as the traffic and the number of traffic signals increased, so also did the need for systematic control of traffic signals. In 1963, traffic-adjusted control of a one-dimensionally distributed system of traffic signals was first conducted in Tokyo. Later, in 1966, an experiment in area control of traffic signals was conducted with the approximately forty signal controlled intersections in the Ginza area of Tokyo (the most famous shopping area in Tokyo).

This was the first introduction in Japan of digital computers to control traffic. In this method, the principal of one-dimensional traffic-adjusted signal control mentioned above is extended to area control in such a way that many different patterns may be taken. A unique feature of this method is that it takes into consideration directional control patterns for east-west and north-south directions. However, fundamentally it is based on traditional ideas of

---

*Based on a case description (including Table 6.4) specially prepared by Hiroshi Inose, Department of Electronic Engineering, University of Tokyo, Hiroyuki Okamoto, Technical Councillor, Traffic Bureau, National Police Agency, and Yoshibumi Miyano, Head, Traffic Facilities Section, Traffic Division, Tokyo Metropolitan Police Department. The authors wish to express their gratitude to the Japan Society of Traffic Engineering, the Committee for Designing the Tokyo Area Traffic Control System, and the Study Committee on the Relationship between Traffic Flow and Exhaust Emission, for their collaboration and contribution.

TABLE 6.4 Traffic Control throughout Japan (as of March 1976)

City	Population DID[a] (in thousands)	Area DID (km²)	No. of signalized intersections[b]	No. of information displays[b]	No. of TV cameras[b]
Sapporo	823	88.3	317	3	4
Sendai	439	51.4	174	8	4
Yamagata	143	19.3	40	3	2
Tokyo	8,793	549.3	2,815	60	19
Mito	92	13.8	100	5	3
Utsunomiya	188	26.4	115	10	2
Urawa	784	82.4	163	18	4
Chiba	365	48.6	180	0	0
Katsunan	783	86.6	57	10	9
Yokohama	1,935	205.8	581	27	10
Kawasaki	907	88.2	280	3	2
Niigata	276	33.6	40	3	3
Nagano	136	19.9	40	3	2
Shizuoka	476	57.2	140	6	2
Hamamatsu	216	30.0	40	5	2
Kanazawa	251	25.0	90	7	1
Fukui	132	15.8	55	5	3
Gifu	264	28.2	115	0	1
Nagoya	1,854	191.4	635	13	5
Kyoto	1,301	101.7	283	5	5
Osaka	5,396	429.6	987	38	19
Kobe	1,157	75.0	374	8	9
Wakayama	254	33.7	99	0	1
Okayama	219	27.0	124	5	10
Hiroshima	504	48.7	253	11	14
Takamatsu	171	26.0	69	3	1
Matsuyama	183	22.0	40	3	2
Fukuoka	720	82.0	300	12	20
Kitakyushu	880	112.3	344	9	17
Kumamoto	347	41.6	178	18	6
Kagoshima	307	32.4	131	4	7
Total			9,159	305	189

[a]DID, densely inhabited district.
[b]Numbers of signalized intersections, information displays, and TV cameras are those within the computer controlled areas.

pattern selection and arterial control, and therefore it could not be considered adequate to deal with the traffic congestion in the Tokyo Metropolitan Area.

The Tokyo Metropolitan Police Department, on the basis of the results of the experiments mentioned above, secured the cooperation of the Japan Society of Traffic Engineering and continued the development and research for more sophisticated systems. The result was the entirely new Area Traffic Signals Control System, with much higher control capability, which was put into operation in May 1970 and covers an area of about five square kilometers in downtown Tokyo.

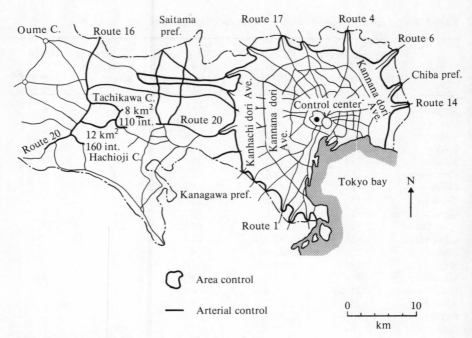

FIGURE 6.10   Control Area in Metropolitan Tokyo.

The Japanese traffic control systems that started in downtown Tokyo were dual systems: human control centered around the traffic information center, and mechanical control centered around the sophisticated traffic signal control. However, experience gained in applying these dual systems indicates that, to develop a traffic control system of the highest capability, these two systems should be integrated, each complementing the other.

The concept of integration has now been firmly established in the construction of the traffic signal control systems in this country. On the basis of hardware technology, which is centered around traffic measuring technology and electronic computers, the concept that this would be technologically possible has now been proven true, and the effect of each system has been proven to be large.

With this background, the National Police Agency, as a project of its "Five Year Plan for Facilities Construction for Traffic Safety Installations" beginning with the fiscal year 1971, decided to implement integrated traffic control systems as national policy in 28 cities throughout the nation. In Tokyo, the success of the wide area traffic signal control system, opened in May 1970 as well as the facilities policy being established by the National Police Agency, led the Tokyo Metropolitan Police Department to establish the Traffic Control Center, and to expand the control area further. In November 1970, the Committee for Designing the Tokyo Area Traffic Control System was established under Dr. Inose's chairmanship to design a wide area traffic control and surveillance system for the entire Tokyo Metropolitan Area, which is to cover in its final stage 8,000 signalized intersections. By that time, the number of automobile sensors is predicted to reach 16,000.

Figure 6.10 shows a map of the future control area in Metropolitan Tokyo. The area inside the bold line plus such satellite cities as Hachioji and Tachikawa are subject to two-dimensional area control; the major routes indicated by the bold lines are subject to one-dimensional arterial control. In this way, the concept of the Tokyo traffic control system facilities was decided upon, and in the fiscal year 1973, the Traffic Control Center was budgeted. By 1975, the system reached the scale described in the next section.

The goals toward which this system was originally aimed were safety and the smooth flow of traffic. Accordingly, the control principle was also organized around these fundamental objectives. However, since around 1970, air pollution caused by autombile exhaust emissions, and environmental noise have become social problems. Hence the traffic control policy in Japan is now directed toward the suppression of these kinds of adverse traffic effects, and is pointed in the direction of vigorous control of traffic flow with the aim of decreasing the total traffic volume. Following these changes in the sociopolitical climate, the methods of the traffic control system have also been changed to fulfill these requirements. It is a big advantage of systems built around electronic computers that such changes are possible and easily implemented. By utilizing their adaptability to change as described in paragraph 6.3.1.3, this sytem has become an important factor in advancing the national policy mentioned above.

6.3.1.2 Implementation. The major features of the system are as follows.

1. *On-line traffic data collection*

The outputs of all the traffic detectors are transmitted to the control center by means of data transmission lines leased by the Nippon Telegraph and Telephone Public Corporation (NTT). After processing by computers, the data are utilized for:

- on-line real-time control of traffic signals
- wall-map displays for surveillance and possible intervention by traffic officers, for broadcasting to the public, for controlling remote roadside displays and variable traffic signs, and for answering telephone inquiries

2. *Dynamic selection of signal control modes*

One of the following five control modes is automatically selected in accordance with the traffic conditions:

- stop mode, to minimize the number of stops when traffic is relatively light
- delay mode, to minimize the delay time when traffic is average
- capacity mode, to maximize the traffic capacity when traffic is close to saturation
- queue mode, to suppress the spread of queues when the queues build up
- jam mode, to recover from jamming when the traffic is oversaturated

3. *Dynamic formation of tree-offset pattern*

In accordance with the traffic conditions, an optimum tree is formed, which minimizes the total waiting time in the area. In switching from one offset pattern to another, smooth transfer is employed to avoid traffic confusion.

4. *Dynamic formation of cycle and split patterns*

In accordance with the traffic data at each intersection, the cycle and split values are automatically assigned. The entire area is dynamically divided into cycle subareas in accordance with the traffic situations, in such a way that all the intersections in a cycle subarea are assigned the same cycle length, and that the difference between it and the cycle length of the neighboring cycle subareas is below a predetermined value.

5. *Multilevel fall back structure*

Under normal conditions, the entire system is under the control of central computers in a redundant configuration. When the computer system fails, the centrally located standby computer or hardware controller takes over. In case of its failure, the local controller in each intersection individually controls signals in the intersection. Graceful degradation of the control functions is thus provided with this multilevel fall back structure.

6. *High reliability design and computerized maintenance*

Redundancy concepts are introduced in the system design so as to meet a reliability objective of less than ten hours of downtime in ten years. Diagnostic programs for the outstation devices are provided to check the conformity of answerback signals from each local controller with the CPU command, and to verify the validity of traffic data from each vehicle detector compared with past data and data from neighboring detectors. Diagnostic programs are also provided for monitoring the operation of the equipment in the control center.

Vehicle detectors are installed at about 150 m from each of the stop lines at critical intersections, but, for reasons of economy, on only one lane per approach. To detect long queues, additional detectors are provided at points 300 m, 500 m, and 1,000 m behind the possible bottlenecks on one of the lanes. Detectors are also installed on all lanes at strategic points on important arteries to measure their cross-sectional traffic volumes. Most of the detectors are ultrasonic radars.

The traffic data obtained by the detectors are transmitted over NTT's leased lines to the control center. The data are then processed and utilized for controlling the traffic signals as well as for displaying the traffic situation.

The data transmission system employs the time-division multiplexing technique rather than the conventional frequency-division multiplexing technique. Although the costs of both techniques are comparable, considerations of reliability, accuracy, maintainability, and compatability with the central processor were found to favor the time-division multiplexing technique. The upward and downward frames for the data to and from the control center, respectively, have different frame rates. An upward frame has a frame rate of 300 milliseconds and carries detector output as well as answer back signals from the local controller, whereas a downward frame has a frame rate of one second and carries hold and advance commands for stepping signal indications.

Traffic information is also obtained in visual form by remote-controlled CCTV cameras located at various points in the streets. Video signals of 4 MHz bandwidth from CCTV cameras are transmitted over leased telephone lines with repeaters every 470 meters for 0.32 millimeter PEF cable, 2,000 meters for 0.9 millimeter PEF cable. Selection of pictures and camera control are performed at a CCTV monitor console. In addition, traffic information is obtained from neighboring traffic centers, police officers and patrol cars by means of wired and wireless telephones.

Figure 6.11 shows the layout of the displays, consoles, etc. in the control center. The traffic situation is shown by two methods: link indication and congestion length indication. Link indication gives the traffic situation on links in the downtown areas where the distance between signalized intersections is short (cf. Figure 6.12). The red lights are turned on when abnormal congestion is detected by comparison with the previous indications. Congestion length indication shows the extent of congestion at intersections on the outskirts of the city by means of lights (cf. Figure 6.13), which change with the length of the congestion.

A supervising officer is situated at the console where he can observe the overall traffic situation within the city; when on-the-spot activity is required, he can send instructions for traffic control by radio telephone or by wired telephone to the front-line officers. For traffic information services, eight booths are provided for radio station reporters. In addition, roadside displays are installed to provide congestion and routing information. Inquiries by telephone are answered by operators.

The data processing to be performed at the control center may be classified into the following three categories.

(i) Processing of input data from detectors:
sampling of detector pulses, smoothing disposal of malfunction detectors and calculation of degree of congestion, etc.

(ii) Determination of control parameters:
determination of control mode, cycle length calculation, formation of cycle subarea, formation of tree, split calculation, offset pattern selection and adjustment, etc.

(iii) Control and monitoring of signal indications:
control of signal indications, checking of answer back signals, etc.

Table 6.5 shows the result of job analysis classified by the execution cycle. In the table, the number of intersections is denoted by I, and the number of detectors, major intersections, and major links are assumed to be $2I$, αI and βI, respectively. Assuming that $I = 8,000$, $\alpha = 0.2$, and $\beta = 0.4$, the total number of computer steps, as given by Table 6.5 is 3.12 megasteps/second. Taking into account such factors as the overhead of the monitor program and the allowances for load variations, the computing speed necessary to control 8,000 intersections with a single computer was estimated as 5.2 megasteps/second. Similarly, Table 6.5 shows the results of estimating the required memory capacity.

As a result of the above job analysis, it was found to be preferable to employ a hierarchical structure for the computer system. It is known from Table 6.5 that the data processing of cate-

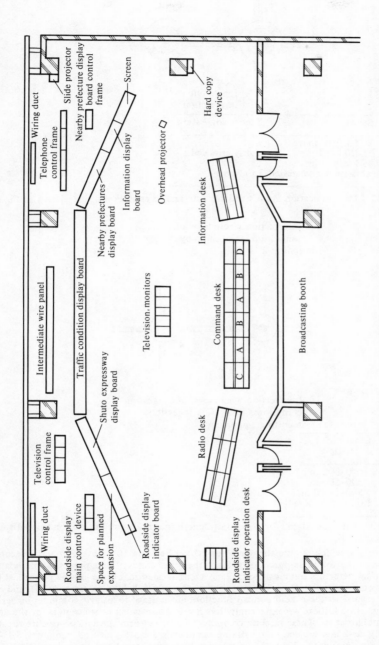

FIGURE 6.11 Implementation of instruments in traffic control. Command desk sections: A, supervisory television operation desk; B, CRT display position, traffic condition display board control desk; C, information display board control desk; D, nearby prefecture display board control desk.

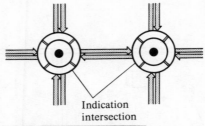

Indication	Speed	Traffic congestion
Off	20 km/h or more	Light
Yellow	15–20 km/h	Fairly crowded
Red	15 km/h or less	Crawl
Red flashing	Abnormally congested	Confusion

FIGURE 6.12   Link indication lamp standards (cf. Traffic condition display board in Figure 6.11).

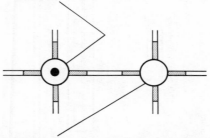

Indication	Length of congestion
Off	0–300 m
Green	300–500 m
Yellow	500–1,000 m
Red	1,000 m or more

FIGURE 6.13   Lamp standards for congestion length indication (cf. Traffic condition display board in Figure 6.11).

gories (i) and (iii), which require the execution period between 50 millisecond and 1 second, occupies 90% of the total computer steps. Since these categories of data processing are related to the individual detectors and local controllers, it was decided that these jobs be performed by computers lower in the hierarchy. The rest of the jobs, which belong to category (ii), and are more or less related to the overall control function, are assigned to a single computer higher in the hierarchy. In addition, separate computers are to be installed in subcenters for the area control of satellite cities. These satellite computers are to perform the data processing of categories (i)–(iii), and are subject to overriding commands from the central computer. NEAC 3,200/50s computers with 65 K word core memory were employed lower in the hierarchy; each of these performs data processing of categories (i) and (iii) for up to 500 intersections. At the present time, the computer highest in the hierarchy is an NEAC 2,200/375 with a core memory

TABLE 6.5 Estimation of Computation Time and Memory Capacity

Program name	Execution cycle	Dynamic step (steps per second)	Core file	Mass storage file
Sampling of detector pulses	50 ms	$200I$	$8.4I$	—
Total ($I = 8,000$)		1.6×10^6	67.2 kW	—
Control of signal indications	1 s	$25I$	$13I$	—
Checking of answer back		$30I$	—	—
Green extension		$70\alpha I$	$25\alpha I$	—
Delay calculation		$75\beta I$	$15\beta I$	—
Others		$0.3I + 240\alpha I$	$40\alpha I + 0.53I$	—
Total ($I = 8,000, \alpha = 0.2, \beta = 0.4$)		1.18×10^6	260 kW	—
Smoothing	100 s	$1I$	$6I$	—
Relative offset to absolute		$3.2I$	—	$64I$
Adjustment of offset		$1\beta I$	—	$9\beta I$
Others		$0.11I + 10\alpha I$	—	$260\alpha I + 3I$
Total ($I = 8,000, \alpha = 0.2, \beta = 0.4$)		0.055×10^6	48 kW	1010 kW
Disposal of malfunction detectors	5 min	$0.39I$	—	$80I$
Calculation of necessary length of cycle		$0.83\alpha I$	—	$60\alpha I$
Determination of split		$0.7I + 4.4\alpha I$	—	$13.3I + 150\alpha I$
Selection of offset pattern		$0.69I$	—	$21I$
Preparation of time tables		$1.67I$	$4.8I$	$60I$
Determination of the mode of control		$0.23\alpha I$	$0.8I$	$150\alpha I$
Others		$1.67I + 8.84\alpha I$	—	$640\alpha I + 18.25I$
Total ($I = 8,000, \alpha = 0.2$)		0.065×10^6	44 kW	3140 kW
Formation of cycle subarea	30 min	$0.04I$	—	$12I$
Formation of tree		$1.67I$	—	$20.8I$
Distribution of traffic flow		$2.2 \times 10^{-7} I^3$	—	$2 \times 10^{-3} I^2$
Prediction of traffic parameters		$7 \times 10^{-5} I^2$	—	$2 \times 10^{-4} I^2$
Others		$55.6\alpha I$	—	$60\alpha I$
Total ($I = 8,000, \alpha = 0.2$)		0.22×10^6	—	499 kW

TABLE 6.6 Reliability Prediction

Computer	Item	MTBF (h)	MDT (h)	Switching time (min)
1. Higher in the hierarchy	Central processing unit	500	2	2
	Magnetic disk-pack control	6,000	2	0
	Magnetic disk-pack	3,000	2	0
	Multiple communication control unit	4,000	2	0
	Peripheral equipment switch	20,000	2	0
	Line switch	100,000	1	-
2. Lower in the hierarchy	Central processing unit	2,000	2 (4)^a	1
	Communication control unit	4,000	2 (4)	0
	High speed communication control unit	8,000	2 (4)	0
	Line switch	200,000	1 (3)	-

^{a}Parentheses denote the cases where satellite computers are located in subcenters.

of 393 K word; this will be replaced by a larger computer before the system reaches its final size.

Table 6.6 shows the predicted reliability data for the computers. It can be seen from this table that for the computer highest in the hierarchy a duplex configuration with a total downtime of 8.3 hours in 10 years is sufficient, because the computers lower in the hierarchy remain operating during the changeover time of 10 minutes. For the computers lower in the hierarchy it can be seen that a downtime of 2.6 hours in 10 years is achieved if one standby computer is provided for every three active computers in the control center, and that a downtime of 2.1 hours in 10 years is achieved if a dual configuration is provided for the satellite computers in unattended subcenters. The downtimes are similarly allocated to the rest of the subsystems, to meet the reliability objective of a total downtime of 10 hours in 10 years.

The present system structure and scope, as of March 1976, are shown in Figure 6.14 and Table 6.7.

6.3.1.3 Operational Experience. As mentioned earlier, while the scale of this system is expanding year by year, various trials are also being carried out regarding the method of control, and new features are being incorporated. For example, lowering the levels of emission gases (CO, HC, NO_x, etc.), and reducing noise and vibrations caused by automobiles have been made additional objectives of the system. Furthermore, with the aim of reducing the volume of personal automobile traffic, a policy of giving priority to public mass transportation facilities, namely buses, has been adopted, and bus priority signals have been built into the system.

Every effort is being made in hardware and in software, to completely maintain the functions of this traffic control system, which is the largest in the world. In particular, to increase the reliability of the local signal controllers, a great deal of effort has been expended to improve the MTBF (mean time between failures).

The Traffic Control Center operates around-the-clock, performing not only its conventional police duties, but also providing traffic information to the general public, a service that is much appreciated.

1. *Prevention of traffic accidents*

Table 6.8 shows a comparison of the occurrence of traffic accidents within and outside the computer control area. It goes without saying that traffic accidents bear a relationship to traffic regulations, traffic laws, and other aspects of the total traffic policy, but this table still shows a rather sharp drop in traffic accidents inside the computer control area.

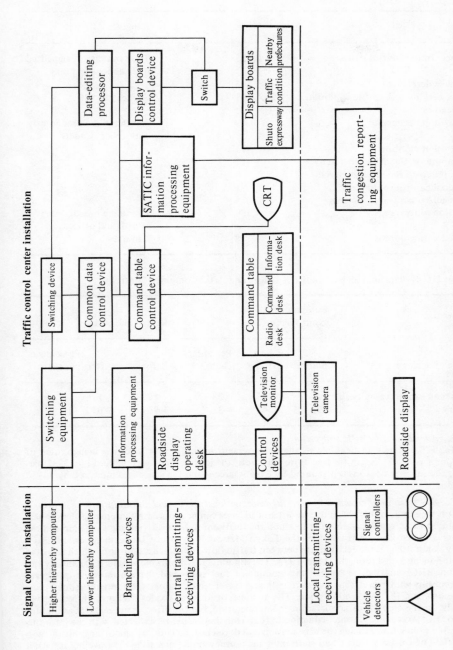

FIGURE 6.14 Schematic drawing of the Tokyo traffic control system (SATIC, Semi-automatic traffic information and control system).

TABLE 6.7 Scope of Traffic Control System (as of March 1976)

Equipment	Scope	Remarks
Concentrated control signals	2,815	Total number of signalized intersections is 8,003
Vehicle detectors	3,346	
Intersections with traffic condition indicators	361	Capacity is 500
Television cameras	19	Another 4 are used jointly with other systems
Television monitors	11	
Computer in the higher hierarchy	1	
Computers in the lower hierarchy	8	
Arterial control computer	1	
Specialized control processors	15	
Area of control center	1,380 m^2	Excluding broadcasting booths and offices
Broadcasting booths	8	6 in use

TABLE 6.8 Traffic Accident Statistics Before and After Computer Control

	Computer controlled area			Rest of the area		
Type of accident	Before	After	Percentage change	Before	After	Percentage change
Fatal and serious casualties	47	36	−23	127	116	−9
Injury	616	424	−31	1,267	1,092	−14

2. *Alleviation of traffic congestion*

Table 6.9 shows an example of the improvement in traffic flow after the inauguration of computer control. As can be seen from this table, the journey time was reduced by 13–31 percent, the number of stops was reduced by 24–45 percent and the speed was increased by 14–18 percent.

3. *Suppression of emission of harmful gases, with savings in fuel consumption*

It is well-known that the emission of harmful gases can be reduced by improving the running condition of automobiles. In order to study the relationship between running condition and the volumes of harmful gases emitted, the Tokyo Metropolitan Police Department established a study committee on the relationship between traffic flow and exhaust emission, which reported the following conclusions in April 1974. The amounts of carbon monoxide (CO) and hydrocarbons (HC) emitted, as well as the amount of fuel consumed, have a more or less linear relationship with travel time, so that reducing travel times should substantially reduce these emissions, and the fuel consumption. The amounts of nitrogen oxides (NO_x) produced depend largely on the amount of acceleration and deceleration, but the effect is not as great as that for CO or HC. Accordingly, some reduction in NO_x emissions could be expected with the introduction of a policy for reducing the acceleration and deceleration. Thus, by smoothing out the normal flow of roadway traffic and shortening the journey time (increasing the traveling speed), it is possible to reduce the CO, HC, and NO_x emissions and at the same time reduce fuel consumption.

TABLE 6.9 Journey Time, Number of Stops, and Speed Before and After Computer Control

		Upstream			Downstream		
		Before	After	Percentage change	Before	After	Percentage change
Harumidori Avenue	Journey time (s)	1275	861	−31	983	763	−22
	Number of stops	18.1	10.0	−45	11.7	7.7	−34
	Speed (km/h)	12.3	18.3	+48	14.3	18.4	+29
Eitaidori Avenue	Journey time	372	290	−22	401	350	−13
	Number of stops	5.7	3.4	−40	5.7	3.8	−34
	Speed (km/h)	14.8	18.6	+26	13.6	15.5	+14
Chuodori Avenue	Journey time	652	559	−14	704	596	−15
	Number of stops	10.3	7.9	−24	12.3	7.5	−39
	Speed (km/h)	15.5	18.2	+17	14.4	17.0	+19

TABLE 6.10 Estimated Annual Reductions in Emission of Pollutants and in Fuel Consumption for Tokyo

	Before computer control	After computer control
Journey time per km (s)	242.1	189.2
Stops per km	3.91	2.47
Average link speed (km/h)	14.9	19.0
Acceleration and deceleration distance per km (m)	563	356
Steady cruising distance per km (m)	437	644
Running mode per km (model)		
Idling (s)	110.1	66.9
Acceleration (s)	58.7	37.1
Cruising (s)	39.4	58.0
Decelerating (s)	43.0	27.2
Total emission per year		
CO ($\times 10^3$ tons)	34.8	22.7
HC ($\times 10^2$ tons)	64.7	44.8
NO_x ($\times 10^2$ tons)	17.6	15.7
Total fuel consumption per year ($\times 10^3$ kl)	252	215

These conclusions run counter to the estimates generally believed earlier and obtained from stationary engine test results, that increasing the speed would reduce CO and HC emissions, but would increase NO_x emissions. These new conclusions constitute strong evidence in support of the efficacy of smooth traffic flow policies.

In the light of these new results, we have made efforts to improve the running condition of traffic on major routes within the city — in other words, to reduce the journey time — and benefits have begun to accrue. Table 6.10 shows estimated reductions of emission gases in

Tokyo before and after the introduction of computer control. It can be seen that, in one year, CO would be reduced by 12,100 tons, HC by 1,990 tons, and NO_x by 190 tons. Also, a saving in fuel consumption of 37,000 kiloliters would be realized.

4. *Cost-benefit considerations*

The above data can be used to estimate the economic benefit to be gained as follows. The major factors for the economy are the reduction in journey time and the savings in fuel consumption, the yearly benefit of which is expressed by the following equation

[yearly benefit (yen)] = [journey time reduction (s/km)] ×
[benefit brought by journey time reduction (yen/s/vehicle)] ×
[total road length (km)] × [average traffic volume in 12 h in the daytime (vehicle/day)] ×
[yearly working days] + [yearly reduction of oil consumption (l)] × [oil price (yen/l)]

According to this equation, the benefit gained per computer controlled intersection in the year 1974 is estimated at 25,219,000 yen (US$84,000). In contrast, in the year 1974, 1999 intersections were under computer control, requiring a total investment of 6,526,554,000 yen (US$21,755,000) for the implementation of the local controllers and the equipment in the control center. This means that an investment of 3,490,000 yen (US$11,600) per intersection provides the above annual benefit, which is more than seven times as large as the initial investment.

5. *Suppression of noise and vibrations*

Residents of areas adjacent to major routes suffer from noise and vibrations caused by large volumes of motor traffic. It is known that automobile noise and vibrations are related to the starting and stopping of automobiles, and to the running speed. One approach to improving these conditons is to reduce starting and stopping as much as possible, and to keep vehicles at a controlled speed. To achieve this, the smallest possible cycle is chosen for the signals under computer control, and the mutual offset of the signals is adjusted so that vehicles traveling at the controlled speed can make as few stops as possible. In this way, if a vehicle travels above the control speed, the number of stops increases, and traveling becomes more difficult, thus discouraging violation of the control speed limit.

Another effective method of reducing the effects of noise and vibrations on residents is to keep vehicles as far as possible from people's homes. For this purpose, a policy was adopted that designated the lane farthest from the sidewalk as the lane to be used by large vehicles.

The Tokyo Metropolitan Police Department has combined these two policies, and given the compound policy the name of the road where it was first tried out – the Kanjo Nana-go Sen method (abbreviated the Kan-Nana method).

Table 6.11 shows examples of the results of the Kan-Nana method. It can be seen from this table that the speed was reduced by 24-25 percent, the number of stops was reduced by 65-77 percent, and the noise level was lowered by 3-4 phons.

6. *Bus priority signals*

Table 6.12 shows the results of one experiment where bus priority signals were installed at five intersections in a one kilometer section of a downtown street within the computer controlled area. It can be seen that the waiting time of buses at traffic signals was reduced by

TABLE 6.11 Suppression of Noise and Speed by the Kan-Nana Method

	No. 2 Keihin highway (inbound)			Keiyo highway (inbound)		
	Speed (km/h)	Number of stops	Noise (phon)	Speed (km/h)	Number of stops	Noise (phon)
Before	59.7	17	65	58.4	13	65
After	45.1	6	62	43.7	3	61
Change	−14.6	−11	−3	−14.7	−10	−4
Percentage reduction	24	65	5	25	77	6

TABLE 6.12   Effects of Bus Priority Signals

	Inbound		Outbound	
	Waiting time (s)	Average speed (km/h)	Waiting time (s)	Average speed (km/h)
Normal signal	60.9	15.0	48.4	16.3
Bus priority signal	33.9	17.2	22.9	19.1
Percentage improvement	44	15	54	17

TABLE 6.13   Improvement in Reliability of Local Controllers

Period	Number of controllers under survey	MTBF (h)	Improvement[a]
May 1971-March 1972	435	3,790	–
April 1972-March 1973	644	5,589	1.5
April 1973-March 1974	1,042	4,309[b]	1.1
April 1974-March 1975	1,620	6,872	1.8
April 1975-December 1975	2,193	9,149	2.4

[a]Compared with initial period.
[b]This reduction in MTBF was due to the unusual lightning damage during the year.

44-54 percent and the speed increased by 15-17 percent. The success of this measure has led to a plan to gradually install bus priority signals in the computer controlled area.

7. *Improvement in reliability of local controllers*

Table 6.13 shows the improvement in the reliability of local controllers under computer control. The number of local controllers covered by these statistics grew from 435 in 1971 to 2,197 in 1975, while the MTBF increased approximately 2.4 times; in other words, the failure rate has been improved to approximately one failure per year.

The following actions were thought to be particularly valuable in improving the reliability of the signals:

● under standardized instrument specifications, factory inspections were tightened, and production quality supervision was improved
● the lightning arrestor was improved, and failures that used to occur due to lightning in the summer were reduced
● the external noise suppression circuit was improved; this reduces the number of incidents in which a local controller goes off-line from the computer control

To improve the reliability of the controllers even more, it was decided to use LSI devices for the main parts of the controllers. Special LSI devices were developed for use in traffic signal controllers, and prototypes have now been installed in the streets where they are undergoing tests. It is hoped that, by these steps, the MTBF of the local controllers can be improved to 30,000-40,000 hours.

8. *Operating condition of Traffic Control Center*

The Traffic Control Center is manned by 28 police officers. In addition to instructions concerning traffic enforcement sent by radio telephone to patrol cars, policemen on motorcycles, and front-line police stations information regarding the condition of traffic and traffic accidents is sent out regularly by the Center.

The telephone information desk receives inquiries about traffic from the entire community. Generally, they handle an average of about 600 questions a day, but when there is snow, stormy weather, strikes, or some other abnormal condition, 2,000–2,500 inquiries are received.

6.3.2 THE OSAKA SYSTEM*

6.3.2.1 Case History and Basic System Design Principles. The city of Osaka has been suffering from the severest road traffic situation among Japanese cities during the last ten years, because of insufficient capacity in the road network (cf. Hasegawa 1972, 1975). Therefore, in 1973 the Police Department of the Osaka Prefecture began to install a traffic control system that is to cover almost all areas in the Osaka Prefecture, including the urban and intercity expressway systems. The Osaka Traffic Control System has been studied by several research committees, all of which are under the direction of Professor Eiji Kometani of Okayama University.**

This report deals with the realization of the concept of graceful degradation in road traffic control systems, which is a part of the achievements of the research committees. As is well-known, this type of system design concept does not have general principles that can be applied to general cases, and this report confines itself to the special case as above.

Road traffic control systems in ordinary surface streets and avenues with traffic signals have very distinguished features compared with ordinary process control systems. Road traffic control systems generally contain various objects and parameters that are so complex they are essentially difficult to handle with a computer system. For instance, the capacity of a road section is not only a function of the road itself but also a function of the degree of congestion, with statistical fluctuations in some cases. It should also be noted that the system is not going to control the cars in the controlled area but to control the human beings driving the cars. What is worse, it is almost impossible to have a unique set of objective functions in the control system and to have objective and effective measures for the evaluation of the traffic control system.

Although there are certain difficulties associated with road traffic control systems, these same difficulties may yield possibilities for simplifying the control systems. One favorable feature is that the sensitivity of the control often shows special behavior. In general, a sudden change of the signal control parameters causes more congestion when selecting signal parameters that are not suitable for optimizing the traffic pattern in the area. When the signal parameters are changed suddenly though moderately, congestion upstream from that signalized intersection grows very rapidly, and once congestion rises above a certain limit, it takes a very long time to recover the smooth traffic flow. This is because traffic congestion generated in this way has, in general, a rather low sensitivity to traffic signal control. This phenomenon has actually been recognized in measurement of real traffic flow in Osaka (Mishina and Sato 1970, Committee of Urban Traffic Control Systems 1970, 1971, 1972a), and is one of the foundations of the graceful degradation design of the road traffic control system. Thus, it is very important that the signal parameters be kept unchanged even when a system failure occurs. To achieve this, the correspondence between the hardware system failures and the control level degradation must be studied thoroughly. If it is achieved, traffic flow in the control area may remain undisturbed for some time.

When a system failure occurs and traffic in the area is not heavy, but the failure lasts for a fairly long time so that the traffic becomes heavy, traffic congestion will be generated in the area. If a failure occurs when the traffic is heavy and then lasts for a long time so that the traffic becomes light, the average queueing time for the vehicles at each intersection becomes long. Therefore, the mean time to repair (MTTR) should be kept as short as possible. It should

---

*Based on a case description specially prepared by Toshiharu Hasegawa, Department of Applied Mathematics and Physics, Kyoto University, Kyoto, Japan.

**The author would like to express his appreciation to the members of the various committees, especially to Professors Eiji Kometani and Tsuna Sasaki of Kyoto University, and Professor Toshio Fuijsawa of Osaka University, for their guidance. He would also like to thank the Traffic Division of the Police Deparment of Osaka Prefecture, the Hanshin Expressway Corporation (cf. Section 7.2), and the Tateishi Electronics Company, which is the prime contractor for the system described here and in Section 7.2.

be noted here that there is a kind of trade-off between MTTR and the mean time between failures (MTBF). From the systems point of view, some of the subsystems need not have very good MTBFs if they have a short MTTR.

It is not always possible to estimate the time required to repair a failure. When a long repair time is expected, it is necessary to select fixed time signal settings suitable for traffic patterns derived from an off-line analysis. It should be noted here again that this type of graceful degradation is only one of many graceful degradation systems, even in traffic control systems.

6.3.2.2 Implementation and Operation.

In the traffic control system in Osaka, the controlled road network is divided into segments, arteries and other streets. A segment consists of signalized intersections and links between them. In a segment, traffic is usually so heavy and complex that it is almost impossible to derive reliable information about the traffic flow by any feasible instrumented detecting method. Information from closed circuit television systems although important, is difficult to use quantitatively. The object of traffic control in a segment is to prevent traffic flows at intersections from blocking each other and to ensure that the traffic flow does not jam into deadlock. Segments are located in the central parts of the city.

Arteries are main streets that connect medium zones of the city; they are determined by the origin–destination distribution between the medium zones to channel the flow of traffic between them freely. Each artery has one segment at most and is rather fixed at any time except for an emergency, such as blockage of the road by an accident. In the case where the artery is blocked, alternative arteries, which are selected beforehand, take over the role of the troubled artery.

The other streets in the controlled area work as buffers to the segments and arteries, i.e., the difficulties in the segments and arteries are shifted to these streets. This introduces distributions of queue lines in the traffic flows so that the overall delay becomes less than that in the case of concentrated queue lines.

The classification of the above three parts of the network was made possible by the extensive studies on traffic flow in Osaka. Not only were ordinary traffic surveys carried out, but also traffic surveys using aerial photographs (Committee of Urban Traffic Control Systems 1972). Figure 6.15 shows the classification of the streets and intersections in the Osaka City area in the initial stage. Alternative arteries are determined beforehand to work as substitute arteries when their corresponding arteries have their capacity reduced for some reason.

The traffic control system in Osaka covers more than 1000 intersections with about 2000 detectors. In the final stage, it is expected to cover about 4000 intersections.

One of the most important features of the traffic control system in Osaka is the hierarchical hardware system closely connected with the control level degradation. The functions of the traffic control are classified into two groups according to their importance, frequency of operation, types of computer calculations, etc. The lower level computers detect the presence of vehicles, and compute traffic flow parameters such as volume, time occupancy, average speed and queue length, etc. The upper level computer carries out the other functions of the control such as controlling the lower level computers, making decisions on overall control strategies, etc.

Not only the hardware system but also the control level has hierarchy. In this system the hierarchy of the control level is called system degrade. According to the failures of the hardware subsystem, the level of control is degraded. Figure 6.16 shows the relation between hardware failures and the control levels. This is the foundation of the graceful degradation design in the traffic control system of Osaka. When the control level goes down because of a hardware failure, the degraded control level depends on the location of failure. For instance, a failure of the signal transmission system causes a degradation to level 6, as shown in Figure 6.16. If this degradation takes place when the system has been working at level 1, the effect on the corresponding traffic signals could be great. However, the number of traffic signals affected is rather limited and the total effects in the whole traffic control system may not be large. In contrast, a degradation from level 1 to level 2 does not affect the traffic signals much, although all of the traffic signals in the controlled area will shortly be affected to some extent, according to the change of the traffic pattern in the area. Thus, when the upper level computer goes down, lower level computers continue to control their corresponding traffic signals with the last signal parameters given by the upper level computer. To achieve this, the traffic signal parameters decided strategically should be converted to the real time settings at the lower level computers.

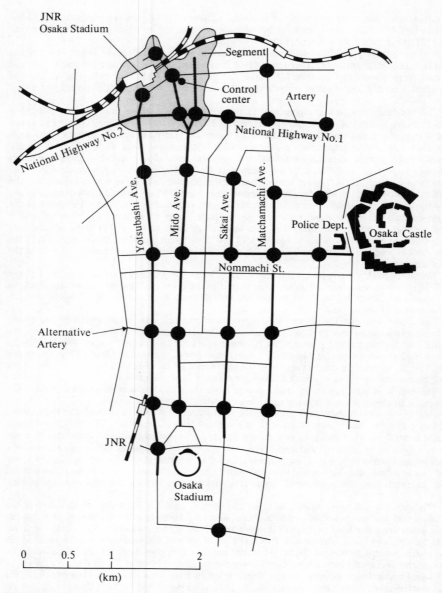

FIGURE 6.15 Controlled area in the initial stage in Osaka.

The hardware system of the traffic control system in Osaka is shown in Figure 6.17. The upper level computer is a FACOM 230-35 made by Fujitsu Ltd., with a core memory of 65 K word. The lower level computers are OMRAC-Ts made by Tateishi Electronics Co., and are special purpose computers for traffic control. The upper level computer has various peripherals but the lower level computers consist of CPUs, core memories of 65 K word each and com-

Level 1. Segment control, detour guidance control, display of traffic parameters, man-machine interface control, off-line jobs, etc.
Level 2. Actuated artery synchronization control, jam control, actuated one point control, etc.
Level 3. Actuated artery synchronization control, simple artery synchronization control, actuated one point control, etc.
Level 4. Same as Level 3.
Level 5. Linkage control of adjacent signals, actuated one point control, fixed cycle control, etc.
Level 6. Actuated one point control, fixed cycle control, etc. (2nd order fail control).
Level 7. Individually fixed cycle control (3rd order fail control).
Level 8. No traffic signal light.

1,upper CPU or channel controller failure; 2,lower CPU or data scanner failure; 3,first order fail controller failure; 4,local master controller failure; 5,linkage cable failure; 6,terminal signal controller failure; 7,failures of each signal transmission system.

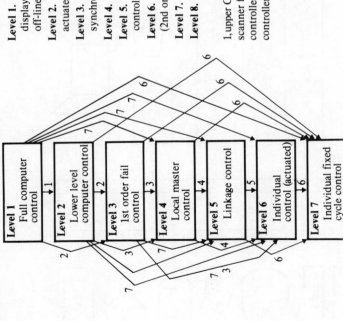

FIGURE 6.16 Hardware failures and control levels.

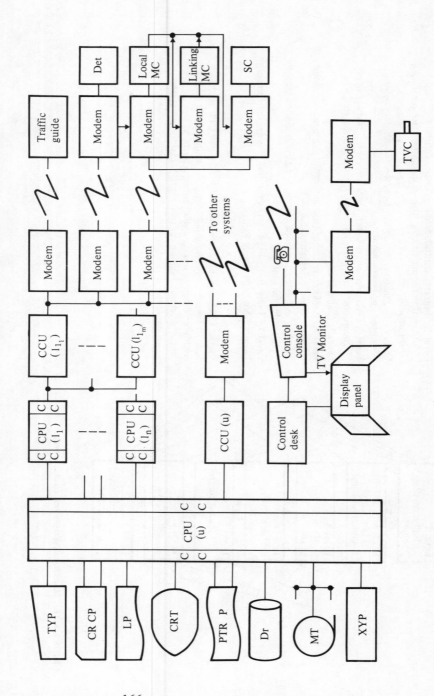

FIGURE 6.17 Hardware system of the traffic control system in Osaka: CC, channel controller; (u), upper level; (l), lower level; Det, detector; MC, master controller; SC, signal controller; TVC, TV camera.

puter channels. The latter (including their initialization) are controlled by the upper level computer. As a matter of course, they may be initialized by a paper tape reader when the upper level computer is down. As the upper and lower level computers are connected through their channels, the capacity of this computer complex is fairly large because of the absence of a communication neck between the upper and lower level computers.

The first order fail control shown in level 3 of Figure 6.15 has two categories. In conventional traffic control systems, this first order fail control has only one mode, i.e., when the computer system is down, all of the controlled traffic signals are controlled with the prefixed signal parameters. This means that a sudden change of parameters is to be expected. In conventional systems in Japan, the extent of the parameter change may exceed 10 percent. This size of parameter jump is large enough to cause disasters in the city of Osaka, judging from our surveys (Committee of Urban Traffic Control Systems 1972b).

Therefore, in order to have graceful degradation between level 2 and level 3, the communication control unit (CCU) of the lower level computer is designed to backup its computer. The CCUs of the lower level computers store in their core memory the information for signal cycle, split, and offset of each traffic signal in their corresponding control area. The information for the signal parameters is supplied from the upper level computer to the lower level computers every five minues and is converted to real-time settings. The real-time settings for the signals are stored in the core memory of the CCU. In case of failure of a lower level computer system, each traffic signal in its controlled area is controlled by the stored data in its CCU. Because of this backup system, the signal parameters are not changed and remain optimal for the time being. Of course, if the lower computer remains down for a fairly long while, the signal parameters do not remain optimal. Therefore, when the required time to effect repairs is expected to be long, such as in the case of a scheduled down for the computer system, this type of first order fail control is not adopted. Instead, the second category of the first order fail control is adopted. In this case, each traffic signal is controlled by one of the preset fixed time signal settings according to the prevailing traffic pattern. These preset fixed time signal settings are prepared for several traffic patterns to have simple artery synchronization control. The selection of the signal setting should be done carefully. This may, however, be possible since the selection may be done after fairly long observations. When an unexpected system failure takes place, the control of the first category starts automatically.

At control level 1, all of the systems are operating normally and the highest control strategies are being carried out. When the upper level computer system or the channel between upper and lower level computers is down, the control level degrades to level 2. At this level, vehicle-actuated artery synchronization signal control, traffic jam control, actuated single intersection control and others are carried out by the lower level computers. For the time being, there will be no conflict between the controlled areas of the lower level computers. There is, however, expected to be mismatching at the boundaries of controlled areas if the failure of the upper level computer lasts for a long time. At any rate, degradation from level 1 to level 2 is gradual.

At level 3, a lower level computer is down and the first order fail control is carried out. As mentioned above, when the expected time required to repair the failure is not known, the CCU of the lower level computer controls the traffic signals in its area with the data stored in the core memory of the CCU to keep the control parameters as close as possible to those already in use. When the time required to repair the failure is expected to be long, one of the preset parameter sets for fixed time signal settings are used to control the traffic lights.

At level 4, a control level similar to level 3 is followed, not with the present data but with the predetermined parameters stored in the local master controller. At levels 3 and 4, actuated artery synchronization control, simple artery synchronization control, and actuated one-point control are performed. The main difference between level 3 and level 4 is the mode of degradation. Degradation to level 3 can be performed gracefully but to level 4 may not be so graceful. At the same time, the storage locations of parameters are different. In level 4, the local master controllers, which were used before the installation of the present control system, are utilized as the backup systems. Therefore, this level of control is performed, for instance, when the whole system of the control center is down, and the control grade becomes the same as the control grade before the installation of the present control system. From levels 4 to 6, the control methods are the same as before.

This traffic control system has been operated since October of 1972, and there has been only one case when the signals were controlled by the CCU of the lower level computer. It is reported that the expected graceful degradation was achieved for that case. The upper level computer, FACOM 230-35, is a medium-sized business computer and has failed several times, although the failures have not affected the traffic control significantly. Figure 6.18 shows the display panel and the control consol in the control center.

6.3.2.3 Evaluation of Operational Experiences. In road traffic control systems, it is sometimes possible to have a system with a rather high cost-performance relation by shortening the MTTR, even with subsystems that do not have such excellent MTBFs. For instance, in the Osaka system, the upper level computer system does not have to be a multiple computer system with respect to the MTBF, because of the graceful degradation achieved when it fails. In this system, a hardware monitor panel is installed in the system, which tells the attendants of the control machine room the location of the hardware failure, in order to lessen the MTTR. The MTTR of a vehicle detector is of the order of hours if the failure is of a detectable kind, and the MTBF of a detector is longer than 30,000 hours. The MTBF of the controlling computer system is not known because of the frequent upgrading of the system according to the extension plan. The MTTR of the computer system is short: it is reported that repair times longer than a day, including a repair of a failure in the information transmission system, are seldom.

Various problems still remain concerning the system failures. On the night of April 12 (Saturday), 1975, the signals of 27 intersections in the central business district (CBD) were all fixed at red for about an hour because of an unhappy coincidence of undetectable errors in the hardware. This period could have been less than three minutes if a proper procedure, as suggested by Figure 6.16 had been taken by the attendants of the control center. The more reliable and failure free the system becomes, the more important the training of the personnel at the control center becomes. It may be desirable to have accident simulators for training against failures in the future, as a kind of man–machine interface.

Detecting subsystem failure is another important problem. In the Osaka system, monitoring the subsystems is performed with software and hardware as often as possible. As a matter of course, it is generally impossible to have a complete failure detection system.

The effect of the installation of a traffic control system in the fiscal year 1972 (Police Department of the Osaka Prefecture 1974) for six controlled routes, has been to make average improvements in journey time, number of stops, delays at intersections, and sectional speed of 17.5 percent, 29.2 percent, 27.7 percent and 24.2 percent respectively. For the fiscal year 1973 (Police Department of the Osaka Prefecture 1975) for 12 controlled routes, these improvements were 17.6 percent, 21.3 percent, 26.4 percent and 23.1 percent. In the fiscal year 1973, the average queue lengths of 10 important intersections had decreased by 38.7 percent.

Evaluating a traffic control system is very difficult. For instance, we sometimes measure average journey time, average delay and stoppage by traffic signals, average speed at some section, and the number of accidents before and after the installation of the traffic control system. It is almost impossible to define the exact correlation or dependence between these factors of the traffic flow and the installation of the traffic control system. It should also be noted that, even if a traffic control system produces favorable improvements in these factors, it is not necessarily a good one, since, if the former traffic control system was very poor, any new traffic control system should produce fairly large improvements. The number of accidents is highly dependent on the extent of congestion, on road safety equipment and on police regulation.

We have come to the conclusion after almost ten years of experience that the evaluation of a road traffic control system should depend not only on the control strategies or control systems but also on how reliable the system is for the people in that area. At present, the traffic control systems in the Osaka area are the most reliable information systems in that area. People there seem to take them as indispensable social services.

REFERENCES

Committee of Urban Traffic Control Systems. 1970, 1971, 1972a. Survey report of area traffic control in Osaka, Parts 1, 2, and 3. Institute of Urban Systems Research, Osaka.

FIGURE 6.18 Osaka traffic control center.

Committee of Urban Traffic Control Systems. 1972b. Report of aerial photograph survey. Institute of Urban Systems Research, Osaka.

Hasegawa T. 1972. The traffic control system in Osaka. First USA-Japan Computer Conference Proceedings. Tokyo: AFIP and IPSJ, pp. 494-495.

Hasegawa T. 1975. Graceful degradation in road traffic control systems. Second USA-Japan Computer Conference Proceedings. Tokyo: AFIP and IPSJ, pp. 64-68.

Mishina T. and T. Sato. 1970. Traffic planning and control for EXPO 70. Traffic Engineering, November: 12-19.

Police Department of the Osaka Prefecture. 1974. Report on the operational effects of the traffic control system of the Police Department of the Osaka Prefecture for the fiscal year 1972.

Police Department of the Osaka Prefecture. 1975. Report on the operational effects of the traffic control system of the Police Department of the Osaka Prefecture for the fiscal year 1973.

6.4 USSR EXPERIENCES: MOSCOW AND ALMA-ATA*

The fast growth of industry and agriculture in the USSR required more efficient transportation, especially road transport, which amounts to 80 percent of the general transport volume and to about 50 percent of the passenger traffic. This included extending the street network (from 177,300 km in 1950 to 645,000 km in 1975) and increasing production (from 270,000 units in 1970 to 1.2 million in 1975) and registration of vehicles. During the next few years further increases are expected. Consequently the necessity to develop automated road traffic control arose first in large cities and then on freeways. Projects were undertaken by several institutions and offices to achieve this aim. This contribution describes the following computerized control systems:

- ISUMRUD, a system for automated control of arterial streets
- START, a system for centralized control of all street traffic in Moscow
- GOROD, a system for automated street traffic control in large or medium sized cities

6.4.1 THE ISUMRUD AND START SYSTEMS OF MOSCOW

6.4.1.1 Case History and Basic System Features. Moscow covers an area of about 875 km^2 within the boundaries of a 110 km freeway circle surrounding the urban district. It has more than eight million inhabitants. The daily traffic density is swelled by a large number of tourists, transit passengers and commuters. Traffic demand is served mainly by subways, but also by buses, trams, cabs, and private cars. The street network of Moscow has a structure of diametral and circular thoroughfares. On a general length (main roads) of about 3,500 km it contains more than 770 signalized intersections, 160 of them being concentrated in the inner city within the boundaries of the so-called Garden Circle, which covers only 3.3 percent of the total area.

In order to raise the efficiency of road traffic control, in 1965 the office Mosgortransproject started work on a computerized traffic control system. This system became operational in 1967 under the name of ISUMRUD.

The intersection chosen for practical implementation was the square at Serpukhov gate, a complicated intersection including seven road junctions and split into six single intersections (Figure 6.19). The system configuration shown in Figure 6.20 was developed to meet the following basic requirements:

- centralized traffic signal control by means of a control computer
- investigation of control algorithms
- operative traffic control by means of dispatching techniques

---

*Based on a case description specially prepared by M. P. Pechersky and B. G. Khorovich, Mosgortransproject, Moscow, USSR.

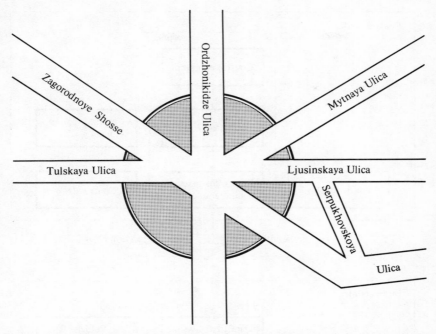

FIGURE 6.19   Intersection Serpukhov gate, at which the ISUMRUD system was installed.

The system was designed to be centralized, i.e., each change of a signal phase can be only realized with an instruction from the control center. The impulses generated by vehicle detectors are also transmitted to the control center. During a breakdown of the central control computer or in accordance with a dispatcher's decision, a central interval (coordinated) control program is activated. In the case of failing signal transmission the local control program stored in the local controller is started automatically.

The control computer calculates the signal plans for all intersections (for this purpose, the changing points for each phase of every intersection have to be calculated and given to the transmission units), examines the local controllers and works through a check procedure during each cycle. The data or signal transmission is carried out using time multiplexing on telephone lines, at a frequency of 100 Hertz.

The operation of the ISUMRUD system provided the experience necessary for developing the more sophisticated and larger systems START and GOROD.

The START system was designed to control road traffic in the Moscow area within the borders of the Moscow freeway circle. It was planned to solve the following main tasks:

- automated control of traffic lights and changeable traffic signs
- centralized traffic-actuated control
- gathering and processing of statistical data

6.4.1.2   Structure of the Area Traffic Control System START. Compared with the ISUMRUD system (Figures 6.19 and 6.20), START is characterized by the following differences (cf. Figure 6.21):

- use of a multiprocessor system instead of a single computer
- direct connection between the control desks and the computing system
- the local controllers do not contain blocks for modifying information

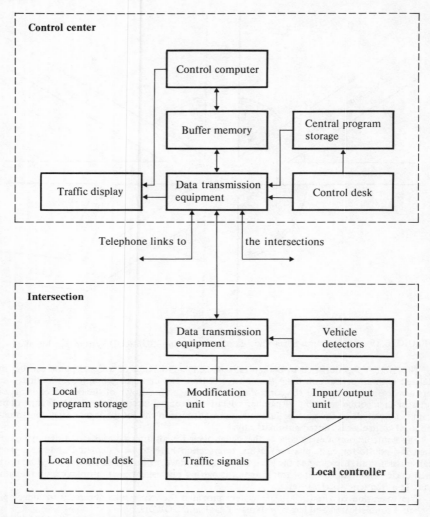

FIGURE 6.20 Block diagram of the ISUMRUD system of Moscow.

The START system can be divided into two subsystems (cf. Figure 6.21):

- the automated traffic control system
- the dispatching control system

The first subsystem contains a central control computer (M-6000) and several lower level computers of the same type. These lower level computers are used for primary data processing tasks and the formation of output instructions. The central and peripheral computers operate in redundancy.

Telephone channels with a transmission rate of 100 Baud are used for the information transfer by time multiplexing to the various peripheral devices such as local controllers, change-

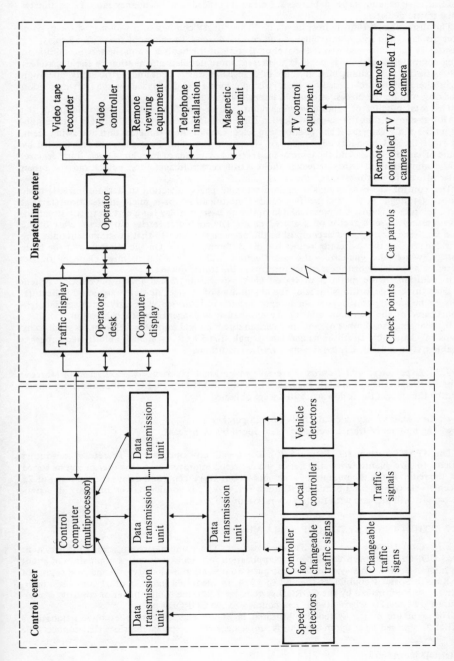

FIGURE 6.21 Block diagram of the START system of Moscow.

able traffic signs, and traffic detectors for measuring speeds and occupancy rates, i.e., estimates of the traffic densities.

The dispatching control subsystem (cf. Figure 6.21) consists of a wall-map display, computer displays, and remote surveillance and control devices. By means of the traffic display, the operator is kept informed about the technical state of the hardware and the traffic conditions. Using computer displays it is possible to present more detailed information to the dispatcher, such as traffic flow parameters, the state of changeable traffic signs and traffic lights, operation of peripheral units, etc. Closed circuit television cameras and displays allow the traffic situation to be observed on the busiest streets and intersections.

The total number of remote viewing points during the first stage of introduction of the START system within the so-called Garden Circle amounted to 24, each of them being equipped with 1–5 TV cameras. The operators can select each of these cameras, turn and swing them, and store the TV broadcast by means of videorecorders. The TV signals are transmitted by means of coaxial cables and the control instructions by means of telephone lines. Furthermore, radio communication exists between the control center (dispatching point) and the urban traffic office, the patrol cars, and the check points.

The control algorithms consist of two essential parts: selection of a (coordinated) basic strategy (strategic level), and traffic-actuated modification for a single intersection (tactical level). In order to choose the precalculated optimum basic strategy for a certain traffic situation, it is necessary to set a vector of measured values. The computer carries out the selection of the optimum basic strategy and accomplishes the necessary change to that program containing the cycle length and the switching points for the different phases. On the tactical level the basic strategy, which also guarantees the coordination, will be adjusted to meet changing traffic. Therefore a special algorithm searches for gaps in the traffic queues.

In addition to the methods of traffic light control mentioned above, the START system contains further alogorithms utilizing the possibilities of changeable traffic signs and indicators. Thus an impending traffic jam can be recognized and indicated very early using special algorithms based on the maximum density in areas that show danger of traffic congestion.

By changing the number of lanes in a certain direction and imposing different restrictions by means of changeable traffic signs and lane signals, the START system is able to accomplish a variable traffic control adapted to actual traffic conditions.

6.4.1.3 Experiences and Results. Extensive operational experience was obtained in Moscow with the ISUMRUD system (cf. Figures 6.19 and 6.20). From studies before and after implementing the system the following results were obtained:

- the average delays were decreased by 25 percent
- the number of traffic accidents was reduced by 10 percent

The START system (cf. Figure 6.20) is not yet fully operational. Therefore, it is rather difficult to give quantitative estimates of the expected improvements. However, it can be stated that the reduction in the number of accidents (about 10 percent) and delays (of the order of 25 percent) will save about 1.5 million Roubles per year. It is considered certain that this economic benefit will increase with further operation of the system.

6.4.2 THE GOROD SYSTEM OF ALMA-ATA

6.4.2.1 Case History and Basic System Features. The computerized traffic control system called GOROD* has been developed for application in urban cities with a population of about one million or less. This system was first put into operation in the city of Alma-Ata, capital of the Kazakh Soviet Republic, at the end of 1974. In the initial stage, 38 of 61 controlled intersections were controlled by the GOROD system. By 1980, the total number of controlled intersections grew to 120, where 60–80 were coupled to the GOROD system.

The structure of the system is illustrated in Figure 6.22. A control computer (initially a NAIRI 2, replaced later by an M6000) is supplemented by a special interface that connects the

---

*GOROD means city.

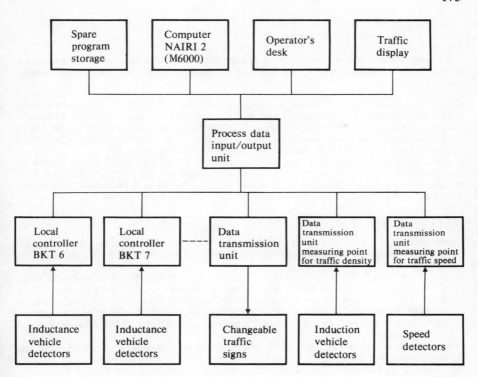

FIGURE 6.22 Block diagram of the GOROD system of Alma Ata.

intersection equipment with the control center, which contains a wall-map display and an operator desk. Every intersection is equipped with inductive vehicle detectors, local controllers, and changeable traffic signs.

Information given by the vehicle detectors (on average five per intersection) proceeds to a transmission unit, which is integrated in a controller, and through two-core telephone cables to the control center. The signal transfer is carried out by means of a time-multiplexing method. These signals are received and passed by means of a process data input and output unit to the computer, which processes the data to prepare instruction output using the same transfer unit to traffic lights. In this way the local controller BKT-6 is only a translator for these instructions. In the case of a mistake or interruption, the local controller automatically activates an internal fixed-time control strategy. The control center also gets information from the local controllers, namely detector signals and information about the implementation of the computer instructions and the technical state of the control equipment. The local controller BKT-7, moreover, allows the realization of traffic-responsive control by applying a gap-searching algorithm with back transfer of the detector signals.

This system of automated control of traffic signals is planned to accomplish the following tasks:

- analyze traffic density at characteristic crossing points of main roads
- select an especially suitable coordinated control strategy, either time or traffic dependent
- improve the selected strategy as a function of the prevailing traffic conditions at every intersection

- control traffic signs
- enable automatic system supervision

In order to raise the reliability of the system in the case of a breakdown, an additional program memory and controller are triggered off, making fixed-time coordinated control possible.

6.4.2.2 Experiences and Further Projects. During the years 1975-1977 the effectiveness of the system was checked before and after implementation using the floating-car method. The results showed a reduction of the waiting time in front of the traffic lights to about 70 percent and an increase in the average travel speed of at least 16 percent. The annual economic benefit from these two factors alone was calculated to be at least 800,000 Roubles.

In addition to the arterial and area traffic control systems ISUMRUD, START, and GOROD sketched briefly here, a freeway traffic control and surveillance system called ARDAM is being developed in the USSR. The first steps to implement ARDAM on a section of the Moscow-Riga freeway 42 km long have been taken (cf. Bibliography for more details).

BIBLIOGRAPHY

Frimstein, M. I. 1975. Structure and analysis of the ARDAM system. Giprodor NII, Automated Highway Traffic Control System ARDAM, Report No. 15, Moscow 1975, pp. 16-22 (in Russian).

Kaplun, G. F., M. P. Pechersky, and B. G. Khorovich. 1969. ISUMRUD — automated street traffic control system. Gorodskoye Khosyaistvo Moskvy, No. 8: 39-43 (in Russian).

Pavlenko, G. P. 1976. Design and structure of the arterial part of the traffic control system GOROD. Proceeedings of the First Scientific-Technical Conference of the Member Countries of the Council of Mutual Economic Assistance on traffic safety problems, Moscow, p. 667 (in Russian).

Pechersky, M. P. and M. Gurevich. 1969. Area traffic control in Moscow, Traffic Engineering and Control, No. 5: 560-562.

Pechersky, M. P. and B. G. Khorovich. 1976. Area traffic control system START in Moscow. Proceedings of the First Scientific-Technical Conference of the Member Countries of the Council of Mutual Economic Assistance on traffic safety problems, Moscow, p. 698 (in Russian).

Saidenberg, Ya. I., F. G. Usmanov, and E. B. Khilazhev. 1973. An automated traffic computer control system in a large city. In Automated Control System and Hardware Experiences, Collection of Papers. Zapadno — Sibirskoye Knizhnoye Isdatelstvo. Omskoye Otdelenye, pp. 17-20 (in Russian).

Shinkarev, N. I. 1976. Scientific Technical Progress in Transportation. Moscow: Shanie (in Russian).

Vasilyev, A. P. 1975. Basic tasks and stages of automated arterial street traffic control. Giprodor NII, Automated Highway Traffic Control System ARDAM, Report No. 15, Moscow, pp. 5-15 (in Russian).

Vasilyev, A. P. and M. N. Frimstein. 1976. Development of an automated highway traffic control system. Proceedings of the First Scientific-Technical Conference of the Member Countries of the Council of Mutual Economic Assistance on traffic safety problems, Moscow (in Russian).

6.5 EXPERIENCES OF A DEVELOPING COUNTRY: NAIROBI, KENYA*

6.5.1. CASE HISTORY

Nairobi is the capital of Kenya and has a population of more than 650,000 people.

The central business district (CBD), with an area of only 2.5 km^2 employs about 55,000 people. Owing to the concentration of employment within the CBD, it is estimated that about

---

*Based on a case description specially prepared by L. W. Situma, Traffic Engineer, Nairobi City Council, Nairobi, Kenya.

180,000 trips are generated by the area. 80 percent of these trips occur during the peak hours between 6.00 am and 6.00 pm. Terminal facilities for all these trips are about 6,600 on street parking (2,700 metered) and 8,900 off-street parking bays (4000 within private developments).

In 1969 the City Council of Nairobi, in conjunction with the Kenya Government and the World Bank, decided that traffic congestion within the CBD had reached a critical point for the economic development of Nairobi. This awareness led to the establishment of the Nairobi Urban Study Group (NUSG). The terms of reference for the NUSG were to propose suitable guidelines for short- and long-term development plans for Nairobi. While the NUSG was being set up it was further agreed that, as a matter of urgency, installing traffic signals at critical junctions within the CBD could temporarily alleviate the situation without disrupting the final recommendations of the NUSG.

Uhuru Highway and Kenyatta Avenue — two major streets in Nairobi — were chosen as first priority streets for signalization (cf. Figure 6.23). Kenyatta Avenue is the main artery of the CBD. The Avenue qualifies for signalization because of its closely spaced intersections, multiple turning movements, and the additional complexity of service roads on either side. The Haile Selassie Avenue/Uhuru Highway junction is the southern gateway to and from the industrial area, airport, southern suburbs and other southern parts of the Republic (cf. Figure 6.23). In 1974, peak hour traffic volumes entering Nairobi, at University Way junction, Kenyatta Avenue junction, and Haile Selassie Avenue junction, were, respectively, 4,900, 5,400, and 6,000 vehicles per hour.

From the Central Area Traffic Management Study, it was established that the cause of traffic congestion at most junctions within the CBD was a complete lack of intersection capacity to cope with the current traffic volumes through the intersections.

In 1972, the capacity of the major weaving intersections — the roundabouts at Kenyatta Avenue and Haile Selassie Avenue — was already exceeded and average approach delays of 4–5 minutes per vehicle during peak hours were very common. Figure 6.24 illustrates this fact for a particular intersection. On Kenyatta Avenue itself, capacity was severely diminished by the lack of responsive traffic management. There were as many as 28 different traffic movements at each major intersection, created mainly by allowing interaction between the service roads and the center carriageway. The accident rate was also high and pedestrian movements at peak hours were very frightening at the crossroads on Kenyatta Avenue.

Two approaches were taken into consideration in dealing with the problems sketched here:

- redesign the intersections, especially the roundabouts of the Uhuru Highway and the crossroads on Kenyatta Avenue
- install a computerized traffic control system

Although there are several types of control systems in use throughout the world, it was decided from the beginning that a coordinated movement signal system should be installed in Nairobi. There were several constraints in both civil design and signal systems alternatives that were considered by the Council. Amongst the constraints the following were notable:

1. For design purposes, the data available were not adequate, and hence subjective assumptions became prevalent in the calculations.
2. The high capital cost of the equipment and construction, the Council's poor financial state of affairs, and inflation dictated the need for a system with the highest benefit/cost ratio.
3. The CBD of Nairobi is a floral area and any scheme designed to eliminate the trees and flowers is strongly resisted by the residents. The Nairobi City Council and the Kenya Government were opposed to the removal of the roundabouts, which are very beautifully planted and are an established feature contributing to the overall landscaping of Nairobi.
4. As Kenyatta Avenue is in the heart of the CBD, each alternative was constrained by the need to maintain adequate parking facilities and environmental aesthetics. Fortunately, all the parking facilities on Kenyatta Avenue are in the service roads, and thus the capacity of the main carriage-way is free from the effects of parking.
5. The need to maintain the aesthetics at any cost and the lack of adequate road reserves limited the widening of approaches to the junctions.

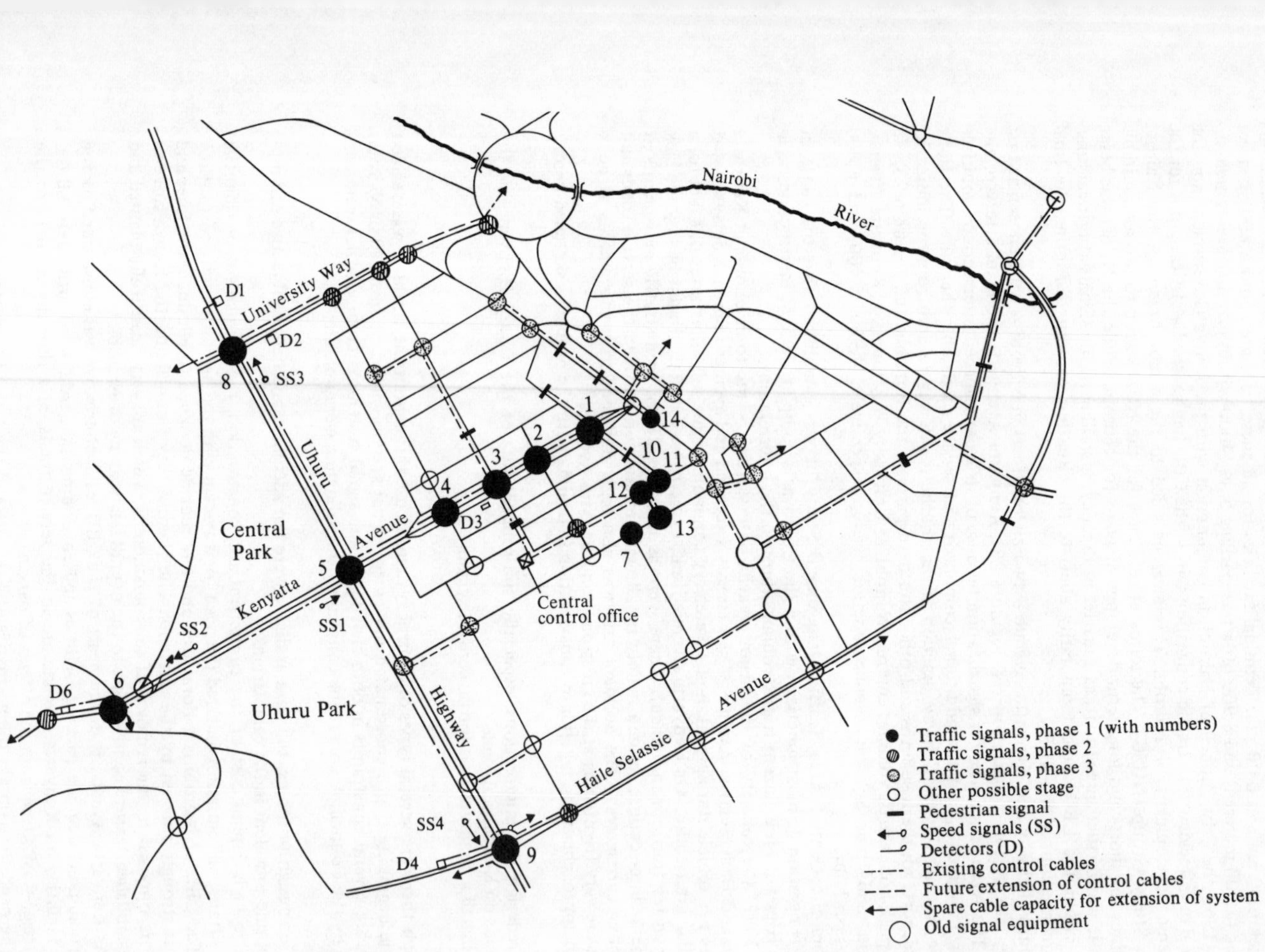

FIGURE 6.23 The traffic signal system in Nairobi.

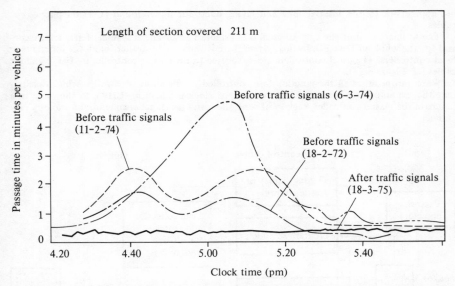

FIGURE 6.24 Approach delays to roundabout on Kenyatta Avenue east of Uhuru Highway of Nairobi (cf. Figure 6.23) before and after the installation of the traffic control system.

Considering these constraints, the following objectives were accepted in planning the traffic control sytem and redesigning the intersections:

- to increase junction capacities by increasing the capacities of individual approaches to each intersection
- to reduce delay times and hence traffic congestion at each junction
- to reduce the rate of accidents at all junctions and enhance traffic safety

In view of the various constraints, the basic concept of the selected alternative is a combination of signalized roundabouts, one-way approaches on Kenyatta Avenue, and a flexible progressive signal system. Owing to the identified constraints, this was considered to be the only option that would accommodate the traffic congestion on the two streets.

6.5.2 IMPLEMENTATION AND OPERATION

All the management of the research, development, and construction work needed for the implementation of the scheme was carried out by the Traffic Section of the Nairobi City Council.

After a considerable analysis of the collected data on traffic delays and congestion, it was decided to signalize a total of 14 junctions. Ten of these junctions on Kenyatta Avenue, Kimathi Street and Simba Street are in such close proximity to each other that it was possible to achieve a high degree of intergreen movement.

When the entire scheme was finally approved and the necessary finance obtained, an international tender was prepared for the purchase of the signals equipment. The tender for signals equipment was awarded on the basis of the following contraints: (1) the capital cost of the equipment must be low and there must be proof that such equipment is in use in other parts of the world; (2) the equipment supplied was to be readily adaptable for inclusion into a system of central computerized area traffic control. M/S Siemens of the Federal Republic of Germany was awarded the contract to supply and install the signals equipment.

The civil engineering work was undertaken by both a local contractor, and direct labour under the Works Superintendent in the City Engineer's Department. The work involved widening approaches to the intersections and laying ducts for underground (UG) telephone-type cables.

The ultimate goal of the City Council is to establish an advanced system of area traffic control for the whole of the CBD. In this system there is a master controller for all the junctions or local controllers. The local controllers are connected to the master controller by UG telephone cables (cf. Figure 6.25).

When in operation all the junctions are controlled by the master controller, which operates on either an automatic control (ELSA) or a taped clock mechanism (UHR). In the event of a failure in the master controller, each local controller can operate independently on standby programs.

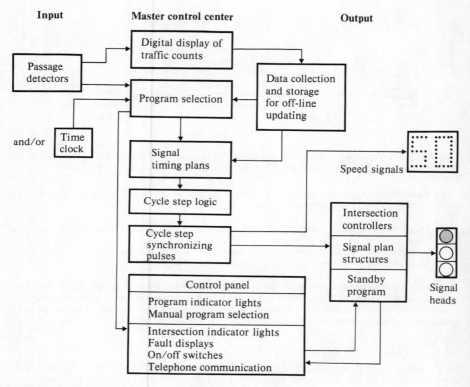

FIGURE 6.25 Block diagram of the Nairobi control system.

A KSV-02/12 electronic master contoller is used that is capable of controlling 20 intersections via six signal plans. In our case only 13 junctions (numbered 1–13 on Figure 6.23) have been connected to the master controller. All controllers were installed on City Hall Way, Simba Street, and Kimathi Street around the Hilton Hotel. Detector loops and speed indicators were installed at points shown in Figure 6.23.

As a result of the experience and knowledge gained elsewhere, and the traffic volume data collected, a series of pre-timed signal programs were given trial runs. Based on the established travel pattern it was decided that only four signal programs would cater for the identified traffic fluctuations.

6.5.3 OPERATIONAL EXPERIENCES

Before the installation of traffic signals, there were three police officers on duty on Kenyatta Avenue during peak hours. The installation of traffic signals reduced the role of man from controller of traffic to supervisor of the system. The police are now there to enforce the traffic act.

Before the installation of traffic signals, one could experience a delay of up to 6 minutes at the junctions on Uhuru Highway. After the installation of traffic signals, delay times were down to 1.5 minutes or less (cf. Figure 6.24).

There were at least two accidents per week on Kenyatta Avenue and Uhuru Highway before the installation of the traffic control system. This accident rate could be reduced to about one accident a month. It is generally agreed that the signals have reduced the number of accidents considerably.

As already mentioned, the Nairobi City Council and the Government are very keen to preserve the aesthetics of the CBD. The system installed has preserved all the flowers and trees along Kenyatta Avenue and Uhuru Highway. In fact, when the signals are in operation a Christmas tree effect is produced on Kenyatta Avenue and Uhuru Highway.

For convenience and safety, the central control office is lcoated in the City Hall. Should there be any failures or destruction in the control center the local controllers will continue to operate independently. In this case there would be no green wave, and the operating programs at each junction would have to be changed as before.

The system has been in operation since 1975, and in general has been problem free and very reliable.

In the initial stages, the general public were very sceptical about the success of the signals. After one year's operation the signals proved so successful that there is now public demand for an immediate extension of our signalization program.

The capital cost for the equipment was US$263,375 (K Shs 2,106,980), and the civil engineering works cost a total of US$222,780 (K Shs 1,880,000).

As this was a pioneering scheme, it has not been possible to quantify the operative and administrative costs. At the time of writing, the maintenance costs of the equipment are about US$8,190 (K Shs 60,000 paid to M/S Siemens (K) Ltd), for a maintenance contract. Figures for accident costs were hypothetical figures obtained from the literature and were not based on local values.

The difficulties and complexities of converting the potential benefits into monetary values, and the lack of any experience in the developing countries, have made it impossible to calculate a realistic benefit/cost ratio for this scheme. However, one may assume that under the existing conditions the benefit/cost ratio of the scheme was above 1.00.

7 International Freeway and Road Traffic Control Systems Experiences

In this chapter the experience gained with freeway traffic control systems is analyzed by means of specially prepared case descriptions dealing with advanced projects in the USA, Japan, and France.

The first case description deals with the Dallas Freeway Traffic Corridor Control Project, and presents experience gathered in the USA with integrated traffic control systems that enable coordinated control of both

- area traffic in an urban street network
- freeway traffic

The second contribution, on the Hanshin Expressway, describes one of the most advanced freeway traffic control systems of Japan, which is already partly integrated with the Osaka area traffic control system discussed in paragraph 6.3.2.

The final two case descriptions present the results obtained in France with the following projects:

- the Paris freeway corridor control projects
- the project Operation Atlantique

The last control project does not deal with urban freeways. It has been included in this chapter, however, since the control principles used and the results obtained are also considered interesting for freeway traffic control problems in an urban region.

7.1 THE DALLAS FREEWAY TRAFFIC CORRIDOR CONTROL PROJECT*

7.1.1 CASE HISTORY**

7.1.1.1 Background. In 1967 the Federal Highway Administration (then the Bureau of Public Roads) began the development of a research study to investigate the design, implementation, and operational requirements of a traffic surveillance and control system for an urban freeway corridor. The study was designed to demonstrate the effectiveness through the installation and operational experience of the local transportation agencies. Thus, the Dallas North Central Expressway Corridor Surveillance and Control Project was formally begun in July 1968 with an administrative research contract between the Texas A & M Research Foundation and the Department of Transportation's Federal Highway Administration (1968). Assurances of cooperation were received from local operating agencies although no legally binding agreement or contract was executed to specify details as to the character of such involvement. The Texas Transportation Institute (TTI) is the research agency for the Texas A & M Research Foundation.

The North Central Expressway Corridor was selected for several reasons, the most important being the willingness of the City of Dallas (city) and the Texas State Department of Highways and Public Transportation (then the Texas Highway Department) to accept the responsibility for the hardware installations and the operations and maintenance requirements of the systems. This specific corridor is one of the most heavily traveled areas in the State, and the major artery, the North Central Expressway, is an old freeway with old design standards (Figure 7.1). Traffic congestion is increasing because of the increasing traffic demand caused by major land development activities in the north sector of the city and Dallas County. The possibility of major construction of new transportation facilities in this sector is very low because of higher priorities to complete the Interstate System and other major arterials in other areas of the city and state.

The alternatives facing the local agencies for improving mobility in this important corridor of the metropolitan area were to improve the capacity of the roadway system, to provide alternative modes of travel, and to improve the operational efficiency of the existing roadway system. All three alternatives are being implemented at various levels of effort.

Capacity is being added in short sections of the arterial street system to improve bottleneck sections and to improve approaches to critical intersections. These are long range improvements and generally respond to increased traffic loads caused by land development in the vicinity. Although these improvements contribute to the total travel in an urban corridor, their impact on peak period corridor movement of through traffic is slight, since the improvement does not span the length or breadth of the corridor.

Improvements to the transit service are being made in several areas of the corridor. A park-and-ride facility has been installed adjacent to the freeway. A bus ramp for preferential entry to

---

*Based on a case description specially prepared by W. R. McCasland, Research Engineer, Texas A & M University, Texas Transportation Institute, Freeway Surveillance and Control Department, Houston, Texas, USA. The research reported in this paper was developed under administrative contracts (DOT-FH-11-6931 and DOT-FH-11-7825) with the Department of Transportation's Federal Highway Administration in cooperation with the City of Dallas and Texas State Department of Highways and Public Transportation. Appreciation is extended to those agencies for their support in the research endeavor, to the research staff of the Dallas Corridor Project, and, in particular, James D. Carvell — Project Manager — whose work is the basis of this paper.

***Disclaimer*: The technical contents of this paper have appeared previously in the reports in the reference list by the Texas Transportation Institute for the Federal Highway Administration (RF-590-1, RF-590-2, RF-590-final, RF-836-1, RF-953-2, RF-953-3, RF-953-4, RF-953-6, RF-953-7, RF-953-8). The contents of this paper reflect the views of the author who is responsible for the facts and accuracy of the data presented herein. The contents do not necessarily reflect the official view or policies of the Federal Highway Administration. This paper does not constitute a standard, specification, or regulation.

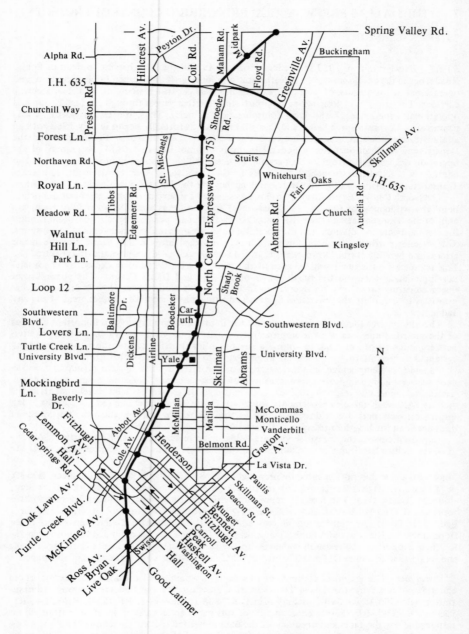

FIGURE 7.1 North Central Expressway Corridor, Dallas, Texas: ● interchange; ■, study office.

the freeway has been constructed. Preemptive control of traffic signals along major bus routes was installed in 1977. An express bus service has been added to certain routes serving the outlying areas of the city. Even with the success of these measures, which are designed to encourage drivers to leave their automobiles and ride buses, the vehicular demand in the freeway corridor continues to increase.

Improvements in the operational efficiency of urban roadways are being made through the application of traffic surveillance, control, and motorist information systems. These improvements, which also benefit the transit systems, are the subject of this report.

7.1.1.2 Summary of System Design. Along the ten-mile section of the North Central Expressway, thirty-four entrance ramps are instrumented for metering control during peak traffic periods, and fifteen frontage road (diamond interchange) intersections and sixty-eight surface arterial intersections in the freeway corridor are being instrumented for computer control. Three rotating drum type variable message signs have been installed along one arterial to advise motorists of traffic conditions on the freeway and to direct drivers to an alternate route. A telephone call system has been installed in the control center to improve traffic information for pre-trip planning. Three portable matrix signs will be employed along the urban freeway to provide additional information relative to incidents and diversion routes. A central control system consisting of a cluster of minicomputers will monitor and supervise the control of all these devices.

7.1.1.3 Constraints. The Dallas Corridor Project is unique in that the research agency is the motivating force in the development of the control systems but the responsibility for installation and operation rest with the local operating agencies of the city and the state. This arrangement has both advantages and disadvantages.

Technical considerations. The research agency provided the conceptual hardware design, computer specifications development, corridor control strategies, and computer programming necessary to implement the strategies. This permits the research project to explore many hardware and software alternatives that many city and state organizations cannot do.

For example, after first considering the strong central control computer concept, a system design of a hierarchy of digital computers with minicomputers located at each multiphase intersection was selected for the Dallas Project. The state and city took the conceptual designs and prepared the plans and specifications, awarded contracts, and supervised the installation.

A major disadvantage of this arrangement was the time required to develop the conceptual design into an operating system. Part of the problem was in the definition of system responsibility in the installation, and differences between contractors had to be resolved in numerous conferences. Another problem was that this demonstration project had to be promoted for construction priority by a research agency with no real authority in the decision making of the local agencies.

Economic factors. The economic considerations contributed to delays in the installation of the project. First, the source of funding within an agency was not specified at the beginning of the project. Therefore, all new federal and state funding programs were explored at length to determine if they were eligible for use in the corridor project. Second, the time delay compounded the effects of inflation. Third, some improvements to signal hardware, such as new signal poles and electrical wiring, were not anticipated. Finally, some costs were underestimated because of the complexity of the computer and surveillance systems and of the manpower and facilities required to operate and maintain the equipment.

Legal Constraints. The time schedule was of extreme importance to the research agency because of the contract requirements to which it was legally committed. However, the operating agencies were not legally committed to providing an operating system by a specified time. This lack of contractual arrangements with the city and state at the beginning of the project proved to be detrimental to the overall development of the project in terms of time and research resources required to complete the study.

7.1.1.4 Objective. The objective of the Dallas Corridor Project is to "optimize" traffic flow through an urban freeway corridor by sensing traffic patterns and conditions, adjusting traffic signal timings, and directing traffic to routes with better operating conditions. Initially, the term "optimize" referred to reducing vehicular travel time in the corridor. This is still a basic goal, but with an added incentive to minimize vehicle occupant travel time in the corridor through preferential treatment of priority vehicles.

It has been the approach of this project to design flexibility into the control hardware and software to test the implementation of other control strategies, such as reducing total travel to satisfy the Environmental Protection Agency's proposed transportation regulation. This approach, considered essential for a research and demonstration project, is probably too costly for most operational installations.

7.1.1.5 Basic concept. The basic concept of the strategy to be tested in the Dallas Corridor is to improve total corridor traffic movements by making the most efficient use of the major arterials during peak periods. Ramp control has been applied to the urban freeway to minimize merging conflicts and control demand at a level below capacity. Surface street intersection signal systems are controlled by a central digital computer system with a program that promotes progressive movement in the dominant flows and provides offpeak signal timings that minimize individual vehicle delays at intersections. The adjustment of peak traffic demands by diversion to alternate routes with better operating conditions is incorporated in the total corridor strategy.

7.1.2 IMPLEMENTATION AND OPERATION OF THE SYSTEM

7.1.2.1 Management. There is a need for close involvement by all agencies during phases of the development and installation of digital computer control systems. There is a tendency to design a computer system that will or can do all things for all people. Research projects usually employ this approach, but the reality of construction and maintenance costs and operational needs must be represented to temper the design.

There is a tendency to approach the installation of a computer control system as just another construction project. Experience has proven that extra care must be taken in the placement of the detection system in the pavement and the installation of the electronic equipment in the environmental cabinets in the field. Computer systems require better accuracy and more reliable operation than conventional signal control systems. More information is routinely collected from these systems and most control strategies require a higher level of traffic data for formulating timing patterns. Those persons responsible for the conceptional design should assist in promoting the necessary quality control in the installation of the equipment. Acceptance tests should be made more demanding and exacting to elicit this quality, which will pay for itself in lower maintenance costs and better operating results.

Therefore, representatives of the design, construction, and operation activities should meet early in the project to determine the requirements of the system and to establish a review process at critical times in the development process. A well-timed meeting of all agencies involved could save several weeks in the schedule and reduce costs.

The Dallas Corridor Project established three types of groups to assist in the development of the project.

1. *Administrative review group.* Members of this committee represented the policy-making level of the various agencies and considered decisions on funding, project priority, and level of involvement. The meetings were infrequent (4–6 month intervals).
2. *Technical review board.* Members of this committee represented the operational level of the various agencies. Details in design, construction, and operations were discussed in meetings that were held each month.
3. *Task force.* This committee, formed to consider specific areas of work, consisted of some members of the Technical Review Board and other agency staff members who have special expertise necessary to solve problems.

The installation of a computer traffic control system requires extra supervision and acceptance testing, but these measures are ineffective unless the contract is awarded to a qualified company. Computer controlled signal systems are relatively new and often companies without experience in building similar systems must participate in the bidding process. It is essential to examine their capabilities in terms of facilities and personnel to be assigned to the job. It is

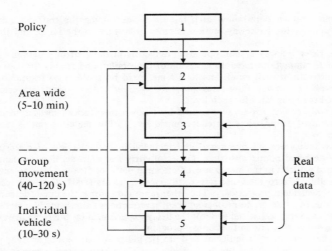

FIGURE 7.2 Corridor control levels.

particularly important to select a general contractor who can manage the installation with a systems approach, i.e., a knowledge of the interface requirements of the telecommunications, computer, and traffic hardware systems.

7.1.2.2 Concepts and methodology used for operations and control (McCasland et al. 1971, Whitson and Carvell 1974, 1975). The Dallas Corridor Study has developed a hierarchical approach to corridor control with several levels of control ranging from policy determinations to minimization of delay for an individual vehicle. These levels are integrated to form a comprehensive strategy that is synergistic in nature (Figure 7.2).

Level 1 concerns policy decisions. Matters of designating control area limits, allocating funding for implementation, determining implementation schedules, and establishing basic goals of the traffic control system are handled at this level.

Level 2 considers the corridor decomposition. This corridor is subdivided according to geographic constraints, geometric considerations, traffic patterns, and travel conditions.

Level 3 specifies control variable ranges. With regard to intersection control, cycle length, offsets, and phase sequence will be evaluated during this control interval. Driver communications systems will also be controlled at this level.

Level 4 implements the overall corridor control strategy. The time frame for this level is loosely called cycle-by-cycle. This level sets the metering rates for ramp control and the green duration, queue clearance, and phase skipping for arterial intersection control. Freeway incident detection also resides at this level.

Level 5 actuates the control program for individual vehicles. This is the lowest level planned for the Dallas Corridor Control System. Override options on the freeway ramp metering system for merge and queue clearances reside at this level. Skip phasing and phase extension at intersections will be used by this level.

Freeway ramp control – theory of operation. The basic theory of entrance ramp control is to change the rate of flow on the ramps in accordance with the traffic flow conditions on the freeway. The control rates are established on a one-minute basis, but the control program attempts to release the ramp vehicles at intervals that will also improve merging operations. This is done by sensing and projecting vehicle gaps in the merging lane and adjusting the release time of ramp vehicles to coincide with the arrival of the gaps. The adjustment of release times produces irregular intervals between ramp vehicles, but generally adheres to the predetermined one-minute flow rate.

The merging rate is determined each minute by examining the traffic conditions of the freeway system, the local freeway section in close proximity to the ramp, and the ramp itself (Figure 7.3).

There are three freeway system operation detectors in the control program: flow rates upstream of each ramp, flow rates downstream of each ramp, and speeds through a bottleneck section downstream of each ramp. Speeds are measured by speed traps formed by two detectors. Five levels of freeway flow rates and two levels of critical speeds are established with predetermined metering rates for each level.

Local freeway conditions are represented by a downstream speed measurement. When this speed measurement drops below a predetermined level, a low metering rate is called from the control program.

Two ramp indicators are monitored and included in the selection of metering rates. For inadequately designed ramps with sharp angles of entry and short acceleration lanes, the detection of stopped vehicles in the merge area of the ramp with the freeway lane preempts the ramp control by calling a zero metering rate until the area clears (DM). For well-designed ramps, merge control override is unnecessary. The other ramp indicator is the queue detector (DQ). The purposes of this override are to protect adjacent intersections from being blocked by the queue of vehicles at the ramp signal and to limit the delay experienced by the ramp drivers. This override calls a high metering rate in the control program.

Because there are many conditions and traffic indicators to be considered, priorities are established. For example, if the queue override calls for a high metering rate at the same time that the critical speed indicator calls for a low metering rate, the control program will use the predetermined priority settings to decide which rate to employ. The usual priority set is: first, merge; second, queue; third, critical speeds; and fourth, system (Figure 7.4).

After metering rates have been established, time limits are set to establish projection windows, which are segments of time during which acceptable gaps are utilized. The segments will vary in length but are usually about 40 percent of the metering rate. For example, if a metering rate of three vehicles per minute allows vehicles to be released at intervals of 20 seconds, the projection window would be 8 seconds.

Upstream detectors measure gaps between vehicles and the speeds of the gaps. Release times for the ramp vehicles are calculated from this speed–gap data and the geometrics of the freeway entrance ramp designs. If the release time coincides with the projection window, then the ramp vehicle is controlled by gap acceptance; otherwise, the vehicle is released according to the time established by the metering rate. Metering rates will vary from 2 to 12 vehicles per minute for single lane entrance ramps.

7.1.2.3 Hardware Design (McCasland et al. 1971).

Control system. The Dallas Corridor Control Project features a distributed control logic approach to traffic control. The control equipment consists of off-the-shelf items available in the form of computer time sharing components and data communications gear. To gain the most versatility, with the combined objective of using off-the-shelf computing equipment, a signalized intersection has the appearance of a teletype insofar as the telecommunications serial code structure is concerned. Thus, a myriad of standard equipment is available "upstream" of the intersection level of the hierarchical control system.

The decentralized hierarchy of digital computers link together to form a system of interrelated components functioning ultimately as a sophisticated control system (Figure 7.5). The hierarchical structure tolerates subsystem failures without total loss of control. The urban system control computer (an IBM 1800 system in the Dallas design) is located at the command control location where supervisory personnel are present. The telecommunications (I/O) computer (Nova 820, 16K words) is a front-end computer for the urban system computer and drives the display board. It directs incoming and outgoing data from several corridors over multiple data sets utilizing voice grade telephone lines at rates of 2,400 baud.

The corridor strategy computer (Nova 820, 32K words) is located near the center of the subnetwork (corridor) over which it has direct control. It is equipped with a variable sequence arterial multiplexer (Nova 820, 16K words), which handles continuous communications over low speed telephone lines with at least 96 intersection computers.

The intersection computers (2K and 4K words minicomputers) interpret incoming serial

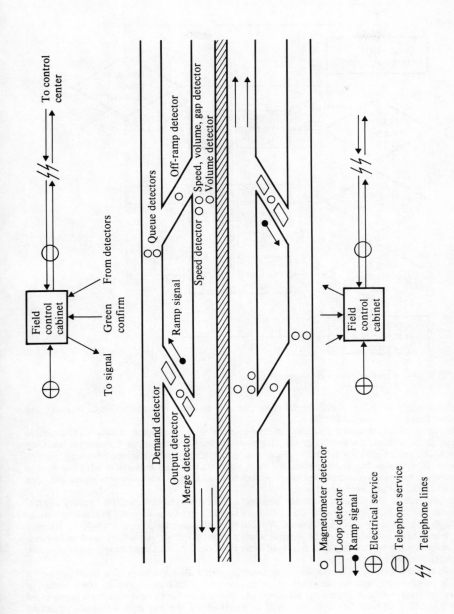

FIGURE 7.3 Typical freeway installation.

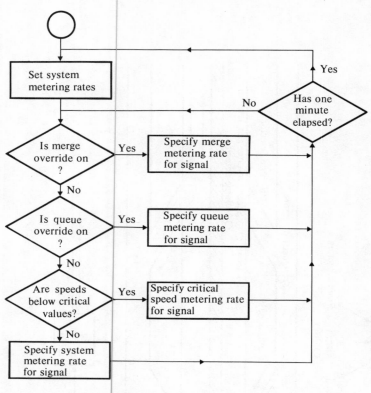

FIGURE 7.4 Ramp metering rate selection.

data commands to change traffic signal status, receive physical intersection data, formats, and transmit these data in serialized form.

This system was designed prior to the introduction of microprocessors, which are desirable at least at the intersection level of control logic. A software-driven arterial multiplexor was used because the Data General 4060 multiplexor was not available and a telemetry message consisting of 9 information bits was required. It now appears that an 8-bit message is sufficient for the information interchange and the hardware multiplexor would serve the telemetry needs, although only 64 lines could be accommodated.

In addition to the variable sequence intersection control system there is a freeway control computer (Nova 820, 16K words) that monitors traffic conditions on the Central Expressway and operates the ramp metering system. A fixed sequence arterial computer (24K words minicomputer) has been added to the control system, which will drive the two- and three-phase traffic signals by supervising existing fixed time field equipment. Communication is over low speed telephone lines.

Surveillance system. The detection system for the Dallas Corridor is a combination of vehicle loop sensors and magnetometers on each entrance and exit ramp to the freeway, at regular intervals along the freeway, and on each approach to each major signalized intersection in the corridor. Special bus detection will be installed on the bus routes at each signalized intersection.

A closed circuit TV system of nine cameras is installed on the freeway to assist in incident identification and motorist information operation.

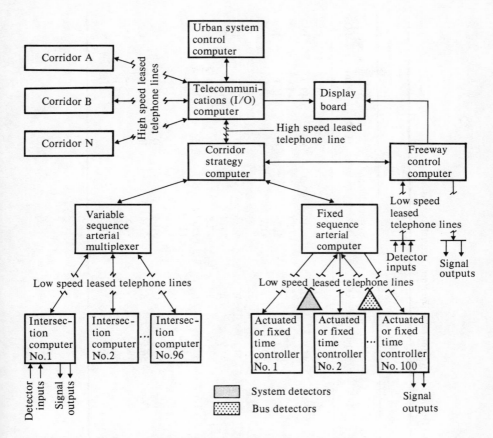

FIGURE 7.5 Corridor control system design for the Dallas Corridor study.

Operation of the system. The computer control system is designed to adjust automatically to changing traffic conditions. However, the detection system, even as extensive as it is, cannot provide all information as quickly as other surveillance sources. Therefore, an operator interacts with the control system.

The motorist information systems at present are all manually operated, with decisions made from visual surveillance, computer data, and radio and telephone reports. There are no immediate plans for automating these systems.

Finally, the maintenance of a traffic control system with 40 or more minicomputers and over 600 vehicle detectors requires a significant level of manpower. It is in this area that improvements must be made before large scale systems of this type will be implemented. Replacement of minicomputers in the field with hardwired units will improve the maintenance requirements, but the major effort must still be in the development and installation of an accurate and reliable vehicle detection system.

7.1.3 EVALUATION OF OPERATION

The evaluation of the Dallas Corridor Control System is limited here to the freeway ramp metering system and the frontage road control system. Measures of the effectiveness of these

TABLE 7.1 Typical Operating Conditions Before and During Ramp Control on the North Central Expressway

Section	Period	Inbound (am)			Outbound (pm)		
		System demand (Vehicles/2 h)	System speed (mph)	LOS	System demand (Vehicles/2 h)	System speed (mph)	LOS
Lemmon to Mockingbird	No control 1971	9998	40.8	E	8472	30.5	F
	Control 1971	9752	45.9	E	8410	47.9	D
	Change (%)	−2.5	+12.5	D	−0.7	+57.0	
	Control 1975	10498	41.1	E	8620	35.1	E
	Change (%)	+5.0	+0.7	D	+1.7	+15.0	
Mockingbird to Loop 12	No Control 1971	7148	28.1	E	6086	40.2	D
	Control 1971	7196	37.9	E	5914	41.8	D
	Change (%)	+0.7	+34.8	F	−2.8	+4.0	
	Control 1975	7734	33.9	E	7502	36.9	E
	Change (%)	+8.2	+20.6	E	+23.3	−8.2	
Loop 12 to Forest	No Control 1971	6072	51.5	D	6106	53.4	C
	Control 1971	6118	56.5	D	6134	52.6	C
	Change (%)	+0.7	+9.7	C	+0.5	−1.5	
	Control 1975	7067	46.9	E	7047	47.1	D
	Change (%)	+16.4	−8.9	B	+15.4	−11.8	

two systems and estimates on the potential benefits of the systems yet to be completed have been made.

7.1.3.1 Ramp Metering.
Table 7.1 summarizes the changes in operating conditions one year after ramp control was initiated and four years after control was initiated. Significant improvements in speeds were noted in 1971, but increases in the traffic demand in 1975 have caused the speeds to approach the no-control level (Carvell 1972, Carvell et al. 1975). Tables 7.2 and 7.3 present the accident experiences on the North Central Expressway. Accident rates declined significantly during the control periods while the total number of accidents in the non-control periods increased.

The benefits of ramp control are estimated in terms of increased annual costs if the ramp control were removed. Table 7.4 summarizes the costs to be approximately $1.75 million per year. The system costs are approximately $600,000 for the installation and $75,000 per year for the operation and maintenance.

7.1.3.2 Intersection Control.
Figure 7.6 identifies the three signal control subsystems that will be placed under corridor computer control. The benefits to be expected from improved signal control are reduced travel times, lower numbers of stops, and fewer accidents. Other factors, such as energy conservation, protection of the environment, and improved quality of travel, are recognized but not included in the estimated benefits of the system. Table 7.5 presents estimates of the benefits in terms of annual costs if the computer control signals had not been implemented (Whitson and Carvell 1975). The total of approximately $10 million per year was calculated on the basis of expected reductions of 8.2 percent in delay and 26 percent in numbers of stops. The systems cost approximately $2 million to install and $175,000 per year for operation and maintenance.

REFERENCES

Blumentritt, C. W. 1969a. Analysis of the data communication requirements for the Dallas North Central Expressway Control System. Research Report RF-590-1, from the Texas Transportation Institute, Houston, Texas, USA.

Blumentritt, C. W. 1969b. Analysis of the communication system requirements for pilot study of signalized intersections. Research Report RF-590-2, from the Texas Transportation Institute, Houston, Texas, USA.

Carvell, J. D. 1972. Evaluation of ramp control operating on the North Central Expressway. FHWA Report RF-836-1, from the Texas Transportation Institute, Houston, Texas, USA.

Carvell, J. D., et al. 1975. Operational benefits of the Dallas Corridor Study. Report prepared by the Dallas project staff for the City of Dallas.

Dudek, C. L. and J. D. Carvell. 1973. Feasibility investigation of audio modes for real-time motorist information in urban freeway corridors. Research Report RF-953-8 from the Texas Transportation Institute, Houston, Texas, USA.

Dudek, C. L. and R. H. Whitson. 1973. Toward automatic incident detection on the North Central Expressway. Research Report RF-953-7 from the Texas Transportation Institute, Houston, Texas, USA.

Federal Highway Administration. 1968. Optimizing flow in an urban freeway corridor, Department of Transportation Contract No. FH-11-6931.

McCasland, W. R., et al. 1971. Optimizing flow in an urban freeway corridor. Research Report RF-590, Final Report from the Texas Transportation Institute, Houston, Texas, USA.

Stockton, R. and C. L. Dudek. 1973. Motorist diversion strategies for the North Central Expressway. Research Report RF-953-2 from the Texas Transportation Institute, Houston, Texas, USA.

White, B. 1974. Actuator program for frontage road control. Research Report RF-953-6 from the Texas Transportation Institute, Houston, Texas, USA.

Whitson, R. H. and Carvell, J. D. 1974. Frontage road control strategy for the North Central Expressway. FHWA Report RF-953-4 from the Texas Transportation Institute, Houston, Texas, USA.

TABLE 7.2 Accidents One Year (1970–1971) Before Versus One Year (1971–1972) During Control (No control in off-peak directions)

		Accidents (am)[a]				Accidents (pm)			
	Weekday total	12M to 12N	Peak period (7–9 am)	Control direction (inbound)	Non-control direction (outbound)	12N[b] to 12M	Peak period (4–6 pm)	Control direction (outbound)	Non-control direction (inbound)
Before 6/10/70–6/9/71	838	243	104	68	36	595	156	94	62
During 6/10/71–6/9/72	823	244	100	49	50	579	133	74	60
Change (%)	−1.8	0.0	−3.8	−27.9	+38.9	−2.7	−14.7	−21.3	−3.2

[a]M = midnight, N = noon, midday.
[b]Includes peak period accidents.

TABLE 7.3 Accidents One Year (1970–1971) Before Versus One Year (1973–1974) During Control (Control in both directions)

		Accidents (am)[a]				Accidents (pm)			
	Weekday total	12M[a] to 12N	Peak period (7–9 am)	Peak direction (inbound)	Off-peak direction (outbound)	12N[b] to 12M	Peak period (4–6 pm)	Peak direction (outbound)	Off-Peak direction (inbound)
Before 6/10/70–6/9/71	838	243	104	68	36	595	156	94	62
During 6/10/73–6/9/74	978	238	76	54	22	740	140	86	54
Change (%)	+16.7	−2.1	−26.9	−20.6	−38.9	+19.7	−10.3	−8.5	−12.9

[a]M = midnight, N = noon, midday.
[b]Includes peak period accidents.

TABLE 7.4 Summary of Estimated Increased Annual Costs (Disbenefits) as a result of Discontinuing Ramp Control

Costs due to time delays caused by increased congestion	1,267,800
Vehicle operating costs increase	252,900
Accident costs	49,100
Costs due to time delays to transit riders	170,800
Total	$1,740,600

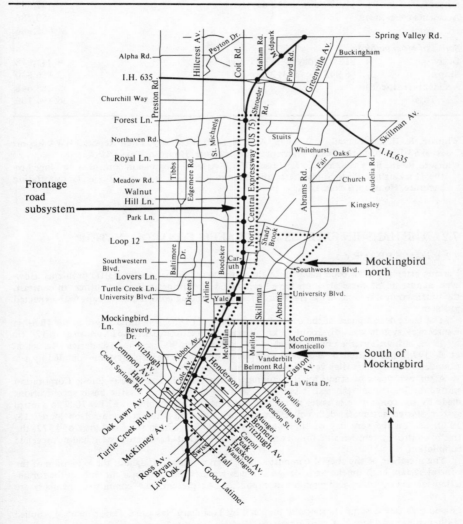

FIGURE 7.6 Signal control subsystems: ●, interchange.

TABLE 7.5   Total Annual Costs by Subsystem

Frontage Road
Delay:	$3,413/day × 312 days/yr	=	1,064,856
Stops:	$1,601/day × 312 days/yr	=	499,512
Accidents (rear-end):			76,954
Total			$1,641,322

Mockingbird North
Delay:	$3,396/day × 312 days/yr	=	1,059,552
Stops:	$1,579/day × 312 days/yr	=	492,648
Accidents (rear-end):			53,786
Total			$1,605,986

South of Mockingbird
Delay:	$12,954/day × 312 days/yr	=	4,041,648
Stops:	$6,945/day × 312 days/yr	=	2,166,840
Accidents (rear-end):			285,648
Total			$6,494,136

Whitson, R. H. and Carvell, J. D. 1975. Arterial control strategy. Final Report FHWA Report RF-953-17A from the Texas Transportation Institute, Houston, Texas, USA.

Whitson, R. H., B. White, and C. J. Messer. 1973. Arterial progression control as developed on the Mockingbird Pilot Study. Research Report RF-953-3 from the Texas Transportation Institute, Houston, Texas, USA.

7.2   THE HANSHIN EXPRESSWAY TRAFFIC CONTROL SYSTEM*

7.2.1   CASE HISTORY

Various attempts have been made throughout the world to cope with traffic problems. However, almost all of these attempts have been aimed at existing traffic difficulties. In contrast, the traffic control system on the Hanshin Expressway is being constructed to cope with expected problems.

The basic design phase of the control system began in 1967, when the length of the Hanshin Expressway system was only about 25 km. It was planned to lengthen the Expressway to 120 km by 1976, although at the end of March, 1976 it was about 90 km. The basic design (Kometani et al. 1974) was started by a Special Committee on Traffic Control Systems founded by the Foundation of Express Highway Research, Tokyo, Japan.

After two years of study by the committee, the Hanshin Expressway Public Corporation started to implement the traffic control system in accordance with the basic specifications made by the committee in order to be ready in time for the opening of EXPO 70. The control system started automatic data collection in 1969 and semi-automatic control on March 15, 1970, i.e., on the very day of the opening of EXPO 70. Within the fiscal year of 1972, the traffic control system became fully controllable, although its hardware system had not reached completion.

The members of the special committee have been working throughout the expansion of the control system both on the basic design through research and development, and on the formulation of management and maintenance policies. At present, the committee members are

---

*Based on a case description specially prepared by Toshiharu Hasegawa, Department of Applied Mathematics and Physics, Kyoto University, Kyoto, Japan (cf. Acknowledgment to paragraph 6.3.2).

working on the continuity of the system with respect to its control strategies and hardware system in the course of the future expansion of the Hanshin Expressway.

The average number of cars flowing into the Hanshin Expressway in one day is about 400,000 at present. The construction expenditure for the traffic control system has been about 15,000 Japanese Yen per meter of the expressway (about US $80,000 per mile). This amount of money is negligible compared with the construction expenditure for the expressway.

The fundamental object of traffic control on the Hanshin Expressway is to ensure, as far as possible, a smooth flow of traffic for the users of the expressway. Natural congestions on the throughway are extremely undesirable because they result in unnecessary decreases in the volume of traffic going through the congested areas. The inflow control, which controls the number of vehicles intending to use the expressway, is then adopted to avoid natural congestion as much as possible by optimizing the inflow traffic. In the case of accident or emergency, vehicles approaching the scene of the accident are requested to leave the expressway. At the same time, vehicles intending to use the congested route of the expressway are advised not to use it. At present, variable sign boards located on and around the expressway are used to give information or direction indication to drivers.

The traffic control system on the Hanshin Expressway ranks as one of the largest and most capable expressway traffic control systems in the world. With the extensive cooperation between the area traffic control system in Osaka and the traffic control system on the Meishim Expressway, the traffic control systems on the Hanshin Expressway are going to play an important role in the regional traffic control in the Osaka district.

7.2.2 IMPLEMENTATION AND OPERATION OF THE SYSTEM

7.2.2.1 Method of Traffic Control. The method of traffic control on the Hanshin Expressway has, roughly speaking, two phases according to the state of traffic flow in the expressway system. One is the inflow traffic control phase, which is designed to avoid natural congestion. The other is an emergency control phase in case of an accident and is designed to eliminate the effect of the accident as soon as possible. These two phases are not independent of each other but are closely related.

Inflow control. The inflow control phase is also divided into two categories: linear programming control (LP), and sequential control. LP control is for stationary traffic flow. A traffic flow is called stationary when there is no traffic jam in any section of the expressway and there is no sudden jump in traffic volume incoming to or outgoing from the expressway system. Under LP control, the volume of inflow traffic at every on-ramp of the expressway is controlled to maximize the total number of incoming cars without causing a traffic jam in any section of the expressway.

The origin and destination (OD) of the expressway traffic should be derived for LP control. As the origins of the expressway traffic are supposed to be on-ramps and the destinations of the traffic are supposed to be off-ramps, it is far easier to derive the OD of expressway traffic than of the ordinary avenues and streets system. However, there is no way to derive the OD at any instant as on-line information. Therefore, we should use off-line analysis and the available on-line data. In order to maintain the high accuracy of the derived OD, estimates of the ODs are made only when the traffic flow is assumed to be stationary. On-line data include traffic volumes of inflowing cars at all on-ramps and outgoing cars from all off-ramps. The entropy method of estimating OD using these data has proved to be a powerful tool (Sasaki 1968, Sasaki and Myojin 1968). In this method, the parameters of the estimation equation, which describes the probability that a car entering the expressway from a particular on-ramp goes off the expressway at another particular off-ramp, are determined by off-line analysis. The real OD patterns have been investigated by the Hanshin Expressway Public Corporation many times, by asking drivers of inflowing cars about their trips on the expressway at that time or at other times. The parameters of the estimating equation are derived from these data and are revised at least once a year. The probability estimated by this equation is called the OD transition probability. In deriving the estimating equation, travel times required to go from a particular on-ramp to a particular off-ramp via expressway and via ordinary surface streets are measured.

It is assumed here that all cars entering the expressway from a particular on-ramp and leaving from a particular off-ramp take the same route or path. Route matrices for each on-ramp are then defined, and these determine which sections of the expressway are passed by a car entering the expressway from a particular on-ramp and leaving at a particular off-ramp. A section of the expressway is the road between two successive merging points, branching points, on-ramps and/or off-ramps, i.e., within one section there is only one entrance and one exit.

From the estimated OD transition probabilities, the traffic volumes of the on- and off-ramps, and the route matrices, it is possible to estimate the traffic volume in each section of the expressway. Let a vector X denote the estimated traffic volumes in the expressway sections. As a matter of course, X cannot be a good estimate owing to the fairly large estimation error. The main contribution to the estimation error is the fact that the time required for a car entering the expressway from an on-ramp to reach a particular off-ramp has been neglected in the estimation so far. Let e denote an error vector for the estimation, i.e., by denoting measured volumes in every section of the expressway as Y, we have,

$$Y = X + e$$

In order to have a good estimation of e, the traffic flow on the expressway should be kept uncongested.

The objective of the traffic control on the Hanshin Expressway when the traffic flow is stationary is to maximize the number of inflowing cars to the expressway without causing traffic jams on any section of the expressway. Therefore, by denoting the capacities of every section in a vector form as C, we should control the inflow traffic in such a way that

$$X + \hat{e} \leq C \tag{7.1}$$

where $\hat{e}$ is an estimated e. Under condition (7.1), and that providing the maximum allowable inflow traffic volume on each on-ramp does not exceed the actual demand (in other words the number of cars arriving at that on-ramp intending to use the expressway), the total number of cars allowed to use the expressway is maximized. This type of maximization can be reduced to a linear programming problem. This is the reason why this type of control is called LP control.

LP control has proved to be powerful when the traffic flow on the expressway and the number of inflowing cars is stationary. However, at another period in the day, the traffic flow and the inflow traffic volume may become nonstationary, i.e., they may vary to a fairly large extent in a short interval. In this case, LP control cannot work because of the restrictions for good estimates of X and e. This situation occurs typically during the morning and evening rush hours. Therefore, when a sudden large change of the inflow traffic volume is detected, sequential control is adopted.

Sequential control is used when it is estimated that there are one or more sections where congestion is expected to occur in the near future, and that LP control cannot work, or when congestion has already taken place in some section. As is well known, once a section of a road has become congested, the number of cars passing that section decreases tremendously. Therefore, the most important object of the traffic control on the Hanshin Expressway is to prevent congestion in any section or to remove the congestion as soon as possible if it does occur. Then, if necessary, on-ramps are closed, depending on the extent to which they influence the troubled section of the expressway. Each section of the expressway has a traffic flow that consists of cars entering the expressway from various on-ramps, the contribution of each on-ramp inflow being different in general. Then, in case of congestion on a particular section, the on-ramp where the inflow traffic contributes most to the traffic flow of that section and thus its contribution to the traffic flow appears very quickly is closed. If this on-ramp closure is expected to be insufficient, the next most influential on-ramp is closed, and so on sequentially.

Criteria to decide when and how many ramps should be closed or limited are derived by various off-line analyses and simulations. According to the criteria, the computer system tells when and where this sequential ramp closure control should start. The sequential ramp closure control is powerful when congestion is not too severe. This type of control, however, is not powerful enough for severe congestion such as when a bad accident occurs on the expressway. Therefore, when the capacity of a certain section decreases to almost zero because of an accident or otherwise, not only inflow control but also outgoing recommendations or orders should be made to the drivers.

Emergency control. When a traffic accident or a disabled car blocks the traffic flow on a section of the expressway, the traffic capacity of that section decreases tremendously and the steady flow on the expressway is impaired. In this case, also, the inflow traffic at the on-ramps upstream of the troubled section, should be controlled so that the expected jam can be avoided, will be less obstructive or will be removed quickly. In addition to the above sequential ramp closure control, drivers on the expressway are recommended by the traffic controller of the Hanshin Expressway Public Corporation or ordered by the police in charge of traffic control to leave the expressway from an indicated off-ramp. This additional control is not performed in all traffic accidents or blockages due to disabled cars, because in some cases the cause of the decrease of the traffic capacity may be cleared in a short time.

Emergency control, which is a combination of sequential control and outflow control, is very important and powerful and is closely related with detour control or guidance, not only in the Hanshin Expressway network but also in the whole road network in the Osaka–Kobe area. Detour control or even guidance on the Hanshin Expressway network itself is very difficult and complex. This is because the Hanshin Expressway network is surrounded by the expressway system operated by Nippon Dohro Kodan (Japanese Highway Public Corporation) with several interchanges, and the shape of the expressway network of the Hanshin Expressway Public Corporation itself is rather complex. It is impossible to have optimized detour control in a network like this. Another difficulty in detour control is estimating the time required to clear the cause of the trouble. After long and detailed discussions and investigations, it was decided not to rely on estimates of the time required to clear the cause of trouble because of the very large variance in estimations.

7.2.2.2 Hardware System. The traffic control system on the Hanshin Expressway consists of the following hardware: vehicle detector system, detector information transmission system, information processing system, information display system, traffic control device system (variable sign boards), and a closed circuit TV (CCTV) system. This traffic control is one of the largest and most advanced systems in the world and has been expanding since 1969. It also employs loop detectors, including short loop detectors, multiple loop detectors, and long loop detectors, together with ultrasonic detectors that include Doppler detectors and pulse detectors. The functions of the detectors are roughly two: one group of detectors detect traffic volume and time occupancy and the other detects average speed, traffic volume, and time occupancy. Long loop detectors are used to detect space occupancy or queue length. The detectors are located at on-ramps, off-ramps, throughways, upstream of the on-ramp booths, and downstream of the off-ramp. Some of the throughway detectors detect the average speed of the vehicles together with the time occupancy and the traffic volume at that point. The detectors near the on- and off-ramps detect the queue length. There are 356 detectors at 178 detecting points.

Figure 7.7 shows the hardware system of the traffic control system. Inflow traffic volume at each on-ramp is measured at the toll booth with very high accuracy, because this detector is also used for checking the amount of money collected at the booth. The devices in the control center shown in Figure 7.7 are located in the control center in Osaka. Subcenters are located remotely along the expressway routes. The subcenter in Kobe is not connected via transmission lines operated and maintained by the Hanshin Public Corporation but via data lines of the common carrier with the control center in Osaka. After the completion of the construction of the Nishinomiya–Osaka line of the Hanshin Expressway, Kobe subcenter will be connected via self-operated lines. Kobe subcenter, unlike other subcenters in the system, has its own data logging system, CCTV system, and TV monitoring system together with a small control console.

The information flow to and from the control center is shown in Figure 7.8. No new techniques for the information transmission system are required for the traffic control system. Information concerning the traffic flow to and from the broadcasting station is very useful. Many drivers on the expressway listen to the traffic information broadcast and sometimes follow the information given by the radio fairly well. Most radio stations around Osaka have their own traffic information collecting system with light planes or helicopters, which often bring useful information to the control center of the Hanshin Expressway. Radio links are used to keep close contact with the patrol cars of the Police Departments of the Osaka and Hyogo Prefectures (Kobe area), and the patrol cars of the Corporation.

Thirty-four TV cameras are controlled from the control center. Tilting, panning, and zooming for some of them is carried out remotely from the control desk. The TV signals are transmitted over ordinary telephone cables in the frequency mode.

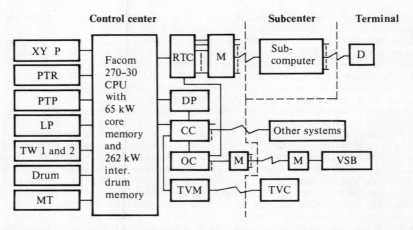

FIGURE 7.7 Hardware system of the traffic control: RTC, real time controller; M, modem; DP, display panel; CC, control console; OC, output control; TVM, TV monitor; Subcomputer, front end processor; D, detectors; VSB, variable sign boards.

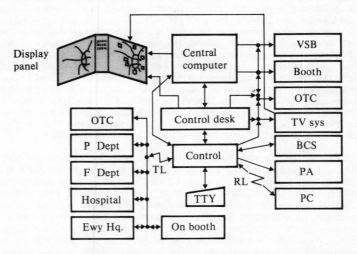

FIGURE 7.8 Information flow to and from the control center: OTC, other traffic control systems; P Dept, Police Department; F Dept, Fire Department; Ewy Hq., Hanshin Expressway Public Corporations main office; Control, output controller; TTY, typewriter; On booth, on-ramp booth attendant; Booth, booth attendant instruction; TV sys, TV cameras; BCS, broadcasting stations; PA, public address system; PC, police cars.

The control center of the traffic control system on the Hanshin Expressway has roughly the following functions: data collection and processing for traffic control; reception and processing of emergency calls from the emergency telephones located along the expressway; automatic traffic control; manual traffic control; automatic and manual exchange of information with other traffic control systems; monitoring failures of the whole traffic control system; maintenance of the system; transmission of instructions and information; modification and improvement of the traffic control system, and others.

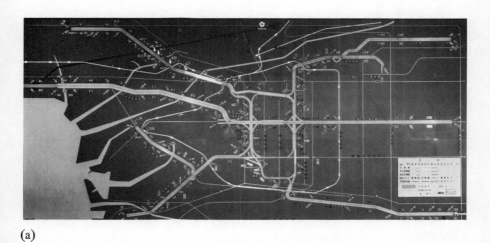

(a)

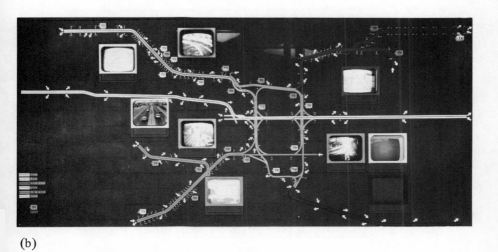

(b)

FIGURE 7.9 Display panels in the Hanshin Expressway Control Center in Osaka.

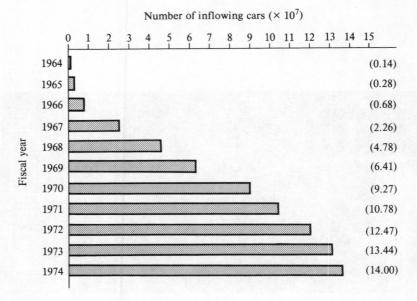

FIGURE 7.10   Traffic entering the Hanshin Expressway.

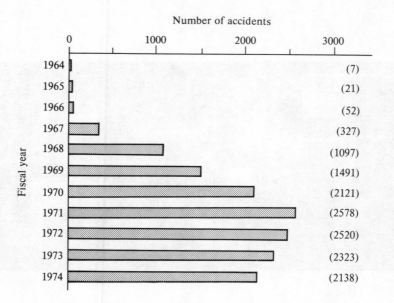

FIGURE 7.11   Time-dependent development of the number of accidents.

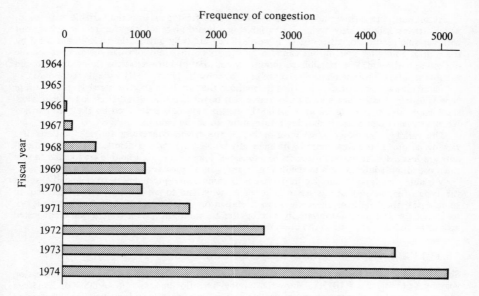

FIGURE 7.12 Time-dependent development of the frequency of congestion.

The control center has a control room and a machine room. The control room has a control console and three display panels: a graphic panel, a digital information panel, and a monitor panel. Figure 7.9(a) shows the graphic panel, which gives a qualitative assessment of the traffic flow in the entire system of the expressway. The section number of the expressway, the rate of inflow control in each on-ramp, and the operation of the variable sign boards are also shown in the graphic panel. Time occupancies, traffic volumes, and the average speed of vehicles on the sections that have the largest five time occupancies are given automatically with the section numbers of the digital information panel (not shown here). These three parameters can also be given manually for up to ten sections. The digital information panel also shows queue length, rate of inflow, traffic control at on-ramps, and speed limit. Figure 7.9(b), shows the monitor panel, which has eight TV monitors and lamps that indicate the location of the emergency calls. TV cameras are controlled automatically, e.g., when the control system detects any irregularity in the sequence of parameters received, a TV camera closest to the troubled section is automatically turned on with an alarm bell. The Kobe subcenter has its own TV monitor and emergency call panel with three TV monitors and six cameras. For quantitative information, it has a digital information panel.

The hardware monitor panel is installed in the maintenance room next to the machine room. Every block of the hardware system in this traffic control system is monitored as much as possible and in case of hardware failure of a detectable kind, this monitor panel rings an alarm bell and indicates the location of the failure.

7.2.2.3 Operation of the Control System. As there are no traffic signals on the throughway of the Hanshin Expressway, the control of cars depends mainly on the on-ramp inflow control and on the guidance or orders given to drivers by the variable sign boards to leave the expressway. The largest problem in giving effective traffic control is the Japanese traffic regulations. The Hanshin Expressway Public Corporation does not have the power to control drivers with legal force. Any recommendation by the Hanshin Expressway to the drivers may be ignored. Only the Police Department has the right to force drivers to follow the orders. Consultation with the Police Department has been underway and it is expected that this problem will be solved in the future.

At present, the inflow and outflow control of traffic depends on the variable sign board, which drivers follow fairly well. The variable sign board gives information to the drivers not with variable pictures or signs but with words. They are fully automatic devices controlled by the central computer system, but can also be controlled manually through the control console in the control room. It is possible to monitor what kind of information is shown on a certain variable sign board by the controller at the control console. There are 91 variable sign boards.

There are two ways of showing the information: the film type and the matrix type. The film type is usually for a place where the variable sign board is installed rather close to the drivers, since large film is hard to control, while the matrix type is usually used on the throughway. When a variable sign board is showing information, an amber light on top flashes.

The variable sign board gives information to the drivers concerning traffic situations of a varying nature. Fixed sign boards with manually changeable signs controlled by booth attendants are located near on-ramps for fixed situations. Many other sign boards are located to give the most suitable information to the drivers in and around the expressway.

Variable sign boards can give information, automatically and if necessary manually, about the type of situation that has arisen, and where, why, and to what extent it has occurred. In some cases, the length of the congestion is shown on variable sign boards. The length is not measured by detectors automatically but by the TV monitor system watching the congested area or by patrol cars going in the other direction.

7.2.3 EVALUATION OF OPERATIONAL EXPERIENCES

Since the opening of the expressway, the number of cars using the expressway has been increasing year by year. Figures 7.10–7.12 show the number of inflowing cars, the number of accidents and the frequency of congestion, respectively. The number of accidents has been decreasing since 1972 but the frequency of congestion increased every year except 1970. 1970 was the year of EXPO 70 and was the time when there was a significant improvement in the road network in the Osaka district. One of the main reasons for the increase in the frequency of congestion is the daytime reconstruction or repair of the expressway. The decrease in the number of accidents is not confined to the Hanshin Expressway; it is occurring on almost all road networks throughout Japan.

As mentioned earlier, the number of cars flowing onto the expressway in one day is about 400,000. The average number of cars in the Osaka area is about 300,000 per day and the maximum traffic volume in a section per day is about 140,000. The Kuko line (Airport line) has this maximum sectional traffic volume. The length of the expressway in the Osaka area is about 65 km. An average delay to a car in a typical morning rush hour on the Kuko line has been found by extensive investigations to be about 5 minutes. The average number of cars delayed in the rush hour is about 10,000. Because of the aforesaid difficulties, the inflow control performed now is quite coarse. The inflow traffic is controlled by controlling the number of open booths at each on-ramp. From investigations and mathematical studies, it is expected by using fine control to reduce the numbers of cars delayed to 8,000 for 1 minute in the morning rush hour. The committee members have been working on finding the optimal strategies in practice and the optimal hardware for the inflow traffic control.

The effects of variable sign boards have been investigated by questioning the drivers on the expressway (Hanshin Expressway Public Corporation 1973). These investigations revealed that about 99 percent of the drivers who answered the questions notice the variable sign boards. Of the 20 percent of the drivers who answered the question: "What would do if you notice a sign board telling you that the route you are going to take is congested when you are trying to take the expressway?" 40 percent said they would transfer to the ordinary surface streets. If the length of the congestion is known to be more than 5 km, more than 80 percent of the drivers said they would transfer to the surface streets. The responses were higher than expected.

The reliability of the hardware system and the information derived from the detectors and given to the drivers are the most important things in traffic control. The system has been checked and checked again in order to cope with new problems and the development of the expressway system. For instance, the accuracy of all the detectors and the information transmission system was thoroughly checked by an independent institution in 1971. It was revealed that the reliability of any type of detector is more than 96 percent in detecting time occupancy.

The availability of the system is now very high. The estimation error of the traffic volume in each section, estimated every 5 minutes, is usually less than ± 10 percent.

This report has presented a rough sketch of the traffic control system on the Hanshin Expressway and brief aspects of its operation. This traffic control system will be included as one of the most important parts in the regional traffic control system in the Osaka district.

The main part of this regional traffic control system will be the area traffic control in the Osaka Prefecture operated by the Police Department. Another important part will be the traffic control system on Meishin Expressway (an intercity expressway that connects Nagoya and Kobe), Kinki Motorway and Chugoku Motorway (also intercity expressways) operated by Nippon Dohro Kodan. This regional traffic control system is one of the largest challenges for traffic control in the world.

REFERENCES

Hanshin Expressway Public Corporation. 1973. Report on the surveys on the effects of road information sign boards of Hanshin Expressway and on the detours of the Expressway.

Kometani, E., T. Hasegawa, and S. Inada. 1974. On the operation of the traffic control system on the Hanshin Expressway. Traffic Control and Transportation Systems. Amsterdam: North-Holland/American Elsevier, pp. 237-248.

Sasaki, T. 1968. Probabilistic models for trip distribution. Proceedings of the Fourth International Symposium on the Theory of Traffic Flow: pp. 205-210.

Sasaki, T. and S. Myojin. 1968. Theory of inflow control on an urban expressway system. Transactions of the Japan Society of Civil Engineers, No. 160: 88-95.

7.3   FRENCH EXPERIENCES: THE PARIS FREEWAY CORRIDORS AND THE PROJECT *OPERATION ATLANTIQUE*

7.3.1   THE PARIS FREEWAY CORRIDORS*

7.3.1.1   Case History and Basic Design Principles. This contribution deals with the traffic control systems installed in the various traffic corridors around Paris. The map shown in Figure 7.13 illustrates that the main traffic streams flowing into and out of Paris have been channeled into four traffic corridors:

1. The northern corridor, which was implemented first, serves a highly industrialized area and le Bourget airport. Its main traffic facilities are the A1 freeway and several national roads (cf. Figures 7.13 and 7.14).

2. The southern corridor carries both the commuter traffic and all traffic to and from the Mediterranean coast. Moreover, Orly airport is located in this corridor. The corridor is characterized by three freeways A6, B6, and C6 and several parallel national roads (cf. Figure 7.15). The A6 freeway is the oldest freeway of the corridor; it leads to Lyon and Marseille in the south of France. The B6 freeway was built next to serve Orly airport. The C6 freeway was built last to serve the regional food market of Rungis.

3. The corridor to the west of Paris, which covers the largest area, includes the A12 and A13 freeways as its main facilities.

4. The eastern corridor on the other hand is the smallest; it was implemented as recently as 1977.

The problems that led to the decision to install traffic control and guidance systems will be illustrated by selected figures for the northern and the southern corridor.

Southern corridor (cf. Figure 7.15). There are queues every day and especially every weekend towards Paris. These can reach a length of 20 kilometers on a week day morning and 50 or

---

*Based on a case description specially prepared by Eugene Sacuto, SRE d'ile-de-France, Unité Exploitation Opérationelle Quest, P.C. de Boulogne, France.

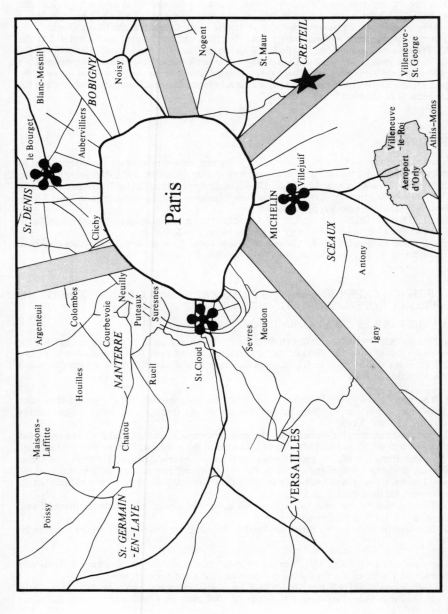

FIGURE 7.13 The four traffic corridors of Paris: ✱, control center of the corridor; ★, regional control center.

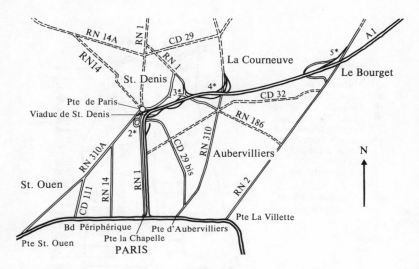

FIGURE 7.14 The northern freeway corridor of Paris (metered ramps are denoted by asterisks).

more kilometers on a Sunday night after a long weekend. These queues can double or triple the travel time of a motorist who commutes every morning to Paris.

Northern corridor (cf. Figure 7.14). During each peak period, queues 5-9 kilometers long occur on the A1 freeway towards Paris. To solve these problems, it was decided to test various kinds of control principles and equipment in the different corridors. The objective of having different control policies in the same region is to be able to identify an optimal solution that could then be implemented on a large scale.

Two different control principles were thoroughly analyzed:

- speed control
- ramp metering connected with route guidance

The first principle has been tested extensively on the western corridor. The second principle was preferred on the northern and southern corridors.

7.3.1.2 Implementation and operation. The research work that led to the implementation of the control systems was financed by the government. All equipment was installed by the local agencies of the public works. Some technical agencies give technical advice on a national or regional level.

Corridor control centers have been installed in each corridor. They are supplemented by a regional control center as indicated in Figure 7.13.

Radio transmission has been tested for connecting the control centers with the changeable direction signs, etc. However, we have experienced difficulties, especially in dense urban areas such as around Paris. Therefore communication cables were used.

In the western corridor, 45 variable speed limit signs are installed, which recommend driving speeds for specific traffic situations. Moreover, variable direction signs, as shown in Figure 7.16, are used to control the amount of traffic entering the A13 freeway. Three different states are distinguished:

- normal traffic conditions on the freeway, i.e., the freeway can be entered in the conventional way (Figure 7.16(a))

208

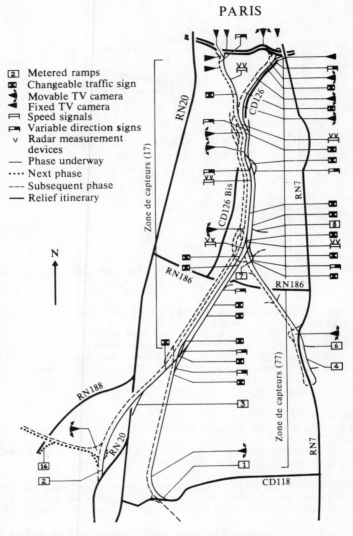

FIGURE 7.15   The A6–B6–C6 traffic corridor in the south of Paris.

- access to the freeway is controlled by ramp metering if the traffic conditions on the freeway are characterized by high traffic volumes and densities (cf. Figure 7.16(b))
- the entrance ramp is closed if severe congestion is occurring on the freeway (cf. Figure 7.16(c))

Examples of variable direction signs as they are used in the southern corridor are shown in Figures 7.17–7.19. Figures 7.17(a), 7.18(a), and 7.19(a) show the variable direction signs that the drivers meet first when approaching the freeway entrance ramp. Figures 7.17(c), 7.18(c), and 7.19(c), on the other hand, illustrate how the alternative routes are shown to the drivers in

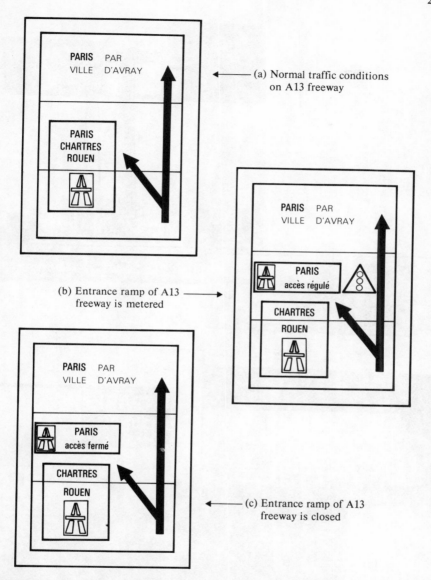

FIGURE 7.16 Variable direction signs as used in the western traffic corridor of Paris at Côte de Picardie (crossing between national road RN 185 and road CD 182; cf. Figure 7.13).

front of the entrance ramp. Variable direction signs of this type are widely used in the southern, i.e., so-called A6–B6–C6, corridor (cf. Figure 7.15). Figures 7.14 and 7.15 indicate the freeway entrance ramps that are metered in the northern and the A6–B6–C6 corridors. For the northern corridor, according to Figure 7.14, about 65 percent of the traffic demand is composed of traffic coming into the A1 freeway from the entrance ramps numbered 2-5. The objectives

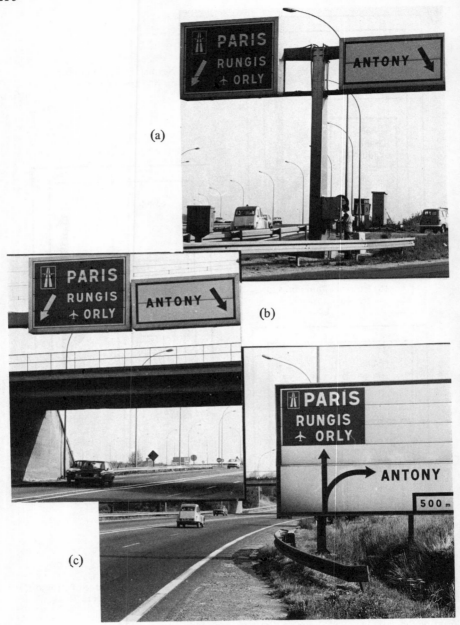

FIGURE 7.17 Changeable direction signs as used in the southern corridor of Paris – normal traffic conditions: (a) traffic signs at the entrance ramp, which is open; (b) before the merge on a gantry; (c) before the signs on the gantry.

FIGURE 7.18 Changeable direction signs as used in the southern corridor of Paris — congested freeway — an alternative route is indicated: (a) traffic signs at the entrance ramp; (b) before merge; (c) first sign met by driver.

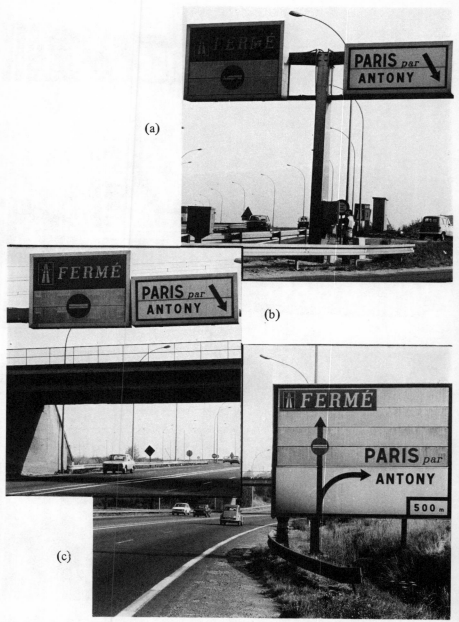

FIGURE 7.19 Changeable direction signs as used in the southern corridor of Paris — entrance ramp is closed for all traffic: (a) traffic sign at the entrance ramp; (b) before merge; (c) first sign met by driver.

of the traffic control experiments in this corridor, which are the first of this type in France, are to reduce the queues on the A1 freeway by ramp metering.

The timing of the traffic lights on each entrance ramp has been carefully set by considering:

- flow reduction needed to stop congestion on the freeway
- traffic demand on the ramp
- number of cars that could be stored on the ramp
- left-over capacity of surface street facilities

Usually, cycles of 65 seconds have been used with 15–30 seconds of green time. The traffic light settings on the surface streets have been altered to increase their capacity. Illegal parking has been eliminated during the experiment.

The variable direction signs of the northern corridor of Paris have been used at peak hours to divert traffic from entrance ramps to alternative routes towards Paris.

7.3.1.3 Operational Experience. The experience gained in the traffic control experiments will first be judged with respect to the effectiveness of the tested control principles, i.e., speed control, and ramp metering and route guidance.

Speed control experience. The experiments carried out in the western corridor indicated that the use of variable speed limit signs will contribute to a reduction in the number of accidents but will not reduce the length of the queues; in other words, variable speed control is a safety tool but does not help to reduce congestion.

Ramp metering and route guidance experiences. The experiments carried out in the northern corridor showed the following results:

- The total travel time of drivers on the corridor network has been reduced by 13 percent.
- The queues have been reduced by 45 percent on the freeway.
- An inquiry was made to evaluate the opinions of the drivers on this experiment. 54 percent of the drivers using the surface streets thought that their travel time had been increased.
- 33 per cent of the drivers using the freeway thought that their travel time had been reduced.
- More than 35 percent of the drivers using the entrance ramps involved have followed at least once the advice shown on the variable direction signs.
- 78 percent of the drivers feel that the experiment was absolutely necessary or useful.

The results and the response of the drivers have been so good that it has been decided to keep using every morning the traffic lights on the ramps and the variable direction signs on the surface streets.

The experiments carried out in the Paris corridors were faced with numerous constraints and difficulties of various kinds.

The most important technical constraints are the need for reliable information to give orders to the variable direction signs, and the need for a control center to shelter the person who receives information concerning traffic and sends orders to the signs. Technically, this means that not only do variable signs have to be implemented, but also the communication systems should be designed in such a way that information can be received and orders sent back, so that the variable signs can actually be traffic responsive and not just man responsive.

The other constraints are political. It is obvious that there is a conflict of interests between the freeway drivers and the surface street motorists. For better traffic conditions on the freeway, one solution would be to close all of the entrance ramps inside the Paris area. By doing this we would have a very high level of service on the freeway, although congestion on the surface streets would become intolerable.

The improvement of freeway traffic conditions is often compensated by a decrease in the level of service on the surface streets network. By restricting traffic on the entrance ramps, we make it better for long distance traffic but worse for commuting traffic.

There is also another kind of constraint that is special to France. The police corps working at the command post of the freeway corridor are only in charge of freeway traffic; the surface street traffic is taken care of by the urban police corps, who do not receive orders from the command post. Therefore some kind of agreement is required between the two police corps.

Within these same constraints, there is a real need for coordinating the actions of the various government agencies in order to be consistent for the driver.

Freeways are very expensive to build and the average income tax payer does not understand how an important freeway can be congested six months after being opened to traffic. Therefore the policy should be: entrance ramps must be controlled, metered, or closed sometimes to keep traffic flowing on the freeway.

BIBLIOGRAPHY

Direction des Routes et de la Circulation Routiére. 1975. Corridor operations: technical guide, index of possible actions (in French).

Direction des Routes et de la Circulation Routière. 1976. The corridor: an urban motorway and its associated network rationally exploited (in French).

Dressayre, K. and J. -P. Grunspan. 1975. The western corridor or 6000 hours lost per day. Transport, Environnement, Circulation, No. 12 (in French).

le Dieu de Ville, D. and Laurens, B. 1976. Corridor A6–B6–C6: The end of the nightmare. Transport, Environnement, Circulation, No. 19 (in French).

Ledru, M. and J. Andrianmanantou. 1974. Urban corridors: an avant-garde technique for exploiting traffic. Equipement, Logement, Transport (in French).

Poulit, J., J. L. Durand, and M. Ledru. 1975. The exploitation of rapid peri-urban highways and their associated network, The Corridor Operation. Revue Générale des Routes et Aérodromes, Formation Permanente, Fascicule No. 10 (in French).

Service de l'Exploitation Routière et de la Securité. 1976. Corridor Operation, equipment and the exploitation of rapid urban highways. Index of possible actions (in French).

Service Régional de l'Equipment d'Ile de France. 1976. Results of experiments to regulate traffic in the A1 corridor.

7.3.2 THE PROJECT *OPÉRATION ATLANTIQUE**

7.3.2.1 Case History. During the great seasonal migrations and especially during the summer holiday departure and return periods, the major roads of the French network become saturated, and traffic jams increase in number.

The national or main road RN 10 (cf. Figure 7.20) is a typical example of this phenomenon. It serves the Atlantic coast between Tours and the Spanish frontier, and carries traffic bound for Spain and Portugal, and even Morocco. During July and August, 1974, this major road between Tours and the Spanish frontier (550 kilometers) was completely saturated, with traffic jams lasting 11 hours. From the economic point of view, this saturation represented a loss of more than ten million francs to the community.

Confronted with this situation, the public authorities, in particular the Department of Roads and Road Traffic (Ministry of Equipment), decided to put into effect a project to regulate the traffic on the main RN 10 roads. This began in 1975 as Opération Atlantique with the following aims:

- regulating the traffic by better temporal distribution of the demand
- regulating traffic by better spatial distribution of demand

The first aim was quickly put aside, because on the one hand it raised the problem of staggering holidays (a problem going beyond that of road traffic alone), and on the other hand, the "sheeplike" character of the French people would not lead one to expect positive results.

The basis of the traffic control was therefore distribution in space, which was made possible by the good quality and the recognized importance of the secondary road network in France. The actual foundation of Opération Atlantique rests on this principle. Moreover, this project

*Based on a case description specially prepared by M. Gien and M. Pouliquen, Ministère de l'Equipement, Centre d'Études Techniques de l'Equipement, Bordeaux, France.

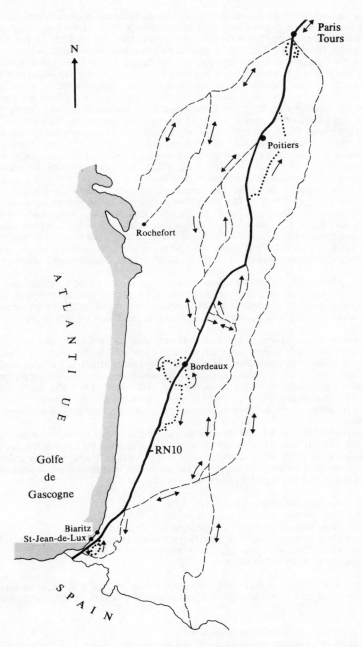

FIGURE 7.20 The project Opération Atlantique: —, RN 10 main route; - - -. relief routes; ...., bis routes.

marked a turning point in regulating traffic at seasonal peak periods. Until that time, the principal actions carried out in France had been relatively local. Opération Atlantique, on the other hand, regulates the traffic on the 550 kilometers of the main road RN 10 as a whole.

The spatial distribution of traffic in Opération Atlantique, is accomplished by two types of itineraries: the "bis" or "second" itineraries and the "délestage" or "load relief" itineraries.

A bis itinerary is a secondary route intended for motorists who like peace and quiet, who attach a relatively low importance to time, and who are therefore willing to run the risk of taking a little longer to do their journey in order to avoid the crush of the main roads on peak days. The bis itineraries are proposed to motorists when the traffic on the principal itinerary is becoming very heavy, and if there is no risk of the bis route itself becoming saturated.

The relief routes enable the traffic to be spread out over a zone covering the main route and the relief route in such a way that the time required to cover the distance between the two extremities of the zone is the same whichever itinerary is followed. Obviously the management of such a zone must be carried out in real time, which requires an adequate compilation of data and means of data processing and management to be set up.

The data compilation necessary for effective utilization of the bis and relief itineraries is accomplished at selected points. This allows the opening of the itineraries to be anticipated, or action to be taken directly on the traffic by measures such as manual operation of traffic lights.

Taking into account the geographical dimensions of the operation, it was decided to set up four control posts (PC), namely, three sector coordination centers (CCS) north, center, and south at Poitiers, Bordeaux, and Saint-Jean-de-Luz, respectively, and a principal coordination center (CPC) at Bordeaux, located in the regional center of information and road coordination.

7.3.2.2 Implementation and Operation. Traffic data are collected in a real-time operation mode in the CCSs. A decision on load relief strategies requires a reading on actual rates of flow and occupancy every six minutes. The CPC receives the data from the three centers every six minutes, which enables it to have an overall idea of the traffic on the RN 10 and the secondary network.

The automatic acquisition of traffic data is carried out by sensors of two types (radar with the Doppler–Fizeau effect and loops) permanently connected to the computer of each CCS by specialized Post and Telegraph lines. At each detection point, there is a cabinet in the form of an interchangeable block containing a detector, transmitter–receiver, power supply, and alarms. To fulfill the functions of visualization, control, and check, the CCSs are equipped with MITRA XV/21 computers (32 K words). For transmitting data between the CCSs and the field equipment, a frequency-multiplexing system is used, operating in the frequency range 300–2600 Hz.

The cyclic transfer of batches of data to the CPC is carried out by modems. A wall-map display shows the different routes and the various counters and indicator lights. It displays the traffic situation valid for the last six minutes in the form of traffic volumes and occupancy rates, the presence of queues, and the routes recommended by changeable direction signs.

The main functions of the CCS computers are:

- acquiring data and reports (panels)
- chronicling events
- elaborating measures (flow – road occupancy rate and queues)
- processing, taking into account possible requirements or commands (choice of itinerary)
- displaying information (synoptic panels and printing evaluations)
- managing periodic transmissions

The choice of itinerary is made every six minutes in accordance with the latest data acquired. The CPC is equipped with a Mitra XV CII computer (32 K words), which fulfills the following tasks:

- controls cyclic liaison for data transfer
- updates the readings (report–synoptic panel–consoles)
- computes the volumes of congestion and management
- manages the relations between peripherals

The traffic status along the whole route is indicated in the CPC by display equipment such as wall-map, alphanumeric, and graphic computer displays.

At CCS and CPC level, apart from the usual tasks of supervising the functioning of the system, the human role with regard to the automatic system consists of:

- opening and closing the bis itineraries according to data supplied by the automatic system
- possibly modifying the strategy of load relief by adjusting the road occupation rates or modifying their threshold
- collecting and broadcasting information, whether or not it comes from an automatic system

Finally, it should be pointed out that, during a period of very dense traffic, air surveillance of the whole route completes the automatic system. Each plane is directed from the CCS concerned in accordance with the information supplied by the automatic compilation of data.

Motorists are notified of the beginning of each bis or relief itinerary by a set of four or five indicator panels. These panels are variable with revolving prisms. When the itineraries are closed, they show a neutral grey face, so that motorists continue to use the principal route (cf. Figures 7.21, 7.22).

When they are open, the panels inform the motorists where the perturbations are located (for the relief itineraries), and the towns reached by the bis or relief itinerary. The itineraries themselves are marked out by specific arrows (green arrows for the bis and yellow arrows for the relief itineraries).

All these sets of signs are remotely controlled from the CCSs. The automatic functioning is controlled by the computers, which decide on the opening or closing of the panels.

As can be seen, Opération Atlantique astutely combines automatic and human means:

- the automatic means give principally a general idea of the traffic, and permit the management of the relief itineraries in real time
- the human means take over from the automatic means when these are not adapted, e.g., opening the bis itineraries, local action (manual operation of the traffic lights), or information concerning the length of the traffic jams.

7.3.2.3 Operational Experiences. In 1976, the Opération Atlantique composed five bis routes and seven relief routes (cf. Figure 7.20).

60 data compiling points (measuring flows and road occupancy rates) were spread out along the RN 10 and the secondary network. Table 7.6 illustrates that implementing the traffic control system in 1976 led to a reduction of congestion of the order of 30 percent compared with the situation in 1975. In France the degree of congestion is measured in hours-kilometers (h-km); 1 h-km corresponds to a line of traffic that is one kilometer long and that lasts for one hour. Table 7.7 compares the traffic congestion occurring for the three important holiday departure and return periods expressed in h-km. The total decrease is 2,139 h-km, i.e., 35.3 percent less in 1976 than in 1975.

For the above periods, the flow statistics show an increase in traffic demand for the departures (end June–beginning July and end July–beginning August) of about 9 percent, and an increase for the return traffic of 17.6 percent (end August).

The evaluation of the traffic congestion presented above already shows the effectiveness of the Opération Atlantique. The following economic evaluation supports this diagnosis from two points of view: (1) saving of time, and (2) saving of energy.

The economic benefit of saving time is calculated on the following basis:

- It is considered that 1 h-km of traffic jam corresponds to about 100 h lost for the community (the dynamic length of the vehicles in a traffic jam is estimated at 10 m, which corresponds to 100 vehicles in a traffic jam 1 km long).
- One hour lost in a traffic jam is estimated during the peak departure periods at 30 francs (10 francs per person transported, the vehicle occupation rate being three persons on average).

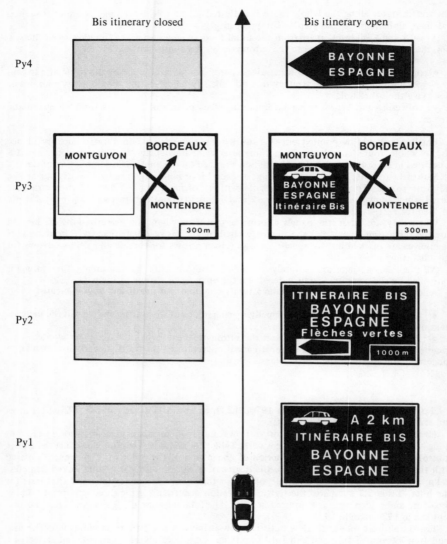

FIGURE 7.21 Beginning of a bis itinerary.

Over the whole of July and August, a comparison of the traffic congestion shows a decrease of 2,400 h-km. The saving in francs therefore works out to be 2,400 × 100 × 30 F = 7,200,000 F.

It should be noted that 4,000,000 francs were invested in 1976 to set up the operation, to which should be added 1,500,000 francs for a publicity campaign, i.e., a total of 5,500,000 francs. It can be seen, therefore, that from a strictly economic point of view the operation was made profitable in the space of one summer.

Certain studies in France have indicated that the excess consumption of a vehicle moving in a traffic jam for one hour is four liters (~ 1 gallon). The saving in energy due to the Opération

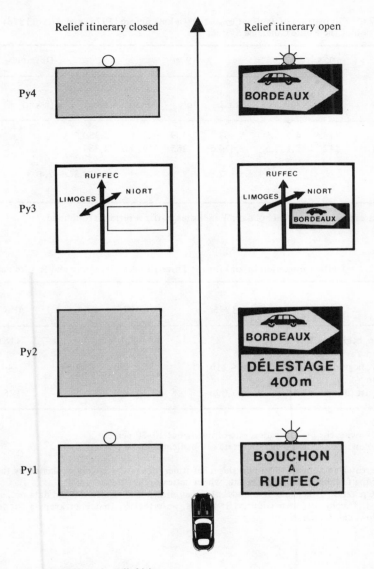

FIGURE 7.22   Beginning of a relief itinerary.

Atlantique during one summer is therefore substantial, since it appears to be 2,400 × 100 × 4 = 960,000 liters.

The reaction of the public to the introduction of the route guidance system is measured by the rate of obedience of the motorists in following the bis and relief itineraries. This rate is the ratio between the motorists following these itineraries and the motorists appearing at the entry sites.

For 1976, the average rates were as follows:

TABLE 7.6 Comparison of Traffic Congestion (h-km) Before (1975) and After (1976) the Introduction of the Route Guidance System

Traffic congestion[a]	1975			1976			Difference	
	July	August	Total	July	August	Total	(h-km)	(%)
Y(h-km)	749.5	4,713.5	5,463	2,039	1,311.5	3,350.5		
W(h-km)	542	2,132.5	2,674.5	385	1,999.0	2,384		
Y + W (h-km)			8,137.5			5,734.5	2,403	29.53

[a] Y indicates traffic leaving on holiday; W indicates traffic returning from holiday.

TABLE 7.7 Traffic Congestion (h-km) for the Three Holiday Departure and Return Periods

Period	1975	1976	Difference (%)
End June, beginning July (direction Y)	85.5	240.5	+181.2
End July, beginning August (direction Y)	4,116	2,298	−44.17
End August (direction W)	1,090	1,383.25	+26.9

- following the bis itineraries, rate of obedience 10–30 percent
- following the relief itineraries, rate of obedience 40–80 percent

The greater amount of traffic on the relief itineraries can be largely explained by the shock effect of the first panel at the entry site, which indicates "traffic jam at . . . ".

The Opération Atlantique is a model of the application of electronics and data processing to regulate traffic in the open country. It proves, nevertheless, that man keeps a fundamental role in this kind of operation.

8 Findings and Summary

Chapters 4-7 aimed to analyze the contribution of computer control technology to improving the efficiency of existing urban highway transport systems, by presenting a survey of

- basis systems concepts (Chapter 4)
- concepts and methods of control (Chapter 5)
- international experience gained so far with computerized area and freeway traffic control systems (Chapters 6 and 7)

Three basic systems concepts were introduced in Chapter 4: (1) traffic signal control; (2) comprehensive automobile control (CAC); (3) automated highways.

Computerized traffic signal control systems are in operation in many urban cities around the world. The CAC concept has reached the development status of large scale demonstration projects, and its general introduction can be expected for a small number of countries during the next ten years or so. The automated highway concept is still a subject of fundamental research and it is uncertain whether it will become feasible before the year 2000.

The existing control methods and concepts were analyzed in Chapter 5 for the following four categories of systems:

- route guidance systems (cf. Section 5.2)
- urban street traffic flow control systems (cf. Section 5.3)
- freeway traffic control systems (cf. Section 5.4)
- vehicle control systems (cf. Section 5.5)

International experiences with the first three systems categories were presented in Chapters 6 and 7. From these experiences and the methodological considerations of

Chapter 5 the following conclusions may be drawn focussing on technical, efficiency, economic, social, and institutional issues.

8.1 ROUTE GUIDANCE SYSTEMS

8.1.1 TECHNICAL ISSUES

Computer-aided route guidance by means of variable route signs represents a fairly well developed technology for systems of the one-to-one type. So far no route guidance system has been installed for complex many-to-many situations as they exist in large urban street networks. This is caused by several methodological, hardware, and human-engineering problems:

- in the case of many origins and destinations the route guidance algorithm and the corresponding travel time estimation methods become rather complex
- a human driver can become frustrated if he is faced with a forest of changeable route signs

The first problem can very likely be solved by the distributed microcomputer-based control concept described in paragraph 5.2.2.4. The second, i.e., the human-engineering problem, can by no means be avoided in many-to-many systems. Thus, this problem represents an incentive for the development of CAC systems. But, for the near future the CAC system cannot be considered as an alternative for systems based on variable route signs. Effective route guidance without changeable route signs would require that all vehicles or at least a sufficiently large part of them, are equipped with the necessary onboard equipment. Such a target can only be reached stepwise, and requires at least one decade for research, development, testing, and series installation.

8.1.2 EFFICIENCY ISSUES

Experiences in France demonstrate the capability of route guidance systems to reduce traffic congestion, expressed in queue length multiplied by waiting time, by about 30–70 percent under special traffic conditions such as: holiday departure and return periods; traffic during Olympic Games (in Grenoble). Reduction in congestion resulted in significant energy savings; for the project described in paragraph 7.3.2 fuel savings on the order of about 10 million litres per year could be estimated.

8.1.3 ECONOMIC ISSUES

The benefits achieved in the form of fuel and time savings balanced the initial installation costs in less than one year of system operation (cf. paragraph 7.3.2).

8.1.4   SOCIAL ISSUES

The feasibility and efficiency of route guidance depends almost completely on its public acceptance, i.e., on the willingness of drivers to follow the displayed route recommendations. Initially it was argued that only a small percentage of drivers would behave cooperatively. But, the French experiences made it clear that one can expect an acceptance rate of recommended alternative routes on the order of 40-80 percent.

8.1.5   INSTITUTIONAL ISSUES

The decision to create comprehensive route guidance systems has in general been made by national or regional governments in cooperation with local authorities. This is especially true for such innovative systems as the Japanese Comprehensive Automobile Control (CAC) System, which contains route guidance as its main features (cf. Sections 4.2 and 5.2). The project to develop technology for a CAC system was one of 12 major projects conducted with funds provided by the Ministry of International Trade and Industry National Research and Development Program. This program was launched by the Japanese Government in 1966. Projects receiving support from this program must satisfy, among others, the following conditions (cf. reference MITI 1975 to Chapter 4).

- the proposed technologies must be urgently needed, must be expected to play a leading role in technological progress, and cannot be developed by the private sector alone
- research and development must be carried out cooperatively by government, universities, and industry

There is no other country or national government that has assigned such a high priority to the development of route guidance systems based on the CAC concept.

8.2   AREA TRAFFIC CONTROL SYSTEMS

8.2.1   TECHNICAL ISSUES

A well-developed and sufficiently proven technology is available for implementing and operating centralized area traffic control systems. A whole arsenal of methods for determining optimal signal control strategies has been created during the last 10-20 years. To date, time-dependent or traffic-responsive selection of fixed-time signal plans, developed by off-line optimization techniques like TRANSYT, have resulted in the best improvements in traffic operations.

Existing centralized computer control and surveillance systems differ remarkably in scale and complexity. They range from mini systems, covering one complex or a few intersections, to giant systems, in which several thousands of intersections are coupled with the control center. Much progress could be made in improving the reliability of the latter systems by installing a hierarchy of computers in the control center and by introducing so-called graceful degradation design principles (cf. Section 6.2).

The most expensive and very often the most unreliable and troublesome system part is the communication subsystem required for linking the control center with local detectors and controllers. This is a characteristic of all types of centralized control systems, and provides the main incentive for creating a new systems generation, i.e., the generation of distributed computer control. The main feature of this new control systems generation is minimizing the amount of data that has to be exchanged between the control center and the local units, by providing the individual intersections with their own data processing unit in the form of a local microcomputer controller. This will lead to a further increase in reliability and flexibility, a reduction of the overall systems costs as well as to a simplification of computer software.

It is certain that the introduction of new control systems structures, i.e, of a new systems hardware generation, will initiate further development of the control strategies and the systems software in general. The installation of microcomputer based systems started recently in some places. It is expected that this traffic control system generation will find very broad application during the 1980s.

8.2.2 EFFICIENCY ISSUES

Experience so far with centralized area traffic control systems has demonstrated their capability to prevent accidents, alleviate congestion, save energy, reduce air pollution levels, and give priority to public transport means.

The following are typical improvements obtained for an operating system:

- a reduction of journey times by about 10-30 percent
- a reduction in the number of stops by 25-45 percent
- an increase in the mean speed by about 15-45 percent
- a reduction in the number of accidents by about 10-30 percent
- improvements in traffic flow conditions resulting in fuel savings of about 15 percent (cf. Table 6.10)
- estimates made for the Tokyo System demonstrated its capability to reduce the emission of carbon monoxide (CO) by about 35 percent, hydrocarbons (HC) by about 30 percent, and nitrogen oxides (NO_x) by about 10 percent (cf. Table 6.10)
- experiments made with bus priority signals demonstrated the feasibility of reducing the mean waiting time of buses at traffic signals by about 40-55 percent, and increasing the mean speed of buses by up to 15-17 percent

8.2.3 ECONOMIC ISSUES

The economic efficiency of computerized traffic control systems is in general assessed on the basis of the time savings achieved. Thus the results of the corresponding cost/benefit analysis depends strongly on the values to be attached to small time savings by various categories of road users. An evaluation basis for which a general consensus can more easily be obtained is to represent fuel savings by their monetary equivalent. Even if time and fuel savings alone are considered, it is found for almost all existing computerized area traffic control systems, that the initial installation costs are balanced by the benefits accumulated in the first six months of operation or even in a shorter time period. This estimation does not take into account the decrease of the number of accidents and of air pollution levels.

Such a favourable economic assessment exceeds the most optimistic expectation of any cost/benefit analysis in public works or business enterprises by at least one order of magnitude.

8.2.4 INSTITUTIONAL ISSUES

The role of city councils, governmental agencies and industrial companies in creating large scale computerized area traffic control systems was and still is different in individual countries, and has changed significantly within those countries during the last 10-15 years.

In the early years, in some countries, national government agencies had no or only slight involvement in the decision making process and the organization of the research and development work. The projects were started as joint efforts between the city and corresponding industrial firms, as mentioned in Section 6.1 with respect to the cooperation between the city of San Jose and IBM. In other countries, e.g., in the UK, governmental agencies played a leading role from the beginning.

However, the major research and development efforts that were initiated during the last 5-10 years became subjects of national research plans and international cooperation. This is especially true for the development and evaluation of advanced software packages such as UTCS, ASCOT, SIGOP, TRANSYT, RTOP, and others in the UK, USA, and Canada.

The most active involvement of governmental agencies in the decision making and project management process can be observed in Japan. Here the creation of large scale area traffic control systems represents a major task of the "Five Year Plans for Facilities Construction for Traffic Safety Installations", which has been set up by the National Police Agency. The first five year plan began in 1971 and a second one was started in fiscal year 1976. In these plans the creation of computerized traffic control systems in 28 cities throughout Japan is considered as a national policy, as pointed out in paragraph 6.3.1. A second interesting institutional factor concerns the project organization principles used in Japan. They are characterized by close cooperation between

- the corresponding cities (in general represented by the police departments)
- industrial companies
- universities

The whole research and development work is reviewed by special committees where, in the case of the Tokyo and Osaka systems, university professors have been appointed as committee chairmen.

8.3 FREEWAY TRAFFIC CONTROL SYSTEMS

8.3.1 TECHNICAL ISSUES

The feasibility of the following main tools of freeway traffic control has been proved in operational systems:

- ramp metering
- speed and lane use control, motorist information, and automatic incident detection
- merging control

But several technical problems are still the subjects of fundamental research and development work. Among these are

- developing dynamic, i.e., ramp feedback, control algorithms and the corresponding on-line parameter and state estimation methods (cf. 5.4.1.2 and Figure 5.16)
- developing spatially distributed, microcomputer based ramp metering systems (cf. Figure 5.17).

8.3.2 EFFICIENCY ISSUES

The achievable traffic improvements may be characterized as follows:

1. Merging control has the capability to improve traffic operations within the on-ramp region by

- reducing mean delays of on-ramp vehicles by about 25 percent
- reducing accident frequency by about 25–30 percent
- increasing ramp capacity by about 10–15 percent

In spite of these significant benefits there is only a small number of merging control systems in operation; all of them are located in the USA. Two reasons can be mentioned for this situation:

- high installation costs and the complexity of the control task
- a real need for merging control arises only in very complicated situations such as poor visibility conditions, very short acceleration lanes, and high accident rates

2. Speed and lane use control in general becomes necessary for long tunnels, freeway bottlenecks (e.g., caused by construction work), or a freeway section that experiences high accident rates. Speed control can lead to a reduction of accident rates by about 25–65 percent; variable speed control is primarily a safety tool but does not help significantly in reducing congestion (cf. 7.3.1).

Reversible lanes have contributed considerably to an increase in capacity and speed.

Computer controlled priority lanes for buses and car pools have so far been used on a broad scale only in the USA, where a reduction in the mean times for bus trips of about 35 percent has been achieved. However, this benefit has to be paid for by a significant increase in automobile trip times.

Motorist information systems, which use Bulb Matrix Message Signs, and automatic incident detection systems have demonstrated their potential in

- preventing accidents
- reducing the time required for detecting accidents and providing urgent medical and other assistance.

So far the high installation costs of these systems have only justified their implementation for the most heavily travelled freeways such as in Los Angeles.

3. Ramp metering, i.e., coordinated control of the inflow rates at the individual entrance ramps, represents the main tool of freeway traffic flow control. International experience gained so far has demonstrated the capabilities of ramp metering systems

- to reduce congestion with regard to both duration and size
- to increase mean travel speed by up to about 55 percent (cf. Table 7.1)
- to reduce the number of accidents by about 8–30 percent (cf. Table 7.2).

There is an unusual phenomenon of freeway ramp control that is not widely understood. Ramp control improves the traffic flow not only on the freeway itself, but also on the adjacent surface streets. At first glance, this seems impossible, since reducing inflow rates at freeway ramps must divert some vehicles from the freeway, which, one would assume, could only produce congestion on the surface streets. However, ramp control improves the efficiency of the freeway itself, actually enabling it to carry more vehicles and resulting in a net benefit to the whole freeway corridor.

8.3.3 ECONOMIC ISSUES

A cost/benefit analysis based on an evaluation of time savings leads to favorable results similar to those summarized for area traffic control systems in Section 8.2.

It is a concern in such cost/benefit analyses that the installation of a freeway traffic control system may lead to only temporary travel time reductions or to none at all (cf. Table 7.1 and paragraph 7.2.3). This may be caused by a significant increase of traffic demand during or after the installation of the control system, which leads to traffic speed decreases once more to the "no-control level" (cf. Table 7.1). In such a situation the benefits are estimated in terms of the increased annual costs that would result from removal of the traffic control system. These increased annual costs, i.e., the so-called "disbenefits", would balance the installation costs in less than one year of operation (cf. Table 7.4).

It should be mentioned here that there is a school of thought that does not accept these types of cost/benefit considerations for the following reasons (cf. paragraphs 6.3.2.3 and 7.2.3):

• traffic control and information systems should be understood as an indispensable social service, i.e., in the same way as the role of street lighting and other road safety equipment is presently understood
• the installation of these systems consequently has to be considered essential if there is a real demand for them
• since the costs of the computer control system are very low compared with those of the freeway itself, it seems justified to consider a computerized traffic control system as an indispensable, integral part of a heavily travelled freeway

8.3.4 SOCIAL ISSUES

Limiting freeway access by ramp metering can cause public opposition from certain groups of drivers, who could feel restricted in using the freeway, "which has been constructed by their taxes" (cf. French experience mentioned in paragraph 7.3.1).

This factor calls for a special attention to public relations, e.g., the opening of a ramp metering system and its operating principles have to be publicized.

8.3.5 INSTITUTIONAL ISSUES

The decisions to create large scale freeway corridor control systems were in general made by the corresponding national governmental agencies in cooperation with city governments and local public authorities, e.g., the Expressway Public Corporations in Japan.

The case descriptions of Chapter 7 underlined the need for close involvement of all agencies during all phases of the development and installation of the computer control system. For this purpose, special reviewing committees, consisting of representatives of the various agencies (governmental offices, local authorities, research institutes, and universities), were established for some of the projects described.

In the case of the Dallas Freeway Corridor Control System, a three-level structure of project management was adopted. The two most important management levels were (cf. paragraph 7.1.2):

- policy-making level, represented by the administrative review group, which considered decisions on funding, project priority, and level of involvement, and held meetings at 4-6 month intervals
- operational level, represented by the Technical Review Board, which discussed details of design, construction, and operation, and held meetings once a month

Another institutional aspect that should be mentioned here concerns the operation of freeway corridor control systems under special traffic regulations. In France, for example, the freeway police corps are only in charge of freeway traffic, the adjacent surface streets being taken care of by the urban police corps. In the Japan Expressway, public corporations operate the freeway control systems but they do not have the legal force to control the drivers. Area traffic control systems, on the other hand, are operated by the police departments.

These and similar institutional factors may become important if coordinated operation of the freeway and the area traffic control systems are to be established, leading to the need for close cooperation between various police corps and public authorities.

8.4 VEHICLE CONTROL

8.4.1 TECHNICAL ISSUES

The creation of computerized onboard control systems concerns above all three task categories:

- engine/powertrain control
- headway control
- onboard units for route guidance in the framework of the CAC concept

All three control systems are still at the research and development stage where the major technical barrier for their installation in series-manufactured motorcars is the development of sufficiently small, reliable, robust, and cheap sensors and actuators.

8.4.2 EFFICIENCY AND ECONOMIC ISSUES

Computerized onboard systems are expected to deliver three main types of effects:

- reductions in fuel consumption by 10 percent or more
- essential reductions in emission rates of pollutants
- increases in safety

So far the available test results have not provided sufficient information for a reliable cost/benefit consideration.

8.4.3 SOCIAL ISSUES

There are many social factors that could influence such a basic innovation process as the introduction of microcomputers in automobiles. One key problem concerns the question; "What additional amount of money is the driver willing to pay for the above-mentioned benefits?" This question cannot yet be answered reliably.

8.4.4 INSTITUTIONAL ISSUES

The incentives for accelerated development of microcomputer based onboard systems have been generated by the following institutional factors:

- enactment of emission laws in the USA and other countries
- possible future fuel economy legislation
- governmental policies for increasing traffic safety

These institutional factors have contributed to increased efforts by the automobile and electronic industries to introduce microelectronics into cars.

Thus, inspite of the above-mentioned uncertainties and barriers, it is not unlikely that the computerized onboard control systems analyzed in Section 5.5 will become a reality during the next decade.

Part Three

Control and Monitoring of Public Transport Systems

Part 3 analyzes the role of computers in improving the efficiency and attractiveness of existing public transport systems. A survey of basic systems concepts is given in Chapter 9, while Chapter 10 reviews important concepts and methods of control and surveillance. International experience gained with computerized dial-a-ride, bus monitoring, and rapid rail transit systems is evaluated in Chapters 11–13.

9 Basic Systems Concepts

Existing public transport systems are faced with two problems, as discussed in Section 1.5:

- decreasing attractiveness in comparison with the private automobile
- decreasing efficiency resulting from reduced ridership and/or increased operational costs

How can computer control technology help to lower the negative effects caused by these problems? Figure 9.1 illustrates that the following requirements have to be considered:

- providing more flexibility with regard to changing traffic demand
- improving operational reliability and adherence to published timetables
- assisting passengers in using the system, and improving the quality of service
- reducing the number of personnel required, and providing better working conditions for the remaining staff
- making optimal use of available vehicle fleets, and minimizing operational costs

These requirements lead to two types of concepts: (1) improving the operation of existing systems; (2) introducing operational innovations, i.e., changing the basic operating principles of existing systems.

The first concept includes three basic systems categories:

- train operation and traffic control systems
- bus/tram operation and traffic control systems
- passenger guidance and information systems

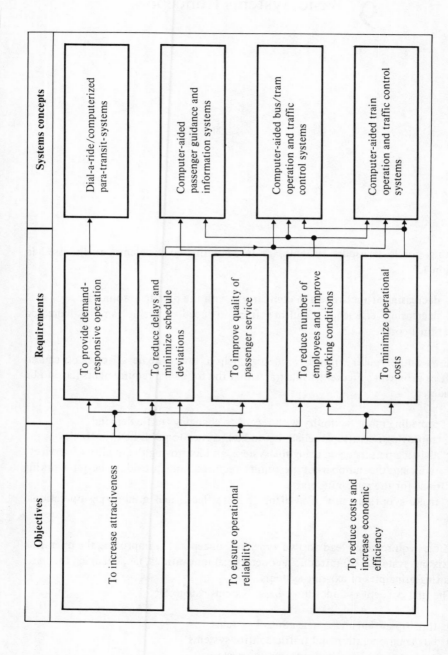

FIGURE 9.1 Concepts of public transportation control systems.

The second concept leads to the so-called

- dial-a-ride concept (cf. Slevin 1974)
- computerized para-transit systems (cf. Kirby 1974)

9.1 TRAIN OPERATION AND TRAFFIC CONTROL SYSTEMS

The role of computers in increasing the attractiveness and efficiency of rapid rail transit systems concerns three categories of tasks (cf. Figure 9.2):

1. *Operation planning*

- computer-aided preparation of train timetables, crew schedules, and rolling stock rostering and maintenance plans

	Rail rapid transit	Bus/tram transit
Operation planning		• Route planning
	• Timetable preparation, crew, rolling stock, and maintenance planning	
	• Evaluation of operation data regarding changing travel patterns, schedule disruptions, accidents, etc.	
Operation/ traffic supervision and control	• Automatic identification of vehicle locations, and timetable disruptions • Dispatching control (minimizing timetable deviations) • Passenger guidance and information • Electric power supply control • Control of depots operations and environmental conditions	• Priority control
Vehicle protection and control	• Control of individual moving states (acceleration, cruising, precise station stopping) • Control of complete train running process	(without special significance for conventional systems)

FIGURE 9.2 Tasks of computerized planning and control systems.

- computer-aided compilation of statistics, evaluation of actual train operation data regarding schedule disruption, accidents, changing traffic demand patterns, etc.

2. *Operation and traffic supervision and control*

- automatic monitoring of train locations, determining deviations from schedules, and deriving measures for minimizing schedule deviations
- passenger guidance and information
- monitoring depot operations
- electric power supply supervision
- control and surveillance of environmental conditions in subway tunnels, etc.

3. *Train protection and control*

This task involves automatic control of the movements of individual trains regarding

- the individual moving states, e.g., by speed and acceleration control, precise station stopping
- the whole train running process, e.g., by energy-minimal train control

The present monograph mainly covers the second and third task categories, where special attention is paid in Chapters 10 and 13 to automatic train traffic supervision (ATS), and automatic train operation (ATO).

Passenger guidance and information systems are important for all modes of urban transportation. Therefore, they are treated as a special and more or less independent system in the following (cf. Sections 9.3 and 10.4).

9.2 BUS/TRAM OPERATION AND TRAFFIC CONTROL SYSTEMS

The contribution of computers to improving the attractiveness and efficiency of urban bus and tram systems concerns two types of complex tasks (cf. Figure 9.2).

1. *Operation planning*

- bus route planning
- preparation of bus and tram schedules, crew schedules, maintenance, and rolling stock rostering plans
- evaluation of operation data and preparing statistics

2. *Operation and traffic supervision and control*

- automatic identification of vehicle locations and timetable deviations
- dispatching control, i.e., operational modifications of bus/tram operation

regimes in order to minimize timetable deviations and to meet special demand requirements
- priority control at signalized intersections
- passenger guidance and information

This Section focusses on the second task category, with two of the tasks, automatic vehicle location and dispatching control, and priority control, being analyzed in detail in Chapters 10 and 12, respectively.

9.3 PASSENGER GUIDANCE AND INFORMATION SYSTEMS

Passenger guidance and information systems are a major area for computer applications. For urban transit, two categories of tasks are of primary interest (cf. Figure 9.3):

1. *Automatic fare collection*

- ticket vending and checking
- accounting sales

2. *Automated passenger information*

- providing schedule information
- displaying and broadcasting information on operations and conditions, e.g., departure platforms, delays, etc.

Moreover, other services like automated seat reservation and automatic luggage registration or automatic left-luggage offices either already are or are about to be highly computerized processes.

The computerized passenger service systems that are relevant to urban transit are analyzed in Section 10.4.

9.4 THE DIAL-A-RIDE CONCEPT AND COMPUTERIZED PARA-TRANSIT SYSTEMS

The basic aim of introducing operational innovations into existing urban transportation systems is to provide the possibility for adapting transport supply to changing traffic demand. This means that the operation of a fixed route and fixed schedule transport system must be changed to a more flexible operational mode, characterized by changeable or completely variable driving routes and schedules. The resulting

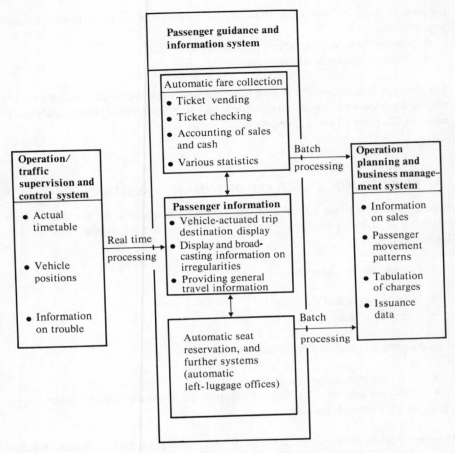

FIGURE 9.3 Tasks of passenger guidance and information systems.

urban transport concept became known under such names as dial-a-ride, dial-a-bus, bus–taxi, demand-bus systems.

Figure 9.4 illustrates the desired operation mode. The prospective passenger is expected to ring the control center, either from his home telephone or by direct line free telephones at fixed stops, explain his desired destination, the point of origin, and give the number of people who want to make the journey (cf. step 1 in Figure 9.4). The control center identifies the location of an appropriate bus (cf. step 2), and gives the passenger the approximate arrival time, trip duration, and expected fare (cf. step 3). If the passenger accepts the offer (cf. step 4) then his request is written onto an appropriate bus tour schedule and passed to the driver, either by hand, if the bus starts its tour from the control center, or by two-way radio.

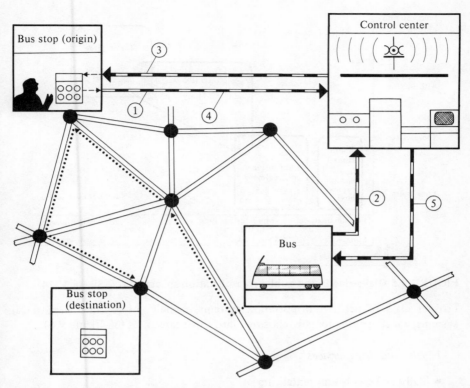

FIGURE 9.4 The dial-a-ride concept (cf. reference Peckmann 1975, to Chapter 11).

Such a concept obviously covers a wide variety of transport demands, broadly filling the gap between the conventional fixed-route bus service and the taxi or private car (cf. Figure 9.5). Compared with an ordinary taxi service it has the advantages that (1) more than one party may take the vehicle at any given time; (2) the routing can be programmed to yield a combination of high use and reasonable waiting and trip time; and (3) the fares may be substantially lower.

What role will modern computer technology play in the implementation of such a demand-responsive transport concept? If only a few buses are operating in a network with a small number of origin and destination zones, then the bus dispatching work can be done by a human controller. However, for many buses, origins, and destinations, a sophisticated computer system is required to assist the dispatcher in real-time scheduling and routing of the various trips.

In addition to introducing operational innovations into conventional bus services the dial-a-ride concept described here can also be used to improve the operation of a taxi fleet.

From this point of view the dial-a-ride concept must be considered as a part of a

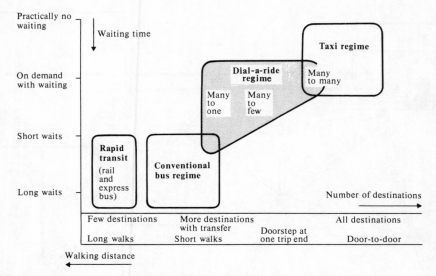

FIGURE 9.5 Dial-a-ride in the public transportation spectrum (cf. Slevin 1974).

broader class of urban transportation systems, namely, of so-called para-transit systems, which may be roughly classified into three categories (cf. Figure 9.6):

1. *Hire and drive services*

- daily and short-term rental cars

2. *Hail or phone services*

- regular taxis
- dial-a-ride systems
- jitneys

3. *Pre-arranged ride-sharing services*

- car pools
- subscription bus

Figure 9.6 illustrates the basic features of these para-transit systems and the specific place of dial-a-ride within the framework of para-transit.

REFERENCES

Kirby, R. F., et al. 1974. Paratransit: Neglected Options for Urban Mobility. Washington, DC: The Urban Institute.
Slevin, R. 1974. The First British Dial-a-Ride Symposium. Conference Proceedings, Center for Transport Studies, Cranfield, UK.

| | Hire and drive services | | | Hail or phone services (bus, taxi) | | Pre-arranged ride-sharing services | |
	Daily rental car	Short-term rental car	Conventional taxi	Dial-a-ride	Jitney	Car pool	Subscription bus
Spatial factors (Location trip-length)	• Airport access • All types	• All types	• Short trips • CBD oriented • Small cities • Airport access • Line-haul feeder	• Short trips • Intra-suburb • Inter-suburb • Intra-CBD • Small cities • Line-haul feeder	• Tourist corridors • House-to-work corridors • Business/commercial corridors	• Long trips • Home-to-work corridors • Line-haul feeder	
Temporal factors (Time of day)		• All hours	• All hours	• Midday • Early morning • Late evening	• Peak hours • Midday	• Peak hours	
Demographic (Traveller/trip purpose)	• Out of town visitors • Business trips • Recreation trips	• All trip purposes of residents living in medium to high density areas	• Professional and managerial • Out of town visitors • Limited mobility groups	• House wives shopping and personal business • College students • Business trips • Limited mobility groups	• Tourists • Workers • Business trips • Shopping trips	• Workers of all income groups • Shopping	• School trips

FIGURE 9.6 The dial-a-ride concept as part of para-transit services (cf. Kirby 1974).

10 Concepts and Methods of Control and Surveillance

It is only partly possible to assign the control and surveillance methods necessary for implementing the basic systems concepts discussed above to the individual levels of the general control tasks hierarchy introduced in Chapter 3 (cf. Figure 3.4). Figure 10.1 illustrates that the three levels of the general control tasks hierarchy according to Figure 3.4 include the following functions for public transport systems:

Level I: Route Guidance. This level is obviously of special interest only for the dial-a-ride concept, which needs methods of real-time routing and scheduling for its implementation. For conventional public transport systems, route guidance only becomes important in exceptional situations, e.g., for very large timetable disruptions, accidents or vehicle breakdowns.

Level II: Traffic Flow Control. The following tasks have to be assigned to this level:

- passenger flow control, including ticket vending, ticket checking, information display and broadcasting, and other services
- control and surveillance of train traffic, including tasks such as automatic train traffic supervision and dispatching control
- control and surveillance of bus/train traffic, including functions such as automatic vehicle monitoring, dispatching control and priority control

Level III: Vehicle Control. This level is only of special importance for problems of automatic train protection and operation.

10.1 ROUTE GUIDANCE IN DIAL-A-RIDE SYSTEMS

The demand-responsive operation of any dial-a-ride system is characterized by three features:

	Bus systems	Tram systems	Rapid-rail systems
Level 1 Route guidance	The dial-a-ride concept (cf. Section 10.1)	Not applicable — except in emergency situations	
Level 2 Flow control (vehicles *and* passengers)	Automatic fare collection and passenger information (cf. Section 10.4)		
	Control and surveillance of bus/tram traffic and operation, including (cf. Section 10.2) • Automatic vehicle monitoring • Dispatching control • Priority control		Control and surveillance of train traffic and train operation, including (cf. Section 10.3) • Supervisory control • Automatic train operation • Automatic safeguarding systems
Level 3 Vehicle control	Not applicable		

FIGURE 10.1 Survey of basic urban transit control and surveillance concepts.

- basic operating principles
- basic control systems structures
- control and dispatching methods

10.1.1 BASIC OPERATING PRINCIPLES

Six basic operating principles may be distinguished, which range from very simple modifications of fixed-route systems to rather complex operation modes (cf. Figure 10.2):

• The varied stop principle corresponds to a jitney-type operation, i.e., the vehicles are operating on fixed car routes, but they are allowed to make demand stops between the regular stops.

• The route deviation principle permits the drivers to leave a nominal route in order to pick up passengers or drop them off.

Operation mode	Principle	Features
Varied stops	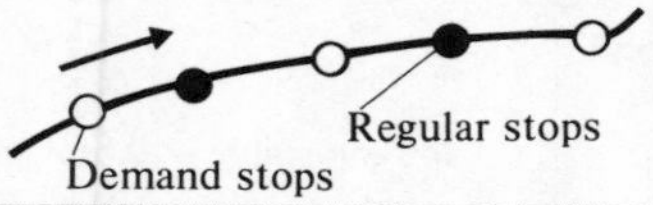 Regular stops / Demand stops	Jitney-type service
Route deviation	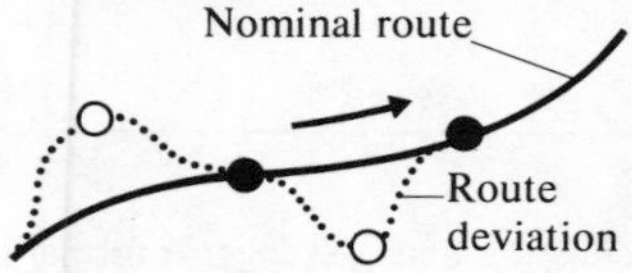 Nominal route / Route deviation	Tourist corridors
Point deviation	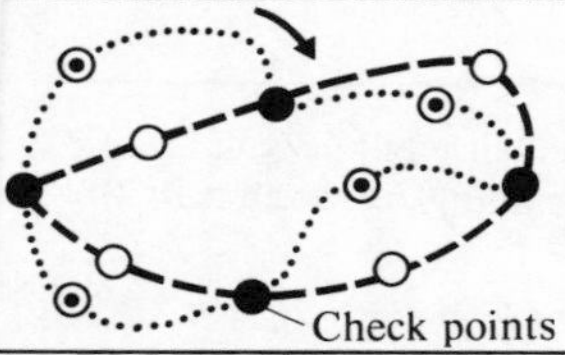 Check points	Check points: railway stations, fixed-route bus stations, etc.
One-to-many	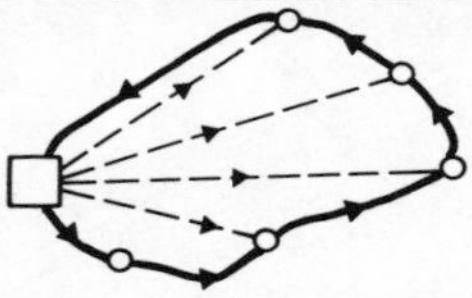	Scattering traffic— afternoon peak periods (work-to-home)
Many-to-many	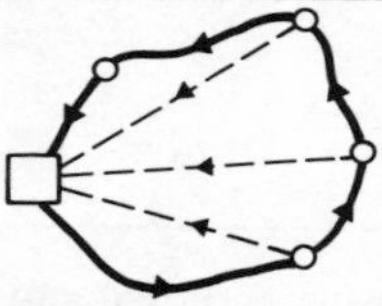	Gathering traffic— morning peak periods (home-to-work)
Many-to-many	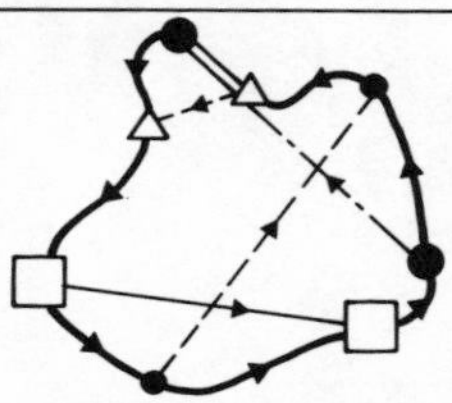	Gathering traffic— early morning, late evening, midday (shopping)

FIGURE 10.2 Survey of basic operation modes of demand-responsive para-transit systems.

- The point deviation principle does not use a nominal route, but certain checkpoints that the drivers are expected to serve at prescribed times. These checkpoints could be railway stations, fixed-route bus stations, etc.
- The one-to-many or scattering principle is characterized by one main origin and many destinations.

- The many-to-one operation mode aims to provide a demand responsive gathering service for a situation that is characterized by many origins and only one destination.
- The many-to-many principle provides a demand-responsive service for many origins and many destinations, and represents the most general and complicated operation mode.

It is possible to have a dial-a-ride system that uses only one of the principles summarized here. However, more advanced systems are able to change from one operation mode to another in the course of a day, for example, according to the following cycle:

1. Early morning (diffuse trip patterns) — many-to-many
2. Morning peak hours (trips from homes to work) — many-to-one
3. Midday (diffuse trip patterns) — many-to-many
4. Afternoon peak hours (trips from work to homes) — one-to-many
5. Late evening and night (diffuse trip patterns) — many-to-many

Moreover, the spatial decomposition of a complex many-to-many system can become desirable and useful. Here the whole service area is divided into several zones that touch each other at a single point, e.g., at a rapid rail or fixed-route bus station (cf. Figure 10.3). In this way several many-to-one or one-to-many systems are obtained, instead of a rather complex and complicated many-to-many structure.

10.1.2 BASIC CONTROL SYSTEM STRUCTURES

The required control system structure depends heavily on the chosen operation mode, the vehicle fleet size, and the number of origins and destinations. Three categories of system structures may be distinguished (cf. Figure 10.4):

1. *Completely decentralized systems.* Here no control center exists and the decisions on varied stops, route deviations or point deviations are made by the drivers themselves (cf. the system of Mansfield, Ohio, USA in reference Guenther 1970, to Chapter 11). It is obvious that such a system structure is only feasible for the very simple operation modes shown in Figure 10.2.

2. *Non-computerized central control systems.* At present, this is the most widely used system structure. The individual vehicles are linked to the control center via a two-way radio communication system (cf. Southall 1974, Robinson 1974). A telephone is used for communication between prospective passengers and the dispatchers in the control center.

3. *Computerized central control systems.* The introduction of computer-aided dispatching control systems becomes necessary if the number of operating vehicles gets larger than about 10 or 20. The computer is used to assist the dispatcher in

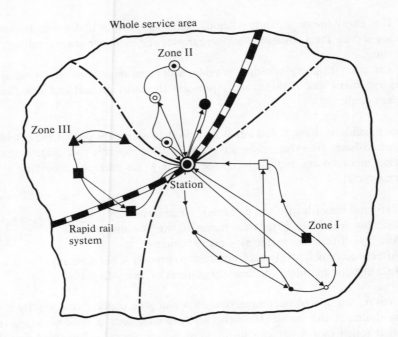

FIGURE 10.3 Division of a large service area (many-to-many system) into several zones (many-to-one, one-to-many systems).

- estimating travel and waiting times
- determining optimal driving routes
- calculating fares
- preparing various statistics, etc.

(cf. case description 11.1 for more details).

Moreover, the use of a digital computer provides the possibility of handling more effectively the information exchange processes between the control center and the vehicles and bus stop installations. For this purpose, the two-way voice radio communication system mentioned above is supplemented by digital data links that are used for the following tasks:

1. Partial automation of the processes of information exchange between the control center and equipment installed at the various bus stops. These bus stop equipment installations include, in special systems concepts (cf. reference Peckmann 1975, to Chapter 11), such functions as:

- keyboards for inputting the trip requests (time of departure, destination, number of persons)

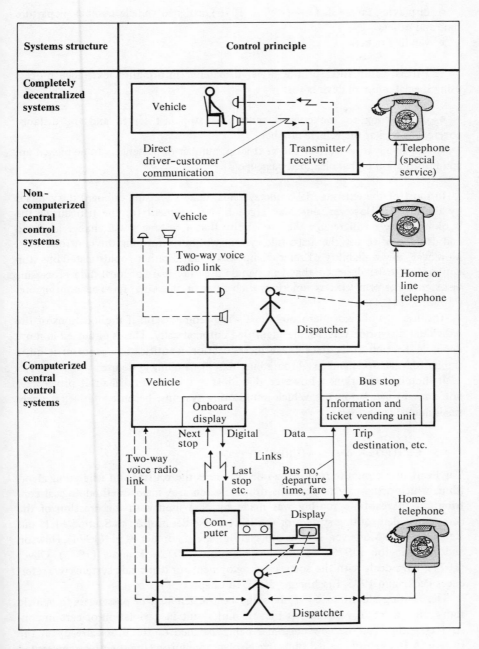

FIGURE 10.4 Basic control system structures.

- displaying fares (cf. Copper et al. 1976), assigned vehicle number, departure time and others
- vending tickets

2. Partial automation of the driver–dispatcher information exchange process using special onboard devices for:

- transmitting to the control center the stop last served and the distance covered, as well as the vehicle number
- displaying to the driver the next stop, number of passengers to be picked up, and information on handicapped passengers

In most of the existing dial-a-ride systems, tickets are sold onboard the vehicles by the drivers. However, one may argue that a payment for the trip during the booking process will reduce the probability that a passenger will change his mind and decide not to use the dial-a-ride system without informing the control center. Moreover, the availability of microcomputers enables rather sophisticated bus stop equipment to be designed that can handle a certain part of local data processing tasks, and perform various functions such as error checking, fares accounting etc. (cf. Section 10.4).

The use of onboard microcomputers will become useful if the locations of the individual dial-a-ride vehicles are identified automatically. This is discussed in more detail in Section 10.2, since the vehicle location identification problem is quite similar for dial-a-ride and fixed-route bus/tram dispatching systems.

It should be mentioned, however, that most of the current dial-a-ride projects do not have to use automatic vehicle monitoring systems, because of the relatively small number of vehicles.

10.1.3 CONTROL AND DISPATCHING METHODS

The heart of a computerized dial-a-ride system is the routing and scheduling algorithm, which assigns passenger trips to vehicle tours. A major methodological contribution to real-time routing was made by N. Wilsen with the creation of the computer aided routing system or CARS (cf. case description in Section 11.1, and Rebibo 1974). Moreover, several other authors, e.g., Breuer et al. (1974), Howson and Heathington (1970), Mason and Mumford (1972), Manlow (1973), Oliver (1974), have dealt with the software development for dial-a-ride systems (cf. references Peckmann 1975, Etschberger 1975, to Chapter 11).

The routing algorithm determines the best assignment of passengers to vehicle tours from a set of assignments that do not result in a violation of certain constraints, either for the new customer being assigned or for those already in the system. A frequently applied objective involves minimizing the function represented by the following expression (cf. Rebibo 1974):

$$Q = T_{NP} + T_{sp} + T_d = \text{minimum}$$

where T_{NP} is the increase in the length of a tour caused by new passengers (including the time required to serve the pick-up and delivery stops), T_{sp} is the increase in delivery times for customers already scheduled, and T_d is the delivery time for the customer being scheduled.

The service constraints concern maximal admissible values of waiting time, travel time, and total time from call-in to delivery. Moreover the number of available buses plays the role of a constraint variable (cf. Section 11.1 for more details).

The routing algorithm can be significantly simplified by reducing the number of admissible origins and destinations, e.g., by restricting the pick-up and delivery points to specially equipped, fixed bus stops (cf. reference Peckmann 1975 to Chapter 11).

10.2 BUS/TRAM MONITORING AND CONTROL

Bus/tram control can be achieved in two different ways:

- monitoring and dispatching control of a fleet of vehicles (buses or trams) by a specially equipped control center
- priority control of buses and trams in signalized networks, within the framework of computerized traffic light control according to Part 2 of this work

The first task includes two functions (cf. Figure 10.5):

- automatic vehicle monitoring, i.e., identifying the locations of the individual vehicles and comparing them with the locations described in the schedules
- dispatching control, i.e., deriving a suitable control strategy that could reduce schedule disruptions and deal with other irregularities

Conventional systems try to solve these tasks manually using three types of information sources:

- bus drivers phoning the dispatching center from special telephone report points
- pointmen (stationary supervisors) reporting on schedule adherence by phone from strategic points
- mobile supervisors who observe line conditions throughout the system

The information exchange processes can be improved by linking the control center with the individual vehicle drivers via voice radio links. However, in the case of large bus/tram transit systems operating several hundred or even several thousand buses and trams, the voice communication process can become complicated (cf. case descriptions of Chapter 12 for more details). Therefore, there is a need for

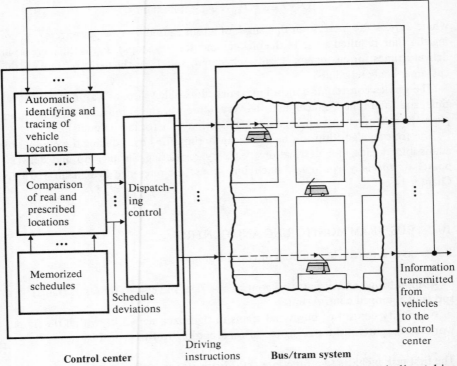

FIGURE 10.5 Basic scheme of automatic vehicle monitoring and dispatching control.

automation of at least a large part of the processes of information exchange between control center and vehicles. The availability of powerful process computers has provided the possibility of solving these automation tasks in which automatic vehicle monitoring represents the key problem.

10.2.1 METHODS OF AUTOMATIC VEHICLE MONITORING

Two categories of methods or techniques are required for implementing automatic vehicle monitoring systems:

- methods of automatic vehicle location
- methods of vehicle polling, i.e., techniques for organizing the automatic data exchange between the digital computer located in the dispatching center and the individual onboard devices

1. *Automatic vehicle location.* The automatic identification of the locations of vehicles operating in an urban area represents a problem of general importance, i.e.,

it not only concerns fixed-route public transport systems but also the management of fleets of law enforcement cars, para-transit vehicles (taxis, dial-a-ride buses), medical emergency vehicles, delivery and repair service cars, postal service cars, and others.

Therefore, significant research effort has been made to develop usable multivalent location identification methods (cf. Buck and Salwen 1974, Gould 1973, Hansen 1976, Ritter and McCoy 1977, Roth 1977, Symes 1977). Three basic principles are considered here (cf. Figure 10.6): (a) dead reckoning; (b) proximity or signpost method; (c) radio location methods.

(a) Dead reckoning: Dead reckoning methods locate a vehicle by computing its distance and direction of travel from a known fixed point, e.g., from special bus stops.

For fixed-route systems the only equipment necessary is a precision odometer for distance measurement; variable-route systems also require some compass-type device. A radio link is used to transmit the measured distance and — if necessary — certain azimuthal information to the control center.

The advantages of this location principle are low installation costs. Certain problems may result from the accumulation of small errors in distance measurement, which can lead to large position errors if the distance from the above-mentioned fixed point becomes large. To compensate for this error, the distance measurement system must be updated on a regular basis. This can be done manually, by comparing the vehicle's route with a known or feasible route, or by combining dead reckoning with the second method, i.e., the signpost principle.

(b) Proximity or signpost method: This method provides the locations of vehicles by determining the relationship between the individual vehicles and fixed locations strategically placed throughout the network (cf. Figure 10.6, part (b)). These fixed locations are indicated by so-called signposts, i.e., by special passive or active devices in the form of magnetic sensors buried in roadways, ultrasonic and optical radiators, or conventional radio transmitters or receivers. There are two types of vehicle–signpost interactions:

- The signpost equipment identifies the passage of a vehicle and transmits a corresponding signal to the control center via land-lines. The identification of the passage of a vehicle can be achieved by an active onboard unit that emits some sort of continuously low level coded signal, or by a passive bus unit, e.g., a reflecting plate.
- An onboard device identifies that a signpost has been passed, which is continuously emitting certain frequencies. This can be used to update the above-mentioned onboard distance measuring system. The vehicles are regularly polled by the control center to record the updated distance measurements.

(c) Radio location methods: This category of methods attempts to locate a vehicle by direct measurements of radio signals traveling between the vehicle and

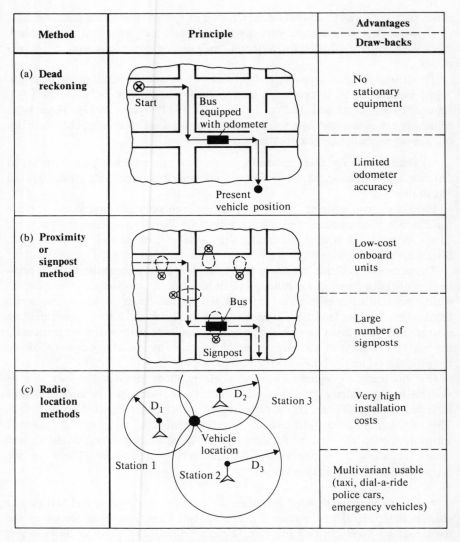

FIGURE 10.6 Automatic vehicle location methods (cf. Ritter and McCoy 1977).

some fixed stations (cf. Figure 10.6 (c)). Such methods have been used for years in a variety of military and commercial applications. Most widely used are trilateration methods, which use the measured distances D_1, D_2, and D_3 between the vehicle and at least three stations to identify the vehicle location. There are fairly well developed methods available, such as Omega, Decca, and Loran, which are used in standard marine and aeronautical navigation systems. Research work is underway to adapt these systems for vehicle location problems occurring in urban areas (cf.

Ritter and McCoy 1977, Symes 1977, Hansen 1976). However, the resulting systems costs are rather high and very likely are not justified for application to fixed-route public transit systems alone. Therefore, such radio location methods only become important for bus/tram monitoring systems in the framework of so-called multi-user area-coverage automatic vehicle monitoring systems, which can serve not only public transportation systems but also other urban systems, e.g., various urban emergency systems (cf. Symes 1977).

At present, only the dead reckoning and proximity methods have been used in bus monitoring systems, as illustrated in the case descriptions in Chapter 12.

2. *Vehicle polling techniques.* Three principles of vehicle polling may be distinguished: (a) synchronous polling; (b) commanded or random access polling; (c) volunteer polling.

(a) Synchronous polling. In this technique each vehicle transmits location data at a preselected time within a polling sequence. The equipment on the vehicle keeps track of the start of the polling sequence and internally determines when the appropriate time to respond occurs. The start of the polling sequence must be periodically transmitted to each vehicle for correction purposes.

(b) Command or random access polling. Here the control center sends a request to each vehicle whenever location data are required. This random access technique is the most flexible but requires substantially more time than synchronous methods.

(c) Volunteer polling. This method is vehicle oriented, i.e., each vehicle has to check first whether the channel is "clear", and, if so, transmit the location data to the equipment in the control center. After determining that the channel is clear, before transmitting it is usually necessary to employ some technique to give a random delay to each vehicle to ensure that certain vehicles dominate in the use of the channel (cf. Hansen 1976, for more details).

Which polling method appears to be most advantageous depends on various factors, such as the size of the vehicle fleet, and the performance of the onboard equipment. In case of a large number of vehicles, the synchronous polling principle must in general be used in order to meet the complicated time requirements. The application of the other two polling principles can become desirable for relatively low polling frequencies, which may result either from a relatively small number of vehicles, or from providing the vehicles with their own data processing unit, i.e., with an onboard microcomputer.

An onboard microcomputer could memorize the bus schedule and compare the prescribed location with the actual one. If there are no significant deviations then there is obviously no need to inform the control center of the vehicle's location. Thus the data exchange between the control center and the vehicles can be restricted to those cases where schedule deviations exist. The onboard system has then to poll the computer in the control center, which leads obviously to the above-mentioned volunteer polling method.

Summarizing, it is clear that two basic hardware structures have to be taken into account (cf. Figure 10.7):

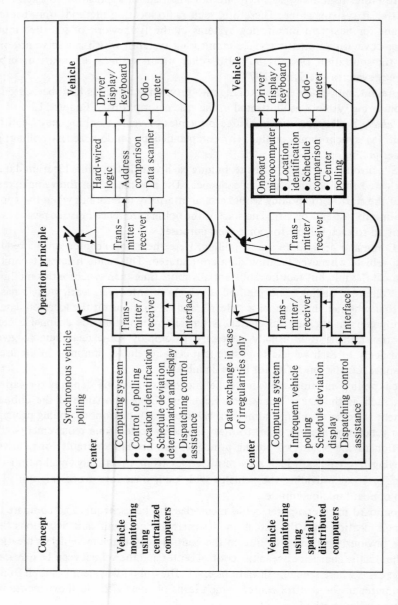

FIGURE 10.7 Vehicle monitoring based on centralized or distributed computing systems.

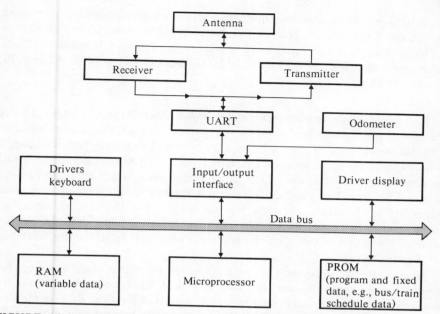

FIGURE 10.8 Block diagram of an onboard microcomputer suitable for vehicle monitoring systems (UART = universal asynchronous receiver transmitter = LSI chip) (cf. reference McGlynn 1976, to Chapter 3).

- Centralized computer-aided vehicle monitoring systems in which a stationary computer controls the polling processes and compares prescribed and real schedules as illustrated in Figure 10.7, and explained further in the case descriptions in Chapter 12. This type of system, which uses synchronous vehicle polling methods, has been implemented in most current computerized bus monitoring systems.
- Distributed computer-aided vehicle monitoring systems, which are characterized by a distribution of the computing power, i.e., by the installation of microcomputers onboard the vehicles. These microcomputers can carry out the comparisons between real and prescribed schedules for their own buses (cf. Figure 10.8). This leads to a dramatic reduction in the amount of data that has to be exchanged between the control center and the vehicles, as well as to less expensive installation requirements for the control center. Therefore, it is likely that future bus/tram monitoring systems will use this distributed control concept. The Toulouse system is probably the first bus monitoring system that makes use of onboard microcomputers (cf. Section 12.3, Figure 12.6).

10.2.2 METHODS OF DISPATCHING CONTROL

The techniques that can be used to deal with schedule disruptions discovered by the automatic vehicle monitoring system vary according to the type of disturbance and within certain constraints. Four categories of control strategies may be taken into account: (1) headway control, by advancing and retarding buses; (2) reducing recovery time; (3) turning buses short; (4) closing large gaps (cf. Ullmann 1974, and references Finnamore 1974, Pikarsky 1973, Giraud 1976, Frey 1976, Perrin 1976, attached to Chapter 12).

1. *Headway control by advancing and retarding buses.* On many routes the buses are operated at equable headways. These bus routes very often experience so-called dynamic instabilities, as illustrated by Figure 10.9. The spacing between buses is a function of passenger volumes and arrival times. Any change in the number of passengers carried affects the vehicle's time performance, which can lead to a small delay of the bus. This small delay can cause the bus to take onboard even more passengers at the next stop while the following bus will remain less occupied. Finally one gets the well-known phenomenon that after a long wait an overdue crowded bus arrives followed by two or three nearly empty buses running in tandem (cf. Figure 10.9(c)).

What measures can be taken to control such an unstable process? The scope for controlling the speed of a bus in traffic is limited. Therefore, the main tool for headway control is to change the time the bus waits at the stop. By a coordinated modification of the waiting times of the corresponding buses the bus flow can be stabilized. The price to pay for this result is a small increase in the total travel time (cf. Ullmann 1974).

2. *Reducing recovery time.* Under certain conditions it will not be possible to eliminate schedule deviations until the bus reaches a terminus. By reducing the scheduled recovery time it may be possible to ensure that the bus leaves the terminus in time.

3. *Turning buses short.* For very large delays this measure can become the only feasible tool to achieve schedule recovery. It may be necessary to combine such short turnings with an interchange of passengers and crew to another bus.

4. *Closing large gaps.* This is achieved by the following measures:

- injecting a reserve bus
- removing a bus from a convoy in the opposite direction on the same service
- removing a bus from a convoy in another service

Further control strategies as well as those summarized here are analyzed in the case descriptions in Chapter 12.

At present, the dispatching control tasks are in general solved by the dispatcher and not by the computer. Therefore the role of the computer is concerned with automation of the vehicle monitoring process, and assisting the dispatcher to derive proper control actions.

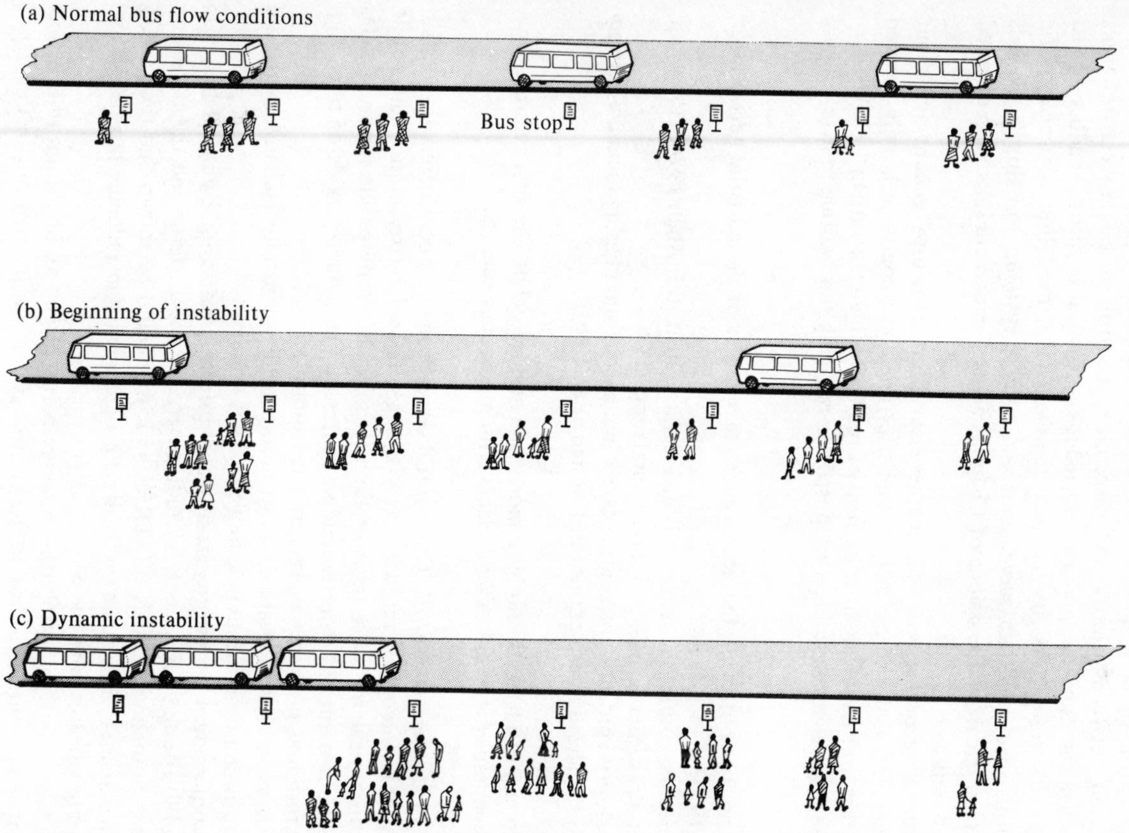
(a) Normal bus flow conditions
Bus stop
(b) Beginning of instability
(c) Dynamic instability

FIGURE 10.9 Dynamic instability of bus headways.

10.2.3 PRIORITY CONTROL

In Chapters 5 and 6, priority control of public transit vehicles is dealt with as a special problem of computerized traffic light control. However, there is a close interrelation between priority and dispatching control since both control tools aim to solve the same task, namely, to reduce delays and to maintain adherence to schedules. Therefore priority control represents a natural link between traffic light control and bus/tram surveillance and control systems. For this reason it is useful to supplement discussions of Chapters 5 and 6 with a consideration oriented to public transit.

The need for priority control becomes clear if one takes into account that about 10-30 percent of the whole bus or tram trip time corresponds, in heavily traveled urban areas, to the waiting time in front of red traffic lights (cf. Allen 1976).

Three approaches have been proposed for reducing this waiting time (cf. Figure 10.10):

- priority control of the access to certain routes or parts of the network (cf. TRRL 1975)
- priority control of single and linked traffic lights (cf. Allen 1976, Hoppe and Vincent 1972, Mertens 1973, Vincent and Hoppe 1970)
- priority control by operating buses in platoons and phasing them collectively through signalized intersections (cf. Herman et al. 1970)

The last approach is still nothing more than an idea, while the first two concepts have been implemented at several places with remarkable success.

1. *Priority access control.* This control concept may be considered as a certain type of inflow control (cf. Chapter 5), characterized by restricting automobile access to special routes or parts of the network. Thus only public transit vehicles are allowed to enter the corresponding routes during certain periods of time, e.g., during rush hours, or under special traffic conditions.

Priority control is established by automatically controlled (changeable) road signs for prohibiting automobile movements at junctions.

A combination of priority access and priority traffic light control has been successfully tested for a route including parts of the Bitterne and the Bursledon Roads in Southampton, UK (cf. TRRL 1975). It could be shown that the priority control schemes discussed above can help significantly in reducing bus delays and improving schedule adherence.

To date, priority control using computers has played an important role, mainly as part of computerized area traffic control systems as pointed out in Chapters 5 and 6. There is only a very limited number of special priority computers in operation. An example is provided by the Zürich tram priority control system described by Mertens (1973). Most of the existing priority control schemes use conventional

Concept	Operating principle	Status
Priority access control	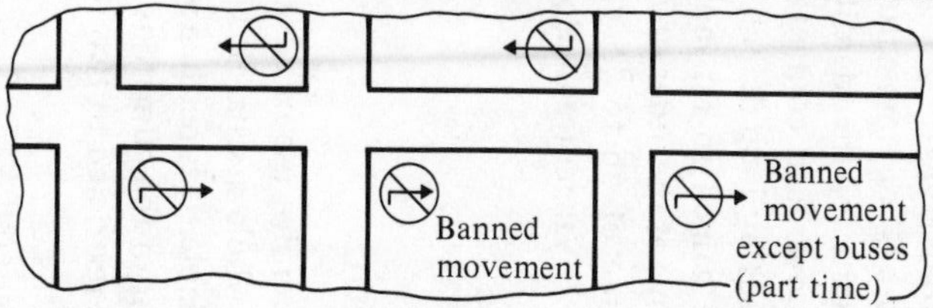Banned movement Banned movement except buses (part time)	Efficiency successfully proven in real applications (cf. TRRL 1975)
Priority traffic light control	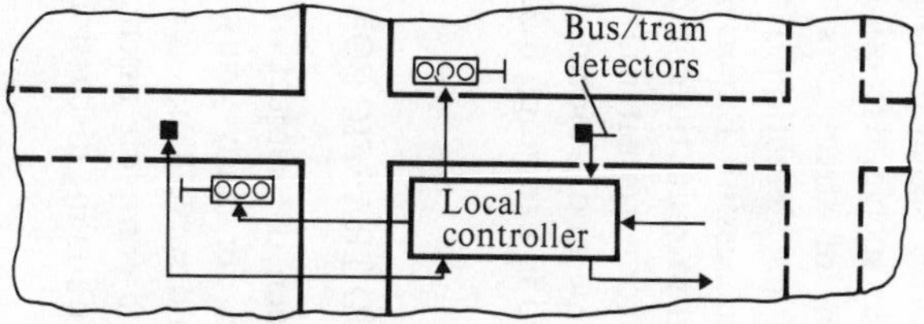Bus/tram detectors Local controller	Successfully applied at many places (cf. Allen 1976, Mertens 1973, Hoppe and Vincent 1972, Vincent and Hoppe 1970)
Priority traffic light control and bus platooning	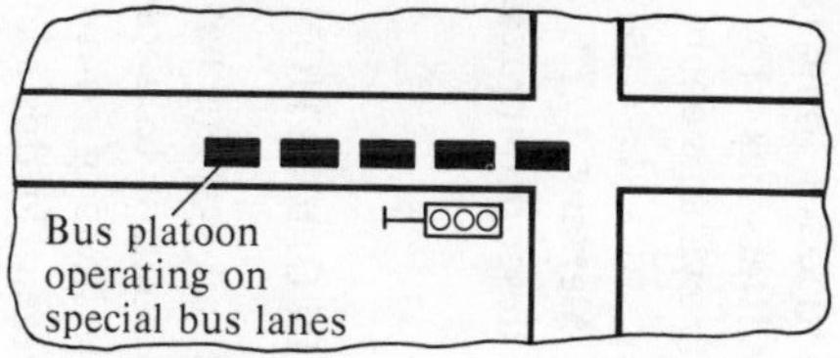Bus platoon operating on special bus lanes	Feasibility not yet proven (cf. Herman et al. 1970)

FIGURE 10.10 Concepts of public transit priority control.

electronic systems. However, this situation may change with the introduction of microcomputers as local controllers, which will provide high flexibility in implementing various types of priority control concepts.

2. *Priority control of traffic lights.* There are two possibilities for introducing priority features into traffic light coordination and control schemes (cf. Allen 1976)

● Passive priority, i.e., computing fixed-time signal plans in such a way that special attention is paid to operational parameters of public transit vehicles. One problem, for example, is to design a progressive traffic light system in such a way that a tram or a bus can synchronize with the automobile traffic in passing a series of intersections uninterrupted or at least with a minimum of stops. The efficiency of this control principle is unfortunately very sensitive to unavoidable changes in passenger service times occurring at the various bus or tram stops.

● Active priority, i.e., modifying signal plans in a bus- or tram-actuated operation mode. This concept requires special detection units for the identification of public transit vehicles. Detection of street cars via overhead contacts has been standard practice for some time, and detection of buses by loop detectors and on-board transmitters has proved successful. Experience gained so far indicates that active priority systems could lead to reductions in delays of the order of 70 percent, corresponding to a decrease of total transit route time of up to 15 percent (cf. Allen 1976, Mertens 1973, Hoppe and Vincent 1972, Vincent and Hoppe 1970, and the case description presented in paragraph 6.3.1).

10.3 TRAIN OPERATION AND TRAFFIC CONTROL

Urban rapid rail systems have been the subject of extensive use of advanced automation and control technologies for many years. A whole arsenal of various concepts and methods of control and surveillance is available. It is neither possible nor useful to present in this section a complete discussion of the numerous techniques and approaches. Therefore, the following discussion is restricted to two topics:

● highlighting the present and future role of computers
● discussing new methodological options of control and surveillance that are or will be provided by the introduction of computers

There are three categories of control and surveillance tasks under consideration for computer application (cf. Figure 10.11, and Boura et al. 1970, Burrow and Thomas 1976, Hahn 1971, Horn and Winkler 1978, Kubo and Kamada 1974, Lehmann 1972, Mampey and Paulignan 1976, Mathson 1977, Schenk 1974, Strobel and Horn 1973, Strobel et al. 1974, Takaoka 1973, Vollenwyder 1972, Wunderlich and Kockelkorn 1975, Watanabe and Urabe 1973, Winkler 1979, Yamamoto 1970):

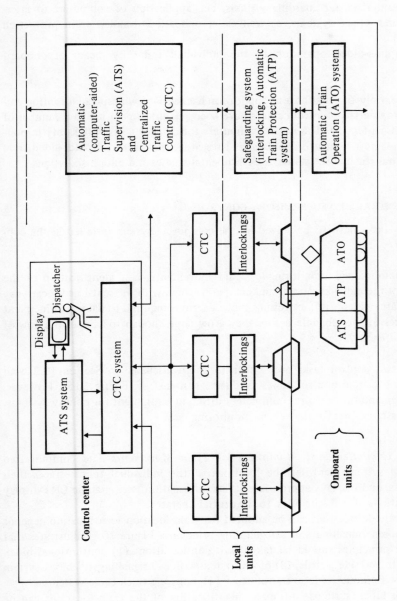

FIGURE 10.11 Survey of control and surveillance tasks (cf. Table 10.1 for definitions of acronyms).

- centralized computer-aided supervisory control. i.e., monitoring and displaying the status of the total transit system, and supervising dispatching control
- computerized safeguarding systems, i.e., application of computers to interlockings and block systems, as well as in onboard automatic train protection equipment
- computer-aided operation of the individual trains by means of onboard computers

To date, only the first area of application has received widespread attention and several computerized control centers have been put into operation in a number of cities (cf. Chapter 13). The use of computers for control tasks with safety responsibility remains a subject of fundamental research. However, computer-aided train operation has reached the status of experimental studies and pilot installations.

10.3.1 CENTRALIZED SUPERVISORY CONTROL

Two main tasks have to be solved by the computing system installed in the control center:

- computer-aided monitoring of train movements, i.e., identification of the locations of the individual trains and comparing them with the scheduled locations
- dispatching control, exercising control actions required to damp out incipient disturbances to the schedule before they grow large enough to hamper the flow of traffic significantly

These two control tasks are similar to those discussed in Section 10.2 with respect to bus/tram control and surveillance systems (cf. Figure 10.5). However, their implementation requires completely different equipment for railway systems. This is illustrated first for the vehicle monitoring task.

10.3.1.1 Computer-aided Monitoring of Train Movements. The computerized monitoring system identifies the locations of the individual trains, traces their movements, and displays appropriate status information, e.g., graphic CRT display representations of real schedules, to the central operator.

The methodology that can be adopted for train location identification depends on the chosen signalling and safeguarding principles. Figure 10.12 illustrates that four basic principles should be taken into consideration: (1) conventional block system with wayside signals; (2) block system with cab signalling; (3) block system equipped with inductive loop transmitters; (4) long-loop track cable systems.

For the block systems, automatic identification of the train locations can be achieved by checking whether the individual block sections are occupied by a train or not. The block occupancy is detected by measuring electric currents that are

Signalling and safety systems	Train location identification	Information transmission from Track to train	Information transmission from Train to track	Section length for speed control	Maximizing line capacity	Applied at (examples)
(1) Conventional block system 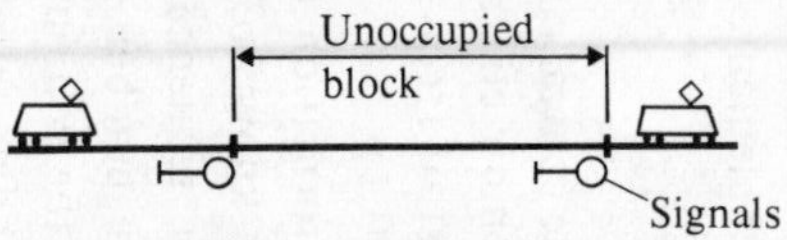	Passive by means of rail currents, section length > 250 m	At discrete locations	Not feasible	Not available		Berlin S-Bahn (during 1930s)
(2) Cab-signalling system 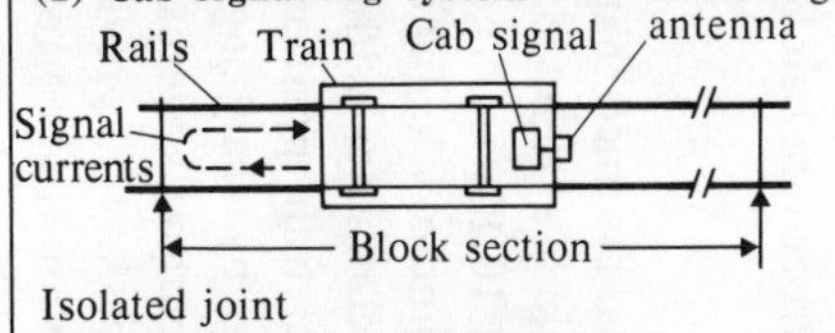		Continuously via Rail signal currents	Not feasible	> 250 m	Not feasible	Subways of Stockholm, Rotterdam, London (Victoria Line)
(3) Short-loop track cable system 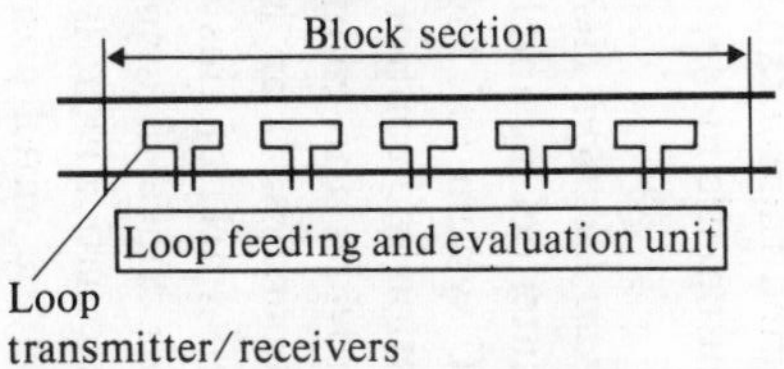		Short loops	Possible at discrete locations	~ 80 m		Subways of Berlin, Munich, Vienna
(4) Long-loop track cable system 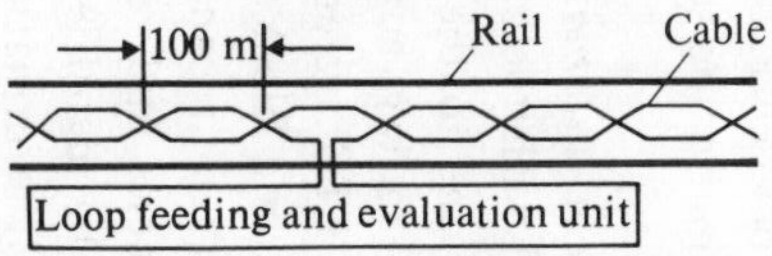	Active by cable crossings, section length 25 m	Long loops	Possible at all locations	25 m	Possible	Subways of Hamburg, Helsinki, Amsterdam Metro

FIGURE 10.12 Survey of basic safeguarding and signalling systems according to Bopp et al. (1975) (cf. Chapter 13).

supplied to the isolated rails; when the rails are short-circuited by the train wheels then an occupancy signal can be detected. The accuracy achievable in location identification depends on the block section length; the error is in general larger than 250 meters (cf. Figure 10.12).

The long-loop track cable system is characterized by two basic features:

- The track cable contains crossings, spaced 100 meters apart, that can be read by an onboard detector. This results in exact position measurements. Their accuracy can be improved further by interpolating the 100 meter distances with onboard odometer measurements.
- The position values obtained thus can be transmitted — together with other information — to the stationary equipment, e.g., computers, via the long-loop track cable. As a result a very high resolution in position measurements of about 25 meters is obtained (cf. Figure 10.12)

The availability of a long-loop track cable system permits an automatic train monitoring system to be created, which is similar to the bus monitoring system discussed in paragraph 10.2.1. The digital data link provided by the track cable permits a stationary control computer to poll the individual trains and to require the transmission of train position and other relevant data. In this way the computer can trace the movements of the individual trains.

However, in most existing computerized train tracing systems a block-section oriented train location approach is preferred for two reasons:

- the high installation costs of long-loop track cables are very often not justifiable
- the high location accuracy provided by the track cable system is required for direct control of train speeds but not for the task of central supervision of the total transit system.

A block-section oriented train location procedure can be implemented in either an automatic or a semiautomatic mode.

1. *Semi-automatic train location.* Here it is assumed that the central computer is coupled with the local interlocking and blocking systems via remote control devices that belong to the CTC system (cf. Figure 10.12, and the list of frequently used terms in railway automation and control technology in Table 10.1).

The CTC system provides the central computer with information on switch settings and block occupancies. If an initial train position is known to the computer, then it can keep track of this train by analyzing the sequence of block occupancies and releases. Since in many existing systems the number of a train cannot be identified automatically, it is necessary to provide the computer manually with information on the initial position and the corresponding number of the respective train. This leads to a semi-automatic train location procedure.

TABLE 10.1 Frequently Used Terms from Railway Automation and Control Technology (cf. Figure 10.11)

ATS	Automatic train supervision or automatic traffic supervision; The ATS system monitors the status of the total transit system and exercises dispatching control functions
ATO	Automatic train operation The ATO system performs the functions normally performed by the train driver
ATP	Automatic train protection The ATP system protects trains from collisions and other accidents
ATC	Automatic train control The ATC system includes both ATO and ATP; the term ATC is sometimes used instead of ATP
CTC	Centralized traffic control The CTC system is a remote control and communication system, which permits local interlockings and all traffic to be controlled from a control center by a human operator or in a computer-aided mode
COMTRAC	Computer aided traffic control The COMTRAC system is an advanced Japanese ATS system
API	Automatic passenger information The API system controls information displays and broadcasting units at stations, in trains, etc.
AFC	Automatic fare collection and accounting The AFC system consists of ticket vending and punching equipment, accounting computers, etc.

2. *Automatic methods.* A completely automated train tracing process can be achieved by transmitting the train numbers directly from the trains to the central computer. For this purpose, the train number has to be stored in the memory of an onboard device. This device reports the train number to the central computer at discrete points, e.g., at passenger stations, via inductive loop receivers.

10.3.1.2 Dispatching Control. Computer-assisted dispatching control is by definition a man–machine interacting process. Existing systems differ in the distribution of the various tasks between the dispatcher and the computer (cf. Chapter 13).

Two concepts may be distinguished:

- The task of the computer is restricted to the determination and graphical presentation of the real schedule and of the deviations from the prescribed schedule. It is up to the dispatcher alone to take proper action in the case of irregularities (cf. paragraph 13.3.2).
- The computer deals with the irregularities and proposes decisions to the dispatcher. For this purpose the proposed optimum modified schedule is displayed to the dispatcher on a CRT. If the dispatcher approves the computer's proposal

then it is executed automatically. Otherwise the dispatcher can interact with the computer via a light pen or a CRT display keyboard (cf. Figure 10.11).

The tools available for dispatching control may be classified as follows:

1. Schedule modifications for short delays, e.g., headway adjusting by changing running speeds and/or dwelling times at stations (cf. Mampey and Paulignan 1976). The objective of this control principle is to minimize deviations from the original schedule.
2. Schedule readjustment for large delays due to incidents (vehicle breakdowns, track maintenance, etc.), by changing the locations where trains cross and pass each other as well as the assigned station platform (cf. Strobel 1974).

The schedule modification and readjustment tasks include a complicated real-time optimization problem that is characterized by a manifold of constraints and objectives. Several authors proposed the use of the objective function (to be minimized):

$$Q = \sum_{i=1}^{n} g_i D_i(N) = \text{minimum}$$

which represents the weighted sum of delays $D_i(N)$ over the trains $i = 1, \ldots, n$ at the end of a suitable chosen disposition time interval, i.e., at $T = N \Delta t$, Δt being a constant time increment (cf. Strobel 1974). By means of the weighting factors g_i, certain priorities can be assigned to specific trains.

This possibility may be of special interest in the case where the computer-controlled district covers not only urban rail transit lines but also long distance traffic. A detailed discussion of the complex and complicated disposition and priority rules that have to be considered in dealing with larger schedule modifications has been published by Wunderlich and Kockelkorn (1975).

10.3.2 COMPUTER-AIDED TRAIN OPERATION

The computerized train operation subsystem performs functions normally carried out by the driver. But in almost all cases it is not intended to introduce completely automated, i.e., unmanned operation, of the trains. The tasks of the automatic train operation (ATO) subsystem are three:

- to relieve the driver from routine tasks so that he can pay more attention to the supervision of onboard devices, as well as to the track and external signals
- to solve tasks more precisely than the driver can do
- to deal with functions that a human driver is not able to perform

Most existing ATO equipment is designed as conventional, i.e., hard wired, electronic devices. However, there is a growing interest in using onboard computers for two reasons (cf. Strobel 1974, Watanabe and Urabe 1973, Kubo and Kamada 1974, Vollenwyder 1972):

- enlargement of the scope of the tasks and the efficiency of ATO
- availability of cheap and robust microcomputers

The first experiments were made with onboard analog computers (cf. Vollenwyder 1972). The use of a general purpose minicomputer has been studied in Japan with respect to the Shinkansen; a system called ATOMIC (automatic train operation by mini computer) was developed and experimentally tested (cf. Kubo and Kamada 1974). Current research and development activities concern onboard microcomputers (cf. Strobel and Horn 1973, Strobel et al 1974, Winkler 1979, Horn and Winkler 1978).

The following functions are considered to be interesting and important for implementing computerized ATO systems:

- energy optimum train operation with priority to schedule adherence
- control of the individual moving phases
- automatic recording of operational data

10.3.2.1 Energy Optimum Train Operation. Two interrelated energy optimum control problems can be formulated (cf. Figures 10.13 and 10.14):

1. A train shall be operated in such a way that it covers the distance between the two stations, A and B, with a minimum of traction energy consumption taking into consideration the requirements for:

- a fixed arrival time T_B at station B
- a varying departure time T_A from station A, which may be caused by changing passenger volumes
- a set of various constraints such as speed limits, maximum admissible values of acceleration and deceleration

Which driving regime must be used in order to achieve this optimum operation mode?

2. A train shall pass a sequence of stations with a minimum of energy consumption taking into consideration the requirements for:

- a fixed arrival time at a certain station, e.g., at the terminus station of a line
- small variations of the order of several seconds for the running times between stations as well as for the waiting times at stations

How must the train running times between the individual stations be chosen in order to meet the desired optimum?

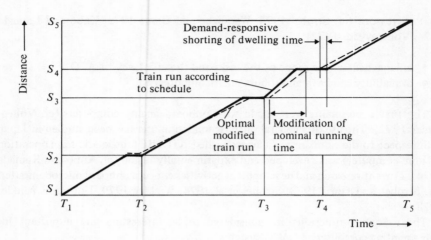

FIGURE 10.13 Energy-optimal real-time modification of the nominal schedule ($S_1, \ldots, S_5$ = Stations 1-5) (cf. Horn and Winkler 1978, Winkler 1979).

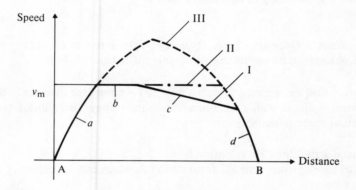

FIGURE 10.14 Energy-optimal train operation between stations A and B: a, acceleration; b, constant speed; c, coasting; d, braking; I, energy-optimal train run; II, time-optimal train run if the speed limit $v \leqslant v_m$ is considered; III, time-optimal train run if no speed limit exists (cf. Strobel et al. 1974, Strobel and Horn 1973, Horn and Winkler 1978).

Both optimization problems have been studied extensively by Strobel et al. (1974) and his associates (cf. Horn and Winkler 1978, Winkler 1979).

The results, obtained in both simulation studies and experiments with onboard microcomputers carried out on suburban commuter railway lines in Dresden, can be summarized as follows:

1. Average energy savings in the range of about 10-20 percent are achievable — in several cases the savings reached about 25-30 percent.

2. The optimum driving regime contains only four phases:

- maximum acceleration
- constant speed
- coasting
- braking at maximum admissible operational deceleration rates

3. The switch-over points from one moving phase to another vary according to running time and therefore have to be computed in a real-time operation mode by the onboard computer using the three measured state variables:

- running time still available
- target distance
- running speed

Odometer measurements have proved to be sufficiently accurate if they are recalibrated at selected points.

4. The existence of only four driving regimes permits the optimal control strategy to be implemented in the form of a man-machine system, as illustrated by Figures 10.15 and 10.16.

5. The algorithm has proved its capability to compensate for delays provided they do not exceed certain limits.

A computer-aided energy optimum train operation system of the type described here was put into regular service on the Berlin S-Bahn in 1980.

However, most of the functions of existing systems are in general restricted to the schedule adherance task (cf. Watanabe and Urabe 1973 and Chapter 13).

10.3.2.2 Control of Moving Phases.
This category of control tasks includes two functions:

1. Speed control, i.e., keeping the train speed automatically at those values that are prescribed, respectively, by the onboard optimum control system discussed above, or by stationary safe-guarding systems.

2. Precise station stopping, i.e., automatic stopping of the train at the station platform in such a way that the entrance doors will meet with fixed station platform signs where passengers are expected to queue up. The implementation of an automatic station stopping system requires two special measures, as illustrated in Figure 10.17:

- installation of certain wayside markers in the track that give the precise train position and enable the onboard distance measurement device to be recalibrated

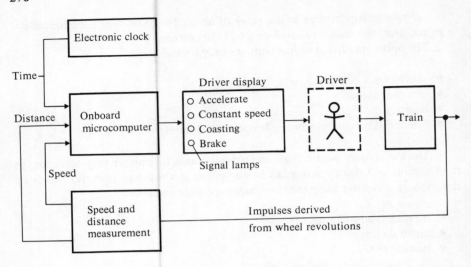

FIGURE 10.15 Block diagram of an energy-optimal computer-assisted train operating system (cf. Strobel et al. 1974, and Figure 10.16).

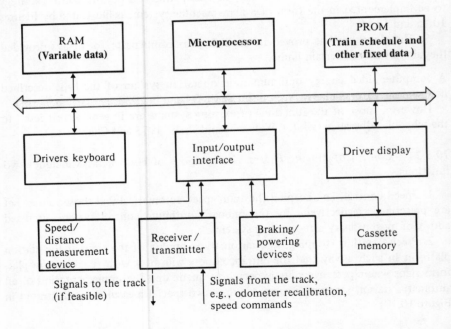

FIGURE 10.16 Basic components of an onboard microcomputer system (cf. Winkler 1979).

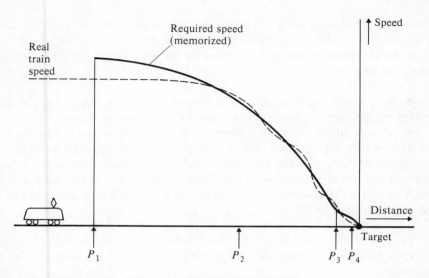

FIGURE 10.17 Preset position stopping; P_i, position markers; P_1, process has started; P_2, recalibration of onboard distance measuring device (odometer); P_3, brake release for better driving comfort; P_4, final braking.

- memorizing the stopping profile, in the form of the relation between the required speed and the target distance, in the onboard ATO system

The two ATO functions sketched briefly here are performed in many existing urban rail systems by means of hard-wired electronic onboard devices. The availability of an onboard microcomputer would provide the possibility to integrate the functions into the computer tasks.

10.3.2.3 Automatic Recording of Operational Data. An onboard computer can be used for several recording and checking tasks, such as preparing train operation statistics, automatically diagnosing faults with regard to the ATO system itself, as well as concerning other onboard systems (cf. Figure 10.16).

10.3.3 COMPUTERIZED SAFEGUARDING SYSTEMS

Railway safeguarding systems are composed of three basic parts:

- stationary interlocking and blocking systems
- onboard automatic train protection (ATP) devices
- communication links between trains and stationary units that transmit relevant safety data

Almost all existing safeguarding systems are designed on the basis of electromagnetic safety relays, i.e., neither hard-wired electronic units or digital computers have found significant broad application up till now. This has to be credited to simple physical reasons:

- In a well-designed relay-based system it can be warranted that the safety relays will switch to the power-off position if a failure occurs.
- When a failure occurs in a transistor it can either be short-circuited, interrupted, or only degraded, i.e., out of control. So for a complex system with many integrated circuits, including a complex number of transistor functions, it is impossible to analyze all possible failures and their consequences, detect them, and take such measures that the system will always be transferred to a restricted but safe state (cf. Mathson 1977).

In spite of these well-known problems, there is a growing interest in replacing relay-based systems by computerized systems. This tendency is supported for two reasons:

- The modernization of existing older railway safety and signalling systems by means of modern relay interlocking and blocking equipment is very expensive, requires the erection of large new buildings (because of the large size of relay devices), and only permits a limited degree of automation to be achieved.
- Space flight and military computer applications have demonstrated the possibility of implementing control systems that are able to meet very high reliability and availability standards.

Current research and development activities aim to develop fail-safe computerized systems on the basis of two concepts (cf. Mathson 1977, and Figure 10.18):

- achieving fail-safe redundancy by doubling or tripling the number of computers and adding an external fail-safe comparator or majority voter logic (cf. Figures 10.18(a) and (c))
- arranging two totally different program versions in one computer with totally opposite ways of structuring data (cf. Figure 10.18(b))

Using a structure in accordance with Figure 10.18(b), a computer based interlocking system has been developed for the Swedish State Railways (cf. Mathson 1977). This system was equipped with third generation minicomputers. Moreover, a fail-safe onboard microcomputer system has been developed and successfully tested in Sweden (cf. Mathson 1977). The application of a three-computer system as shown in Figure 10.18(c) has been adopted by SEL for their SELTRAC System: SELTRAC has been designed for two different categories of systems (cf. Dobler and Eltzschig 1975):

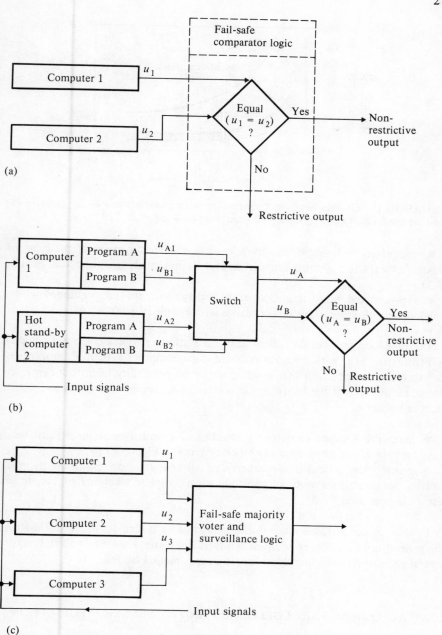

FIGURE 10.18 Approaches for achieving computer control systems with fail-safe features (cf. Mathson 1977, Dobler and Eltzschig 1975).

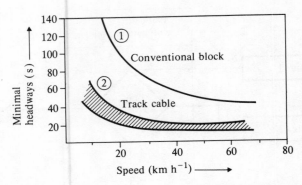

FIGURE 10.19 Minimal headways achievable with conventional block systems (1) and long-loop track cable systems (2) according to Bopp et al. (1975).

- completely automated unconventional rapid transit systems of the type discussed in Part 4 (cf. systems Transurban and Kompaktbahn mentioned in Section 18.3)
- subways and other municipal rapid rail systems, that are equipped with a long-loop track cable system in accordance with Figure 10.12 (row (4))

The long-loop track cable system links the stationary triplex computer complex of Figure 10.18 (c) with a duplex onboard microcomputer system. Proponents of this system concept claim that it will provide the possibility of introducing operational innovations of the following kinds into the operation of subways and similar rapid rail systems:

- maximize the line capacity by enabling safe train operation at short headways that are not achievable by any block system (cf. Figure 10.19)
- provide the potential for supervising all trains almost continuously, thus creating the conditions needed for introducing a completely automated, i.e., driverless, train operation

In spite of the progress made during recent years, it seems justified to state that the general introduction of computers into railway safety systems will still require several years of research, development, and experimental testing.

10.4 AUTOMATIC FARE COLLECTION AND PASSENGER INFORMATION

10.4.1 COMPUTERIZED FARE COLLECTION SYSTEMS

The automation of fare collection systems is generally justified for economic reasons that involve manpower savings as well as fraud prevention. There is no fundamental theoretical difference between automatic fare collection systems as applied to different "utilities" such as urban transit systems, inter-city rail transport networks, toll roads, parking lots. The specific system design, however, may be very different according to the particular characteristics and operational rules of the utility concerned.

For urban transit systems, a distinction can be made between two basic functions of a fare collection system (cf. Kent 1969/70, New Developments in Ticket Issuing 1977, von Rohr 1977, Rottenburg 1977, Takehara and Okada 1977):

- sale, checking, and punching of tickets
- full accounting of sales and cash received, preparing sales and traffic statistics, etc.

10.4.1.1 Sale and Checking of Tickets. The availability of microcomputers opened the way for the creation of a new generation of both ticket vending machines, and ticket checking equipment.

In the past, the application of automatic ticket vending machines was in general restricted to simple fare structures and a rather limited number of different tickets. The installation of microcomputers in self-service vending machines provides the potential for dealing even with complicated fare structures such as occur in suburban or long-distance traffic. This has been demonstrated already in several successful applications of microcomputer controlled ticket vending machines. The Japanese National Railways, for example, designed such equipment for selling passage tickets and limited express tickets for the Shinkansen (cf. paragraph 13.4 and Takehara and Okada 1977). An application to a municipal rapid rail system is described by Kopp (1974) with respect to the Frankfurt S-Bahn in the FRG.

A simplified block diagram of a microcomputer controlled ticket vending machine is given in Figure 10.20, which illustrates the functioning and the use of the equipment. The passenger who intends to use the machine is asked to follow a three-step procedure: (1) select, (2) pay, (3) take.

1. First the passenger has to select the desired destination or zone by pressing a corresponding button at the Destination Keyboard. Moreover, some additional information such as Child's Ticket, Adult's Ticket, Return Ticket, etc., must eventually be inserted.

2. The microcomputer uses these data as well as the memorized information on the origin–destination distances and tariff regulations for computing the fare. The fare is then indicated at a "Still-to-pay" display. If the passenger puts in a coin its value is automatically subtracted from the fare and the difference value, representing the "still-to-pay" fare, is displayed. The passenger has to add coins or notes until the display shows "0.00".

3. The microcomputer then activates a wire-dot printer, which prints the ticket completely or supplements the corresponding data, respectively. Several systems

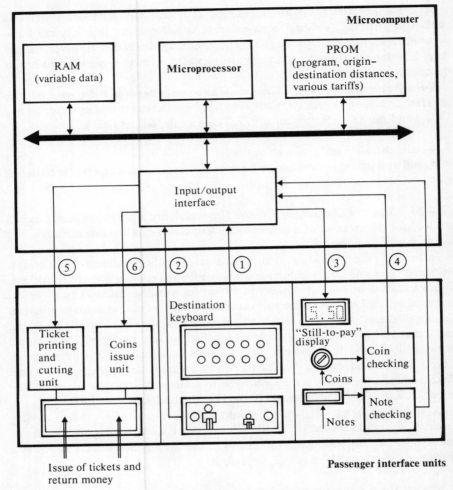

FIGURE 10.20 Simplified block diagram of a microcomputer controlled ticket vending machine.

use tickets that contain a thin magnetic film for memorizing the relevant data, such as money paid, date, departure station, etc. If the value of the coins and notes inserted by the passenger exceeds the initially displayed fare, then the corresponding return money will be determined by the microcomputer and issued by a special unit together with the ticket itself.

It should be mentioned that the ticket vending process described here may be adapted to other design principles. The various design concepts differ with regard to the following features:

- Using magnetic coded credit cards instead of bank notes. It is considered to be very complicated to achieve a reliable note checking process if more than one note type has to be accepted by the vending machine. On the other hand, current Japanese experience has demonstrated the feasibility in principle: Japanese National Railways are going to introduce microcomputer based ticket vending machines that will permit passengers to use three types of notes — 1,000 Yen, 5,000 Yen, and 10,000 Yen (cf. Takehara and Okada 1977).
- The issue of return money can become a reliability problem and is connected with additional installation and maintenance costs. Therefore, some systems concepts decline the issue of return money. Two alternatives can be considered: the passenger is asked to put into the machine not more than the indicated fare; the fare actually paid is magnetically stored on a thin magnetic film placed on the ticket. The second concept is called the "stored-fare ticket" principle and has been implemented in the fare collection system of San Francisco's Bart (cf. Section 13.2). This stored-fare ticket concept requires a special ticket checking system in the form of automatic entrance and exit gates.

These gates are equipped with reading devices that can identify the magnetically coded information, i.e., fare, departure station, etc. At the exit gate the used fare is subtracted from the stored fare. This gate will not open if the stored fare does not cover the fare required for the distance actually travelled. In such a case the passenger is asked to increase the stored fare to an indicated level by using an automatic ticket vending machine located inside the station. Both the entrance and the exit gates are supervised and controlled by a process computer; the installation of microcomputers at these gates is under consideration.

Only a few urban rapid rail transit systems use such a sophisticated but expensive computerized ticket checking system based on automated entrance and exit ramps. The Bart system of San Francisco and the Paris Metro have adopted this approach (cf. references Jacoub 1975, Estournet 1973, to Chapter 13).

Most urban transit authorities restrict themselves to simplified ticket checking procedures, i.e., sampling ticket inspection by inspection staff is in general preferred. This requires, however, the availability of automatic ticket punching or devaluation equipment that has to be used by the passengers themselves. Such equipment has to put a mark on the ticket, e.g., by printing on it the date, the time of day, and the location of the equipment (the number of the station or of the vehicle).

It should be mentioned that ticket devaluation units represent another area of application for microcomputers, which can update in the printing units the date and the time of day. Moreover, in the case of magnetic coded tickets, the microcomputer can serve as a real ticket checking device.

10.4.1.2 Accounting of Sales. The introduction of microcomputers in automatic ticket vending machines, as well as in the corresponding equipment located in

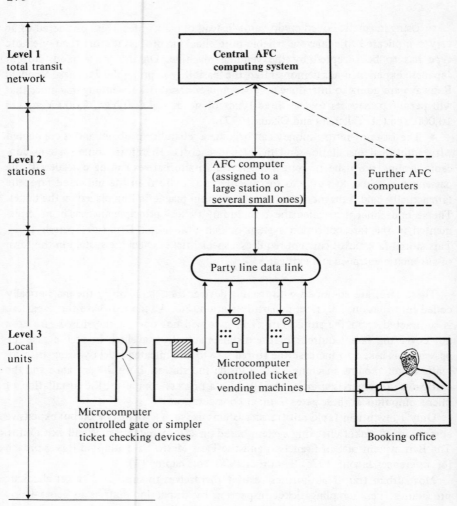

FIGURE 10.21 Hierarchically structured computerized fare collection system.

booking offices, provides the additional advantage that the accounting of sales and cash received can partly be automated.

For large urban transit systems it can, however, become desirable to automate the whole accounting process. This leads to hierarchical computer structures as illustrated in Figure 10.21.

At the lowest level, i.e., ticket vending machines, automatic gates, etc., the microcomputers perform the tasks of automated data collection and aggregation. They transmit the resulting information to a general purpose computer that may be in charge of a large station or of several smaller ones. This and the other AFC computers at level 2 are connected to a central computing system that deals with the overall evaluation of the complete transit network.

A hierarchical system with a structure similar to that sketched in Figure 10.21 has been installed for the Paris Metro where the application of microcomputers at level 3 was considered (cf. Jacoub 1975, to Chapter 13). In addition to ticket selling and checking this system involves the following tasks:

- prepares full accounts of sales and cash received for about 350 stations in an area of about 110 km^2
- prepares sales and traffic statistics
- monitors clocking in and out of station personnel
- reports on the general operation of the system itself

The whole automatic fare collection (AFC) system covers about 1600 gates and 360 booking offices; it will be expanded to a total of over 4,500 gates and vending machines (for more details cf. Jacoub 1975, to Chapter 13).

10.4.2 PASSENGER INFORMATION AND FURTHER SERVICE SYSTEMS

Figure 10.22 aims to illustrate the role of computers with respect to passenger information, fare collection, and further service systems. Two kinds of information systems are of interest:

- providing general schedule information
- providing operational information on train or bus destinations, departure platforms and times, and arrival delays, etc.

Two variants of schedule information systems may be taken into consideration:

- computer-assisted schedule information provided in information offices
- computerized automated schedule information provided by self-service equipment

In the past, self-service schedule information systems have been installed by means of relatively big general purpose computers. These systems could be justified only for the most important railway stations. The advent of the microcomputer provides the possibility to create relatively cheap automatic schedule information equipment, which may very well be justified for urban transit and suburban transport systems that employ various alternative transport modes.

The second task, i.e., providing operational information, is closely related to the corresponding train or bus operation and traffic control system. The train destination display unit, located at station platforms, for example, may be controlled

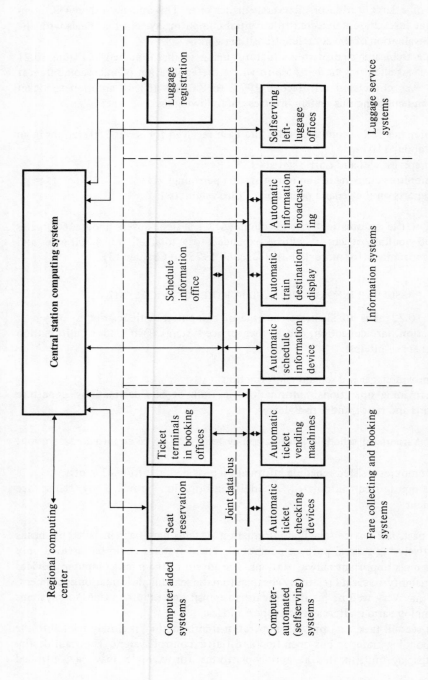

FIGURE 10.22 Concept of a computerized passenger service system for a large railway station (cf. AEG 1978).

directly by the approaching trains as illustrated in detail in the case descriptions presented in Sections 13.2 and 13.3. Another possibility is to control these displays by stationary computers located in the control center.

There are two other processes that will very likely become important areas of application for computerized systems, especially for microcomputers, namely,

- self-service left-luggage offices at railway stations
- computer-aided luggage registration (cf. Freidl 1975)

Both problems are, however, only of special interest for long-distance travel and not for urban transit.

REFERENCES

AEG. 1978. Passenger Service in the Railway Station of the Future. Study by AEG-Telefunken, Nachrichten-und Verkehrstechnik AG, N 312 V. FRG: Kassel (in German).

Allen, D. C. 1976. Signal priority for transit vehicles. Research Report No. 34 from the University of Toronto, Canada, and York University, UK, Joint Programme in Transportation.

Bopp, K., G. Ciessow, H. Linde, et al. 1975. Further development of present rapid rail transit systems. Study by AEG-Telefunken sponsored by the Federal Ministry of Research and Technology, FRG, September 1975 (in German).

Boura, J., M. J. Savage, et al. 1970. The role of computers in train regulation. Paper presented at the First International IFAC/IFIP Symposium on Traffic Control. Versailles, France.

Breur, M. W. K. A., et al. 1974. The bustaxi, a study of different algorithms for allotting passengers to vehicles. Paper presented at the Second International IFAC/IFIP/IFORS Symposium on Traffic Control and Transportation Systems, Monte Carlo.

Buck, R. and H. Salwen. 1974. Channel measurements for automatic vehicle monitoring systems. Report No. DOT-TSC-OST-73-33 from the Department of Transportation, Urban Mass Transportation Administration, Washington, DC, USA.

Burrow, L. and T. Thomas. 1976. Measurement and communication in automated transport. Report by the Urban Transport Research Group, Warwick University, Coventry, UK.

Copper, J., F. Faulhaber, and D. Ghahraman. 1976. Fare calculation and vehicle communication modules for paratransit systems. Paper presented at the Third International IFAC/IFIP/IFORS Symposium on Control in Transportation Systems. Columbus, Ohio, USA.

Dobler and Eltzschig. 1975. The SEL operation control system SELTRAC – development stage and further extension. In Nahverkehrsforschung '75, Federal Ministry of Research and Technology, Bonn–Bad Godesberg, FRG (in German).

Freidl, W. 1975. Computer-controlled sorting of luggage and express goods. Signal und Draht, 67 (9/10): 187–190.

Gould, A. V. 1973. Automatic vehicle monitoring applied to bus operations. IEEE Transactions on Vehicular Technology, VT-22 (2): 42–45.

Hahn, H. -J. 1971. Automatic operation of urban rapid transit and underground railways. Brown Boveri Rev., 12: 553–565.

Hansen, G. R. 1976. Automatic vehicle monitoring systems study. Report JPL 5040-26 by the Jet Propulsion Laboratory, California Institute of Technology, Pasadena, California, USA.

Herman, R., T. Lam, and R. Rothery. 1970. Further studies on single-lane bus flow: transient characteristics. Transportation Science, 4 (2): 187–216.

Hoppe, K. and R. A. Vincent. 1972. Public transport priority in signal-controlled networks. UITP Revue, No. 4: 266–268.

Horn, P. and A. Winkler. 1978. On the energy-optimal train control and schedule modification. Die Eisenbahntechnik, 26 (8): 324–328 (in German).

Howson, L. und W. Heathington. 1970. Algorithms for routing and scheduling in demand responsive transportation systems. Research Publ. GMR-944 from General Motors Corporation, Warren, Michigan, USA.

Kent, J. 1969/70. Automatic fare collection for railways. Proceedings of the Institution of Mechanical Engineers, 184: 129–134.

Kopp, H. 1974. Automatic ticket vending machines on the Frankfurt S-Bahn. Signal und Draht, 66 (6/7): 123–132 (in German).

Kubo, S. and S. Kamada. 1974. Automatic train operation by minicomputer, Schienen der Welt, April: 303–308 (in German).

Lehmann, S. 1972. Automatic control of locomotive driving and braking, and the automation of operational processes in trains, Elektrische Bahnen, 43 (2): 33–38 (in German).

Mampey, R. and J. F. Paulignan. 1976. Traffic control policies for an underground system. Paper presented at the Third International IFAC/IFIP/IFORS Symposium on Control in Transportation, Columbus, Ohio, USA.

Manlow, M. J. 1973. A control algorithm for dial-a-ride. Traffic Engineering and Control, May.

Mason, F. J. and J. R. Mumford. 1972. Computer models for designing dial-a-ride systems. Paper presented at the Society of Automotive Engineers, Automotive Engineering Congress, Detroit, Michigan, SAE Paper 720216.

Mathson, R. 1977. Computers replacing safety relays in railway signalling. In, Van Nante Lemke, ed., Digital Computer Applications to Process Control. Amsterdam: North-Holland Publishing Company, and IFAC.

Mertens, F. H. 1973. A control system giving priority to trams. Traffic Engineering and Control, March: 514–518.

Neubauer, R. 1974. Informatics in Transit. Eisenbahntechnische Rundschau, 23 (10): 405–410 (in German).

New Developments in Ticket Issuing. 1977. Coach J. and Bus Rev., 45 (9): 45–46.

Oliver, B. W. 1974. A control algorithm for a demand responsive bus project. Proceedings of the Fifth Annual Seminar on Bus Operations Research, University of Leeds.

Rebibo, K. K. 1974. A computer controlled dial-a-ride system. Paper presented at the Second International IFAC/IFIP/IFORS Symposium on Traffic Control and Transportation Systems, Monte Carlo.

Rebibo, K. K., R. L. Scott, et al. 1974. Summary of an automated scheduling system for demand-responsive public transportation. Report UMTA-VA-06-0012-74-2 from the Mitre Corporation McLean, Vermont, USA.

Ritter, S. and J. McCoy. 1977. Automatic vehicle location – an overview. IEEE Transactions on Vehicular Technology, VT-26 (1): 7–11.

Robinson, D. 1974. Mobile radio systems and their application to dial-a-ride systems. In Slevin ed., pp. 108–123 (cf. Chapter 10).

von Rohr, J. and H. Becker. 1977. Experimental operation of mobile automatic ticket vending devices at the Duesseldorf rheinbahn. Nahverkehrspraxis, 25 (8): 326–330 (in German).

Roth, S. H. 1977. History of automatic vehicle monitoring (AVM). IEEE Transactions on Vehicular Technology, VT-26 (1): 2–6.

Rottenburg. J. 1977. Automatic fare-collection systems. Rev. UITP, 26 (1): 69–73.

Schenk, O. 1974. Automatic disposition in accordance with train operation control. Signal und Draht, 66 (8): 134–138 (in German).

Southall, A. 1974. Telecommunications for dial-a-ride. In Slevin P, ed., pp. 103–107 (cf. Chapter 10).

Strobel, H. and P. Horn. 1973. On the energy-optimal control of longitudinal vehicle movement in the presence of state space constraints, Zeitschrift für elektrische Informations- und Energietechnik, 3 (6): 304–308 (in German).

Strobel, H., M. Kosemund and P. Horn. 1974. On the determination and practice-referred trial of control strategies for computer-aided optimized train operations. Paper presented at the Fourth International UIC/AAR/AICCF Symposium on Railway Cybernetics, Washington, DC, USA.

Strobel, H., et al. 1974. A contribution to optimum computer-aided control of train operation. Paper presented at the Second IFAC/IFIP/IFORS Symposium on Traffic Control and Transportation Systems, Monte Carlo.

Symes, D. J. 1977. Multiuser area-coverage automatic vehicle monitoring program. IEEE Transactions on Vehicular Technology, VT-26 (1): 187–191.
Takaoka, T. 1973. Recent tendency in automatic train operation equipment. Hitachi Review, 22 (5): 218–224.
Takehara, J. and Y. Okada. 1977. New ticket vending machines developed by JNR and Osaka Municipal Transportation Bureau. Japanese Railway Engineering, 17 (1): 17–19.
TRRL. 1975. Bus demonstration project – a linked system of traffic signals with priority access to the route for buses. Summary Report No. 8, Southampton, Department of the Environment, Transport and Road Research Laboratory.
Ullmann, W. 1974. Computerized dispatching systems for public transit. Research Report from the Wissenshaftsbereich Automatisierungstechnik, Hochschule für Verkehrswesen "Friedrich List", Dresden, GDR (in German).
Vincent, R. A. and K. Hoppe. 1970. Public transport priority at a signal-controlled junction: an experiment in Bern, Switzerland. Traffic Engineering and Control, December: 417–419.
Vollenwyder, K. 1972. Automation of suburban railways and subways by means of onboard computers. Elektrische Bahnen, 43 (6): 133–139 (in German).
Watanabe, J. and M. Urabe. 1973. Computerized automatic train operation system. Toshiba Review, International Edition, February: 7–11.
Winkler, A. 1979. A contribution to energy-minimal train control and schedule modification. Dissertation from the Hochschule für Verkehrswesen "Friedrich List", Dresden, GDR (in German).
Wunderlich, W. and E. Kockelkorn. 1975. Better disposition by means of faster information. Eisenbahntechnische Rundschau, 24 (7/8): 283–294 (in German).
Yamamoto, I. 1970. Analytical methods for train-traffic control problems. Paper presented at the First International IFAC/IFIP Symposium on Traffic Control, Versailles, France.

11 International Dial-a-ride Systems Experiences

More than 100 dial-a-ride systems have been put into operation throughout the world during the last decade. However, most of these systems do not use computers for solving the real-time routing and scheduling tasks.

What problems exist in connection with the implementation of computerized dispatching systems? Under what circumstances may the use of computers be considered useful or necessary? What contributions are dial-a-ride systems expected to make to solving urban traffic problems?

This chapter presents answers to these three questions. For this purpose the experience gained in two advanced demonstration projects, i.e., the dial-a-ride systems of Haddonfield and Rochester (USA), are analyzed first. A general survey of other dial-a-ride projects supplements the findings of these case descriptions.

11.1 USA DIAL-A-RIDE EXPERIENCES: HADDONFIELD AND ROCHESTER*

11.1.1 CASE HISTORY

As metropolitan areas change from a primarily urban environment to include ever increasing communities, conventional transit has found itself serving a rapidly shrinking proportion of the population. The dispersal of housing, shopping facilities, and employment outward from the central business district (CBD) has resulted in a corresponding dispersal of trips away from the high density CBD-oriented trips that are well served by conventional fixed route transit. Today, about 43 percent of the US population lives in the suburbs and 80 percent of all trips both begin and end in the suburbs.

The result has been reduced ridership for transit, increasing transit operating deficit, almost total reliance on the private automobile for urban transportation and limited mobility for those without access to an automobile. In response, the US Department of Transportation, the aca-

*Based on a case description specially prepared by Nigel H. M. Wilson, Massachusetts Institute of Technology, and Eldon Ziegler, Program Manager, Bus and Paratransit Division, UMTA, US Department of Transportation, Washington, DC, USA.

demic community and many cities have been developing public transportation services more suited to today's conditions than the exclusively fixed route systems of the past. One of the most significant of the developments is dial-a-ride.

The early Federal involvement in demand-responsive transit started with the computer aided routing system, Project CARS, at the Massachusetts Institute of Technology from 1967 to 1971. The CARS research effort was focused on the feasibility of dial-a-ride from all points of view. While it was clear that small systems could be controlled effectively by a dispatcher using conventional voice communication over a radio channel, there was a desire to investigate the feasibility of computer controlled systems, which would be necessary if larger dial-a-ride systems were to be implemented. Accordingly, one of the major thrusts of the CARS project was the development of computer control procedures together with a simulation model to assess their effectiveness. The simulation model indicated that the computer algorithm could provide effective service, and the overall CARS project conclusion was that demonstration projects to assess the overall service concept and the utility of computer dispatching should be performed.

Following the CARS project, the Urban Mass Transportation Administration (UMTA), from the US Department of Transportation, began a series of demonstrations and experiments to test dial-a-ride in actual operation. These included special services for the elderly and handicapped in Cranston, Rhode Island and St. Petersburg, Florida and for the poor and unemployed in the Watts area of Los Angeles. These demonstrations and experiments provided operating data and yielded experience on practical problems of service implementation.

The two largest federally funded dial-a-ride demonstrations were in Haddonfield, NJ, and Rochester, NY, and both of these were partially aimed at assessing the effectiveness of computer controlled dial-a-ride. The remainder of this discussion focuses on these two demonstration projects and the description and assessment of computer control based on this experience. Since the Rochester system was built of the experience gained in Haddonfield it makes sense to describe and evaluate Haddonfield before discussing Rochester further.

11.1.2 THE HADDONFIELD SYSTEM

The Haddonfield demonstration project ran from May 1972 to March 1975 serving an area of almost eleven square miles and focused on a rail transit station serving Philadelphia. An average of 10–12 vehicles were used in the dial-a-ride service together with several others used in a more conventional fixed route shuttle service. Daily patronage reached a peak of 1,700 during late 1973 when the average fare was 25 cents. During most of the demonstration dispatching was done manually, but computer dispatching was used towards the end of the project, thus providing a good basis for comparison of manual and computer operations.

The computer control algorithm used in Haddonfield was developed at MIT during the CARS project. It is based on the immediate assignment of customers requesting current service. The assignment includes not only the selection of the appropriate vehicle, but also the insertion of the new pickup and delivery stops in the tour (the existing sequence of planned pickups and deliveries) of that vehicle. During this assignment process, trial insertions are made of the new pickup and delivery points at all possible points in all existing tours. According to a specified criterion (discussed below) the best assignment is selected from those that do not violate any "constraint". Three constraints were used in Haddonfield:

- wait time constraint (constant)
- ride time constraint (linear function of trip length)
- total service time constraint (linear function of trip length)

If for a particular trial assignment of a passenger to a tour, the expected wait time, ride time, or total time of the passenger being assigned or any passenger previously assigned is greater than the associated constraint, the assignment is said to be "infeasible", and is considered less desirable than any "feasible" assignment. The feasible assignments are evaluated using an "objective function", which is the sum of two terms: one measures the increase in total service times for all users currently in the system, including the new customer, and the other represents the increase in vehicle tour length.

Passengers requesting service at some time in the future were assigned (using the same procedure) a fixed time before their desired pickup time. The constraints for these passengers were set more tightly than for immediate request passengers to try to provide them with more reliable service.

This algorithm is relatively straightforward computationally, although it makes some very severe assumptions to achieve this simplicity. The central processor used for the computer control system was a Westinghouse Model 2500 computer, a 16 bit 32K word general purpose machine with a 750 nanosecond core memory cycle time.

Since this computer system was developed solely for this demonstration project, it is inappropriate to attempt a costing of the system, since it is clear that a lower cost system could be achieved based on the experience gained in Haddonfield. The Haddonfield computer system, both hardware and software, was modified considerably for the Rochester demonstration project, and this modified system has replaced it for any subsequent implementation of computer controlled dial-a-ride.

Manual and computer dispatching have been compared for that service, both providing many-to-many service from the hours 9 am to 3 pm on weekdays in the Haddonfield Project. From the customers' point of view these two dispatching methods are identical: except inasmuch as different qualities of service may be provided, customers otherwise do not know whether or not the computer is being used. Table 11.1 shows many-to-many data by hour of day obtained under manual operation, and the data resulting from computer dispatching for both scatter and many-to-many trips. It should be stressed that both sets of data refer to actual trips taken in Haddonfield, not to simulation results. Both the demand rates and the numbers of vehicles in service were essentially identical for both manual and computer operations, hence the productivities* for the two periods were similar.

It can be seen from the table that the service provided by the computer to passengers not starting from the high speed rail line, PATCO (i.e., the service for which manual statistics are available), has a wait time significantly lower than that provided by manual dispatchers. In all other respects computer and manual dispatching performance are very similar. It is interesting to note the excellent service provided passengers originating at PATCO (figures for manual dispatching are not available for these passengers): the mean wait time for such passengers was a full four minutes lower than that for other trips, and the standard deviation was also lower. The slightly higher ride time for passengers starting at PATCO is due to a greater average trip length than that of the remaining passengers: 1.56 miles versus 1.36 miles (airline distance).

These results are encouraging for the continued use of the computer in demand responsive systems. In particular it is clear that the computer can perform at least as well as manual dispatchers, even in small systems (10-12 vehicles) operating at low demand levels (40-60 demands per hour). Furthermore, as discussed in the following sections it is clear that improvements in computer dispatching are feasible and would yield superior performance for computer dispatching.

Based on the Haddonfield experience with computer control, the following aspects of the control algorithm were identified for subsequent investigation:

- inflexibility of constraints
- objective function not a true reflection of passenger utility
- different groups of passengers have different service preference
- an automatic reassignment technique should be considered

The Haddonfield algorithm was designed to minimize total service time (for current and future passengers) within fixed constraints on wait, travel, and total service times. Any assignment in which no constraint is violated is preferred to an assignment involving a violation, independent of the value of the objective function. This policy was developed to reduce the number of passengers experiencing "unreasonably long" service times, with the acknowledged and expected effect of some increase in the mean service times. To achieve this goal the constraints must be set about 100 percent above the mean service times. In practice two problems arise from this approach:

---

*Productivity is here measured in passenger trips per vehicle hour.

TABLE 11.1 Dispatching Statistics

	Time	Number of trips	Wait time (min) Mean	Wait time (min) Standard deviation	Ride time (min) Mean	Ride time (min) Standard deviation
Manual Many-to-many dispatching	9–10 am	402	16.5	9.2	9.9	6.6
	10–11 am	370	15.6	10.8	9.6	6.0
	11–12 am	359	16.7	11.1	10.7	5.8
	12–1 pm	431	19.7	13.3	11.4	7.5
	1–2 pm	451	19.6	12.0	11.6	8.7
	2–3 pm	624	18.4	10.6	12.4	8.6
	Total (9 am – 3 pm)	2637	17.9	11.3	11.1	7.5
Computer dispatching	from PATCO	853	10.6	9.3	13.0	8.3
	others (all day)	2658	14.9	11.1	11.3	8.2

1. Because the short-run demand rate varies widely over the course of the day, and because mean service times are very sensitive to the recent demand rate, a constraint set correctly for some times of the day may be incorrect for many other times of the day. The problem is that the constraints are not dynamically set as a function of the number of passengers currently on the system and the number of vehicles currently in service. This problem could be solved by using a short-memory heuristic to compute the current constraint set.

2. A more basic problem is that assignments that may be far superior from the objective function's viewpoint will be rejected if a constraint is violated. This introduces a perturbation in performance and can lead to short-sighted decisions that tend to waste system resources. This problem cannot be solved by any useful setting of the constraints, and its existence argues for a reduction in the role of constraints in the algorithm. This is possible only if the individual customer utility function can be equally or better represented by some other construct.

The Haddonfield objective function implies that users of the system associate with the service a utility function that is linear in service time. This may be an inaccurate and simplistic representation of actual passenger satisfaction, and hence its use might result in customer dissatisfaction. Although the actual utility function associated with dial-a-ride service has not yet been identified, it is clear that for the distribution of service time measures other than the mean are also important, e.g., standard deviation. It is probable that the uncertainty in service is also an important characteristic. One measure of this is the difference between estimated and actual pickup and delivery times. Once again the means and standard deviations of these distributions should be considered.

It is clear that different customers will have different utility functions. For example, someone who is going to work or transferring to a scheduled bus will be very conscious of the latest arrival time. Another person arriving from a scheduled bus or leaving work will be very conscious of earliest pickup time. Thus, a range of different passenger utilities should be able to coexist simultaneously in the algorithm.

Automatic reassignment was felt to be an important function to explore because of the uncertainty of vehicle operations. Specifically, for vehicle breakdowns or when vehicles fall far behind schedule, a procedure that automatically placed passengers on alternative vehicles could be quite effective.

Several conclusions regarding the computer system design were reached as a result of the Haddonfield experience:

- computer scheduled service can be expected to be better than manually scheduled service, even with small fleets
- computer scheduling techniques need to be extended to include a variety of passenger classes and transfers to fixed route vehicles
- mixed computer-manual scheduling is worse than either alone (the computer needs complete accurate information on passenger and vehicles in order to be effective)
- the computer must have a complete, easily used set of scheduling, dispatching, and customer information as well as fleet status commands and functions
- scheduling techniques that rely on constrained limitations on customer pickup and delivery times result in low productivity and poor service

The system provided excellent service, was heavily used, but was discontinued in March 1975 when Federal funding ended. The experience highlighted a typical problem that can occur when funding arrangements for new types of service are not in concert with existing transit institutions.

11.1.3 THE ROCHESTER SYSTEM

As a follow-on to the Haddonfield demonstration project, UMTA then sponsored a Service and Methods Demonstration Project in the Rochester, New York, metropolitan area. This project, which began in April 1975, is partially aimed at testing and evaluating a follow-on computer control system that is based on experience gained in Haddonfield. The first service area includes most of the suburb of Greece and part of the city of Rochester, with a population of about 72,000 spread over 16 square miles. A second service area was started in April 1976 in the suburb of Irondequoit with a population of 44,000 in an area of 7 square miles. While several different services are offered, depending on the type of market and the time of day, the computer system is provided for many-to-many trips that are made at average fares of $1. Many-to-many ridership in Greece has reached a peak of over five hundred per day with a fleet size of up to nine vehicles.

The heart of the computer control system is the mechanism for assigning passengers requesting service to the vehicles providing it. As in Haddonfield, the approach taken to the assignment problem in Rochester is based on the immediate assignment of all customers requesting immediate service. The control algorithm is designed to make decisions that maximize the satisfaction of all system users subject to a constraint on the number of vehicles providing service. Although user satisfaction is influenced by several factors, those elements potentially under the control of the scheduling algorithm that are considered most important are the following:

- wait time (WT) – the time between a request for service being made and the pickup occurring
- ride time – the time between an individual's pickup and delivery
- pickup deviation (PD) – the difference between the promised pickup time and the actual pickup time

Since these factors are elements of user disutility, the objective function is structured to minimize some function of these service characteristics. The specific disutility function differs for different user groups, enabling the algorithm to provide distinct levels of service. To date, users have been disaggregated into three major groupings: immediate service, immediate transferring service, and advanced pickup service. Immediate service customers are those desiring to be picked up and delivered as soon as possible, but who have no dead-lines associated with the trip. Typical trip purposes associated with this type of customer are shopping and social-recreational trips. An immediate transferring service request comes from a customer desiring to transfer to a scheduled fixed-route bus. In this situation the user's primary concern is to catch the first possible departing fixed-route vehicle. Advanced requests are those from users who call in advance of their desired time of travel and wish to be picked up at a stated time. This category includes users making the same trip at the same time each day (periodic users) and those who simply phone ahead for service.

The objective function used in Rochester is quadratic, based on the theory that user disutility increases quadratically as service deteriorates. The quadratic function can incorporate all three service characteristics defined above.

In order to schedule a passenger's pickup and delivery, trial insertions of the two stops are made for every possible location on the tours of all available vehicles. The quadratic objective function is evaluated for each trial tour configuration. The best pair of insertions (pickup and delivery) is that pair with the smallest value of the objective function from all the tour configurations tried.

The objective function consists of the sum of the following three terms:

- marginal disutility to passengers already on the tour
- disutility of service to the new passenger
- marginal system resources expended to serve the new passenger

The last term is based on the need to conserve system resources in order to provide good service for future customers — clearly a delicate balance exists between providing good service for current customers and mortgaging the short-run future system capabilities.

User groups are distinguished by applying different utility function coefficients in the objective function. The type of service provided to any particular user group depends on the relationship between the weights placed on the service characteristics in the objective function. The fraction of system resources devoted to that group and hence their service quality will depend on the relationship between the weights for that group and the weights for other groups.

Both immediate regular and immediate transferring requests are processed as soon as they are received. The disutility to immediate passengers is assumed to depend primarily upon wait time and ride time. After assignment, the algorithm imposes a hard constraint on the tour of the vehicle assigned to carry the transferring passenger, so that she/he will not miss the scheduled fixed-route vehicle.

Advanced passengers are assigned at a variable time before the scheduled pickup, depending upon the number of vehicles operating in the system. The customer is not concerned with the time from assignment until being picked up but rather with the deviation from the scheduled pickup time. The disutility for these customers is thus a function of pickup deviation and ride time only.

While this describes the basic control procedure that was implemented in the fall of 1975, since then considerable changes have occurred, which make the current control system even more different from the Haddonfield one. The major differences between the Rochester control algorithm and the one used in Haddonfield are:

1. Passengers are divided into groups based on their service preferences.
2. Every passenger in a specific group is assumed to have the same disutility function, which is a weighted sum of wait time, pickup time deviation, and ride time.
3. Advanced requests are assigned an amount of time before their earliest desired pickup time, this time being referred to as the asssignment time offset, ATO. Initially the ATO was set up as a constant (as it was in Haddonfield) but it soon became apparent that this was having a detrimental effect on overall system performance.
4. In Haddonfield, the change in system resources was represented in the objective function simply by the change in tour length (future time committment) of the affected vehicle, and this was also used initially in Rochester. This representation quickly proved inappropriate, however, in that, as the system became more heavily loaded, the linear system resources term was quickly dominated by the quadratic passenger disutility term, resulting in conservation of system resources becoming less rather than more important. A second problem with this formulation was that in underutilized systems empty vehicles tended to remain empty and unassigned, resulting in an uneven distribution of work loads among the vehicles. To counteract these problems the following formulation was developed and implemented in the Rochester system:

> change in system resources — change in tour length [a (tour length of affected vehicle)
> + b (mean tour length of all active vehicles)]

where a and b are constants that are both set at about half the base value weighting in the passenger disutility function. Tour length is measured in the same units as passenger service times.

5. In Rochester, an important element of the demonstration objective is to arrange smooth, fast transfers between dial-a-ride and and fixed route, scheduled transit. This was readily accomplished for dial-a-ride to fixed route transfers by use of the assignment time constraint in the objective function.

6. While passenger reassignment has long been recognized conceptually as an important part of a total control procedure, it was excluded in Haddonfield and also initially in Rochester. It soon became appararent in Rochester, however, that reassignment was badly needed, principally because of the great amount of uncertainty present in the vehicle and driver activities.

In past computer automation efforts, a dedicated computer system has been utilized to support the operation. In the Rochester demonstration, it was decided, because of the experimental nature of the computer system, to take a different approach. All of the computer software development work prior to the initiation of the demonstration in April 1975, was performed on the time-sharing facilities of First Data Corporation, the firm doing the software programming. To facilitate further software development it was decided to operate the production version of the software on the same computer facility that had been used in its development — First Data Corporation's DEC System 10 located in Waltham, Massachusetts. It was decided that a leased telephone line would be used as the dedicated telecommunication link to Waltham (with a speed of 600 characters per second), and that a small computer would be installed in Rochester to coordinate the telecommunications link and to control the local computer terminals.

The next major concern was a backup system. Telephone lines are notably unreliable, so the on-site computer was equipped to communicate with the Waltham facility over direct-dialed telephone lines, if required. To further back up the Rochester equipment, each terminal on site was equipped to access directly the Waltham facility over normal long distance telephone lines if the local communications computer broke down.

One of the unique innovations of the Rochester Project is the application of state-of-the-art digital communication techniques to a land-based transit operation. Prior to the start of the demonstration, the Rochester dial-a-ride operation was already one of the few transit operations in the country to use digital communications. During the demonstration project, digital communications equipment produced by Kustom Data Communications, Inc., was interfaced directly with the computer scheduling system.

The vehicle terminals consist of 32 character by 8 line CRTs and an alphanumeric keyboard. The terminal operates off a modified General Electric voice radio system, equipped for two-channel operation. The control facility equipment consists of a minicomputer control unit, radio interface equipment, and a dispatcher console. The communications control unit is designed to operate as a terminal on the automated dispatching system, and interfaces directly both with the communications computer in Rochester and over telephone lines with the computer facility in Waltham.

The system is capable of transmitting a seven-line message to the terminal in approximately three seconds. A comprehensive computer interface allows extensive error checking and message validation, as well as computer monitoring and control of all communication functions. Finally, the new terminal, with its full alphanumeric keyboard in addition to function and status buttons, provides more interactive capability than was originally thought necessary, and is already beginning to provide a more comprehensive approach to driver/dispatching system interaction.

The equipment also provides an extensive control capability over the voice communication process. Each vehicle is provided with a voice annunciator indicator, which signals the availability of the voice channel for communications with the dispatcher. This annunciator is controlled by a request-to-talk/listening sequence that alerts the driver to raise his radio handset when convenient, and signals the dispatcher when the driver is ready. These functions allow for an orderly use of the voice communications channel.

One additional feature is the capability for the dispatcher and the vehicle driver to communicate directly through the digital communications equipment. Each may send messages to the other. In many cases, this eliminates the need for voice communication entirely, and provides a clearer, more understandable information transfer.

11.1.4 THE FUTURE OF COMPUTER DISPATCHING

The future of computer dispatching is obviously tied to the future of demand responsive urban transport systems in general. Dealing with the broader question first, evidence from Rochester tends to support findings in both Haddonfield and Santa Clara that a many-to-many dial-a-ride system operated by transit labor in major metropolitan areas is hard to justify economically. Both Haddonfield and Santa Clara ceased operation, and many-to-many operations have been reduced in Rochester as part of a series of service modifications aimed at increasing vehicle productivities and thus improving the system economics. This restriction of service, by moving towards a zonal, many-to-one or route deviation operation is clearly one approach worth investigation — as Ann Arbor and several similar Canadian systems demonstrate. Such configuration of service has significant impact on the degree of central control required as well as on the role of the computer within the dispatching process. The greater the degree of structure imposed on the service, the easier the control problem and the less the role for a computer. The result may be, even for large systems of this sort, that the computer is only needed for bookkeeping and communication handling functions*.

An alternative approach is to accept that many-to-many dial-a-ride systems will have low productivities, and to aim at reducing costs by shifting operating responsibilities to the taxi industry via shared-ride taxis. In this case the control problem is essentially the same as for dial-a-ride, but the ready potential for very large fleets makes the problem much more difficult. Certainly one of the reasons taxi operators might be reluctant to move into shared rides is the difficulty of manual control. The computer could remove this obstacle either by making the decisions (as in Rochester), or by performing a combination of bookkeeping and alternative-screening functions (a purely book keeping role alone would probably not suffice). Both computer assistance and computer decision-making need more investigation before the best approach can be ascertained with any degree of confidence.

Looking at computer control for many-to-many systems, it is clear that much work remains to be done. Some of the key areas for future research are:

1. *Best mix of computer, dispatcher, and driver in the decision process.* In Haddonfield, drivers were given only their next stop and the computer was in complete control. It was clear that this did not satisfy the drivers and probably did not result in the best system performance. Initially, in Rochester, drivers were given their next stop plus any subsequent stops that were close by (e.g., within half a mile of the previous stop). This was increased to a minimum of the next two stops for each driver to ensure that the driver would never be idle while waiting for the next stop to be transmitted (this occurred frequently during the initial weeks of operation). Drivers were then allowed to choose which stop to make first, provided that they notified the control center if they elected to change the order selected by the computer. This demonstrated the truth of the axiom "a little knowledge is a dangerous thing," in that drivers would often reverse the order because they could not see the basis for the computer's decision, which might lie further down the tour. Drivers are now being given a minimum of three stops, and this has decreased the number of bad decisions drivers are making. This is clearly only one example of the problems in mixing human and computer decision making.

2. *Bounds on algorithm performance.* It has never been possible to state how close to optimum any dial-a-ride control algorithm is, since, to date, no optimal solution algorithm has been developed. This need is now quite pressing since computer control has been proven technically feasible but has not yet been shown to dominate manual dispatching. Thus, optimization approaches to the bounding of system performance are an important factor in the structuring of demand responsive research priorities.

3. *Dealing with uncertainty.* Rochester experience has underlined the central role that uncertainty plays in dial-a-ride operations. Probabilistic elements include driver performance and availability, vehicle performance and availability, and future demands on the system. Developing an effective control strategy that addresses these uncertainties is crucial. Approaches range from immediate assignment combined with efficient reassignment capabilities to deferred assignment algorithms based on pools of unassigned passengers. It is not yet clear which approach is best, although the Rochester experience indicates that reassignment, at least as it is currently practised, is a fairly constrained process.

4. Quantifying user preference. Dial-a-ride and other demand-responsive urban transport systems make the transport designer face up to the unanswered question of what kind of service

---

*Indeed, the proposed Calgary, Canada, system dispenses with the elaborate and expensive onboard communications equipment typical of the systems under discussion, relegating the computer to strictly a bookkeeping role.

should really be provided. There is, as yet, very little evidence on how passengers trade-off reliability with mean speed, for instance, yet these are the trade-offs that the designer and evaluator of control procedures and dial-a-ride systems generally must make. Having recognized these trade-offs, it is now necessary to get sufficiently reliable data on passenger preferences on which to base future operating systems.

BIBLIOGRAPHY

Massachusetts Institute of Technology, 1971. Computer configurations for a dial-a-ride system. Report No. USL-TR-70-14, from MIT, Cambridge, Massachusetts, USA.
Massachusetts Institute of Technology, 1974. Development of the first personal transit system for the Rochester Metropolitan Area. MIT, Department of Civil Engineering, Cambridge, Massachusetts, USA.
Roos, D. 1964. CARS – computer aided routing system. MIT Urban Systems Laboratory, Cambridge, Massachusetts, USA.
Roos, D. et al. 1971. Summary report of the dial-a-ride transportation system. MIT Report TR-71-03, Cambridge, Massachusetts, USA.
Wilson, N., et al. 1970. Simulation of a computer-aided routing system, MIT Report R 70-16, Cambridge, Massachusetts, USA.
Wilson, N. H. M. 1971. The state-of-the-art of routing algorithms for demand actuated systems. Paper presented at the Sixth Annual Urban Symposium on the Application of Computers to Urban Society. Association for Computing Machinery, pp. 66–76.
Wilson, N., et al. 1971. Scheduling algorithms for a dial-a-ride system, MIT Report TR-70-13 from the Department of Civil Engineering, Cambridge, Massachusetts, USA.
Wilson, N. H. M. and B. T. Higonnet, 1973. General purpose computer dispatching systems. Highway Research Board Special Report No. 136.
Wilson, N., et al. 1971. Advanced dial-a-ride, Interim Report R75-27 from MIT, Cambridge, Massachusetts, USA.
Ziegler, E. 1974. The role of dial-a-ride in mass transit. Highway and Urban Mass Transportation, Washington, DC.

11.2 FURTHER DIAL-A-RIDE PROJECTS

Table 11.2 presents a chronological survey of demand-responsive para-transit systems with more than eight vehicles, which were tested and partly put into regular service in the following countries:

- the USA (cf Bauer 1970, Bartolo and Navin 1971, Ford Motor Co. 1970, Guenther 1970, Lex Systems Inc. 1973, and Section 11)
- the UK (cf. Lutman 1974, Oxley 1970, 1973, Slevin 1974)
- Canada (cf. Atkinson 1973, Bonsall and Simpkins 1973, Ontario Department of Transportation and Communications 1971)
- the FRG (cf. Bredendieck 1975, Etschberger 1975, Peckmann 1975, Meyer and Burmeister 1975, Schwedrat 1975, Meyer et al. 1977)
- The Netherlands (cf. Hupkes 1972)
- Japan (cf. Tanizawa et al. 1972)
- Sweden

TABLE 11.2 Chronological Survey of Dial-a-Ride Systems with more than eight vehicles (up to 1975)

No.	City	Country	Operation mode[a]	Opening year	Number of vehicles
1	Atlantic City	New Jersey, USA	vs	1916	35
2	Little Rock	Arkansas, USA	mtm	1946	75
3	Fort Leonard	Montana, USA	mtm	1958	80
4	Hicksville	New York, USA	mtm	1961	30
5	Peoria	Illinois, USA	mto	1964	17
6	Davenport	Iowa, USA	vs	1967	21
7	Gothenburg	Sweden	mtm	1967	40
8	Flint	Michigan, USA	mto	1968	26
9	Reston	Virginia, USA	mtf	1968	25
10	Merced	California, USA	mtm	1970	18
11	Bay Ridges	Ontario, Canada	mtm	1971	14
12	Detroit	Michigan, USA	mtm	1972	13
13	Haddonfield	New Jersey, USA	mtm	1972/73	19
14	Regina	Sasketchewan, Canada	mtm	1972	17
15	Rhode Island	USA	mtm	1972	28
16	Wilton	Maine, USA	mtm	1972	10
17	Bramalea	Ontario, Canada	mtm	1973	10
18	El Cajone	California, USA	mtm	1973	14
19	St. Petersburg	Florida, USA	mtm	1973	13
20	Ann Arbor	Michigan, USA	mtm	1974	45
21	Cleveland	Ohio, USA	mtm	1974	10
22	Ottawa	Ontario, Canada	mto	1974	26
23	Richmond	California, USA	mtm	1974	13
24	Santa Clara	California, USA	mtm	1974	90
25	Rockville	Maryland, USA	mtm	1975	12
26	Rochester	New York, USA	mtm	1975	9

[a]vs, varied stops; mtm, many-to-many; mto, many-to-one; mtf, many-to-few.

It may be interesting to mention that the first experiments with a jitney-type demand-responsive service (cf. Figure 10.2) were carried out in Atlantic City, New Jersey, USA as early as 1916. However, it took 50 years for the idea of a demand-responsive para-transit service to find broader attention. This is illustrated by Figure 11.1, which shows the time-dependent development of demand-responsive systems that were put into experimental or regular service.

So far about eight countries have made major efforts to develop and test dial-a-ride systems. During the last eight years in the FRG, special attention has been given to creating computer-aided dial-a-ride systems. The systems concepts developed became known under such names as:

- COBSY computer-bus-system (cf. Bredendieck 1975, Meyer and Burmeister 1975, Schwedrat 1975)

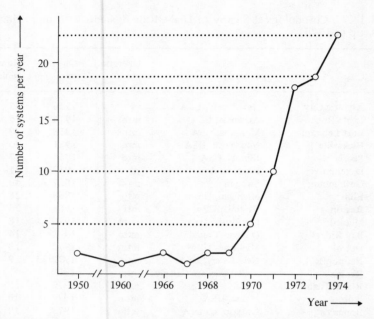

FIGURE 11.1 Number of dial-a-ride systems that were put into experimental or regular service each year.

- RUFBUS, which means hail bus (cf. Etschberger 1975)
- RETAX, rechnergestütztes taxi-bus system, i.e., computer-aided taxi-bus system (cf. Peckmann 1975, Meyer et al. 1977)

A small-scale demonstration project for COBSY was established at the Technische Hochschule Aachen; here COSBY was operated as a University internal demand responsive transportation system connecting the various buildings. A full-scale computerized system was then installed in the Bodenseekreis, a low-density area including the city of Friedrichshafen on lake Bodensee; it has an area of 200 square kilometers and a population of about 105,000 (cf. Etschberger 1975). Moreover, feasibility studies for installing a computerized demand-bus system were carried out for Ahrensburg, a small city with 25,000 inhabitants near Hamburg (Peckmann 1975).

The case descriptions in Section 11.1 and the above-mentioned features of other systems (cf. Table 11.2) permit the following conclusions to be drawn with respect to such systems characteristics as vehicles, service areas, operation modes, and the role of computers.

1. *Vehicles.* The frequency distribution given in Figure 11.2 demonstrates that over 50 percent of the existing systems use less than six vehicles, and less than 20

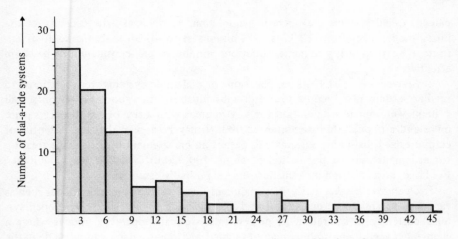

FIGURE 11.2 Frequency distribution of number of vehicles per dial-a-ride system.

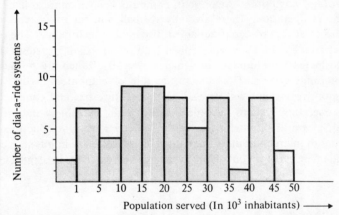

FIGURE 11.3 Frequency distribution of numbers of inhabitants per dial-a-ride system (there are about 16 systems that serve areas with more than 50,000 inhabitants).

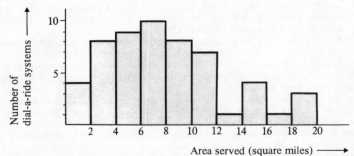

FIGURE 11.4 Frequency distribution of the area served per dial-a-ride system (there are about 13 systems that serve areas larger than 20 square miles).

percent of all systems are operating more than 18 vehicles. The capacity of the buses used ranges from 10-12 seats (minibuses) to 40-50 seats (standard fixed-route buses); in many-to-many situations minibuses are considered to be more effective.

2. *Service areas.* The first applications of dial-a-ride systems were restricted to smaller settlements. Figures 11.3 and 11.4 illustrate that about 50 percent of all systems were put into operation for service areas with a size of less than 15 square miles and a population lower than 35,000. It may be argued that the potential of computerized dial-a-ride systems will permit an enlargement of the service areas up to settlements with a population of about 100,000 or even 200,000 inhabitants. For larger areas division into smaller zones is probably essential.

3. *Operation modes.* Early systems mainly used simple operation modes such as varied stops and route deviation (cf. Table 11.2 and Figure 10.2). Developed systems currently in operation all possess many-to-many operation modes where a change between many-to-one, many-to-many, and one-to-many can be made in the course of the day, as discussed in paragraph 10.1.1.

4. *The role of computers.* At present, computers play a minor role in the dial-a-ride systems in operation. There are only a small number of systems that make extensive use of advanced computer-aided dispatching technology. The reasons are clearly illustrated by the above mentioned features of existing systems. If it is not necessary to increase the number of vehicles over 10-20 then it is not necessary to install a computer system. Thus the future role of computers in the creation of demand-responsive systems will depend on their future size and complexity. However, some operating systems have demonstrated the usefulness and efficiency of computer-aided route dispatching systems. Moreover, the introduction of digital data links and of microcomputers onboard vehicles as well as in bus-stop equipment could provide new aids for controlling even large vehicle fleets in an efficient way.

REFERENCES

Atkinson, W. J. 1973. Telebus project in Regina. In Highway Research Board, Special Project No. 136, Washington, DC.

Bartolo, R. and F. Navin, 1971. Demand responsive transit: Columbia, Maryland's experience with call-a-ride. Paper presented at the Confer-In West, American Institute of Planners Annual Meeting, San Francisco, California, USA.

Bauer, H. 1970. A case study of a demand-responsive transportation system. Research Publication GMR-1034 from General Motors Corporation, Warren, Michigan, USA.

Bonsall, J. A. and B. D. Simpkins. 1973. Dial-a-bus: three years experience in Ontario. Paper presented at the Annual Conference of the Roads and Transportation Association of Canada, Halifax, Nova Scotia.

Bredendieck, R. 1975. A possibility for innovating the computer bus system COBSY. Paper presented at the COBSY Seminar of the Institute für Kraftfahrwesen, Rheinisch-Westfälische Technische Hochschule Aachen (in German).

Etschberger, K. 1975. Description of the demand-controlled street transport system of the Dornier Company – System for the Bodensee Area. In Nahverkehrsforschung '75, Federal Ministry of Research and Technology, Bonn-Bad Godesberg, FRG, pp. 233–239 (in German).

Ford Motor Co. 1970. The Mansfield Ohio dial-a-ride experiment. Final Report from the Richland County Regional Planning Commission and Transportation Research and Planning Office.

Guenther, K. 1970. The Mansfield dial-a-ride experiment. Paper presented at the 11th Annual Meeting of Transportation Research Forum, New Orleans, pp. 215-232.

Hupkes, G. 1972. BUXI: demand-responsive bus experience in the Netherlands. Highway Research Record, No. 397: 38–41.

Lex Systems, Inc. 1973. Santa Clara County transit district – integrated demand-responsive/express bus system. Report on Requirements and Preliminary System Design, Menlo Park, California, USA.

Lutman, P. 1974. Dial or hail it? Experience with scheduled minibus services in Oxfordshire. In R. Slevin, ed., The First British Dial-a-Ride Symposium, September 9-12, 1974. Proceedings published by Center for Transport Studies, Cranfield, UK, pp. 85-87.

Meyer, H. and P. Burmeister. 1975. Study on possible applications of demand-controlled bus systems. Report 3/75 by the Hamburg Consult, Hamburg, FRG (in German).

Meyer, J., M. Peckmann, and E. Wittmann. 1977. Demand-controlled bus system RETAX – system components. In Nahverkehrspraxis, 25 (4): 154-156, 158-159 (in German).

Ontario Department of Transportation and Communications. 1971. Dial-A-Bus, The Bay Ridges Experiment, Downsview, Ontario, Aug. 1971.

Oxley, P. R. 1970. The dial-a-ride bus service. Paper presented at the Planning and Transport Research and Computational Company Seminar, London. Public Road Transport Analysis, pp. 107-110.

Oxley, P. R. 1973a. Dial-a-ride – Demand actuated public transport. Traffic Engineering and Control, 12 (3): 146-148.

Oxley, P. R. 1973b. Dial-a-ride applications in Great Britain. Paper presented at the Third Annual Conference on Demand Responsive Transportation Systems, Ann Arbor, USA. Proceedings published as Special Report No. 136 by Highway Research Board.

Peckman, M. 1975. Description of the demand-controlled bus-system of the Messerschmitt-Bölkow-Blohm Company for Ahrensburg. In Nahverkehrsforschung '75, Federal Ministry of Research and Technology, Bonn-Bad Godesberg, FRG, pp. 240-247 (in German).

Schwedrat, K., 1975. Systems analysis on the development of the demand-controlled passenger transport system COBSY. Paper presented at the COBSY Seminar of the Institut für Kraftfahrwesen, Rheinisch-Westf. Technische Hochschule Aachen (in German).

Slevin, R. ed. 1974. The First British Dial-a-Ride Symposium, September 9-12, 1974. Proceedings published by Center for Transport Studies, Cranfield, UK.

Tanizawa, T., et al. 1972. Demand bus implementation in Onose area, Preliminary Report from Hankyu Bus Company, Osaka, Japan.

12 International Bus Monitoring Experiences

This chapter analyzes the potential and limitations of computerized monitoring and control systems in the improvement of bus services. The experience gained in two advanced European systems, i.e, those of Dublin (Republic of Ireland) and Hamburg (FRG), are considered in detail.

12.1 IRISH EXPERIENCES: THE DUBLIN SYSTEM*

12.1.1 CASE HISTORY

The management of Dublin City Services faced a problem that all urban bus operators have faced for at least a decade. Difficulties arise from an increased volume of traffic, which causes congestion on an inadequate street network, leading to a deterioration in the operating environment for public transport vehicles. In the case of the City of Dublin the problem may be described more precisely in terms of four interrelated phenomena typifying the changing environment in the city.

1. The population of the greater Dublin region has increased by over 240,000 in the last 25 years to approximately 1 million.

2. As well as the rapid increase in population, the area has also experienced a population shift from the inner city to the outer suburbs. The inner city population decreased by 30,000 in the ten-year period 1961-1971. With the main residential suburbs now much further from the city center, the demand for motorized travel has increased dramatically.

3. The number of private cars has increased from 29,000 in 1951, to more than 160,000. Current forecasts predict that this figure will double over the next ten years.

4. Over the same period, the number of car passengers has increased by 33 percent while the number of bus passengers has increased only marginally.

While some alleviation has been gained by traffic management techniques, the absence of any specific aids for public transport has resulted in a poor operating environment for buses.

Prior to 1969 bus control in Dublin was effected using conventional methods. This consisted

---

*Based on a case description specially prepared by B. J. Fitzgerald, Manager, Dublin City Services, Dublin, Republic of Ireland.

mainly in the deployment of supervisory staff at strategic points along the route, usually at city center termini, crew relief points, etc. With increasing congestion, extensive regulation of the service became necessary. Since the degree of congestion may vary in time and location, additional running time is not sufficient to ensure a regular service. The use of stance inspectors as described above helped to correct the more obvious problems in the service, but the technique suffered from numerous deficiencies:

1. This control method was highly labor intensive and required constantly increasing levels of supervision as traffic congestion became more widespread and increased both in intensity and duration.
2. The information available to a stance inspector was limited to visual observations in his immediate vicinity. He had no contact with his colleagues or with a central dispatch office. This led to conflicting and wasteful regulation of the services.
3. The morale of the bus crews and the supervisory staff suffered since no tangible improvements were apparent despite the efforts being made.
4. Divided responsibility for the control of buses led to conflicting instructions for crews.

Thus it was accepted that supervisory staff should be provided with more detailed and up-to-date information on the route being regulated, and the centralized control of a single route or group of routes was recognized as a major objective. The initial stage was to install street telephones at the important stance points in the city; these were connected to a central control room, which was also in contact with the individual bus depots.

Although useful, this represented only an intermediate development stage in the evolution of a modern control system. The information available was still relatively sparse, and response to a particular traffic problem was still too slow to be fully effective. The stance inspector had no means of communicating with crews.

The pilot project of the radio telephone control system was commissioned in 1969. This was a significant departure from the conventional procedure. Each bus on the pilot route was equipped with a radiotelephone that enabled its driver to communicate with a single control inspector. Instructions and regulations now emanated from a single source. The control inspector, who was solely responsible for the entire route, used the voice radio system to obtain location and other data from the drivers on the route. Location data were ascertained verbally from each driver every half hour and plotted on a pre-printed diagram that also showed the expected (or scheduled) position of each bus (cf. Figure 12.1). A location chart could be completed in less than 10 minutes and the controller could then make adjustments to the service depending on the discrepancies between the actual and scheduled positions of the buses. Bus crews benefitted by having immediate communication with their supervisor for emergency purposes or any other reason. The radio telephone scheme was more than a communications network — it was a control system in its own right enabling the controller to monitor and regulate the service as required. In this environment a series of definitive control strategies evolved, which could be applied to correct conditions occurring frequently in a service. These regulations were mainly concerned with effecting a direct improvement in the service although some regulations achieved this indirectly. The more commonly used strategies were:

- transfer passengers and curtail a particular trip
- inject standby buses
- run buses "special" (out of service) to particular points on the route where they could take up service again
- switch crews to enable them to have their meal breaks on time; on occasions a standby bus would be used as a "shuttle" to put several crews back on time in a situation where a number of buses were running late
- hold buses at particular points
- divert buses along pre-arranged routes to avoid accidents, traffic jams, etc.

The pilot scheme, initiated in 1969, operated from a single control room located in the bus authority's Central Head Office. In 1972, the system was extended as a result of the satisfactory

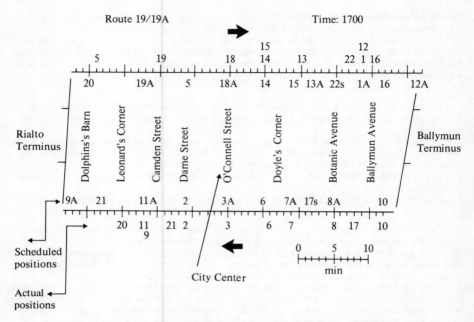

FIGURE 12.1   Control sheet for 17.00 hours: scheduled and actual positions.

operation on one route, and 538 buses of the 900-strong fleet were equipped with radio telephones, using 15 duplex radio channels. At this stage, a local control room was established in each of seven depots located around the city. The control inspectors in each depot were then in sole control of their own route or group of routes, as no services are operated out of more than one depot. This has proved beneficial, insofar as control inspectors are in contact with the same group of operatives, working similar shifts, and are easily accessible in the depot. As a result, a high degree of team spirit has developed between the individual control inspectors and the man they control, and this has helped to create the type of flexibility in the regulation of buses enjoyed in Dublin City. At the end of each day, the completed radio control charts are stored for subsequent analysis by clerical staff to provide management with performance indices for radio controlled routes. Finally, the remaining stance inspectors, the depot maintenance foremen, and a patrol car in each depot were provided with a mobile radio telephone to enable the control inspector to liaise with the principal elements involved in providing the service.

Radio control proved advantageous to operatives, supervisors, management, and passengers:

- Bus crews reported to a single control inspector who was solely responsible for the route. This provided more job satisfaction for both control inspector and crews. With better control of the route, the distribution of the work load was more equitable. Crews could take their meal break on time. The radio provided the crews with an element of protection in the event of vandalism.
- Supervisors now worked in a better environment, with full responsibility for their own routes.
- Management could now monitor the level of service in a scientific way. The radio control statistics led to better rescheduling of the radio controlled routes.
- The improvement in the level of service attracted additional passengers and reversed the decline in patronage. The passengers now enjoyed a better service with more even headways.

Although radio control was a significant improvement on the previous scheme, the potential of the system was constrained by a number of factors:

- Given the limited number of channels available, the system could not be extended to the full fleet of 900 buses using the voice control technique.
- The manual plotting technique was too tedious and left the control inspector little time to plan control strategy.
- A location call every 30 minutes was considered too coarse for control purposes since traffic conditions could change drastically between calls.
- The management statistics, although significantly better than the data available previously, were still considered inadequate and their production was a highly labor intensive activity.

12.1.2 IMPLEMENTATION AND OPERATION OF THE AUTOMATIC VEHICLE MONITORING SYSTEM

On realizing the advantages and limitations of the voice radio control system, Dublin City Bus Services set up the necessary study groups to design a total management system that would provide:

- monitoring
- control
- management information

As a result, the implementation of an automatic vehicle monitoring (AVM) system for all routes in Dublin City Bus Services was recommended. The following objectives were defined for the system:

- real time information for service monitoring
- control capability for the full fleet
- improvement in the level of service
- comprehensive real time management information
- increased efficiency, in terms of better utilization of fleet and manpower

The AVM system was developed as an extension of the existing voice control scheme, using seven depot-based local control rooms, and the existing type of bus position plotting diagram. The overall system design is shown in Figure 12.2.

The central computer generates interrogation messages, which are transmitted via radio links to the full bus fleet; it then validates the replies from the buses, and finally provides the control inspectors' displays, while storing the replies for later analysis to provide the necessary management information. Bus location is by means of an odometer, the starting location being coded in by the driver at the terminus using three thumbwheel switches.

In specifying on-bus hardware, simplicity of design and ease of maintenance were prime objectives: the odometer used provides a simple, reliable, and inexpensive solution to the problem of vehicle location. The configuration described in this paragraph pertains only to the pilot AVM scheme. Extension to the full system will necessitate the provision of additional core memory on the main processor, some minor changes to the software, plus the additional peripherals such as VDUs, serial line interfaces and printers for the remaining six depots.

12.1.2.1 The Central Core Site. The installation at the central core site (cf. Figure 12.2) consists of two computers, which perform separate tasks: the main processor is the Digital Equipment Corporation PDP 11/40, and a PDP 11/10 acts as a front end processor for the main machine.

The front end computer is responsible for the bus polling routines. It generates the interrogation messages and validates the replies for the mobile units. It also repolls vehicles that fail to reply, and then sends the verified data back to the rear end 11/40. The executive is RSX-11A and the user programs are written in ASSEMBLER. The processor configuration includes 16K words of nonvolatile memory.

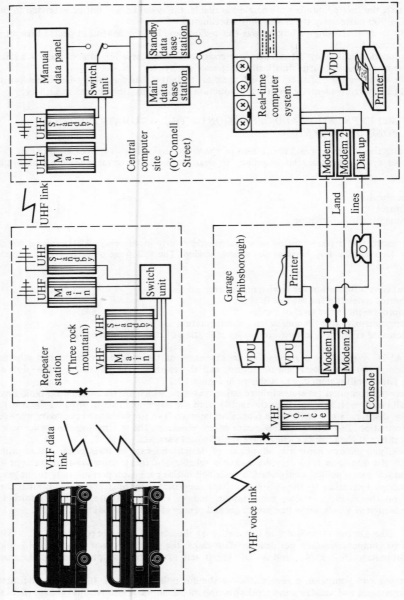

FIGURE 12.2 Structure of the Dublin automatic bus monitoring system.

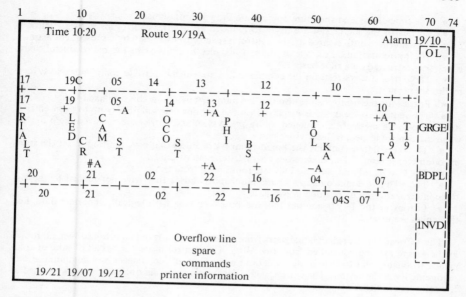

FIGURE 12.3 Route map display.

The main computer processes the data from the mobile units and generates and updates route maps for all routes on display in the system. Scheduled positions are also computed by the rear end machine using a simple simulation technique. It also handles all interactions between the control inspectors and the system, and logs the poll results for subsequent analysis. The rear end PDP 11/40 processor system includes 88K words of core memory, two 3-megaword disk cartridges, one fixed head disk, a 9-channel magnetic tape drive, line printer, teletype and a VDU. The executive used is RSTS-E and all user software is written in BASIC.

The computer system is designed to run continuously without human intervention, automatically reloading software in the event of power failure; in such cases a control inspector is required to adjust the reading for the time of day. At the central computer site only a few tasks need to be done. The statistics tape must be reloaded and a fresh tape loaded for the following day: the day's statistics are processed on the company's main IBM installation. The system accounting programs are run and the lost mileage reports are printed for each garage.

The statistics reports are distributed to the appropriate management personnel the following day. A data base station, which is used to interface the front end to the radio link, encodes the binary data from the bus polling computer with audio tones and decodes the audio coming back from the buses. The base station operates at 800 baud, and the polling rate in the pilot scheme was approximately 400 buses per minute. The base station is duplicated so that a standby station is available should the main link fail. A manual data panel is also provided, which may be used to interrogate an individual bus or may be used to switch all buses to speech in the event of a computer failure. It is also used as a test facility.

12.1.2.2 *The Pilot Garage.*

The installation in the pilot garage (cf. Figure 12.2) consists of two VDUs connected to the computer system via land lines and modems. The VDUs are Hazeltine 2000 video terminals that have a 1998 character dual intensity screen with a field blink facility. They operate at 1200 baud in full duplex mode. The modems are wired for split speed (1200/110 baud) operation on a two-wire circuit. A thermal matrix printer is driven from one of the VDUs and this provides hardcopy for both controllers. The display is updated at 45 second intervals in the pilot scheme.

The screen format has the following features (cf. Figure 12.3):

- The time-of-day and the route number are shown continuously.
- Entries in the alarm slot are made to blink to attract the control inspectors attention. Entries remain there until cleared by the control inspector or overriden by a subsequent alarm.
- The route diagram is set out in units of distance in contrast to the radio control chart, which is constructed on a time basis.
- The cursor always rests on the command line on which controller commands and system replies are printed.
- Entries for the printer appear on the line immediately below the command line.
- The request-to-speak list is displayed on the last line and will display up to eight entries. Request-to-speak messages are queued up by the system and entries are removed from the list when the control inspector deals with them.
- The OL column contains the board numbers of buses that have omitted to set the starting odometer reading — i.e., these buses are said to be out of limits.
- The GRGE column contains buses en route to or from the garage.
- The BDPL column contains board numbers of buses that have not replied to the last three interrogations, i.e., resulted in bad polls.
- Entries in the INVD column indicate buses replying with logically incorrect data, i.e., the replies are invalid.

The software protects individual users from each other. If a terminal breaks down, all routes in the garage may be controlled from one VDU. The control inspector has a full suite of commands to enable him to record all regulations and effect all file changes and communications requirements in the system. The existing radio base stations remain unchanged and were used as the voice link in the AVM system.

12.1.2.3 The on-bus subsystem (cf. Figure 12.4). This comprises a radio telephone, telemetry unit, control heads for both the radio and telemetry unit and an odometer. The radio telephone is a Stornophone CQM 612 VHF/FM radio operating in semi-duplex mode. The channel spacing is 25 kHz and the effective radiated power is 10 watts. The radio control head enables the driver to select up to 12 channels, and also has volume and squelch controls. Push button and foot switch operation of the transmitter is possible. The telemetry unit consists of a memory register to store the on-bus data and a modem to decode the incoming audio information from the radio link and to encode the reply messages. The interrogation message is five characters long including a six bit header. The remaining four characters are in four bit BCD format with start and stop bits to make up six bits per character. The first three characters following the header contain the address or fleet number of the bus being interrogated. Each telemetry unit has identity plugs coded to correspond with fleet numbers of buses.

The last character of the interrogation message is a function code that specifies the response required of the mobile and may have the following values:

- FC = 3   reply with a normal message
- FC = 5   go to the speech channel selected by the driver on the radio control head
- FC = 6   reply only if the alarm code has been set on the vehicle. FC = 6 is used in conjunction with the general address 000 in a special alarm search broadcast every 20 seconds.

On receipt of an interrogation the mobile examines the address and if it coincides with its own identity a reply is sent.

If the interrogation is an ALARM call the mobile will reply only if the ALARM button has been depressed on the data control head. The reply consists of 10 characters including a six bit header. Again the characters comprise six bits including 4 BCD message bits and start and stop bits. The first three characters represent the address of the mobile, which is taken from the identity plugs in the telemetry unit. The next three characters comprise the odometer reading. The seventh message character is the function code, which may have the following values:

- FC = 3   normal state exists on the bus
- FC = 5   the speech facility is requested
- FC = 6   the alarm facility is required

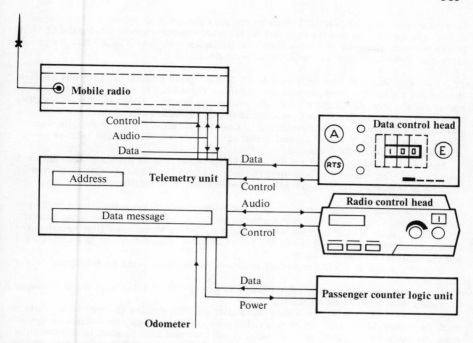

FIGURE 12.4   Bus installation – pilot system.

The last two characters contain passenger loading data, which must be supplied by an external device.

The mobile data unit powers-on in the voice mode and samples the data channel for half a second at two-minute intervals. If data are detected, the data channel is retained until either the mobile is switched to speech or data reception is lost for more than 5 minutes. The telemetry unit is fitted with a rechargeable nickel/cadmium cell to supply power to the logic circuits in the event of a voltage drop in the bus supply.

The data control head enables the driver to interact with the system. Its features include a request-to-speak button, an alarm button, and an entry button that is used to register the odometer reading set by the driver on the three thumbwheel switches. The odometer is a mechanical pulse generator, which sends a pulse to the data box every 44 yards. The pulse count is divided into four and added to the starting value of the odometer set by the driver on the data control head.

From the drivers' viewpoint the AVM procedure is not significantly different from radio control. The odometer readings, which must be set on the data control head at the beginning of each trip, are preprinted on his running board and the times at which these must be entered are indicated. When the driver boards his bus, he switches on the radio, which will be in voice mode immediately, and informs the controller that he is leaving the garage. He consults the running board for the correct odometer reading (always 800 for a garage trip) and enters it on the data control head. The on-bus system changes to the data mode after two minutes and the bus appears in the garage column of the controllers display. When the bus reaches the terminus of the first trip, the driver sets up the appropriate odometer reading and enters it. The bus now appears correctly in the route diagram on the control inspector's VDU. At the beginning of the next trip the driver repeats the procedure and proceeds thus for the remainder of the duty. If the driver can keep to his schedule no verbal communication with the control inspector is necessary unless either of them wishes it. If the driver wishes to speak to the control inspector, he depresses the request-to-speak button on the data control head and waits for the control inspector to reply by opening the voice link. In the event of an emergency, pressing the ALARM button will open the voice channel within 20 seconds. When the bus is pulled into the garage at the end of the day, the driver simply switches off the radio.

12.1.2.4 The data radio link (cf. Figure 12.2). This comprises a directional UHF link between the computer installation and the repeater station on Three Rock Mountain (sited about seven miles from the city center) and a VHF link between the repeater station and the mobiles. On both parts of the link, selection is accomplished with subaudio tones. Three Rock Mountain was chosen as a site for the repeater station to ensure good radio coverage on the data channel. The power output on the VHF channel is limited to 25 watts ERP (effective radiated power) and 5 watts ERP on the UHF channel.

Apart from the complete redundancy in certain areas, such as the duplicated data links and base stations, the system has been designed to fail gracefully. The following features were specified to ensure that the system is reasonably secure from external perturbations:

1. In the event of a mains power failure, the software will reload automatically on restoration of power. Only the time of day has to be changed if the failure is of significant duration.
2. The repeater station is fitted with an auxiliary power source capable of operating the remote equipment for at least 24 hours. Power failure at the remote site is signalled by a warning lamp in the computer center.
3. In the event of a VDU failure all routes within the garage can be transferred to the remaining VDU.
4. If a land line fails the controller effected can switch his VDU to one of the three dial-up lines and continue to control his route(s).
5. If either the data radio link or data base station fails, the standby unit may be selected.
6. In the event of a computer system failure, the mobile units may be switched to speech using the manual data panel, and conventional radio control may be used. In any event the mobile units will revert to voice after five minutes if the data stream from the base station stops, so normally human intervention will not be required. The voice system is independent of the AVM hardware, being a stand-alone communication link based in each garage. (See the system diagram in Figure 12.2.)

12.1.3 EVALUATION OF THE SYSTEM — THE 90-DAY TEST

The system was developed by a consortium of suppliers with Storno Ltd, England, as the main contractor. Various areas of the project such as software, telemetry equipment, and construction were sub-contracted to the other members of the consortium. The Traffic Department of Dublin City Bus Services and the Computer Services Department liaised with the consortium on all aspects of system specification and prototype development and testing.

The functional specification set out the important parameters, the required values for acceptance and the system test procedures. The mean time between failures (MTBF), the mean time to repair (MTTR), the computer system response time, and the location accuracy of the on-bus sytstem were considered especially important. The measurement of the above parameters was the principal objective of the 90-day test.

Ninety-five mobile systems were commissioned in the pilot garage, and five systems were held as spares. The performance of each mobile installation was monitored by recording the failures occurring on each bus and the time to repair the unit. Location checks were carried out on a sample of the scheduled AVM services, and system response time tests were made with the system in normal operation. The test results are summarized briefly in Table 12.1.

Owing to the large commitment of staff to the evaluation of the technical aspects of the system, only a brief examination of the headway pattern on the test routes before and after the introduction of AVM could be undertaken. It was estimated that excess waiting time per passenger would be reduced by 12 percent in the worst case. There was a significant reduction in passenger complaints on the AVM test routes and the bus crews noted the better distribution of loading. Both the crews and the control inspectors received formal training, and in all cases operatives were encouraged to take an active interest in the system as it was being installed.

TABLE 12.1 Results of the 90-day Test

Parameters	Specification	Measured results
Locational accuracy	95% within ± 0.3 miles	96.3%[a]
RTS and ALARM tests	95% to be detected and dealt with correctly	100.0%
Computer system response time	95% of new route requests to be serviced within 6 s	Mean response time was 14 s with a measured maximum of 23 s
MTBF	50 days	MTBF averaged 33 days for test but was much better in the last weeks of the test
MTTR	2 days	MTTR for telemetry units averaged 3.1 days

[a] Approximately 85% of locations were within ± 0.1 miles of actual position.

The introduction of AVM leads to a greater responsibility for the control inspector: having radio contact also with stance inspectors and maintenance staff results in a high degree of job satisfaction. The system provides critical location data with minimum involvement of the crews and this leads to greater independence for the control inspector. In contrast to the radio scheme, the participation of the crews becomes more specific but the level of involvement is reduced considerably. The driver merely sets the odometer reading according to the running board and thereafter uses the system only when he requests communication or is the subject of a regulation initiated by the control inspector.

The control inspectors' attitude to the system was one of immediate acceptance and approval. The awareness of the computer system and the use of computer oriented language, which was new to them, has in no way inhibited their ability as controllers, and indeed there was a significant improvement in the level of control. The crews, within a short time of its introduction, did not hesitate to express their approval for the better environment resulting from the introduction of AVM, and were quite impressed with the more equitable distribution of the work load, which now became possible. The level of acceptance and approval shown by the operatives alone is a significant measure of the success of the system.

The experience gained in the pilot scheme of AVM has proved that the philosophy behind the system is satisfactory, and indicates that this type of system will meet the management objectives set out earlier in this paper.

12.2 FRG EXPERIENCES: THE HAMBURG SYSTEM*

12.2.1 BACKGROUND**

The increase in motorization during the 1960s has led to a decrease in the efficiency of public transit systems. The burden placed on the inner city by individual transportation and the call for the conservation of these city centers led, along with traffic and operational measures, to an increase in the demands made on public local transportation.

*Based on a case description specially prepared by H. Tappert and O. Böhringer, Hamburger Hochbahn, AG, Hamburg, FRG.
**The work that forms the basis of this report was carried out with funds from the Ministry for Education and Science (contract P.6.2./23 HH-HHHB1). The Ministry for Education and Science does not accept any responsibility for the accuracy or completeness of the information or for the observance of the private rights of third parties.

The oil crisis in 1973/1974 underlined the importance of public transportation, and the need to develop new modes of urban transport and improve those in existence. Bus systems will remain one of the main modes of public transport for a long time to come. They accounted for 54 percent of all seat-kilometers accumulated by the different public transport systems belonging to the Verband öffentlicher Verkehrsbetriebe (VÖV) (the Union of Public Transit Companies of Hamburg), which was founded in 1973. However, bus services are greatly influenced by traffic disturbances. The use of new technologies to support bus operation during the last few years has proved to be successful.

Efforts are needed to ensure a safer, more punctual, and more comfortable operation if the bus ridership is to be increased in the future. For this reason the Hamburger Hochbahn AG has used, along with other means, an automatic express bus monitoring system for over ten years. The main features of this system, which was developed by the Hamburger Hochbahn AG in cooperation with an industrial company, and the experience gained so far are described in the following.

12.2.2 IMPLEMENTATION AND OPERATION

12.2.2.1 Basic Design Principles. The structure of the automatic bus monitoring system is illustrated in Figure 12.5. The schedules and routes of the individual buses are stored in the disk memory. Using these data, the computer generates telegrams that are transmitted via a stationary transmitting apparatus to the buses. A transmitter–receiver apparatus connected with a special data processing unit is installed onboard each bus. Before beginning his tour, the bus

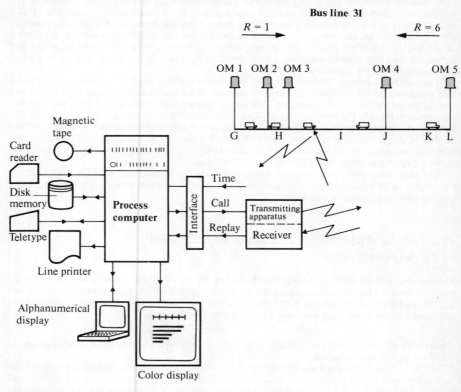

FIGURE 12.5 Structure of the Hamburg bus monitoring system.

driver feeds this instrument with his line and route number. The bus device will only understand the telegram sent out from the stationary data transmitter if the line and route numbers given in the telegram are identical to those stored by the driver in the onboard device. If the numbers are not identical, the driver has to transmit a reply telegram containing the actual position of the bus. The process computer assigns this position to the corresponding bus line and route numbers and displays the actual timetable together with the prescribed one in the form of linear time–distance diagrams on a multicolor computer display (cf. Figure 12.5).

The design principle sketched here, using a data transmission speed of 300 baud, permits the polling of each of the 100 vehicles at time intervals of 30 seconds. If the bus is also equipped with a device for counting passengers, the number of people entering or leaving the bus will also be transmitted to the control center. The principle used for the identification of bus positions is illustrated in Figure 12.5. Let us assume that a bus on line 31 runs between points G and L of its route. The driving direction G to L is marked by setting a variable $R = 1$, and the opposite direction by $R = 6$.

Location code transmitters, OM1–OM5, are installed along the route and permanently transmit two separate frequencies in the six meter range. An antenna in the bus registers when one of these transmitters has been passed. The number of this transmitter will be stored in the data part of the vehicle's device. In addition the 100 meter steps of the odometer are stored in an electronic counter. If the data section detects one of the transmitters OM1–OM5, this counter returns to zero. In this way the distance of the vehicle from the beginning of the route can always be calculated from the position of a location transmitter and the distance of the bus from the transmitter.

To enable a change of the driving direction to be identified, e.g., from the direction marked $R = 1$ in Figure 12.5 to the opposite direction indicated by $R = 6$, a location code transmitter has to be installed at each turning point. If a vehicle is put into operation at a point that is neither the beginning nor the end of a route it is necessary to place a locator at this route entrance point for immediate location identification.

12.2.2.2 Computerized Management System.

The bus location identification system was originally implemented with hard-wired logic only. To enlarge the potential of the system, a process computer, as shown in Figure 12.5, was installed in 1975. The software developed and implemented in the computer is designed to perform the following tasks:

- program controlled polling of the individual vehicles, i.e., generating the polling telegrams and transferring them to the interface
- transfer the actual timetable data from the interface to the computer and error checking of the reply telegrams
- comparison of the prescribed and real schedules (every two minutes), and computation of the timetable deviation
- generation of graphical representation of both the prescribed and real schedule on the color display
- organizing the dispatcher–computer communication
- handling of emergency calls sent by the bus drivers
- informing the bus driver by means of a flashing onboard red lamp if the required departure time valid for individual stop is not yet reached
- control of radio conversation between the bus drivers and the dispatcher
- collection of data from the automatic passenger counters
- identifying defective bus equipment by analyzing all telegrams and reply telegrams with respect to response errors

When irregularities in the bus operation occur, then the dispatcher takes over the management tasks using the information on operational status provided by the process computer. In the case of an accident or other incidents, i.e., criminal activities, the bus driver is expected to press an emergency button. The corresponding acoustical and optical alarm signals as well as the bus line and route numbers will be displayed to the dispatcher within two seconds. The dispatcher can then take the necessary action. For this purpose he can use the information on the operational status presented in the form of graphical timetables on the computer display. He identifies the buses that are ahead of schedule, those with delays exceeding 15 minutes, and those with a break planned for the turnaround point. Using a keyboard the dispatcher can get the graphical timetable representation on the color display for the corresponding busline. Necessary orders are sent to the bus drivers by the radio link.

Experience gained so far indicates that the dispatcher learns to use the system very rapidly and then regards it as an indispensable aid to his work.

12.2.2.3 System Hardware. The onboard system consists of a transmitter–receiver apparatus, a control panel, a data section, and a magnetic antenna for receiving the information broadcast by the location code transmitters. Four frequencies in the two meter range are used for data and voice communication with the control center where the maximal admissible transmitting power is six watts. For data transmission from the computer to the bus, a frequency of $F1 = 153.93$ MHz is used; a frequency of $F3 = 149.33$ MHz is used for the opposite direction. Voice communication is established by means of the frequencies $F2 = 154.107$ MHz (control center to bus) and $F4 = 149.47$ MHz (bus to control center).

Frequency $F2$ is also used for transmission of data, but only in order to break a vocal communication. Passenger counters from Weiss/Prodata are installed in 40 rapid buses. The entrances and exits of the buses are divided so that only one passenger can pass at a time. The sequence in which two light beams are broken is identified by a logic as an entering or leaving passenger. The results are fed into two counters. The contents of the counters are transmitted in the answer telegram to the control center.

The hardware installed in the control center is a specially designed computer transmitting/receiving apparatus interface, the transmitting/receiving apparatus, and the process computer. In selecting the process computer, emphasis was placed on the need for fast data transfer between the magnetic disk and the core memory. The selected machine possesses a 64K word core memory with a cycle time of 800 nanoseconds and a 16 bit word length. A double cassette magnetic disk memory is attached to the process computer as well as a magnetic tape memory that permits data transfer to a universal computer. The alphanumerical display is used for the dispatcher–computer communication. The color display contains a 19 inch screen on which 48 rows with 72 grid places are addressable. One place permits any symbol to be displayed that can be represented by a matrix of 8×6 points.

12.2.3 OPERATIONAL EXPERIENCES

The bus monitoring system described here currently covers about 100 buses of the express bus system. It was necessary to invest approximately 400,000 DM to reconstruct the control center for the introduction of the computerized monitoring system. The development of the software and the hardware required about 110 man-months. These expenses lead to the question of the benefits achieved. The following improvements were obtained:

- The number of persons needed to solve the various supervisory tasks was reduced from 176 to 60. This manpower saving was mainly achieved by automatic bus monitoring and the replacement of local inspection points by emergency cars guided by radio communication.
- The automatic bus location identification permits disturbances to be detected with smaller delays. Thereby it was feasible to save about six reserve buses and the corresponding drivers.
- Delays were reduced.
- The drivers were more disciplined in keeping to the prescribed schedules.
- A small increase of the travel speed of the monitored express buses was achieved.
- The introduction of the emergency call system provided more safety for drivers and passengers.

12.3 FURTHER BUS MONITORING PROJECTS

There is a large number of bus systems that use various types of radio telephone communication as true dispatching aids (cf. Cabeza 1973). However only a few of them are computerized vehicle monitoring systems. Figure 12.6 summarizes selected features of these computerized vehicle monitoring systems, which could reach the status of large-scale demonstration projects or of regular passenger services.

The first computerized bus monitoring system had already been installed by the Chicago Transit Authority (CTA) at the beginning of the 1970s (cf. Shea 1973). The system monitors about 500 of the 3,000 buses operated by CTA. The proximity method, in accordance with paragraph 10.2.1 and Figure 10.6(b), is used for automatic vehicle location. For this purpose about 120 signposts were installed on 60 routes. Each vehicle is polled every two minutes by the central computer. A "Three Phase Expansion Program" has been presented (cf. Shea 1973), which proposes to equip all routes of the CTA network with signposts, i.e., to install about 380 additional signposts and to cover the total bus fleet by the automatic monitoring system.

The Dublin and Hamburg systems, described in Sections 12.1 and 12.2, represent the second and third largest operational projects. Extensive research and development activities were started very early in the UK (cf. Finnamore and Bly 1974, Wheat and Cohen 1974). Radio monitoring has been studied in Bristol and London. An experimental computerized monitoring system was tested on London's bus route number 11. The Bristol and London experiences presented did not lead to promising conclusions regarding the achievable benefit/cost relations. According to their experiences the installation of a costly computerized bus monitoring system is considered justified only for rather complicated traffic conditions such as occur on high frequency services, with a long round-trip time and subject to prolonged traffic congestion of an unpredictable nature (for more information see Finnamore and Bly 1974).

During recent years several significantly large efforts have been made in France. A computerized system called SECAMA, (système expérimental de contrôle automatique du mouvement des autobus) was installed for the bus route 52 in Paris in 1972/1973 (cf. Cassy and Snitter 1974, Perrin 1976). It monitors 35 buses and uses the dead reckoning principle, in accordance with Figure 10.6(a), as the vehicle location procedure.

A relatively large system, which covers a network of ten bus lines with a rolling stock of 96 vehicles, was put into operation in 1975 in Besançon, a French city of about 130,000 inhabitants. Location identification is also by dead reckoning, in which the door openings at bus stops are used for recalibrating the distance measurements taken by the odometer (cf. Frey 1976).

A third French computer-based bus monitoring system has been tested in the city of Lille, in Northern France (cf. New Scientist 1977). Finally, the city of Toulouse started a computerized bus surveillance and control project in 1975, with operational experiments in 1976 (cf. Giraud and Henry 1976). Results obtained so far are considered promising for heavily travelled routes on which perturbations can lead to major deviations from schedules (for more details cf. Giraud 1976). The Toulouse system possesses a special feature: it represents the first known system that uses microprocessor-based onboard devices (cf. Figure 10.8).

City–country	Number of buses	Location identification method	Polling cycle (s)	Year of installation	Further features	References
Besancon–France	96	Dead reckoning with updating by door opening registration at stations	16.5	1975 (December) Beginning of operation	Passenger load registration in classes: 0, 1/3, 2/3, 3/3	Frey 1976
Chicago–USA	500 on 60 routes	Proximity method (120 signposts)	120	1969/70	System supplied by Motorola corporations	Shea 1973
Dublin–Ireland	400	Dead reckoning	60	1969/72 radio control; 1975, computerized	Extension to total bus fleet is planned (900 vehicles)	cf. Section 12.1
Hamburg–FRG	100 (express bus system)	Proximity method	30	1966/69 radio control; 1975, computerized	Passenger counting is used in 40 buses; enlargement of AVM system to city buses is planned	cf. Section 12.2
London–UK	44 on route 11	Dead reckoning	5	1971/73	System supplied by Marconi company	Wheat 1974, Finnamure 1974
Paris–France	35 on route 52	Dead reckoning	10	1972/73 Operational since 1975	Number of passengers is transmitted, system "secama"	Perrin 1976, Cassy 1974
Toulouse–France	16 on route 12	Dead reckoning (as in Besançon)		1975 Operational since 1976	Using onboard microcomputers	Giraud 1976

FIGURE 12.6 Survey of selected computerized bus monitoring systems.

To date computerized streetcar surveillance systems have not received broad attention. Nevertheless, a project to create a computer controlled operation management system for streetcars and underground trams in the city of Hannover, FRG was started (cf. Geist 1975). The proposed system will fulfill three categories of tasks:

- monitor the streetcar operation on surface streets in the same way as discussed above for bus systems
- monitor and control the streetcar operation in downtown tunnels in a similar manner to subway trains
- coordinate the streetcar operations on surface streets and in tunnels, especially with respect to tunnel inflow metering

Summarizing, it can be stated that the creation of computerized public transit vehicle monitoring systems is still at an early stage. The complexity of these systems and the resulting cost, reliability, and other problems justify their application only under special traffic conditions.

The barrier that exists towards widespread introduction of automatic monitoring concepts into public transit systems is very unlikely to be broken unless drastic reductions in cost and complexity can be achieved. It is likely that the concept of vehicle monitoring by spatially distributed onboard microcomputers, as discussed in paragraph 10.2.1 (cf. Figures 10.7 and 10.8), could point the way to overcoming this barrier.

REFERENCES

Cabeza, C. M. 1973. Use of electronic control systems for improving public surface traffic, 40th UITP Congress, Den Haag, The Netherlands (in German).

Cassy, M. and A. Snitter. 1974. Experimental Bus Operation Control System – SECAMA. Paper presented at the Second IFAC/IFIP/IFORS Symposium on Traffic Control and Transportation Systems, Monte Carlo, Monaco (in French).

Finnamore, A. J. and P. H. Bly. 1974. Bus control systems: some evidence on their effectiveness. Paper presented at the Second IFAC/IFIP/IFORS Symposium on Traffic Control and Transportation Systems. Monte Carlo, Monaco.

Frey, H. 1976. Electronic system to assist the management of bus operations. Paper presented at the Third International IFAC/IFIP/IFORS Symposium on Control in Transportation Systems, Columbus, Ohio, USA.

Geist. 1975. Computer-controlled operation management system for streetcars and underground trams. In Nahverkehrsforschung '75, Federal Ministry of Research and Technology, Bonn-Bad Godesberg, FRG (in German).

Giraud, A. and J. -J. Henry. 1976. On-line control of a bus route – studies and experiments in Toulouse. Paper presented at the Third International IFAC/IFIP/IFORS Symposium on Control in Transportation Systems. Columbus, Ohio, USA.

New Scientist. 1977. French buses to keep real time? February 17: 398.

Nicolaev, M. I. and G. I. Kalney. 1974. Automatic traffic control system applied to urban electric transport. Paper presented at the Second IFAC/IFIP/IFORS Symposium on Traffic Control and Transportation Systems, Monte Carlo, Monaco.

Perrin, J. -P. 1976. Automation for bus operation in Paris. Paper presented at the Third International IFAC/IFIP/IFORS Symposium on Control in Transportation, Columbus, Ohio, USA.

Shea, R., ed. 1973. Monitor – CTA final report – An urban mass transportation demonstration project study of automatic vehicle monitoring. Chicago Transit Authority, Engineering Department, Research and Planning Division, Chicago, USA.

Wheat, M. H. and N. V. Cohen. 1974. Bus monitoring and control experiment and the assessment of its effects on bus operation. Paper presented at the Second IFAC/IFIP/IFORS Symposium on Traffic Control and Transportation Systems, Monte Carlo, Monaco.

13 International Rapid Rail Transit Control Systems Experiences

This chapter analyzes the experience gained so far with the introduction of computer control and surveillance systems into urban and suburban railways. Three case descriptions show the state of development reached in the USSR, USA, and FRG:

- the ASU-PP (automated passenger transport control and management) system of the Moscow Metro
- the computerized control systems of the San Francisco BART (bay area rapid transit) system
- the computer-aided train traffic surveillance and control system of the new Munich S-Bahn

Each of these three cases is characterized by certain specialities: the San Francisco BART, for example, which is one of the newest commuter transit systems in the USA, represents the most computerized urban rail transit system of the world. The Moscow Metro also uses highly sophisticated control systems, but does not depend so much on the extensive use of universal digital computers. The new Munich S-Bahn is characterized by the fact that the computer-controlled district not only covers urban rail transit lines but also long distance traffic.

13.1 USSR EXPERIENCES: THE MOSCOW METRO AND THE ASU-PP SYSTEM*

13.1.1 BACKGROUND

Subways represent one of the main and most convenient modes of public passenger transport in the large urban cities of the USSR. The Moscow Metro, for example, accounts for about 35 percent of all urban passenger traffic, and this figure is still increasing. This rise in the demand for

*Based on a case description specially prepared by L. A. Baranov, A. V. Shileiko, and V. I. Astrakhan, Moscow Institute for Railway Engineers (MIIT), Moscow, USSR.

Metro traffic calls for an improvement in both the operation and operational management of the whole rail transit system. Therefore an automated management and control system called ASU (Avtomatisirovannaya Sistema Upravleniya), which makes extensive use of advanced computer and control technology has been developed for the Moscow Metro.

This contribution describes the part of the ASU that directly concerns the passenger transport processes. It is called ASU-PP (ASU-Passashirskikh Perevosok), i.e., automated passenger transport control and management system.

13.1.2 THE STRUCTURE OF THE ASU-PP SYSTEM

The ASU-PP system contributes to a more effective solution of the following three tasks: (1) computer-aided planning and scheduling, (2) organization and dispatching, and (3) operational control of train operation.

The planning tasks include the establishment of working programs for the various parts of ASU-PP as, for example, the train schedules. Investigations carried out in the past led to the conclusion that the train schedules should be changed more often than is presently feasible using manual scheduling methods. At present the timetables of the Moscow Metro are kept unchanged for several years though it is known that during the course of the year seasonal passenger stream fluctuations of more than ± 10 percent occur. It is therefore expected that the introduction of special schedules for the individual quarters of the year will lead to significant economic benefits.

The second main task of ASU-PP concerns computer-aided organization and dispatching of the passenger transport processes. To solve this task, the hierarchical management and hardware structure shown in Figure 13.1 was implemented. The highest level consists of a dispatching center containing a computing complex, a wall-map display, and control panels. At the second level, equipment for automatic control of the peripheral elements is installed. The third level includes those ASU-PP devices that carry out the commands delivered from level 2 (cf. Figure 13.1). The first and second levels are connected by data transmission links. Information exchange between the second and third levels is carried out by means of devices for remote control, remote data transmission, contactless inductive transmitters, and loops (cf. Figure 13.2).

The tasks of the various levels of the dispatching and control hierarchy given in Figure 13.1 may be characterized as follows: The computer complex at level 1 uses actual data describing the existing passenger transport demand at the train changing stations to determine the train schedules and the working programs of all peripheral elements and processes. These programs are stored in a long-term memory. The actual working programs of the peripheral systems are transferred to a working memory every day (cf. Figure 13.1).

The computing system can make operational changes to the train schedules and to the other working programs on the basis of information on the performance of the original working plans. In such a case the corresponding commands stored in the working memory will be updated. The computing system also analyzes passenger stream volumes and computes car circulation plans, energy consumption, and parameters characterizing the operational status of the Metro system. Moreover the computer has to monitor the train positions and the state of all the ASU-PP equipment, and control the corresponding indicator elements on the wall-map display. Figure 13.1 illustrates how the following subsystems are monitored and controlled by the dispatching center:

a. The passenger guidance system controls access to station entrances, changeovers, and escalators or recommends alternative walking routes to the passengers. Turnstiles and automatic gates are used as control devices.

b. Centralized control and train protection devices ensure a safe train operation by automatic cab signalling, speed control, and control of distances between trains.

c. Train traffic control, according to a given schedule, is carried out by the system "Avtovedenie" (automated guidance).

d. Monitoring and control of train movements when entering or leaving depots.

e. Surveillance and control of the electric power supply system.

f. Optimal control of the escalators and their surveillance by TV cameras.

g. Surveillance and control of hygiene conditions within the Metro, e.g., by air conditioning.

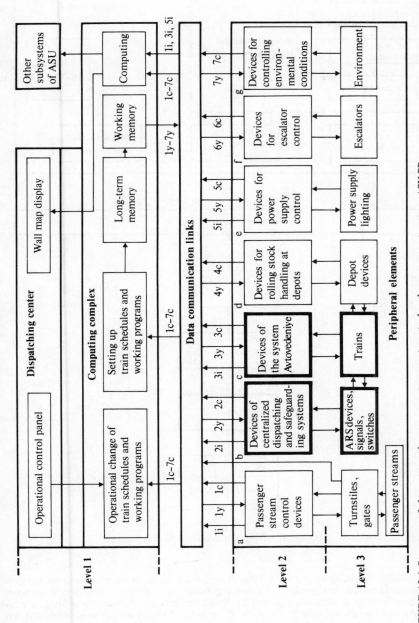

FIGURE 13.1 Structure of the automatic passenger transport control and management system ASU-PP.

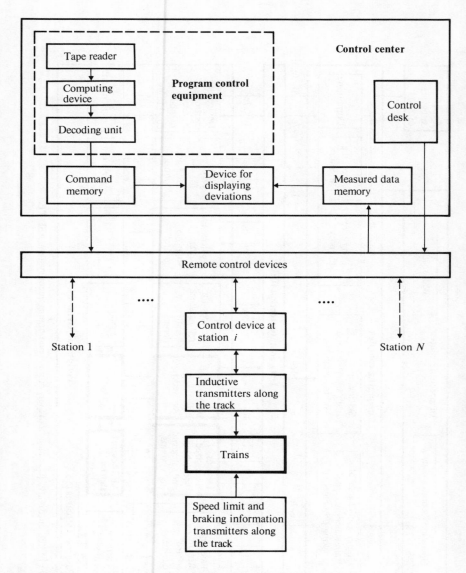

FIGURE 13.2 Structure of the train traffic control system Avtovedeniye.

The dispatcher can speak with the train drivers via a radio communication link and give information via various loudspeakers to passengers riding on the trains, as well as to those using escalators or waiting at stations. Figure 13.1 illustrates that the systems for automatic train protection and operational train traffic control are assigned to the second level of the ASU-PP hierarchy (cf. subsystems b and c). These subsystems are considered in more detail in the following.

13.1.3 THE OPERATIONAL CONTROL SYSTEM

It should be mentioned that a systematic approach for solving the traffic control tasks occurring in Metro traffic only became possible recently. In previous years attention was mainly paid to improving the individual control processes, in particular automated train control.

The USSR was among the first countries to start the development and introduction of automatic train control systems. In 1958/1959 the system Avtomashinist (automatic driver) was tested in suburban traffic. At the end of the 1950s and the beginning of the 1960s the development was started of new equipment for automatically displaying signals, i.e., driving commands at the control panel of the train driver, and for automatic speed control. This system was called ARS (ustroistvo c Avtomaticheskoy Regulirovkoi Skorosti). In 1969 the ARS equipment was first installed on the circle line of the Moscow Metro. All new Metro lines are currently equipped with these devices.

Centralized control systems developed by the Giprotranssignalsvyas institute were first installed on the Leningrad Metro in 1965. In 1970 a central control system created by the Moscow Institute of Railway Engineers (MIIT) and the Moscow Metro was put into operation on the Riga line of the Moscow Metro. The research and development activities were then oriented to the creation of a complex automated train operation control system called KSAUDP (Kompleksnaya Sistema Avtomticheskogo Upravleniya Dvisheniyem Poyesdov), which became operational for test runs on the Krasnopresnenskaya Line in 1974. The KSAUDP was created in a joint effort by MIIT, the Federal Moscow Railway Research Institute, the contracting firm Metrogiprotrans, and the Moscow Metro. The two main tasks of KSAUDP are:

- automated speed control and train protection (system ARS)
- train traffic control (System Avtovedeniye)

During recent years the ARS system has been supplemented by various types of train protection equipment installed onboard the trains and alongside the track. Electric currents of frequency 75–225 Hz circulate through the rails. At these frequencies it is possible to transmit coded commands to the onboard equipment about the admissable train speed in a range of 0–80 km/h. If the command speed is exceeded then the onboard control equipment switches off the traction power and operates the brakes. Moreover a certain number of passive inductive transmitters, installed along the rails and within the stations, transmit speed limits and braking information to the train devices. Using this information, the onboard system ensures that the speed limits are not violated and that the train is stopped at a prescribed point of the station with a position error smaller than ± 2 m.

The main task of the train traffic control system Avtovedeniye is to minimize timetable deviations. The structure of the Avtovedeniye system is illustrated in Figure 13.2. A control center has been established that contains the following: program control equipment consisting of a tape reader, a computing device, and a decoding unit; remote transmitting units for data transmission to and from the stations; memories for storing control commands and data transmitted from the stations; equipment for identifying deviations between the commanded and real control variables; a panel for manual control.

The train schedules that are stored in the program control system are given by three categories of parameters: train headways, stopping times at the stations, running times between stations. Using these prescribed values the control systems installed at the stations control the movements of those trains operating in the line sections assigned to corresponding stations (cf. Figure 13.2). For this purpose the station control system transmits commands to the train such as depart, switch off traction power, close doors, or open doors, by means of a certain number of discrete inductive transmitters installed along the track. The object of this control approach is to keep deviations from the prescribed arrival times to less than five seconds.

To reach this goal each station control unit contains two computing devices: a departure time computer VO (Vychislitel Otpravleniya); a computer VVT (Vychislitel Vyklucheniya Tyagi) for determining the point at which the traction power has to be switched off. The VO computer receives the actual nominal values of headway, stopping time, and running time that are valid for the following line section from the control center. Moreover VO gets information on train departure and arrival times as well as on minimal admissable headways between trains, minimal admissable stopping times, and running time reserves.

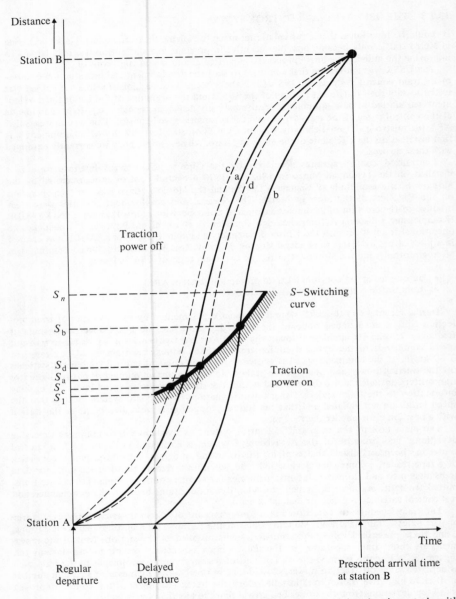

FIGURE 13.3 Train movement trajectories: a, nominal trajectory; b, delayed train; c, train with good running quality; d, train with poor running quality.

The VVT computer receives information from VO on the departure time of the corresponding train, and from the control center the running time valid for the previous section. In addition VVT gets feedback information on the actual train position. These feedback signals are obtained by means of inductive transmitters installed at certain surveillance points along the track.

Using these data the VVT computer determines the position of the train where the traction power has to be switched off. A corresponding command is transferred to the onboard control unit via the inductive transmitter located at that position. The control principle implemented in the VVT computer is illustrated in Figure 13.3, which shows the movement of the train between station A and station B in the form of time–distance curves. Curve a describes the exact nominal time–distance trajectory. The power is switched on between A and the position s_a. On passing the inductive transmitter located at s_a, the traction power is switched off and the train coasts.

Curve b corresponds to the case where the train departs from station A with a delay; here the power remains on until s_b is reached. The switching curve S describes the locations of the admissible switching points resulting from different delays or different train running performances. A train with good running qualities, for example, follows the trajectory c and its power will be switched off earlier, i.e., at s_c. In the case of poor running qualities, the switching point moves to s_d (cf. curve d).

The switching curve S shown in Figure 13.3 has been implemented by the inductive transmitters mentioned earlier. The required number of transmitters results from a quantification of time using one-second steps. The controllable range is defined by s_1 and s_n: if the switching distance s is smaller than s_1 then the train will not reach station B with the required minimum speed; for $s > s_n$, the speed will violate given limits for the maximum admissible speed, and high energy consumption rates result.

In the second case, delays cannot be compensated for by the VVT computer, i.e., by using running time reserves. Therefore a further reduction of the stopping time is required for larger delays, taking into consideration the given limits for the minimum stopping time. If a train arrives at a station earlier than prescribed, VO will not increase the stopping time, rather VVT will reduce the time required for powering. In this way it is possible to save traction energy. If the time required for passenger changing is smaller than the stored minimum stopping time then the train driver may start the train earlier by pressing a button in order to save energy.

The train traffic control system described here is not only characterized by excellent control performances but also by high safety standards. These are achieved by the hierarchical structure of the system shown in Figure 13.2, which leads to fail-safe features. In the case of a breakdown of the central and/or the station control equipment the safety of train operation is ensured by means of the passive inductive transmitters mentioned above, which prescribe the speed limits and start the braking process.

13.1.4 CONCLUDING REMARKS

The introduction of the train traffic control system discussed in this section and of the ASU-PP system as a whole is considered absolutely necessary to improve the operation and operational management of the Metro system, i.e., to fulfill the increasing passenger transport tasks of the Moscow Metro in an effective way. The results achieved and further expected results include, among others, the following:

- an increase in the capacity of densely occupied lines and a decrease in timetable deviations
- an increase in travel speeds without affecting operational safety
- a decrease in operational costs in terms of savings in energy and man power (the number of employees per train was reduced from two to one)
- an increase in the flexibility of train operation

BIBLIOGRAPHY

Lisitsyn, V. M., V. I. Astrakhan, V. M. Maksimov, and V. V. Maleyev. 1971. System for automatic control of subway train traffic. Trudi MIITa (Report of the Moscow Institute of Railway Engineers), No. 370 (in Russian).

Astrakhan, V. I. and V. V. Maleyev. 1975. Automatic system for controlling subway passenger transport. Trudi MIITa (Report of the Moscow Institute of Railway Engineers) No. 492 (in Russian).

13.2 USA EXPERIENCES: THE SAN FRANCISCO BART SYSTEM*

13.2.1 BACKGROUND

The San Francisco Bay area rapid transit (BART) system is a totally automated commuter transit system whose design goals include providing comfortable and highly reliable transportation among the San Francisco Bay Area communities. The system is comprised of 34 passenger stations distributed along 75 miles of wide-gauge, electrified double track (cf. Figure 13.4). Total automation of the train control system was thought necessary to provide fast, frequent, and reliable service. The system had to be designed to meet fully the safety standards of the rapid rail transit industry.

The California Public Utilities Commission (CPUC) has safety jurisdiction over BART. CPUC General Order 127 specifies the rules governing the design, construction, and operation of rapid transit systems in California. BART started limited revenue operation between Fremont and Oakland on September 11, 1972. Operating restrictions were imposed by the CPUC, even on this limited service, owing to difficiencies in the presence detection capabilities of the train protection system. Under this restricted mode of operation, trains were kept two stations apart by a manual signal system controlled by dispatchers on station platforms.

Additional sections of the system were opened for revenue service with the same operating restrictions. Service was extended from Fremont to Richmond on January 29, 1973, a shuttle service between Oakland and Concord began on May 21, 1973, and a shuttle service between Daily City and downtown San Francisco began on November 5, 1973. In July 1974 the supervisory control system was put into operation.

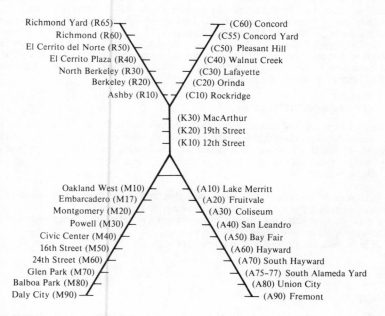

FIGURE 13.4 The BART network.

*Based on a case description specially prepared by Krishna V. Hari, Director of Systems Engineering, BART District, Oakland, California, USA.

13.2.2 IMPLEMENTATION AND OPERATION

In the following, a survey is presented of the basic components of the BART train safety and operation control system restricted to the automated train control system, and the supervisory (train operation) control system.

13.2.2.1 Automatic Train Control. The automatic train control system has two subsystems:

- the train detection and speed control subsystem
- the station stopping subsystem

1. *Train detection and speed control.* Fail-safe train detection is fundamental to proper train protection and control. In the BART train control system, the welded track is short-circuited at intervals and audio-frequency signals are induced into the rails, which operate thus as transmission lines. Signal currents circulate through the rails, and their presence is detected by current sensors mounted on the short circuits (cf. Figure 13.6).

At the control center, the transmitted and received signals are compared by fail-safe circuitry for continuous correspondence. The presence of a received signal that agrees with the transmitted signal indicates an unoccupied block. The absence of a received signal or a received signal that does not agree with the transmitted signal indicates an occupied block. The train protection system is concerned with the local control function centered around each station. A block diagram of the control system is shown in Figure 13.5.

The information for various block transmitters originates in the station, and transmission of the coded signals to and from the wayside locations is accomplished by a time-division multiplexing system requiring only three twisted pairs. The three twisted pairs carry commands, respectively, in the form of binary signals sequentially separated in time to each wayside location, information from the wayside to station, and control signals for synchronization and signalling. A block diagram of the multiplex system is shown in Figure 13.6.

The information to the track block transmitters is the speed commands to the vehicle in coded form. The transmitter is modulated with frequency shift modulation rather than amplitude keying. The proper information for each transmitter is selected by a time-slot selector operating off the synchronization pulse line. Each location has its assigned time slot built into the trackside equipment. The time slot samples the information line, and the sampled information is stored in a memory for the duration of the code cycle. The output frequency of the divider is determined by the information in the memory, and one of the two available frequencies selected for transmission to the track is reversed in phase each cycle for synchronization on the train (cf. Figure 13.6).

The signal from the antenna is passed directly to two narrow-band crystal filters, which respond only to the equivalent crystal-controlled transmitters at the ends of their blocks. Each receiver uses two frequencies that are not used by receivers in location on either side. The outputs of the crystal filters are summed, passed through a threshold detector and then limited in a simple amplifier. The output is constant for an unoccupied track. The threshold detector provides an output only for an unoccupied block. The signal is then detected in a conventional frequency discriminator. The data are fed back to the station in the appropriate time slot.

When the transmitter and receiver are combined the same time slot selector that was used for the transmitter is also used to enable the receiver information to be delivered to the line (cf. Figure 13.6).

The design philosophy of equipment onboard each train is to ensure that the speed of the train cannot exceed the commanded speed. Hence, it is unnecessary for the train protection equipment to determine whether or not the train is obeying the speed commands. Theoretically, for protection purposes, it is sufficient to know that an unsafe speed cannot be transmitted to the train and to know whatever speed is transmitted to the train cannot be exceeded.

The speed command control system, which generates the speed commands to be sent to the various track block transmitters, is located in each station. It is essentially a closed-loop electronic control system; the design was made possible by the development of solid-state, fail-safe "AND" gates. The fail-safe "AND" gates replace the conventional vital relays for encoding speed commands and for fast bit comparison of the transmitted and received information

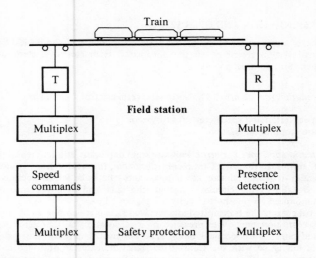

FIGURE 13.5 Block diagram of the automatic train control system. T, transmitter; R, receiver.

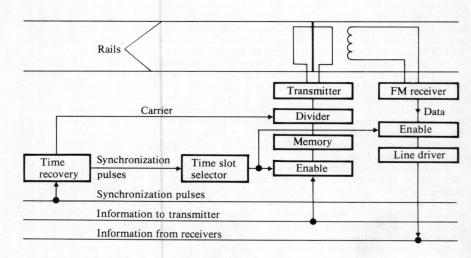

FIGURE 13.6 Block diagram of the multiplexing system.

of a track block. This development enabled the interfacing of the speed encoding and the multiplexing systems.

A six-bit serial comma-free code is used to convey the speed commands to the train. A repetitive sequence of any of the comma-free codes utilized will never be confused with any other code regardless of the time or bit believed as the beginning of the message. Thus no synchronization is required in the vehicle speed decoding system in order to recognize a speed command. Timing in the decoder is not important.

If two adjacent blocks are sending identical speed commands, the two commands are time shifted. The time shifting offers a further opportunity to protect adjacent track circuits from receiving erroneous information over and beyond that provided by frequency separation. The train-carried equipment, being comma free, does not recognize the difference except for a transient loss of one word as the train leaves one block and enters the next, which is easily filtered by the train-carried equipment.

The six-bit, comma-free codes that are utilized in the BART system are 100000, 100001, 101001, 100101, 100011, 101011, 100111, and 101111. Each higher speed command is generated by properly synthesizing the speed code 100000 and its other five time-shifted versions. The synthesizing process is controlled by the various occupancies of the track blocks and various switch positions at crossovers. Conventional vital relay systems were utilized to provide system protection as the train travels from station to station. Figure 13.7 shows a block diagram of the speed command control system.

The synthesized data are encoded into the proper time slot and transmitted to wayside receivers as well as being sampled into the comparator. The data from the wayside receivers are also sampled into the comparator by the receiver time slot. The comparator outputs the occupancy information, which is utilized to contol the speed command synthesis.

The coded signals are picked up from the track by antennas fitted on the control cars and decoded by fail-safe as well as redundant methods (cf. Figure 13.7(a)). The speed decoding equipment drives crystal-controlled oscillators (one for each speed command) whose frequency is compared with the tachometer frequency. The output of speed-maintaining equipment is an analog signal (P signal), which controls the tractive and braking efforts of the train. When the vehicle approaches a station, the P signal is also controlled by a programmed stop signal, the details of which are discussed elsewhere. The vehicle braking effort is generated by both dynamic and friction brakes. Electronic blending systems were required to provide a constant braking effort.

2. *Precise station stopping.* The train speed is controlled by the track circuit system as it approaches the station. In order to make precise station stops, with constant deceleration, accurate information on the distance traveled and the true velocity of the train are necessary. The station platforms are a fixed length of 700 feet. A transmission cable with crossovers is fitted on the power rail coverboard for the length of the platform and is driven by a constant amplitude, constant frequency oscillator. Two sets of small antennas are mounted on the control car (Figure 13.8). The signals from these are compared in phase; one phase change occurs per crossover as the car moves by the transmission cable. Thus very accurate information of distance d traveled and velocity V are obtained on the train.

A hard wired digital computer unit on the train calculates the deceleration $b = V^2/[2(d - d_0)]$, where $(d - d_0)$ is the distance to go before stopping. The output of this computation is passed to a digital-to-analog converter, which generates an analog signal proportional to $V^2/[2(d - d_0)]$, and the signal is passed to a scaling circuit whose output is the required deceleration demand for programmed stopping.

13.2.2.2 Computerized Supervisory and Backup Control Systems. The BART system performance is monitored by a high-speed digital process control computer system and series of digital data transmission links to each of the passenger stations. The status of the field devices is continuously transmitted to the computer (cf. Figure 13.9).

The supervisory control system was required to keep track of each train on the system and to evaluate the performance of each train based on its arrival/departure at each station and at points of convergence. This system will try to improve the performance of a poorly performing train or, at the worst, get the train out of the system to achieve overall system stability. The central supervisory system can in no way overrule the local safety protection system.

Three different subsystems of the supervisory control and backup system are described in the following:

- the train identification system
- the computer augmented block system (CABS)
- the sequential occupancy release (SOR) system

1. *Train identification system.* Each train in the system continually transmits as identification (ID) signal in binary code, consisting of its serial number, final destination, and its length. This signal is a 36-bit code: 10 serial number bits, 4 length bits, 6 destination bits, 4 parity bits, and

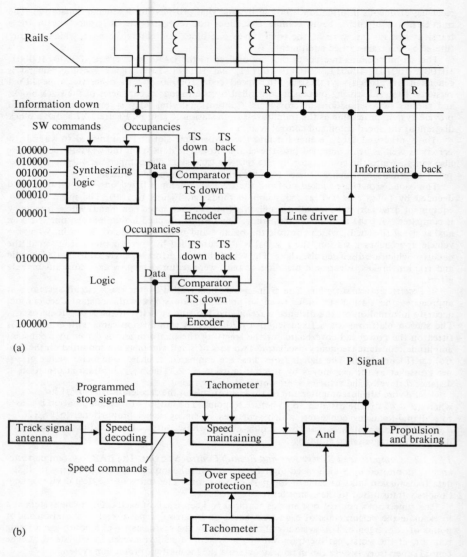

FIGURE 13.7 Block diagram of (a) the speed command control system, (b) the onboard speed control system.

12 format bits. These data are transmitted at 18 Hz, and thus it requires two seconds to transmit a complete ID code. The six destination bits plus the four parity bits form an error correction code. The ID processor in the station will correct any one-bit error in the destination portion of the code. The format bits are used to correctly align the train's ID data in a 36-bit register at the station as the ID data on the train are only partially synchronized with the station. In addition, if more than two errors are detected in the format, the central computer is notified so that appropriate action may be taken.

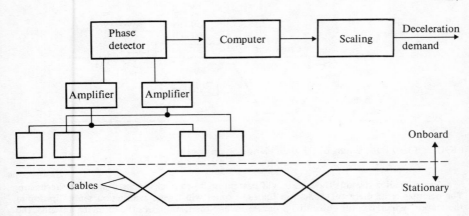

FIGURE 13.8 Block diagram of the station stopping system.

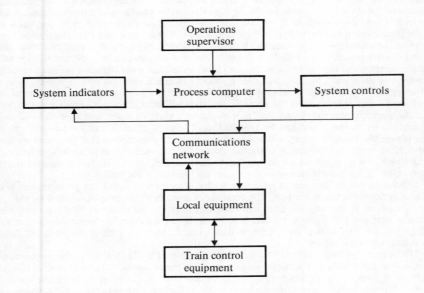

FIGURE 13.9 Structure of the supervisory control system.

The train ID is an FSK signal in the 5-10 kHz region as are all data transmitted to and from the train. The transmitting antennas, one on each side of the train, consist of small loops mounted on the trucks of the train. The receiving antennas are 410 feet parallel wire cables similar to twin-lead transmission lines, but with a two-inch spacing mounted on the coverboard of the third rail. These antennas are located between one and a half and two miles from the station. The data from the train are picked up by the antennas, amplified, and the transmitted through the multiplexing system to the station where the data are processed.

To see how these data are used, consider a typical station with destination number 10, whose track plan is shown in Figure 13.10.

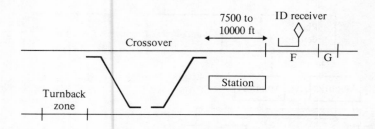

FIGURE 13.10 Functioning of the train identification system.

A train heading towards the station will pass through two track blocks, G and F, in succession. The occupancy of these two blocks in the right order will trigger a direction sense flip-flop in the station, which will turn on an ID processor. When the train passes the receiving antenna, the data are transmitted to the station, where the format is checked and, if correct, the destination and serial numbers are stored in registers.

The destination and serial numbers of the train are transmitted through the data transmission system to the central computer for verification and updating of the computer. The destination portion of the data is used to light the signs at the station platform to inform passengers what the train's final destination is. If, in this case, the destination of the train is 10, a signal is sent to the local control system so that a route across the switches can be set up to the terminal zone.

In each station, three antennas, which extend the length of the platform are mounted on the third rail coverboard. One of these is the program stop antenna, already discussed, plus an ID receiver antenna, similar to the antenna on the wayside, and a performance modification (PM) transmitting antenna. The train has a loop antenna onboard that can recieve signals from the PM antenna. When the train enters the station, the program stop signal is being transmitted, and an all ones signal is sent to the PM antenna. The train will interpret a signal of all ones (no zeros) as a door-open command; when the train comes to a stop, its doors will open. The ID being transmitted by the train will now be switched to all zeros. Two and a half seconds after the last zero is received from the train, a signal is sent to the central computer that the doors are open, and a local timer in the station is started. Normally the central computer will override the local timer. When the time set in the local timer expires, or the central computer sends a release train signal, the PM transmitter will start transmitting a PM signal.

This signal can be any one of six codes, which are used to modify the performance of the train as it travels between stations. The presence of zeros in the code is interpreted by the train as a close-door signal, and the train will close its doors and proceed out of the station. Eight seconds after the door-close command is received by the station, the station destination signs will go out, or if there is another train waiting to come into the station, the signs will display the next train's destination.

The system has two methods of correcting schedules. The dwell time in the station can be modified or the train's performance while running between stations can be modified. If a train is ahead of schedule, the central computer can delay the train's departure from the station as long as desired. It can also release the train early. Trains running between stations will usually run at somewhat less than maximum speed. For instance, an 80 miles per hour (mph) speed command will be interpreted as 70 mph. If a train is ahead or behind schedule, the central computer can send a PM command immediately after it commands the train doors to close. This command is loaded into a register aboard the train, which will then modify the speed commands that it receives from the track circuits. This modification is always toward a lower performance level, so that the safety of the system is not compromised. In this way, schedules can be met and the total system performance optimized. Owing to the restricted operating modes discussed elsewhere, this supervisory system has never been operated under revenue conditions.

If a train is to turn back at a station, when it proceeds out of the station it will cross the tracks and come to a stop in the terminal zone. The train attendant will turn off the lead car and walk to the opposite end of the train, which will become the new head end, and turn it on.

The terminal zone has two antennas, similar to the station platform, but no program stop antenna. When the new head end is turned on, the train will transmit its ID, but with all zeros in its destination code. This signal is sent to the station where the all-zero destination code is interpreted as the train is ready for service again. The train-ready signal is sent to the central computer, which can then send a new destination to the train. When the new destination is verified, the train can then be dispatched.

The destination data are also used where alternate routes are possible. For instance, a train coming in from the Concord or Richmond Line can go either to the Alameda Line or Mission-Market Line. Bits 4 and 5 of the destination code uniquely designate each line of the system. One point of divergence is near the Lake Merritt station. Bits 4 and 5 of a train are decoded as Alameda or Mission destinations as a train approaches Lake Merritt, and are sent to the local control system. This system will then set up a route and align the switches to the appropriate line. In addition, if two trains are converging on the same line, as for instance when a train from the Concord Line and a train from the Alameda Line both request a Mission destination, the first train arriving at the convergence will proceed to its destination first, no matter who makes the first request.

2. *Computer augmented block system (CABS)*. A supervisory control system, which thus far was intended to be used primarily for schedule control purposes rather than for train separation, was then used to maintain two-station train separation. This technique uses the train tracking capability of the supervisory control system to withold dispatch of a following train from a station platform as long as the lead train has not cleared the two-station separation block. This mechanization was named CABS II.

Additional safeguards had to be added to CABS II to ensure positive clearing of trains from station platforms and to stop trains from attempting station run-throughs in order to shorten the train separation from two stations to one. Thus CABS I was put in control of the existing service in July of 1974.

The main thrust of the CABS mechanization was to open transbay service with some reasonable operating headways. Still additional safeguards had to be built into CABS I to enable transbay service, which mainly dealt with the Oakland Wye and The MacArthur merges. Pseudo stations had to be created under CABS I to maintain reasonable operating headways. Thus transbay service became a reality in September, 1974, and the total system was placed in revenue service.

CABS I introduces its own peculiar operating constraints. This add-on package of software/hardware modifications is transparent to the supervisory system schedule control capability. A continual conflict arises between the supervisory control system operating strategies and CABS I. Delayed trains operate to make up time between stations only to be held at the next station. The effects of one poorly performing train are felt throughout the system.

3. *The sequential occupancy release (SOR) system*. While the CABS package was being developed as an interim technique to realize transbay service, resolution of the deficiencies of presence detection of trains was still being pursued. After extensive evaluation of various methods of resolving the problem, it was decided to add a software backup to presence detection known as (SOR).

In its basic form, SOR memorizes the occupancy of a track block and does not release it until the train is detected in appropriate blocks ahead. This process has been implemented system wide, utilizing redundant Alpha LSI minicomputers, both exercising independent control over the primary train protection system.

SOR is an auxiliary computer system, which can inject occupancies (OCC) into the train control system, and thus create hardware interference by extending stopping profiles. The algorithm used can be described with the aid of Figure 13.11.

If the nose of the train in Figure 13.11(a) is just moving across shunt 2, occupancies are placed in the track circuit containing the nose of the train, and in each track circuit behind shunt 2 for a distance of 700 feet. In this example, since all track circuits shown are 700 feet or longer, only two track circuits will be occupied.

In the example shown in Figure 13.11(b), four track circuits would be occupied. Even though the train is well past shunt 2, and its tail does not reach past shunt 3, occupancies must be placed 700 feet back from shunt 2. This results in occupancies being established in all four track circuits as shown. When the nose of the train moves across shunt 1, the occupancies between shunts 2 and 5 can be reset. An occupancy can be reset any time the nose of the train

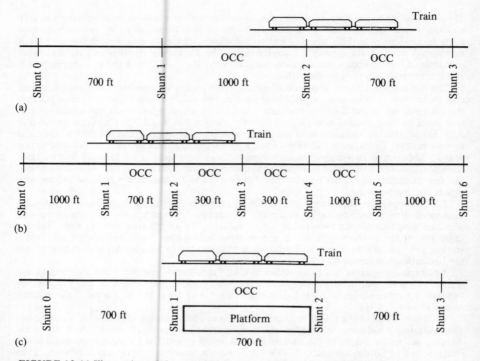

FIGURE 13.11 Illustration of the sequential occupancy release (SOR) algorithm.

crosses a shunt. As the train crosses shunt 1 in the example in Figure 13.11(b), three occupancies can be reset in any two-second interval. Thus, when the train crosses shunt 1, the rear-most occupancy between shunts 4 and 5 can be reset. Two seconds later the occupancy between shunts 3 and 4 can be reset; and finally, the occupancy between shunts 2 and 3 can be reset after a further two-second delay.

There is one major exception to the 700-feet rule. At all station platforms and end-of-line turnbacks, after the train has come to a stop and the dwell has started, all occupancies behind the train can be reset (using the two-second delay rule) leaving only the "real" occupancy of the train itself. As soon as the train pulls out into the next track circuit, the 700-feet rule is reinstated. Thus, as shown in Figure 13.11(c), as the train crosses shunt 2, its occupancies extend 700 feet behind shunt 2. After the train comes to a stop in the platform, all occupancies to the rear of shunt 2 are reset, leaving only the one occupancy in the platform itself. As the train departs and crosses shunt 1, the 700-feet rule is reinstated, causing occupancies between shunts 0 and 1 and between shunts 1 and 2.

13.2.3 OPERATIONAL EXPERIENCES

On January 19, 1975, a BART train operating in automatic mode collided with a maintenance vehicle. This accident led to an extensive investigation by the CPUC with regard to the equipment and procedures of the BART system. Questions were raised about the safety effectiveness of rules and procedures, and the interaction of equipment reliability and safety. BART embarked on a program to develop extensive evaluation of safety procedures and equipment reliability, in addition to developing the capability to detect maintenance vehicles by the primary train detection system. Prototype hardware developed for this purpose underwent extensive field testing, and the system was implemented in June 1977.

TABLE 13.1 A Selection of Critical BART Train Control Performance Reliability Requirements

Problem	MTBF requirements (hours) per duplicate equipment	
	Train Carried	Wayside
No system monitoring or corrective strategy determination	310	310
No corrective strategy execution	310	310
No automatic train movement	270	270
Program stop not within 5 ft of program stop line	9750	9750
No automatic door open	980	980
Speed regulation not within limit	150	150
Incorrect or no destination sign	980	980

13.2.3.1 Reliability Observations. Some important reliability requirements specified by BART in its initial train control procurement are given in Table 13.1.

Many comments have appeared in the public and trade media about the reliability of this system. Thus we shall limit our comments here to some generalized observations.

1. The reliability of the train control system, particularly the onboard vehicle portion, has been much poorer than BART had expected.

2. We see published articles that tie the terms "unreliable" and "the automatic train control system". These sometimes fail to clarify that it is not only the train control system but also several of the other major vehicle systems that contribute to the total BART system unreliability. The vehicle propulsion, braking, and auxiliary electrical systems are equal or more severe sources of operational trouble. This may be due in part to the fact that these latter systems are installed on each vehicle while the train control equipment is present on only two cars of each train.

3. The reliability of the overall wayside system is of less concern than that of the vehicle except during periods of peak thermal stress. The overall in-station environment is benign and the reliability of equipment located in these locations causes relatively few problems. Thus, the primary sources of unreliability in the wayside portion of the system are in the environmentally exposed multiplexing and speed control equipment illustrated earlier.

4. A very large percentage of the unreliability of the control system occurs in the train-carried portion. Most of the problems that do affect revenue service are best classified as system problems in the operating environment rather than traditional component failure problems. An example of this category of incident would be "intermittant" or "no response" to speed commands, which disappear when checked by maintenance, only to reappear, perhaps on the next usage, perhaps not for days or weeks.

5. Many of the component parts that are diagnosed and bench-verified as defective reside in the undercar environment. This is particularly true if the complexity of the equipment in the in-cab (semi-controlled) environment and the undercar environment are compared. Reliability is, all other things being equal, a function of complexity. The observations that on BART this rule does not seem to hold true suggests that other things are not equal. The most evident correlation appears to be with regard to the severity of the environment.

System and equipment environment is a very difficult and expensive problem to correct on existing equipment. The experience of BART should be carefully examined by future designers of control equipment for transit operation. Some of the environmental parameters that appear to be contributing to higher than expected failure rates on BART equipment are: peak temper-

TABLE 13.2 Estimates of Transit Equipment Environmental k Factors

Location	Environment	Access	Factor
In station	Conditioned atmosphere	Controlled	0.5
In station	Unconditioned environment	Controlled	1.0
In station	Unconditioned environment	Uncontrolled	3.0
Wayside	Unconditioned environment	Controlled	1.5
Vehicle	Unconditioned atmosphere	Controlled (in cab)	5.0
Vehicle	Conditioned environment	Uncontrolled (passenger areas)	10.0
Vehicle Under frame mount	Severe unconditioned environment	Controlled	15.0
Vehicle Truck mounted	Extremely severe environment	Controlled	40.0

ature, thermal cycling, corrosive atmosphere, vibration, shock, electric noise, human interface, and the stress of extremely frequent maintenance.

Future designers should consider and either make allowance for or provide control for the environmental impacts (traditionally known in the reliability discipline as environmental "k" factors) summarized in Table 13.2, which are taken from King (1976).

13.2.3.2. Reliability and Safety. Working with the problem of safety in systems that impact or involve the public raises some difficult questions, which cannot be answered here. Two closely related to reliability are:

1. Given that we exist in a world governed by statistical probability, how safe is safe?
2. Since safety of automatic equipment is generally achieved by greater equipment complexity, and since equipment complexity is a determinate of reliability in a way such that a design trade-off exists, how much additional unreliability is acceptable for an increase in safety?

Since some critics of BART's automatic train control system have tended to equate safety and reliability, it should be emphasized here that such equating is not valid. At the same time, we cannot totally separate these concerns for, at the very least, failures do increase the probabilty of an unsafe condition.

BART has studied the safety of all aspects of its design and operations. The statistics of safety per passenger-mile related to failures continue to accumulate favorably with each day of operation.

REFERENCE

King, J. H., Jr. 1976. Reliability Comes To The Transit Industry. Proceedings of the 1976 Annual Reliability and Maintainability Symposium. (Available from IEEE, 445 Hoes Lane, Piscataway, NJ 08854, USA.)

13.3 FRG EXPERIENCES: THE NEW MUNICH S-BAHN*

13.3.1 CASE HISTORY

A few months before the opening of the Olympic Games in 1972 the Munich S-Bahn was put into operation. The heart of the S-Bahn network, which has a total length of almost 400 kilometers, is the section between the stations München-Pasing and München-Ost (approxi-

*Based on a case description specially prepared by K.-H. Suwe, Technische Bundesbahnamtstrat in the BZA (Central Office of the Bundesbahn) Munich, Munich, FRG.

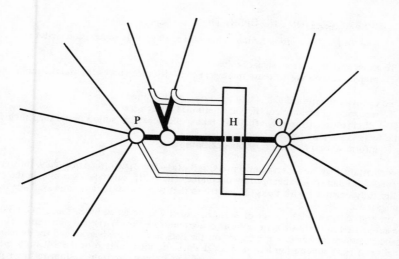

FIGURE 13.12 The Munich S-Bahn network: ■, only S-Bahn traffic; ═, only long distance traffic; − −, mixed traffic (traffic supervision center); H, München main station; O, München-Ost; P, München-Pasing.

mately 15 kilometers); 4.2 kilometers of this section are laid in a tunnel underneath the center of the city. This basic line carries only S-Bahn trains, while the lines branching-off at each end also carry long distance trains of passenger traffic and freight traffic. At each end of the basic line, and at one intermediate point, S-Bahn trains arriving from approaching tracks have to change over to the basic line and vice versa (Figure 13.12). Therefore, there is a high train density on this section and running on schedule is possible only if all trains reach the end points of the basic line without undue delay. Owing to the mixed traffic (S-Bahn trains and other trains) on the branch lines, delays to long distance trains sometimes cause delays to S-Bahn trains. For this reason a center has been constructed at Munich through which we are able to intervene in the operation in advance if a conflicting situation between two trains is at hand. Station inspectors working in the individual signal cabins cannot be charged with this task because they do not have the overall view that is necessary to recognize conflicts in advance in order to make the decisions required for maintaining smooth operation.

The system was designed so that the signal cabins of the S-Bahn network branch-lines could be remotely controlled from one control center in Munich, and the local station inspectors could be withdrawn. This would result not only in the desired centralized traffic supervision but also in a rationalization of the staff. Therefore, train stations of branch-lines were equipped with remotely controllable relay-operated signal cabins. Furthermore, grade-crossings were eliminated or were connected in interdependence with signals. By constructing tunnels or outward platforms the accessibility to platforms was designed so that passengers would not have to cross the tracks (cf. Wehner 1972).

Since the problems associated with remote control of signal cabins by computer were by no means all clarified in 1969, and because of the necessity to take the S-Bahn into operation before the opening of the Olympic Games in 1972, it was decided to design initially a train supervision system with automatic representation of the operational situation through computers (traffic supervision center). This was to be extended later to a substantially automatic remote control center with computers (traffic control center). A further objective was of course to be able to transfer the equipment of the traffic supervision center to the traffic control center.

In addition to the traffic supervision center/traffic control center two further computer control systems were designed for the Munich S-Bahn: the first deals with automatic train running and braking control, while the other controls the train destination indicators.

13.3.2 IMPLEMENTATION AND OPERATION

13.3.2.1 The Traffic Supervision Center. The tasks of the traffic supervision center are:

    a. To receive messages from signal cabins
    b. To prepare (according to these messages) time–distance curves for trains running in the area
    c. To compare theoretical and actual locations of the trains
    d. To recognize in advance conflicting situations between trains
    e. To eliminate recognized conflicts by taking the necessary action or to keep their operational effects to a minimum
    f. To inform station inspectors of the decisions made

In conventional traffic supervision centers, tasks (a)–(f) are performed manually. Information falling under (a) and (f) is exchanged by telephone. The traffic supervisor working in the center is usually fully occupied with tasks (a) and (b) so that there is hardly any time available for the remaining tasks.

In the traffic supervision center of Munich, which was designed by Siemens AG, tasks (a)–(c) are automated so that traffic supervisors can attend to tasks (d)–(f) more intensely. This was a pressing requirement because the traffic on the lines has become heavier and the demands on the quality of train supervision have increased since the introduction of the S-Bahn operation. However, the reasons that led to the design of the train supervision system resulted in the necessity for preventive measures against interference in the traffic supervision center. Despite the fact that in the case of a system failure train operation continues under the supervision of the station inspectors in signal cabins, this failure would have an adverse effect on the quality of operation at least during periods of heavy traffic. Therefore, the center is designed to fail under the principle of graceful degradation. Components whose failure would result in serious interference to the train supervision system, were duplicated (e.g., data converters, process computers). Components, whose failure could be tolerated or that could easily be replaced, were installed singly, or were supported by a small number of spare parts (Figure 13.13).

110 train stations are connected to the traffic supervision center of Munich (cf. Wehner 1972), and of these about 50 signal cabins are located in the S-Bahn area. For the supervision of the whole area we installed three duplex computer systems. In order to be able to fulfill tasks (a)–(c) automatically by computer, the messages that allow computerized tracing of trains within the area are read in the signal cabins and transmitted to the center by means of a remote control system. A route message for each passable connection of two blocks is offered to the remote control system in every signal cabin, independent of the direction of travel of a given train. When a train passes from one block into another that is connected to the first one by a route message, the message becomes effective and is transmitted to the center.

In the case of undisturbed operation the route messages are generated automatically and are transmitted to the center without human intervention. However, if there are interferences in the signal cabin (e.g., a train has to run on substitute signals or on written orders), or if there are interferences in the route message itself, the message may be initiated by the station inspector of the train station concerned if two stepping keys for the two blocks are pushed simultaneously.

In addition to the route messages coming from all stations, train numbers of trains approaching the supervision area together with information on the blocks they occupy are transmitted from several selected train stations to the center. By means of a keyboard, the station inspectors who are situated at the approaching zones and at some train stations where trains could start, key these train numbers into the installation.

The remote control system is a cyclic time frequency multiplexing system, which transmits information exclusively from signal cabins to the center. The reverse direction of transmission is not required for the traffic supervision center. All route messages (active or inactive) coming from all connected train stations are transmitted consecutively in a remote control channel in a cycle of about six seconds. These electrical signals are received in the center by a small computer constituting the central part of the remote control (Dietz Mincal 513); this small computer checks the received messages with regard to transmission errors and detects changes that may have occurred (e.g., did a change of message occur in comparison to the cycle transmitted before?). This data converter transmits to the traffic supervision computer those messages that differ from the previous cycle.

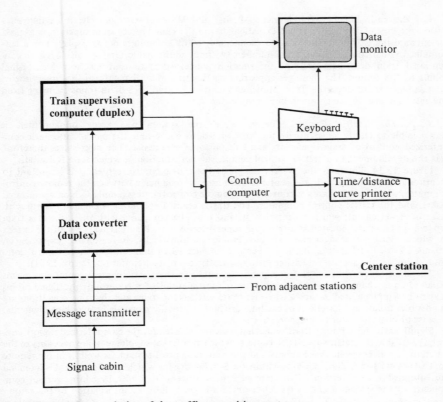

FIGURE 13.13 System design of the traffic supervision center.

The traffic supervision computer is a Siemens AG, Type 304, equipped with drum storage. It identifies the train approaching the supervision area by means of the transmitted train number, and subsequently updates every train by means of the route message. If a route message is received and if one of the two blocks connected by the route messages is occupied by a train (train number) this train number is computer-internally moved to the clear block.

A further basis for updating trains is the train schedule, which, among other things, stores information concerning the direction of a train. During the computer-internal updating of train numbers the agreement between train number location, route message, and direction indication is being checked. Furthermore, the train schedule contains the scheduled running times of trains, enabling the computer to indicate at any time whether a train running in the supervision area is on time or by how many minutes it is out of schedule.

The information prepared in the computer has to be displayed in a way that can easily be understood. For important lines and for lines with heavy traffic it is necessary to provide disposition data and operational records in order to be able to trace irregularities. Therefore we employed time–distance curve printers for the lines München–Augsburg, Treuchtlingen, Landshut and Salzburg/Kufstein. These printers are controlled by a small computer (Dietz, Mincal 513), which was provided for this purpose. In addition to these printers there are data monitors available to supervisors, which give an overall view of the present occupation of all train stations and lines (including also lines that are not equipped with printers). The location of a train is displayed in the block concerned by the train number. The train number is supplemented by a letter giving information on the punctuality of the train in comparison with the schedule. A computerized voice output informs certain traffic superintendents who are working outside the traffic supervision center on the punctuality of a given train if this superintendent dials a number on the telephone adding the train number concerned.

For the basic line between München–Pasing and München–Ost and the remaining lines within the city of Munich, it is not expedient to use the same type of train supervision as used on outward lines. Thus, this area is not connected to the computers but is displayed on a self-contained visual control panel. Train numbers (without additional letters) are displayed on conventional train describers and the information concerned is transmitted from local signal cabins to the center. The messages appearing on this train describer inform the supervisor of long distance trains departing from München mainstation, and of S-Bahn trains changing from the basic line over to the common line in the outer region.

13.3.2.2 The Traffic Control Center. While the system was being implemented it was possible to clarify substantially the basic questions concerning the partial automatic computerized control of train operation, and to gain some experience. The next stage, therefore, was the development of a traffic control center, the construction of which soon followed.

Figure 13.14 illustrates the system design of the traffic control center. It is important to mention at this point, that in the center a distinction is to be made between the remote control computer, which has the tasks of conventional remote control centers (emphasis is on operational tasks), and the traffic control computer. The traffic control computer is charged mainly with tasks of management, which means that its tasks can be compared with the tasks of a train supervision computer located in the traffic supervision center. Practical tests reveal the extent to which management decisions can be obtained automatically at reasonable expense (cf. Schenk 1974). Furthermore, it can be seen from Figure 13.14 that several remote control computers (four on the average) are assigned to one traffic control computer (cf. Delpy 1974).

The changeover from the traffic supervision center to the traffic control center also means expansion and changes in signal cabins. Through installations of relay groups the signal cabins have to be supplemented in such a way that route settings for trains and shunting operations are impossible, if individual sections of the lines are blocked totally or partially (blocking elements) because of operational or technical reasons.

Significantly more information, which has to be transmitted for the computerized remote control, is available for traffic supervision centers, which means that supplements are necessary in the junction circuit between signal cabins and the remote control system, as well as in the remote control system itself. Train stations within the S-Bahn area for which remote control is intended are equipped with a version of the remote control system DUS 600; by adding certain components to the system it allows transmission of a larger volume of information to the center. Train stations connected to traffic supervision centers and not intended for remote control are equipped with a somewhat simpler version of the DUS 600, which does not provide possibilites for expansion.

Instead of the small computer functioning as a data converter in the traffic control center, a message testing device consisting of wired electronics is being used. This device forms the central part of the remote control system and tests the telegrams coming from the DUS 600 with regard to their correct transmission. Undisturbed telegrams are forwarded to the computers in two different versions while disturbed telegrams result in a special message to the computers. For safety reasons the message-testing device is designed with two-channel processing and with safe comparison.

Furthermore, for reasons of safety, the remote control computers are also designed with two-channel processing (cf. Delpy and Suwe 1969). Telegrams coming from the message-testing device and received by the computer are stored and checked with regard to message changes. Detected message changes are transmitted to the traffic control computer (in a given case even in pre-processed form) or they are forwarded to the disptacher by means of indicators.

A line monitor and a station monitor (color monitors) as well as a control- and fault-printer (teleprinter) are intended to serve as indicating devices for dispatchers (cf. Suwe 1975). Through the line monitor, the computer informs the dispatcher mainly on running possibilities of trains, location of trains, train number with indications concerning punctuality, occupied and closed block sections, and the routing of a given train.

While the line monitor covering the total operational range of a dispatcher is provided as permanent indication, the station monitor displays only one train station or one part of a train station. However, this device, which is intended only for emergency cases (e.g., for cases of irregularities or disturbances, or during shunting operations), can be switched to a selected area and provides more detailed information concerning the present traffic situation at a given train

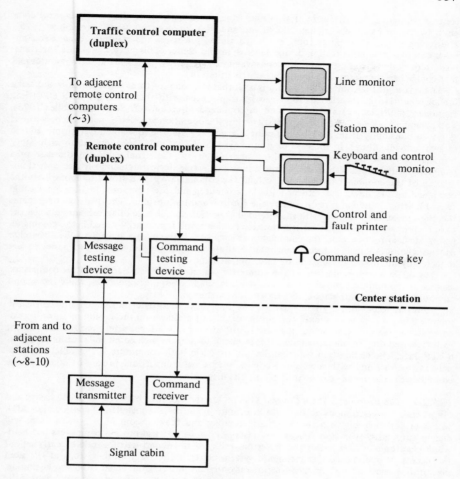

FIGURE 13.14 System design of the traffic control center.

station. This color monitor shows every point position, every point lock, every signal aspect, every interference, etc., as displayed on the control panels in the signal cabins. In the case of irregularities and interferences the dispatcher has to deduce his decision from the messages received with due regard to operational safety. Thus the computerized double processing is expanded to provide a display of the messages on the monitor. Every message that is significant for operational safety is displayed on the monitor in the form of two separately prepared and separately controlled symbols. The content of the message is valid only if the meanings of the two symbols harmonize. This comparison is carried out by the dispatcher as he watches the monitor (cf. Suwe and Zöller 1975).

It is currently being investigated whether these symbols should be permanently displayed in parallel or whether they should be displayed successively. The control and fault printer registers all disturbances detected by the remote control computer. This is necessary because it is not possible to display all disturbance messages on the line monitor and because the station monitor may not be switched to the train station at fault.

The information transmitted from the center to the signal cabins consists of commands that were either prepared automatically by the remote control computer working with the traffic control computer, or that were transmitted from the dispatcher to the remote control computer by means of an alphanumeric keyboard and through a control monitor (cf. Suwe 1975). The remote control computer compiles the command telegrams and transmits them to a computer-external command-testing device located in the center. This device prevents the transmission of command telegrams that were invalidated by faulty processing in the computer and that could therefore result in a dangerous traffic situation.

The layout of the command telegram is such that the code for three keys, which is necessary for an operation in the signal cabin, can be transmitted with the aid of the DUS 600 (time frequency multiplexing) also being used in command direction (cf. Suwe and Zöller 1975). The command-testing device determines whether the operation was performed with or without safety responsibility from the first transmitted group key and the subsequently transmitted one or two individual keys. In the case of group keys that do not point to an operation with safety responsibility, the telegram is transmitted immediately to the signal cabin. In the case of a command with safety responsibility, the command-testing device momentarily prevents transmission of the command to the signal cabin. The command, which is stored temporarily in the command-testing device, is returned to the computer and the operation in the signal cabin caused by this command is registered by the control- and fault-printer. The dispatcher compares the content of the telegram with the intended operation and, in case of agreement between the two, he releases the telegram for transmission (which is still temporarily stored in the command-testing device) by operating the command releasing key. If there is no agreement the dispatcher cancels the temporarily stored telegram. Malfunctions in the command-testing device are detected by double processing and by safe comparison.

One operative characteristic of the computer system is that the traffic control computers, as well as the remote control computers, work in single computer operation, while the second computer serves as a hot standby for reasons of reliability.

However, tests currently being carried out may show that under normal circumstances the remote control computer should control the display of messages on the station monitor in two-computer operation. Commands that are significant for operational safety are for the present to be released only by the dispatcher. It is intended to prepare such commands automatically at a later date. The elaboration will then be performed in both computers in a parallel and independent manner and with subsequent computer-external comparison. In case of failure of one computer the commands can be initiated by the dispatcher.

13.3.2.3   The Automatic Train Control System. On the basic line between München-Pasing and München-Ost a continuous automatic train running control was installed (Type Siemens AG), which in combination with the automatic running and braking control of S-Bahn trains will enable automatic operation. These two installations do not use process computers but hard-wired electronic devices. After the driver has initiated the departure of a train by carrying out the necessary operations, the automatic system controls the train until it stops at the next platform. Because an interference-free operation could not be guaranteed from the beginning, an additional fixed light signal system was installed that can be switched to marker light if the automatic installation functions in an interference-free manner (cf. Wehner 1970).

Prerequisites for automatic train operation are technical devices performing the tasks of the driver. These tasks can be divided as follows:

a. Observance of permissible speeds in individual sections of the line, observance of the line-side signals, and

- in the case of acceleration, determining the theoretical train speed
- in the case of braking, determining the point of brake application taking into consideration the gradient ratio and the braking capacity of the train

b. Execution and control of acceleration and braking until the theoretical speed (possibly stop) is reached at the predetermined destination (in the case of braking).

In conformity with this division of tasks the technical equipment is also divided into two parts (Figure 13.15). The tasks listed under (a) are mainly assigned to stationary control posts

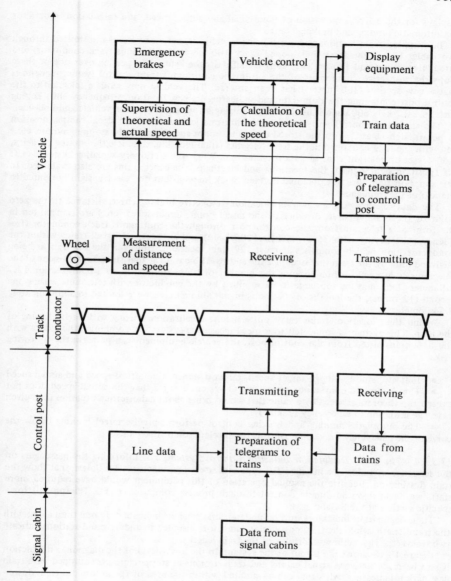

FIGURE 13.15 Structure of the automatic train control system.

of the automatic train control system, while the tasks listed under (b) are carried out by the automatic running and braking control system installed in the vehicle. The components of the automatic train control system that are installed in the vehicle assume the remaining tasks of (a), i.e., localization of the vehicle, data exchange with stationary control posts, control of the displays for the driver, supervision of theoretical and actual speed, and calculation of data for the automatic running and braking control.

Information from stationary control posts to vehicles and vice versa is transmitted through two cables laid in the track and looped every 100 meters (continuous track conductor) and through receiving and transmitting coils installed in the vehicle. In passing over one of these loops the respective component of the automatic continuous train control system recognizes that a new section (100 meters) has been entered. The vehicle now sends a telegram to the control post containing the train data and indicating its new location. Frequency shift keying with 56 ± 0.2 kHz/600 baud is used for transmitting information between vehicle and line-side.

The stationary control post combines the information coming from signal cabins (position of points, signals, etc.) with the stored data (maximum speed, gradients, stopping points, etc.) and with the data contained in the train telegram (train location, train length, braking capacity, etc.) to set up relevant control data for the train. When the stationary signaling system is cut off (position light signal), the locations and lengths of the other trains are also evaluated. In this situation trains are not operated in fixed block intervals but on "electric sight" oriented to the rear of the train ahead.

The data relevant to a given train (location, respective braking curve, distance to the zero point of the braking curve, distance to the target point, target speed, etc.) are transmitted in the form of a telegram from the control post through the continuous track conductor (frequency shift keying with 36 ± 0.6 kHz/1200 baud) and are received by all trains within the conductor loop, which comprises at most 127 train locations. However, the telegram is being analyzed only by the train that occupies the addressed location at that time. Transmission of the telegrams is in cyclic form. Because the number of trains running in an S-Bahn section 12.7 kilometers long may be too large to be supplied by the conductor with telegrams within one second (12 trains), the lengths of the loops in the Munich area were limited to approximately 3.5 kilometers.

Within their location, trains carry out a precise localization starting with each crossing of the loop. This system is connected to a speed measuring system. By combining these data with the data transmitted from the control post, the vehicle equipment can perform the following tasks:

- Indicate, among others, target speed, target distance, theoretical speed and actual speed
- The automatic continuous train control system ensures that the actual speed does not exceed the theoretical speed by a value that would bring about a dangerous situation (in a given case, the initiation of emergency braking)
- The automatic running and braking control performs precise target braking below the curve of the emergency brake and controls the acceleration of the vehicle.

13.3.2.4 The Train Destination Indicator Control System. For informing the passengers on the sequence of trains, we installed flat-type indicators manufactured by Solari, that show the train destination. Because the manual operation of this equipment would have required more staff, we decided on automatic control through process computers (AEG Telefunken) for a specific section of the S-Bahn area.

Train destination indicators on non-underground stations indicate only one train along with the train length, while the train destination indicators installed in underground stations indicate two trains (the first train with display of the train length).

Figure 13.16 shows the layout of the system. In the lower level of the diagram, a distinction has to be made between signal cabins and train stations or stopping points equipped with train destination indicators. All signal cabins situated within the area of the automatic train destination indicator system are equipped with train describers. The first two digits of the train number contain coded information on the destination of the train. The signal cabin, where all train numbers of the controlled area are available, supplies the train numbers of trains entering the area of the automatic train destination indicator control system; these train numbers are transmitted to the center. The transmitter of the remote control system required for this task also transmits updating information about train locations (relays serving to update train numbers as well as switch positions and signal aspects); the control numeral (destination of train) extracted from the first digits of the train number is updated in the computer as the train passes from one section to the next.

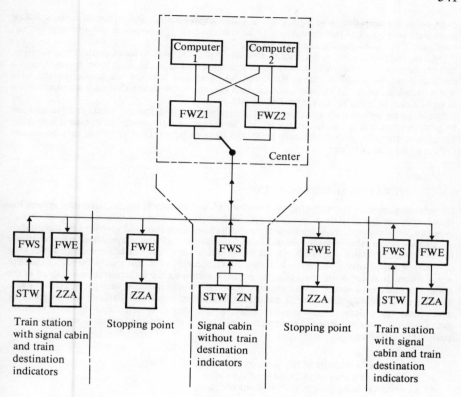

FIGURE 13.16 Automatic control of train destination indicators: FWE, receiver of the remote control system; FWS, transmitter of the remote control system; FWZ, remote control center; STW, signal cabin; ZN, train describers; ZZA, train destination indicator.

Because of the fact that the control numerals of trains running in the area were fed in when these trains entered the area, only the updating information on train locations has to be transmitted afterwards, which means a considerable reduction of the amount of information compared with the alternative possibility of transmitting all train numbers. In addition to the updating information about train locations, the counting results of certain axle counters are also transmitted from the train stations to determine the train length to be displayed on the train destination indicator.

At the stopping points there is no need to collect messages for the computer system. Through receivers of the remote control system, the train destination indicators installed at these stopping points and at regular train stations are connected to the center. The complete installation is controlled by a computer system type AEG 60-10 (16 K words), which for reasons of reliability is designed as a duplex installation. Two remote control centers precede this computer system. Each remote control center is connected to each computer and through the software the computer decides on the remote control center to supply the telegrams. The processing triggered by the received messages is performed in parallel in both computers. However, the resulting commands transmitted to the train destination indicators at the platforms can be issued by only one of the computers (exclusion due to software).

The remote control system works in a cyclic fashion. At first, the transmitters of the remote control system situated in the substations sequentially transmit the messages to the center. This cycle of messages is followed by a cycle of commands; each display setting of a train destination indicator requires two commands to be transmitted: 1, train station/stopping point and platform; 2, train destination and train length.

Station inspectors, in signal cabins that are equipped with remote control transmitters, collect the messages and enter data through a control desk into the remote control system and thereby into the computer. Since these control desks also indicate the display of the train destination indicators, the station inspectors can also check the automatic control and intervene in a given case by entering data into the computer.

At stopping points with a platform inspector the control of train destination indicators can be performed manually and independently of the computer if the centralized control is at fault. To achieve this the train destination indicators have to be switched from the remote control system to manual operation.

13.3.3 OPERATIONAL EXPERIENCES

Experience gained so far with the last described train destination indicator control system has shown that the installation is operating very reliably, i.e., one major failure occurs on average annually. Furthermore, train destination indicators have been accepted very well by the public.

Results from the central control and surveillance of train operation were available for the Munich S-Bahn only with respect to the traffic supervision task. Experience at the level of traffic control center was gained through development work and partly through other DB installations. It is not possible to give precise information on the rationalization achieved in the Munich installation because it was put into operation simultaneously with the Munich S-Bahn. These S-Bahn trains resulted in an additional loading of nearly 50 percent to the five lines, which also carry dense long-distance traffic; in this way, the calculated capacity of every individual line was exceeded. Through the new traffic supervision center it was possible to achieve the following:

- an immediate increase of the line capacity was not necessary
- the degree of punctuality could be maintained
- a considerable increase in staff was not necessary

In the case of traffic supervision centers and traffic control centers it has been shown that a somewhat longer test period is necessary between installing a system and finally putting it into operation; during this test period the system has to participate in the process in real time but does not have to perform the sole control of the process (life test). The expenditure required for a suitably long test period and for fast fault detection will prove worthwhile in the future.

The reliability of process computers (hardware failure) so far proved to be satisfactory. In the case of a duplex half (data converter and process computer) the MTBF values are between 1000–1500 hours and the MTTR values are 3 hours, which means a calculated period of approximately 30 years between two total failures caused by hardware faults. These values are sufficient not only for traffic supervision centers but also for traffic control centers.

In the case of traffic supervision centers computers are not charged with safety functions. In the case of traffic control centers the relay signal cabins remain fully functional and are controlled in the same way as in the case of local operation (key commands). By installation of an additional safety level for blocking elements, the safety functions of signal cabins are expanded; the operator is then relieved from these tasks. As long as auxiliary operations are not automated the tasks of the computer system are: faultless display of messages on the station monitor; faultless output of information on the control and fault printer; and in particular, faultless output of operations with safety responsibility performed by dispatchers. The computer-external safety level (message-testing device and command-testing device) greatly facilitates the necessary safety analysis for the remote control computers working also in single computer operation. Furthermore, the operator can be integrated into the safety methods (comparison of control and fault printer outputs with the intended commands, checking of multiple and repeatedly displayed messages with regard to their interference-free information).

REFERENCES

Delpy, A. 1974. The new systems solution for traffic supervision centers. Eisenbahntechnische Rundschau, 23 (10): 387-395.
Deply, A. and K. -H. Suwe. 1969. The safety and reliability of computerized railway traffic. Eisenbahntechnische Rundschau, 18 (10): 386-395.
Schenk, O. 1974. Automatic disposition of railway traffic control. Signal und Draht, 66 (8): 134-138.
Stapff, A. 1972. Telecommunication facilities of the Munich S-Bahn. Signal und Draht, 64: 167-173.
Suwe, K. -H. 1975. Standardization of the operation and indication in enterprise control centers (BSZ). Signal und Draht, 67 (7/8): 133-137.
Suwe, K. -H. and H. -I. Zöller. 1975. Standardization of remote control systems and traffic supervision centers of the DB. Signal und Draht, 67 (6): 104-112.
Wehner, L. 1970. Signal systems of the Munich S-Bahn. Signal und Draht, 62 (11): 200-204: 62 (12): 109-222.
Wehner, L. 1972b. The technical signal installations on the Munich S-Bahn. Signal und Draht, 64 (11): 181-191.
Wehner, L. 1972a. The traffic supervision center at Munich. Signal und Draht, 64 (1/2): 10-22.

13.4 FURTHER COMPUTERIZED RAPID RAIL TRANSIT SYSTEMS

The existing computer controlled urban railway systems may be divided into two groups (cf. Table 13.3):

1. Metros or similar rapid transit systems that do not share their tracks with any other traffic (cf. Besacier and Stablo 1972, Kavriga 1972, Mies 1974, 1975, Majou et al. 1970, Pins 1972, Papworth and Maxwell 1973, Sassmannshausen 1975, Takemura and Kariva 1973, Washington Subway Automates 1972, Watanabe 1971).

2. Mixed systems where the computer controlled area covers various categories of traffic such as transit trains, suburban commuter trains, and long-distance trains (cf. Blaise and Jauquet 1970, de Heer 1973, Hagland and Berg von Linde 1970, Lemaire et al. 1970, Savage and Harrison 1974, Steinfeld and Geber 1972).

For the first systems category, the major reasons for installing computerized control and surveillance systems are to

- maximize line capacity by train operation at short headways
- decrease the resulting sensitivity against small disturbances caused by changing passenger volumes, etc.

Mainly for these two reasons, almost all municipal rapid rail systems, which are faced with high traffic demands during rush hours, have been equipped or will be equipped with computerized traffic and operation control units. Examples besides the BART and Moscow Metro systems described above are the Hamburg Rapid Transit System, the Paris Metro, the Osaka Subway, Sapporo's Rapid Transit System, the Washington Subway, and the Vienna U-Bahn (cf. references given in Table 13.3).

TABLE 13.3 Survey of Computerized Urban/Suburban Rapid Rail Systems (for explanation of abbreviations, see Table 10.1)

Type of rapid rail systems	Specific objectives of computer control	Examples	Computer control relevant features (references)
Metro-type systems (Network is *not* shared with long-distance traffic)	• To increase line capacity by short-headway operation (down to 90 seconds or less)	Hamburg Hochbahn, FRG	ATS, API (Mies 1975, Sassmannshausen 1975)
		Moscow Metro	ATS (cf. Section 13.1)
	• To decrease sensitivity to disturbances caused by changing passenger volumes, etc.	Paris Metro	ATS, API, AFC (Besacier 1972, Estournet 1973, Majou 1970, Jacoub 1975)
		Osaka Subway, Japan	ATS (Watanabe 1971)
		Sapporo's rapid transit system Japan	ATS, API (Takemura 1973)
		BART at San Francisco, USA	ATS, API, AFC, ATO (cf. Section 13.2)
		Washington Subway, USA	ATS, API (Washington Subway Automates 1972)
		Vienna U-Bahn, Austria	ATS, API, AFC (Pins 1972, Vienna U-Bahn 1976)
Mixed train operation (Network is shared with long-distance and commuter traffic)	• To make optimal use of available tracks and minimize new track construction	Amsterdam, The Netherlands	ATS (de Heer 1973)
		Glasgow, UK	ATS (Savage 1974)
	• To minimize disturbances resulting from the interaction of transit and long-distance trains	Stockholm, Sweden	ATS (Hagland 1970)
		Munich S-Bahn, FRG	ATS, API (cf. Section 13.3)

In almost all the systems the first items that were computerized concerned the tasks of central monitoring and dispatching control (cf. paragraph 10.3.1). The second subject of computer control represents, in general, automatic display and broadcasting of passenger information, since these functions are closely related to the train traffic supervision and control task as illustrated in Figure 9.3 (cf. Table 13.3).

To date, only a few rail transit systems have installad truly integrated automatic fare collecting and accounting systems; such examples are the Paris Metro and the BART system (cf. Table 13.3 and Figure 10.21).

The motivation for providing the second systems category, i.e., networks operating transit trains as well as others with computerized control and surveillance units, results in general from the following two requirements:

- maximizing the overall network capacity by optimal use of the available tracks, thus minimizing costs needed for new track constructions
- minimizing the disturbances that result from interactions between transit, commuter, and long-distance trains

Examples of computer-controlled rail systems of this category are those of Amsterdam, Glasgow, Stockholm, and the new Munich S-Bahn described above. In Amsterdam the computer-equipped control center supervises the train traffic in an area with a radius of about 8 kilometers (cf. de Heer 1973). The control and surveillance tasks of the Glasgow system were initially restricted to the power box area of the Glasgow Central station (cf. Savage and Harrison 1974). The Stockholm system, on the other hand, covers a large suburban area and controls suburban traffic as well as long-distance passenger and freight trains (cf. Hagland and Berg von Linde 1970).

In conclusion, one may state that computerized surveillance and control systems are going to become an integral part of the following:

- high capacity metro-type transit systems operating at short headways
- heavily travelled urban/regional rail networks that are jointly used by various train categories

To date the role of computers has mainly been restricted to central traffic supervising tasks and passenger guidance and information systems. Automatic train operation by means of onboard microcomputers will very likely find broad application during the 1980s. Computer-based safety systems, on the other hand, are still at the stage of basic research and development, and their general introduction is considered unlikely before about 1985.

REFERENCES

Besacier, G. and J. Stablo. 1972. Development of disposition means for the Paris Metro. Revue Générale des Chemins de Fer, 91 (in French).

Blaise, R. and C. Jauquet. 1970. Example of the application of informatics to centralized train control. Paper presented at the First International IFAC/IFIP Symposium on Traffic Control, Versailles, France (in French).

Estournet, G. and P. Griffe. 1973. The automatic fare collection system of the Paris Metro. Revue Générale des Chemins de Fer, 92: 87–97, 130 (in French).

Hagland, C. and O. Berg von Linde. 1970. Computer controlled traffic signalling for suburban traffic railway around Stockholm. Paper presented at the First International IFAC/IFIP Symposium on Traffic Control, Versailles, France.

de Heer, J. J. 1973. Automatic route setting in railway stations by computers in real-time. Schienen der Welt, No. 1: 69–81 (in German).

Jacoub, M. 1975. A large computerized automatic system: the automatic fare collection system of the Paris Metro. Paper presented at the Sixth Triennal IFAC World Congress, Boston, Cambridge, Massachusetts, USA.

Kavriga, V. P. 1972. Construction and development of subways in the USSR. Schienen der Welt, No. 2: 92–108 (in German).

Lemaire, A., H. Autruffé, and R. Quonten. 1970. An automatic train dispatching system for lines with high traffic density. Paper presented at the First International IFAC/IFIP Symposium on Traffic Control, Versailles, France (in French).

Majou, J., P. Audinot, J. Cholley, and C. Magnien. 1970. Experience of R.A.T.P. in the field of automatic train control. Paper presented at the First International IFAC/IFIP Symposium on Traffic Control, Versailles, France (in French).

Mies, A. 1974. The automation of the operation of the Hamburg Metro, Revue UITP, 23 (1): 35–44 (in French).

Mies, A. 1975. Automatic operation of rapid transit traffic by means of computer control of line equipment – I. Objective and basic systems concept. In Nahverkehrsforschung '75, Federal Ministry of Research and Technology, Bonn-Bad Godesberg, FRG (in German).

Papworth, G. R. and W. W. Maxwell. 1973. The railway's share of passenger traffic in large urban areas. Schienen der Welt, No. 2: 207–229 (in German).

Pins, F. 1972. Automatic train operation of the Vienna U-Bahn. Verkehr und Technik, 25 (10): 433–437 (in German).

Sassmannhausen, G. 1975. Implementation of a rapid rail transit control center. In Nahverkehrsforschung '75, Ministry of Research and Technology, Bonn-Bad Godesberg, FRG (in German).

Savage, M. J. and R. P. Harrison. 1974. Real-time systems for train scheduling. Paper presented at the Second IFAC/IFIP/IFORS Symposium on Traffic Control and Transportation Systems, Monte Carlo, Monaco.

Steinfeld, H. and W. Geber. 1972. The regional control center of the Saarbrücken area. ETR-Eisenbahntechnische Rundschau, 21 (1/2): 31–40 (in German).

Takemura, S. and S. Kariva. 1973. Computer total system for railroad operation and Sapporo's rapid transit system. Hitachi Review, 22 (4): 150–160.

Vienna U-Bahn will have automatic train control. 1976. Railway Gaz. Int., 132 (7): 263–264.

Washington Subway Automates. 1972. Railway System Controls, 3: 18–27.

Watanabe, J. 1971. Computerized automatic traffic control equipment delivered to the Osaka subway. Toshiba Review, November: 7–12.

14 Findings and Summary

Chapters 9–13 analyzed the role of computers in improving the attractiveness and efficiency of urban public transport systems by presenting a survey of

- basic systems concepts (Chapter 9)
- concepts and methods of control and surveillance (Chapter 10)
- international experiences gained with computerized dial-a-ride systems, bus monitoring systems, and rapid rail transit systems

The status reached in the development and implementation of the various computer control and surveillance systems is characterized in the following.

14.1 DIAL-A-RIDE (PARA-TRANSIT) SYSTEMS

14.1.1 TECHNICAL ISSUES

A large number of dial-a-ride systems have been put into service during the last 10–15 years. However, only a few of them use computers to assist the dispatcher in solving the real-time routing and scheduling tasks. One reason for this situation is that over 50 percent of existing dial-a-ride systems use less than six vehicles and less than 20 percent of all systems operate more than 18 vehicles. Moreover, computerized vehicle dispatching only becomes a useful tool in those cases where more than 10–20 vehicles have to be dispatched by the control center. Existing computerized systems have demonstrated their potential to improve the quality of service and to increase vehicle productivity.

Advanced systems use digital data links between the central control computer and both the onboard units installed in the individual buses and the stationary automatic ticket vending and information equipment installed at demand-bus stops.

Current technical trends concern the use of microcomputers in the stationary equipment as well as that onboard the vehicles.

14.1.2 SOCIAL ISSUES

Two objectives were aimed for with the introduction of dial-a-ride systems:

- to serve that part of the population that is not allowed or willing to drive such as handicapped persons, senior citizens, school children
- to provide a public transport system to those regions that cannot be served efficiently by conventional fixed-route systems, such as large areas with a low population density; it was expected that a certain part of potential car users could change to the dial-a-ride system

It has been reported that a certain percentage of car owners readily accepted the service offered by the dial-a-ride system. On the other hand, other reports indicate that in several places dial-a-ride systems did not find the expected public acceptance.

14.1.3 ECONOMIC AND INSTITUTIONAL ISSUES

In several countries national governments became involved in funding research and development work as well as in implementing large-scale demonstration projects. There are examples, however, where the dial-a-ride service was discontinued when governmental funding ended (cf. Section 11.1). These experiences highlight typical institutional problems that may occur when funding arrangements for new transport services are not in concert with existing transit institutions. Moreover, it should be mentioned that several other dial-a-ride systems were only in operation for a limited time interval. This has very likely to be credited to the problems of public acceptance mentioned above and the resulting unfavourable benefit/cost relations. Thus the key problem regarding the future role of computerized dial-a-ride systems is probably to identify suitable areas for application.

14.2 BUS AND TRAM TRANSIT SYSTEMS

14.2.1 TECHNICAL ISSUES

The role of computers in improving the operation of bus and street car transit systems is concerned with two categories of tasks: (1) priority control at signalized intersections, (2) central monitoring and dispatching control of the total bus or tram system.

Priority control is widely used for both buses and trams as an integral part of computerized traffic light control systems. Current trends concern the use of microcomputers as part of local controllers to provide much greater flexibility in implementing various priority schemes.

To date, there are only a small number of truly computerized bus monitoring and dispatching control systems. Analogous systems for streetcar networks are still in the process of development. Almost all operating computerized bus systems aim to automate the identification of vehicle locations as well as the determination and display of schedule deviations. The task of dispatching control itself remains with the dispatcher. The size of the existing systems ranges from about 20 to 500 buses (cf. Figure 12.6). The so-called dead reckoning and proximity methods as well as combinations of both are used as vehicle location techniques (cf. Figure 10.6). Considerable research and development efforts have been carried out in the USA to create so-called Multi-user area-coverage automatic vehicle monitoring systems. These systems use radio location methods, such as Loran, and aim to provide the potential for automatic vehicle monitoring, not only for public transit systems but also for

- police cars and other law enforcement vehicles
- medical emergency vehicles
- para-transit cars, etc. (cf. paragraph 10.2.1)

However, to date there is no such system in operation for public transport.

Existing computerized bus systems that monitor large vehicle fleets were faced with two types of problems: (1) high installation costs, (2) insufficient reliability at the beginning of the operation. It is argued that both problems may be lessened by reducing the amount of data that has to be exchanged between the bus equipment and the control computer. This will become feasible when each bus can be provided with its own data processing and memorizing unit, i.e., by installing microcomputers onboard the individual buses (cf. Figure 10.8). This has already been done in the recently installed Toulouse system (cf. Figure 12.6).

14.2.2 EFFICIENCY AND ECONOMIC ISSUES

The benefits gained from computer applications are different in nature for priority control than for automatic vehicle monitoring. Priority systems have successfully demonstrated their capability to reduce delays at signalized intersections and to improve timetable adherence (cf. Chapter 7).

The main benefits that result from the introduction of computerized vehicle monitoring systems are the following:

- Reductions in the number of personnel needed for the various supervision tasks; the Hamburg system, for example, has reduced its man-power from 176 persons to 60 persons, i.e., by a factor of three (cf. Section 12.2).
- Increases in job satisfaction of inspectors and improvements in the discipline

of drivers with regard to exact fulfillment of instructions given by the dispatching center.
- Reductions in the delays in identifying disturbances; this effect led, in the case of the Hamburg system, to eliminating the necessity to retain several of the reserve buses and the associated crews.
- A reduction of the excess waiting time per passenger by about 12 percent and a significant reduction in passenger complaints were observed during the Dublin 90-day test described in paragraph 12.1.3.

It should be mentioned that there are differing opinions about the cost/benefit ratio that can be achieved. British experiences, which were discussed in Section 12.3, did not lead to promising conclusions regarding the relation between possible benefits and required costs. French experiences, on the other hand, were considered to be so promising that a decision was made to install a computerized bus monitoring system even in a relatively small city of about 130,000 inhabitants, namely in Besançon (cf. Figure 12.6).

It may be concluded that the cost/benefit relation is heavily dependent on the man-power savings that can be achieved. If man-power savings similar to those reported for the Hamburg system can be obtained, then the introduction of computerized bus monitoring systems may represent a major tool for decreasing the operational costs of the public transit system and for increasing its overall efficiency.

14.3 RAPID RAIL TRANSIT SYSTEMS

14.3.1 TECHNICAL ISSUES

The present and future role of computers in improving the efficiency of urban rail systems concerns the automation and optimization of three categories of tasks:

- centralized monitoring of the total transit system (automatic traffic supervision – ATS)
- optimal operation of individual trains (automatic train operation – ATO)
- automatic safeguarding and signalling (automatic train protection – ATP)

The last mentioned application is currently a major area for fundamental research, focussing on the replacement of safety relays by computers. A general introduction of computerized safety systems very likely cannot be expected before the second half of the 1980s.

Computerized ATO systems, which are characterized by the use of onboard microcomputers, have reached the status of experimental studies and real installations; a broad application seems to be feasible from a technical viewpoint during the first half of the 1980s.

Finally, the first category of tasks, i.e., automatic traffic supervision (ATS), represents the area for which computers have already found broad application during the last 5-10 years. Computerized ATS systems are already, or are about to become, an integral part of many municipal rapid rail systems. A well-proven technology is available for implementing computer-equipped control centers, which permits the total transit system to be monitored and the appropriate status information to be displayed to the dispatcher.

14.3.2 SAFETY AND ECONOMIC ISSUES

Decisions on the implementation of computerized ATS systems have, in general, not been made on the basis of pure cost/benefit considerations. The case descriptions presented in Chapter 13 make it clear that the introduction of powerful train traffic supervision and control centers is considered essential if traffic volumes and the resulting number of trains exceed certain limits. In Metro-type systems the supervision and control system has to permit safe train operation at short headways while at the same time accommodate disturbances caused by changing passenger volumes, etc. In urban and suburban rail systems, which operate various types of trains on the same tracks, the creation of a computerized traffic control center may become indispensable to maintain train operation on schedule (cf. Section 13.4 and case description in Section 13.3).

Thus, the main incentives for implementing computerized ATS systems are not cost/benefit factors but the traffic demand requirements and the resulting issues of safety and operational reliability.

It is clear, however, that this major purpose of computer surveillance and control is supplemented by several economic objectives such as

- energy savings, e.g., by energy-optimal train operation as discussed in paragraph 10.3.2
- man-power savings by rationalization of the manual traffic supervision process and by automatic train operation (reducing the number of train drivers per train from two to one)

14.4 PASSENGER GUIDANCE AND SERVICE SYSTEMS

14.4.1 TECHNICAL ISSUES

The main area of computer applications in passenger guidance and service systems is automatic fare collection (AFC), i.e.,

- computerized equipment for ticket vending and checking
- computerized systems for maintaining accounts of sales and cash received, as well as preparing various sales and traffic statistics

The spectrum of computerized systems extends from microcomputer controlled ticket vending machines to large-scale hierarchically structured fare collection and accounting systems covering the whole transit network (cf. paragraph 10.4.1).

The availability of cheap, small, and robust microcomputers has initiated further development in the whole area. Broad application of microcomputer based ticket vending and checking equipment can be expected for the various modes of urban transit systems during the next five years. Another promising area for the application of microcomputers is passenger information systems. Moreover, other services, e.g., automated left-luggage offices, are going to become the subject of microcomputer control.

14.4.2   ECONOMIC AND ATTRACTIVENESS ISSUES

The introduction of computers into passenger guidance and service systems has been accelerated by two driving forces: (1) economy; (2) attractiveness.

It has been estimated that it will be feasible to sell more than 70 percent of all railway tickets by microcomputer controlled automatic ticket vending machines (cf. reference AEG 1978 to Chapter 10). This would allow considerable man-power savings. Moreover, further savings would result from computerization of the accounting and the various passenger information processes.

Both fare collection and passenger information systems directly enhance the attractiveness of the transit system. Therefore, the modernization of these services be means of computer technology may help to attract a certain fraction of car owners back to transit riding.

Part Four

New Modes of Urban Transport: Automated Guideway Transit and the Dual-Mode Concept

Part 1 of this monograph illustrated (cf. Sections 2.1 and 3.2) that the partly conflicting expectations, requirements, and objectives of the private car and public transit users, the public transport companies, and the city as a whole cannot be fulfilled in the sense of an optimal compromise by means of the existing urban transport systems. Therefore, the introduction of completely new, automated transit systems is considered to be one basic long-term strategy.

Part 4 analyzes the concepts for creating those new systems (Chapter 15), the concepts and methods of computerized control and automation (Chapter 16), and the international experience gained so far in real passenger service or large-scale demonstration projects (Chapters 17–19).

15 Basic Systems Concepts

This chapter presents a brief survey of the various new systems concepts necessary in dealing with the following two questions:

1. In what sense can the proposed new systems be considered as a further development of existing modes of urban transport?
2. How shall these new systems cooperate with existing ones?

Figure 15.1 illustrates the three basic systems concepts that can be distinguished:

a. *Automated guideway transit (AGT) systems.* This category of new systems can be considered as a specific development of (automated) urban rail-transit systems (Adams 1976, Alden 1972, Alimanestianu 1974, Anderson et al., 1972, Anderson and Romig 1974, Camp and Oom 1972, FuH 1972, Hesse 1972, Haikalis 1968, Kovatch and Zames 1971, OTA 1975). However, one has to take into consideration the following basic differences:

- the use of small vehicles with a passenger capacity ranging from that of a tram to that of an automobile
- the operation of the system is much more flexible, i.e., both a timetable controlled operation, as in the case of conventional fixed-route public transit, and a demand-responsive operation, as in the case of dial-a-ride systems are possible.
- instead of the suspension principle (steel wheel/steel rail) the use of auto-type principles, (rubber-tired wheel/concrete decking) is preferred, in general; more advanced suspension technologies like the air cushion and the magnetic levitation principles are very seldom used (MacKinnon 1974)

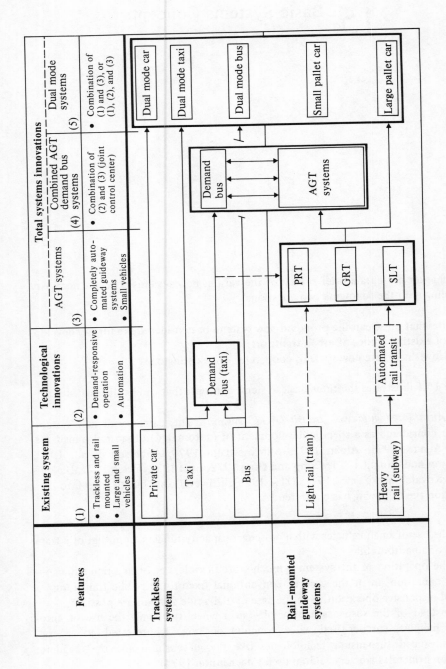

FIGURE 15.1 Survey of basic systems concepts.

The new systems are thus aiming to combine the main advantages of rail transit, i.e., speed, safety, reliability, and capacity, with the merits of automobile and bus traffic, i.e., flexibility, demand-responsive operation, small and comfortable vehicles (cf. Figure 15.1). Obviously certain limits to reaching this target remain with respect to area traffic because of the chosen guideway principle. This disadvantage is avoided in the second basic systems concept:

b. *Combined AGT and dial-a-ride systems.* This concept proposes the joint use of an AGT system and a demand-responsive bus system as described in Chapter 11. Both systems are coupled by a common computer control center, leading to a cooperative operation of the whole system. Nevertheless both subsystems remain relatively independent since the vehicles of one subsystem cannot enter the traffic network of the other subsystem; this means it is necessary to change vehicles in going from one network to the other. To overcome these limits is the target of the third basic concept:

c. *Dual-mode systems.* These systems can be considered as a certain combination of conventional street traffic systems based on private cars, taxis, buses, and automated guideway transit (AGT) systems (cf. Figure 15.1). In area traffic, private cars, taxis, and buses are used in the conventional way, but on densely travelled traffic corridors they are switched to an automated guideway. In the following the concepts proposed and used for the development of automated guideway systems are discussed first.

15.1 AGT (AUTOMATED GUIDEWAY TRANSIT) SYSTEMS

The concepts developed for AGT systems may be classified into three categories depending on the complexity of the traffic network, the vehicle capacity, and the desired operation mode (cf. Figures 15.1, 15.2):*

- personal rapid transit (PRT)
- group rapid transit (GRT)
- shuttle and loop transit (SLT)

The PRT concept aims for system features, which fulfill as far as possible the requirements derived in Section 3.2 (cf. Figure 3.2) for new systems that are expected to provide an alternative to private car use for many city trip purposes. At the other end of the scale SLT systems are characterized by features similar to those of an automated rail-transit system.

15.1.1 SHUTTLE-LOOP TRANSIT (SLT)

Transit systems of this type are frequently called the "horizontal equivalent of an automated elevator" (cf. OTA 1975). They are characterized by (cf. Figure 15.2):

*It should be noted that the terminology used for denoting the various AGT systems is not uniform (cf. list of selected terms given in Table 15.1, and OTA 1975).

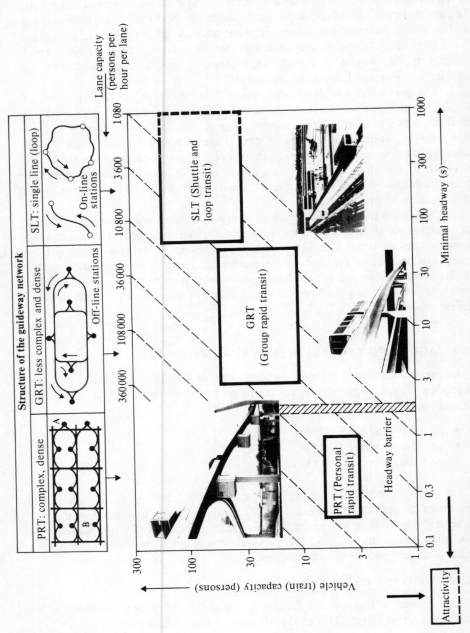

FIGURE 15.2 Classification of AGT systems.

TABLE 15.1 New Modes of Urban Transportation

Acronyms	Definitions
AGT	*A*utomated *G*uideway *T*ransit is a class of urban transportation systems in which unmanned vehicles are operated on fixed guideways along an exclusive right of way
SLT	*S*huttle-*L*oop *T*ransit is the simplest type of AGT system. Vehicles move back and forth on a single guideway (shuttle) or around a closed path (loop). Stations are located on the guideway (on-line stations). There are no or only a few switches (no routing capabilities). Headways are larger than 60 seconds. Vehicles may vary considerably in size and may travel singly or coupled together in trains.
GRT	*G*roup *R*apid *T*ransit denotes an AGT type that serves groups of people with similar origins and destinations. Stations may be located on sidings off the main guideway (off-line stations). GRT networks make use of switching (limited alternative routings). Headways range from 3 to 60 seconds. Vehicles with a capacity of 10–50 passengers may be operated singly or in trains.
PRT	*P*ersonal *R*apid *T*ransit denotes an AGT type that serves one person or groups of up to six usually travelling together by choice. PRT networks are characterized by a large number of off-line stations and extensive use of switching (many alternative routes). Vehicle capacity ranges from 3 to 6 persons; standing is not allowed. Vehicles are operated at headways smaller than 1 second or at most 3 seconds (many authors use the term PRT to denote the whole group of AGT systems).
ICRT	*I*ntermediate *C*apacity *R*apid *T*ransit. This term is sometimes used instead of SLT.
HPPRT	*H*igh *P*erformance *P*ersonal *R*apid *T*ransit denotes an advanced version of GRT sometimes called GRT II. Moderate sized vehicles (8–20 seats) and short headways (3–15 seconds) are basic features.
HCPRT	*H*igh *C*apacity *P*ersonal *R*apid *T*ransit. This term is sometimes used instead of PRT.

- a very simple network structure in the form of a single loop or shuttle line with "on-line stations" lying on the guideway, and with almost no switching
- medium-capacity vehicles or trains consisting of several small vehicles with a passenger capacity ranging from 30 to 150 persons
- relatively large vehicle headways of more than 60 seconds
- a completely automated driverless vehicle operation permitting both timetable-controlled and demand-responsive operation

These features produce the following advantages:

- the simple network structure and the long headways result in relatively low requirements concerning the sophistication of the automated control systems
- the necessary automation technology is fairly well developed, i.e., there is no significant development risk for SLT systems
- about 10 SLT systems are in operation around the world (cf. Chapter 17)

The limits of the SLT concept are obvious:

- only a small number of origins and destinations can be served
- private car use can only be influenced in very limited areas

15.1.2 GROUP RAPID TRANSIT (GRT)

This transit concept is characterized by the following features (cf. Figure 15.2):

- completely automated driverless operation of a larger number of vehicles in more complex networks, with several alternative routes and so-called "off-line stations" lying on bypasses
- use of small vehicles with a passenger capacity of about 10-50 persons, classifying a GRT vehicle not as a mass transit means but as a so-called group rapid transit vehicle
- vehicle operation at headways of 15-60 seconds; for advanced GRT systems a minimal headway of 3 seconds is aimed for, resulting in a maximal systems capacity of about 10,000-40,000 persons per hour per lane (cf. Figure 15.2)
- more flexible operation of the whole system using the potential of a demand-responsive operation with alternative routes for travelling from a given origin to a desired destination

These features of the GRT concept are expected to encourage a certain number of private car owners not to use their cars in the area covered by the GRT network. On the other hand, the creation of GRT systems will require the development of a complex and complicated computer control system in order to achieve the features just mentioned. The risk of unsuccessful development of GRT systems mainly depends on the minimal vehicle headways needed for a certain systems capacity (cf. Figure 15.2). The risk is relatively low for systems with vehicle headways larger than 15 seconds, since in this case the well proven principles of railway safety technology can be used to ensure safe vehicle operation.

The development risk increases when small headways in the range of about 3-15 seconds are required. Here, more advanced techniques of headway control have to be used (cf. Chapters 16 and 18).

15.1.3 PERSONAL RAPID TRANSIT (PRT)

The PRT concept, which aims to provide an alternative to private car use for many trip purposes, is characterized by (cf. Figure 15.2):

- a dense guideway network with a large number of off-line stations providing easy walking access to the system
- very small vehicles similar in size and comfort to a 3-6 passenger automobile
- extremely small vehicle headways, in the range of 1.0-0.2 seconds, which are needed to reach a sufficiently large systems capacity at the low vehicle capacities
- very high flexibility of systems operation, with the potential for timetable controlled and demand-responsive operation, including such features as non-stop origin to destination travel (without the need for changing vehicles), individual use of a vehicle, i.e., the passenger gets "his vehicle" if he wants to travel alone, optimal route selection, etc. (cf. Figure 15.2 and Chapter 19).

The advantage of the PRT concept is that it meets the features derived in Section 3.2 (cf. Figure 3.2) for a hypothetical transit system that would have the potential to reduce private car use within a city. It is becoming obvious that the feasibility of the PRT concept depends on the availability of a very sophisticated computerized control and guidance system. Therefore, there is considerable risk in the successful development of an operational PRT system. This can be illustrated by the so-called "headway barrier" characterized by vehicle headways less than 1-2 seconds (cf. Figure 15.2). To break this headway barrier, a very powerful and reliable headway control system will have to be developed. Moreover, the introduction of new safety standards for transit systems becomes indispensable if a PRT system is to be put into passenger service (cf. paragraph 16.2.3, for details).

The existence of the headway barrier has led to controversial discussions on the feasibility and usefulness of the PRT concept, i.e., of small vehicles (Anderson and Romig 1974, Leutzbach et al. 1974). One school-of-thought recommends avoiding the development of PRT systems and focussing on the simpler GRT concept (cf. OTA 1975). Another school-of-thought claims that the use of small vehicles represents the key solution in the creation of new transit systems competitive with the private car (Loder 1974, Demag 1971, Becker 1973, Fichter 1974, JSPMI 1972, 1975).

Moreover, the following impacts result from a systems concept based on small vehicles (cf. Anderson and Romig 1974).

Costs. Small vehicles will require narrow and light elevated guideways; since the guideway construction costs are the dominating ones, small vehicle sizes will reduce the overall costs in spite of the larger number of vehicles one has to provide. Moreover, a large number of vehicles opens the way for a systematic series production permitting the adaptation of advanced automobile production technologies. Moreover, narrow and light guideway elements can be produced faster and more easily at a central workshop, they can be transported and erected easily, they cover less space, and only seldom do they require removal of buildings and other obstacles.

Architectural aspects. The narrower elevated guideway will result in less visual intrusion, i.e., less negative interference with the city architecture.

Operational advantages. For a traffic network with a large spatial extension and many stops, demand-controlled traffic with short waiting times at the stations is only feasible if a large fleet of (small) vehicles is available (cf. Sher and Anderson 1974).

Passive safety measures. Small vehicles containing no standing room permit the introduction of measures for increasing passive safety such as are used today in the auto (seat belts, etc.).

Individuality and crime. In several countries public transit systems like subways have become a main scene for criminal activities. Therefore, it is often argued that AGT vehicles would provide conditions supporting this trend. However, proponents of the PRT concept claim that this statement does not hold true for the PRT systems, since no passenger is forced to use a PRT car jointly with another passenger if he does not wish to do so.

One result of the considerations summarized here, is that the investigation of PRT systems now represents a major research and development effort undertaken in several countries in spite of the above-mentioned headway barrier and the other complex control problems (cf. Chapter 19).

15.2 COMBINED AGT AND DIAL-A-RIDE SYSTEMS

The attractiveness of the PRT and GRT systems is restricted to trips with origins and destinations close to the AGT guideway network. Therefore it is natural to connect the AGT system with other public transport systems, like urban railways, buses, and streetcar lines. As far as connecting an AGT system with low density areas like suburbs is concerned, the coordinated operation of AGT and dial-a-ride systems using coupled control centers (cf. Figure 15.3) is considered feasible (cf. Navin 1972, Hamilton and Nance 1969). The following sequence (1-5) represents the desired operation principle

1. A prospective passenger who wants to reach, for example, the bus stop B in Figure 15.3 at a certain time has first to inform the control center as to the departure bus stop (cf. A in Figure 15.3), the desired departure time and the destination. To do this, he rings the dispatching center using his home telephone or a special call device at bus stop A.

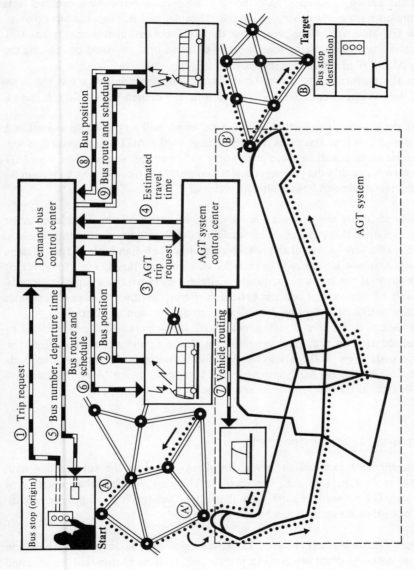

FIGURE 15.3 Concept of a combined AGT and dial-a-ride system.

2. The dispatching center then checks the feasibility of the transport request and informs the prospective passenger on the real departure time of the demand bus at stop A and the estimated total travel time (cf. steps 2-5 in Figure 15.3).

3. If the passenger accepts the offer then the demand bus control center selects a bus, the corresponding driving route and schedule, and informs the bus driver of his task (cf. step 6). At the same time the AGT control center selects an AGT vehicle and the corresponding schedule as well as an optimal route connecting the stations A' and B' of the AGT network (cf. step 7).

4. In the meantime the demand-bus dispatching center selects a bus for transporting the passenger from the AGT station B' to the desired bus stop B (cf. steps 8 and 9).

5. At the scheduled departure time the passenger will be picked up at station A by the demand-bus; at station A' he will change to the AGT vehicle, which is supposed to be already waiting there or to arrive within a short waiting time. The AGT vehicle then transports the passenger under computer control nonstop to station B', where the second demand bus is already waiting.

It is obvious that the operation mode described here has to be considered as the highest level of service imaginable. Its feasibility depends, among other things, on the availability of a very powerful and reliable computer control and information system. This is obvious as it is necessary to keep the confirmation time needed for checking the feasibility of the transport request (cf. steps 1-5) down to a couple of seconds. Proponents of this systems concept believe, however, that the advances of computer technology described in Section 3.4 will provide the solution to this task.

Nevertheless it is likely that this combined AGT-demand-bus concept will be implemented first in simplified operation modes, e.g., the connection of several bus stops with a few AGT stations for collecting passengers in a many-to-one mode and the distribution of AGT passengers by connecting a few AGT stations with many bus stops in a one-to-many operation mode.

15.3 THE DUAL-MODE CONCEPT

This concept aims to avoid the remaining disadvantage of the combined system described in Section 15.2, i.e., the need for changing vehicles when entering or leaving an AGT network. Figure 15.4 illustrates that the following three modifications of dual-mode systems can be distinguished:

1. *Use of existing street vehicles.* The first class of dual-mode systems is based on the use of ordinary street vehicles like private cars, taxis, and buses that are operated by car drivers in low density and non-congested parts of the city. If the route leads through a heavily travelled area, then an AGT-type guideway system has to be used. For this purpose it will be possible to enter the AGT system at certain access points

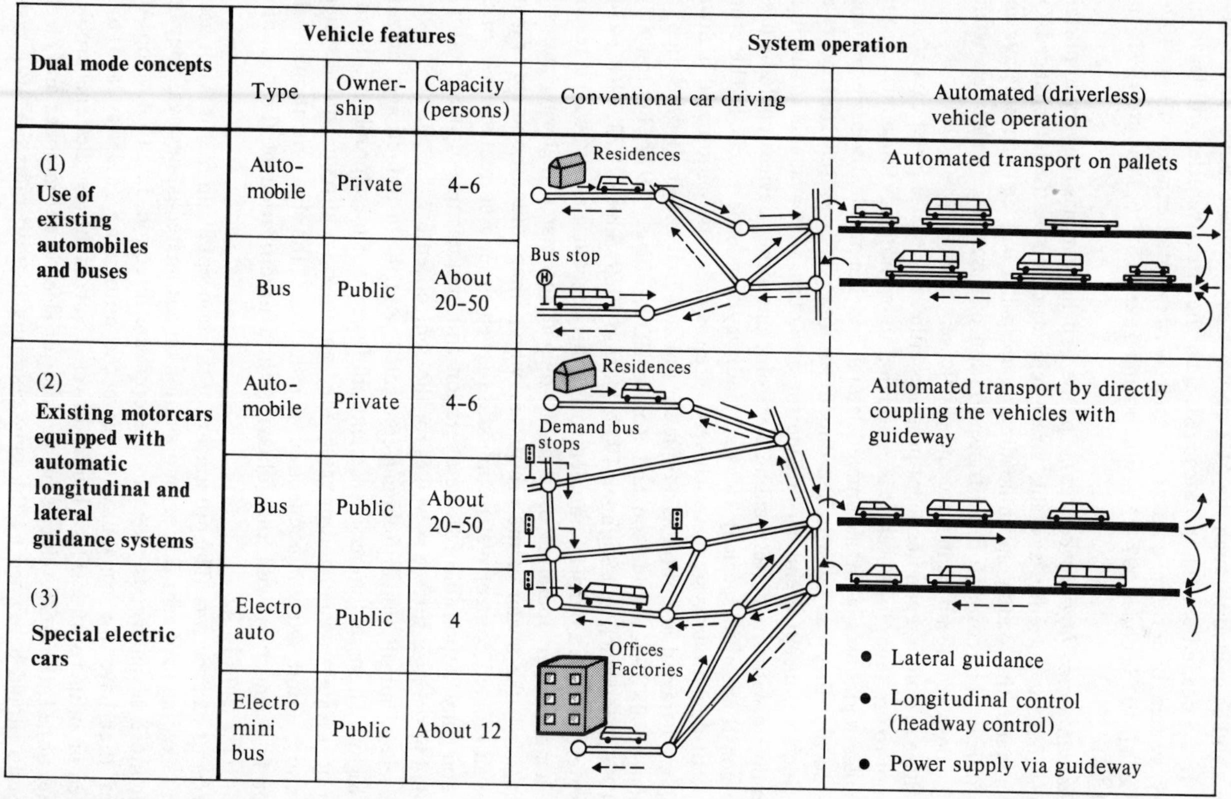

Dual mode concepts	Vehicle features			System operation	
	Type	Owner-ship	Capacity (persons)	Conventional car driving	Automated (driverless) vehicle operation
(1) Use of existing automobiles and buses	Auto-mobile	Private	4–6	Residences / Bus stop	Automated transport on pallets
	Bus	Public	About 20–50		
(2) Existing motorcars equipped with automatic longitudinal and lateral guidance systems	Auto-mobile	Private	4–6	Residences / Demand bus stops / Offices Factories	Automated transport by directly coupling the vehicles with guideway
	Bus	Public	About 20–50		
(3) Special electric cars	Electro auto	Public	4		• Lateral guidance • Longitudinal control (headway control) • Power supply via guideway
	Electro mini bus	Public	About 12		

FIGURE 15.4 Dual-mode system concepts.

where the automobiles and buses can be driven onto special pallet cars. These pallet cars will be guided through the AGT network as described above for the AGT vehicles. If the desired exit ramp is reached, then the street vehicles will leave the guideway system under the control of their drivers (cf. Benjamin 1974, 1976, Dietrich et al. 1975, Loder 1974, Matsunobu and Takashi 1976, Penoyre 1971, and Figure 15.4).

2. *The automated highway concept.* The use of pallet cars has the very important advantage that the currently available street vehicles could be used unchanged. On the other hand, the loading and unloading of the pallet cars can cause delays and other problems. Moreover, for the transportation of buses large pallet cars would have to be designed, resulting in huge and expensive guideways.

For these reasons a second concept proposes either not to use pallet cars at all or simply not to use them for buses. It is considered to be more advantageous to equip automobiles and buses with automated lateral and longitudinal vehicle-guidance systems so that they can be switched directly onto a guideway (cf. Dietrich et al. 1975). Such a concept seems to provide the advantage that available urban highways equipped with the corresponding electronic and other devices could be used, and the construction of specially elevated guideways could be avoided. Therefore, this type of dual-mode system, which very often is called the automated highway system, has attracted attention as a topic for fundamental and applied research (cf. Anderson et al. 1972, Anderson and Romig 1974, Böttger 1972, Fenton et al. 1970, 1976, Kiselewich and Stefanek 1972, Plotkin 1969, Smith 1973, TRB 1974).

However, there is a barrier to implementing this concept, namely, who is responsible if a private car operating on a public-owned automated highway causes an accident? Is it the owner of the private car, who perhaps did not operate or maintain the automated control system onboard his vehicle in a correct manner, or the public company responsible for the operation and maintenance of the electronics embedded in the road surface as well as for the control center?

The related jurisdictional problems connected with the introduction of an automated highway have seldom received attention. Because of these problems, the integration of privately owned and privately used automobiles into dual-mode systems appears to be more likely in connection with the use of the above-mentioned pallet cars, or in connection with the following third modification of the dual-mode concept (cf. Figure 15.4).

3. *Use of electric cars.* The proponents of this concept claim that it is not useful to adapt existing road vehicles for an operation on automated guideways. They consider it more advantageous to go in the opposite direction, i.e., to develop AGT vehicles that have the capability to leave the automated guideway and to be operated on ordinary streets by a human driver. To avoid the jurisdictional problems just mentioned, the AGT vehicles in the form of electric automobiles and electric mini buses should be owned by the same company that operates the AGT system (cf. Figure 15.4). Another advantage of this concept is that the combination of the AGT system with a demand-bus or a demand-taxi system, described in Section 15.2,

can be realized much more effectively (cf. Figure 15.4 and Benjamin 1974, Dietrich et al. 1975).

15.4 CONCLUDING REMARKS

The review presented in this chapter of the basic system concepts illustrated

- the relations between the proposed new modes of urban transport and the existing transportation systems (cf. Figure 15.1)
- the similarities and differences between the individual concepts proposed for the creation of completely new, automated urban transport systems (cf. Figures 15.2–15.4)

The discussion presented makes it clear that from the point of view of automation and control the same basic problems occur in all three system concepts, i.e., the realization of a completely automated operation of a large number of driverless vehicles in guideway networks of different structures. The concepts and methods that have been proposed and used to solve these problems are considered in Chapter 16.

REFERENCES

Adams, C. J. 1976. Review of Conventional Transit Systems with Rail and Automated Fixed Guideway-Transit Systems. Proceedings of the Fourth Annual Intersociety Conference on Transportation, Los Angeles, California, USA.

Alden. 1972. Starrcar system description. Report from Alden Self-Transit Systems Corporation, Milford, Massachusetts, USA.

Alimanestianu, M. 1974. TRAN-SEAT — A high capacity, low cost PRT system with vehicle vertical travel capabilities. In J. E. Anderson and S. Ronig, eds., Personal Rapid Transit II. University of Minnesota, Minneapolis, Minnesota, USA, pp. 517–526 (published by the University).

Anderson, J. E., J. L. Dais, W. L. Garrard, and A. L. Kornhauser (eds.). 1972. Personal Rapid Transit — A Selection of Papers on a Promising New Mode of Public Transportation. Institute of Technology, University of Minnesota, Minneapolis, Minnesota, USA (published by the University).

Anderson, J. E. and S. H. Romig, eds. 1974. Personal Rapid Transit II. Proceedings of the 1973 International Conference on Personal Rapid Transit, University of Minnesota, Minneapolis, Minnesota, USA (published by the University).

Becker, K. 1973. Cabintaxis: a new way for better traffic in our cities. Nahverkehrspraxis, 20 (7): 276–279 (in German).

Benjamin, P. 1974. Analysis of urban dual-mode transportation. In J. E. Anderson and S. Romig, eds., Personal Rapid Transit II, University of Minnesota, Minneapolis, Minnesota, USA, pp. 95–106 (published by the University).

Benjamin, P. 1976. Comparison of dual-mode and other urban transportation systems. In Transportation Research Board Special Report, No. 170, Washington, DC, pp. 28–32.

Böttger, O. 1972. Considerations on the augmentation of road transport capacity by introducing automatic track guidance and headway control. Straße und Autobahn, 23 (1): 8–12 (in German).

Camp, S. and R. Oom. 1972. SPARTAXI for Gothenburg. In J. E. Anderson et al., eds. Personal Rapid Transit. Institute of Technology, University of Minnesota, Minneapolis, Minnesota, USA, pp. 123–140 (published by the University).

Demag. 1971. Systems analysis of personal rapid transit system Cabintaxi (CAT). Report by DEMAG Foerdertechnik, Wetter (Ruhr), FRG (in German).

Dietrich, E. et al. 1975. Dual mode bus systems. A study by order of the Bundesministerium für Forschung und Technologie, Bonn, Friedrichshafen, Munich, Cologne, FRG (in German).

Fenton, R. E., K. W. Olson, and J. G. Bender. 1970. Advances toward the automatic highway. Committee on Vehicle Characteristics, 50th Annual Meeting, 1970.

Fenton, R. E., K. W. Olson, and R. J. Mayhan. 1976. On future automated ground transport – individual vehicle longitudinal control. In Control in Transportation Systems. Proceedings of the IFAC/IFIP/IFORS Third International Symposium, Columbus, Ohio, pp. 211–221.

Fichter, D. 1974. VEYAR; small cars as the key to urban PRT. In J. E. Anderson and S. Romig, eds., Personal Rapid Transit II. University of Minnesota, Minneapolis, Minnesota, USA, pp. 527–534 (published by the University).

FuH. 1972. A new solution for rapid transit? – The project taxitram Elan-Sig. Fördern und Heben, 22 (16): 926–928 (in German).

Haikalis, G. 1968. Supra-Car. Highway Research Record, No. 251: 63–68.

Hamilton, W. F. and D. K. Nance. 1969. Systems analysis of urban transportation – computer models of cities suggest that in certain circumstances installing novel "personal-transit" systems may already be more economic than building conventional systems such as subways. Scientific American, 221 (1): 19–27.

Hesse, R. 1972. German experiences in the planning and development of automatic cabintaxi systems. Intersociety Conference on Transportation, Washington.

JSPMI. 1972. A new urban traffic system: CVS (computer-controlled vehicle system). Japanese Society for the Promotion of Machine Industry, Tokyo, Japan.

JSPMI. 1975. A new urban transit system: CVS (computer-controlled vehicle system). Japanese Society for the Promotion of Machine Industry, Tokyo, Japan.

Kiselewich, S. J. and R. G. Stefanek. 1972. An analysis of interchange operation in an urban automated highway network. Transportation Research, 6: 381–401.

Kovatch, G. and G. Zames. 1971. Personalized Rapid Transit Systems: A First Analysis. Report No. DOT-TSC-OST-71-10, from the US Department of Transportation, Office of the Secretary, Washington, DC.

Leutzbach, W. et al. 1974. On new ways through the city: models for traffic quality – controversies. Bild der Wissenschaft: 80–91 (in German).

Loder, J. L. 1974. Personal automated transportation: A PAT-solution for Australian cities. Transportation Planning and Technology, 2: 221–262.

MacKinnon, D. D. 1974. Personal rapid transit systems at Transpo '72. In J. E. Anderson and S. H. Romig, eds., Personal Rapid Transit II. Institute of Technology, University of Minnesota, Minneapolis, Minnesota, USA, pp. 35–45 (published by the University).

Matsunobu, M. and I. Takashi. 1976 Dual-mode bus system being planned in Hiroshima city. Expressways and Automobiles, 19 (10): 65–69 (in Japanese).

Navin, F. P. D. 1972. The interaction between personalized rapid transit and demand-activated bus transit. In J. E. Anderson et. al., Personal Rapid Transit, University of Minnesota, Minneapolis, Minnesota, USA, pp. 149–173 (published by the University).

OTA. 1975. Automated guideway transit – an assessment of PRT and other new systems. United States Congress, Office of Technology Assessment, Washington, DC, (published by US Government Printing Office).

Penoyre, S. 1971. The road research laboratory's work on dual-mode road vehicles. Traffic Engineering and Control, September: 189–191.

Plotkin, S. C. 1969. Automation of the highways, an overview. IEEE Transaction on Vehicular Technology, VT-18 (2): 77–81.

Sher, N. C. and P. A. Anderson. 1974. Waiting time in small-vehicle systems. In J. E. Anderson and S. J. Romig, eds., Personal Rapid Transit II. University of Minnesota, Minneapolis, Minnesota, USA, pp. 402–416 (published by the University).

Smith, P. G. 1973. Discrete-time longitudinal control of dual-mode automobiles. Paper presented at the Intersociety Conference on Transportation, Denver, Colorado, USA.

TRB. 1974. First International Conference on Dual-Mode Transportation. Report of a Conference sponsored by the Transportation Research Board, Washington, DC.

16 Concepts and Methods of Control and Automation

The automation and control concepts themselves result from the tasks and processes that have to be automated. Therefore, these tasks and the resulting control tasks hierarchy are considered first.

16.1 THE CONTROL TASKS HIERARCHY

Three different groups of problems can be distinguished:

- automated control of the transport processes
- automated passenger guidance and information
- automated surveillance and control of secondary processes, e.g., electric power supply, vehicle and guideway maintenance

The following considerations are restricted to the first category of problems. To illustrate the size and kind of control task, the use of a PRT system is described in more detail.

It may be assumed that a prospective passenger intends to enter the PRT system at station A (cf. the network given in the upper left part of Figure 15.2), and that he wants to travel to station B. For this purpose, he has to carry out the following activities supported by the passenger guidance and information system:

1. From a network map he has to read the number of the desired target station B.
2. He has to put this number into a ticket selling machine using a certain keyboard. This machine then indicates both the station number and the fare. The passenger checks the displayed station number. After inserting the right number of coins or a credit card he will receive a ticket containing the magnetically coded address of the target station B.

3. The passenger then has to put this ticket into an automated gate giving him access to the station platform. A reading device installed in the automatic gate identifies the number of the target station and transmits it to the corresponding control computer.

The control computer checks first whether an empty car is available at station A. If this is not the case then the computer guides the nearest empty vehicle to station A. After stopping automatically at the platform the doors of the vehicle are opened automatically and the passenger can enter the car. As with an automated elevator the doors close after a small delay and the vehicle starts its journey. Now the process control system has to fulfill the following tasks:

- guide the vehicle out of the station and merge it into the vehicle stream moving along the corresponding main line
- keep safe distances between different vehicles and speed limits when passing the main line and the following intersections and merging areas
- select an optimal route for reaching the target station B and guide the vehicle along that route through the network
- after reaching the target station the vehicle has to be stopped at the right position, the doors have to be opened and the passenger can leave the car
- the computer has then to decide whether this empty vehicle should be allowed to wait at station B or be guided to another station where there is a shortage of empty cars

The operation mode described here characterizes a demand-responsive or a demand-controlled working regime. During peak hours i.e., if the traffic demand predominates in certain origin–destination relations, the control system has to switch to timetable controlled operation.

The operation modes and the corresponding control tasks sketched here occur in the same or at least in similar form for all PRT systems developed or planned so far. They are in principle also valid for GRT systems; certain differences arise because the vehicles are not used individually but by groups of passengers travelling to the same station.

In the case of SLT systems (cf. Figure 15.2) a large part of the control tasks mentioned do not exist because of the simple network structure. It follows that, from the viewpoint of the control tasks, SLT and GRT systems may be considered as simpler special cases of the PRT systems. Therefore the following considerations refer mainly to PRT systems.

The control tasks summarized here will, in general, be implemented by a hierarchy of control computers (cf. Figure 3.4) containing, in the case of PRT systems, at least three levels (Figure 16.1): (1) one or several central control computers dealing with the surveillance and control problems at the network level; (2) a set of stationary local control computers handling the traffic flow control tasks occurring

Concepts	Control methods		Results
	Criteria/tasks	Methods	
(1) Central network control (route guidance, empty car disposition) 	• Optimal utilization of empty cars • Optimal use of available network parts • Minimizing travel time • Emergency management	• Route planning • Route selection • Real-time computation	Simulation studies — implemented only for simple network structures (Morgantown–GRT)
(2) Traffic flow control 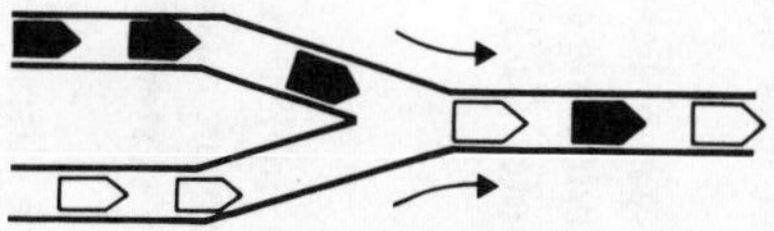	• Maximizing capacity • Minimizing delays	• Synchronous control • Quasisynchronous control • Asynchronous control	Implemented for various GRT and PRT systems
(3) Automatic vehicle guidance *Longitudinal* 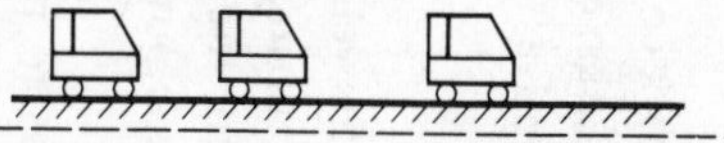	• Maximizing lane capacity • Warranting safety and comfort	• Synchronous control (point follower) • Asynchronous control (vehicle follower)	• Well-proven technologies for headways $T \gtrsim 10$s are available
Lateral 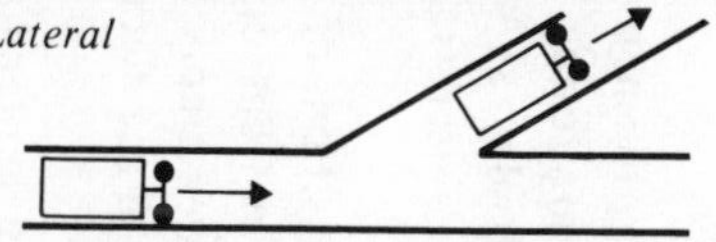	• Warranting safety and driving comfort	• Mechanical-hydraulic systems • Electronic principles	• Well-proven technologies are available

FIGURE 16.1 Control tasks hierarchy.

in merging and station areas, intersections etc.; (3) onboard microcontrol computers or hardwired electronic systems, which are assigned to vehicle guidance tasks.

The onboard control system at level (3) has to fulfill two categories of tasks:

- lateral vehicle guidance including automated course regulation and switching from one course to another;
- longitudinal vehicle guidance including acceleration, speed, position, and headway regulation

The computers at level (2) control the vehicle operation in stations and other conflicting points. A station computer is, for example, connected with the ticket reading device in the automated gates located at the entrances to the station platforms. From these gates the station computer gets information on the number of empty cars required and the trip destinations. If an empty car is not available, then the station computer sends a corresponding message to the network computer. Moreover, the central computer has to be informed about the desired target stations. From the central computer the station computer receives the addresses of the empty cars.

A further task of the station computer is to assign arriving vehicles to the various stops located at the available platforms, sending commands for opening and closing the vehicle doors, controlling the passenger information system, i.e., of various passenger displays showing vehicle destinations, departure times, etc.

The main task of the network computer is optimal routing, i.e., of the traffic-responsive route selection:

- for occupied vehicles travelling between start and target stations
- for empty cars on the way to other stations or to the maintenance depot
- for empty and occupied cars in the case of a disturbance

Moreover, the network computing system has to fulfill a large number of surveillance and monitoring tasks, e.g., displaying the operational state of the PRT system at a wall display as well as on computer CRT displays. Here the following parameters are of interest:

- the positions of the vehicles
- malfunctions occurring in the vehicles, e.g., open doors at non-zero speeds, missing power supply for the driving motor, missing or low braking pressure.
- state of the electric power supply system

The control center has to provide the dispatcher with the capability to override computer instructions in emergency situations as well as to communicate via special telephones with the individual vehicles and stations.

In what ways can the control tasks at the various levels of the hierarchy illustrated in Figure 16.1 be solved? This question is considered first for the lowest level of the hierarchy, i.e., automated vehicle guidance and control. A distinction is made between lateral and longitudinal control problems.

16.2. AUTOMATIC LATERAL VEHICLE GUIDANCE

The lateral vehicle guidance system has to ensure that the vehicles can be moved automatically along a prescribed course, i.e., along the guideway. Moreover, it has to enable the change from one course to another by automated switches lying in front of branching points.

The following objectives must be fulfilled by the lateral guidance system:

- The time delay occurring in passing a branching point must be kept small.
- A course change at branching points should be carried out without moving guideway elements in order to reduce guideway maintenance expenses.

It is mainly for these reasons that the well-proven switching technology used for railways did not find application in GRT and PRT systems. In general, switching principles using onboard devices alone are preferred. These principles* may be classified as mechanical-hydraulic, and electromagnetic course guidance systems, as illustrated by Figure 16.2.

Most GRT systems use a lateral guidance principle similar to that given in column 1 of Figure 16.2 (cf. MacKinnon 1974). Here, the guideway contains side walls, which are touched by certain guide wheels connected with the mechanical-hydraulic servo steering system (cf. Figures 18.5 and 18.6 in Chapter 18 for real systems). If the vehicle is to go straight ahead (cf. vehicles 1 and 3 in Figure 16.2), then the guide wheel will be switched such that it touches the right guideway wall thus guiding the vehicle along this wall. For a left turn the guide wheel has to be switched to the left wall (cf. vehicle 2 in Figure 16.2). To perform this function the vehicle has first to send its number to the stationary computer when it reaches a certain position in front of the branching point. The computer identifies the corresponding driving direction and sends the control command to the vehicle to switch the guide wheel to the right or the left wall, respectively (cf. point A in Figure 16.2). At points B and B' a signal is transmitted again to the computer to indicate whether or not the vehicle has reached the right course after the branching point. If this is not the case then an emergency stop signal will be sent to the vehicle. The scheme in column 2 of Figure 16.2 illustrates a certain modification of the principle just described. Here the guideway contains a center groove in which a guide wheel is inserted (cf. Figure 19.9, which shows a special implementation of this principle).

*Magnetic and air-cushion principles should also be mentioned. However, so far, they have not reached a level of practical significance.

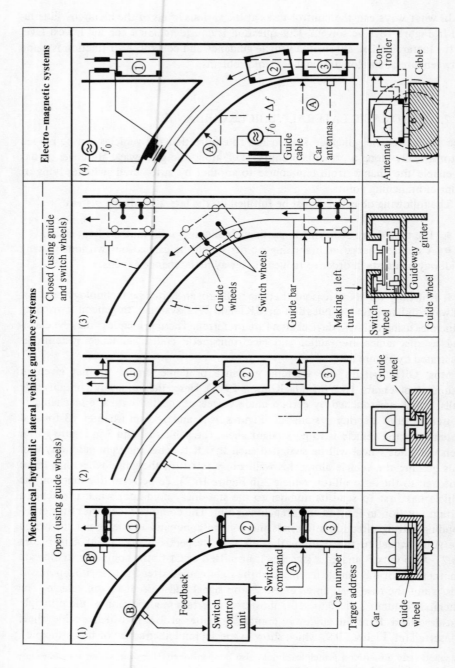

FIGURE 16.2 Selected examples of lateral vehicle guidance principles.

Another principle uses guide rails, as illustrated in column 3 of Figure 16.2. By means of special switchable guide wheels the vehicle follows the guide rail, going straight ahead or turning to the left as desired (cf. Figure 16.2 and Section 19.2).

The electromagnetic principle shown in column 4 of Figure 16.2 is very often recommended for application to dual-mode systems (cf. Figure 15.4, and Fenton 1970, Dietrich et al. 1975, Kästner 1971, Oshima et al. 1965, Strobel and Kästner 1970, Zimdahl 1971). Here, an electric guiding cable is embedded in the guideway surface and supplied by an alternating current of about 4-10 kilocycles per second and 0.1-1 amperes. The electromagnetic field created by this current induces voltages in coils mounted on the vehicle. In this way electric signals are produced that are proportional to the course and course angle deviations. These signals are used by a steering controller and corresponding amplifier to drive a steering mechanism in such a way that the vehicle moves along the guiding cable (cf. Strobel and Kästner 1970 for details). But how can a change from one course to another be carried out? For this purpose, a second cable is installed in the road surface in the area of a branching point. This cable is supplied with a frequency $f_0 + \Delta f$ that deviates sufficiently from the frequency f_0 normally used. If a car is to turn left (cf. vehicles 2 and 3 in Figure 16.2), then a command signal will be transmitted to it by the computer, which switches the filters of the onboard control system to the frequency $f_0 + \Delta f$. The automated steering control system cannot then receive the frequency f_0 sent by the cable leading straight ahead; instead it follows the left turn cable, which sends the frequency $f_0 + \Delta f$. At point A', i.e., after passing the branching point, the onboard filters are readjusted to the original frequency f_0. Another way to carry out course changes is to supply all cables with the same frequency f_0. The power supply is then switched off in the cable section that leads in the direction the vehicle is to not to follow.

The electromagnetic lateral guidance principle described here has been used for a wide variety of applications ranging from low speed electric tractors operating at about 5-10 km/h in stocks and production halls, to high speed test automobiles operating driverless at about 60-100 km/h or even more (cf. Fenton 1970, Kästner 1971). Until now, an application of this principle to GRT and PRT systems has not been reported. Only in research and development work aimed at creating dual-mode vehicles is the electromagnetic lateral guidance system considered a promising principle (cf. Fenton 1970, Dietrich et al. 1975).

16.3 AUTOMATIC LONGITUDINAL VEHICLE GUIDANCE: SPEED AND POSITION CONTROL

The automated control of speed and position of an AGT vehicle represents the first basic task of a longitudinal vehicle guidance system.

16.3.1 AUTOMATED SPEED CONTROL

The speed control system has to ensure that the AGT vehicle follows speed commands transmitted from the corresponding stationary control computer to the

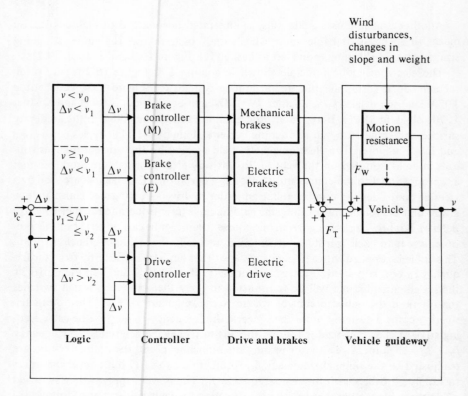

FIGURE 16.3 Speed control system: v_c, commanded speed; v, real speed; F_T, resulting driving (braking) force.

onboard control system. In spite of disturbances such as wind gusts, changing vehicle weights, and guideway slopes, the speed control system has to keep the speed within certain tolerance limits considering given constraints concerning maximal admissible values of acceleration, deceleration, and jerk, which are dependent on safety and passenger comfort requirements.

The basic structure of the speed control system is illustrated in Figure 16.3 (cf. Hinman 1973, 1975, Brown 1972, 1973, 1974, Fling and Olson 1972, Ishii et al. 1974). One observes a characteristic distribution of the whole control task to three control loops, i.e., to a driving control loop and to two braking control loops — one for the electrical and the other for the mechanical brake. The mechanical brake is, in general, used in emergency situations only. Under normal operating conditions the electric motor is used for both functions, i.e., for creating the driving as well as the braking forces. This leads to the simplified block diagram shown in Figure 16.4, in the case of a DC motor which fulfils the requirements mentioned above in the following manner (cf. Hinman 1973, 1975 for details):

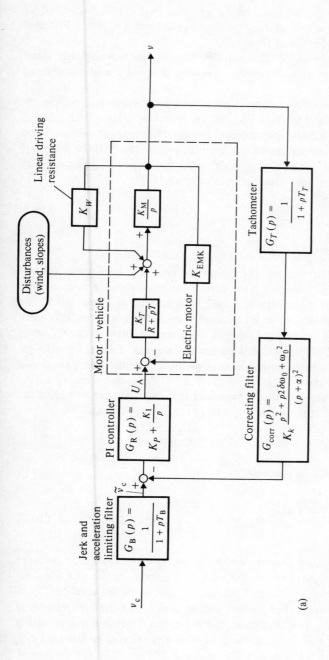
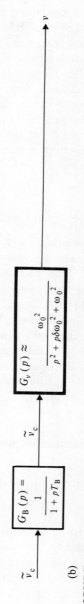

FIGURE 16.4 Speed control system (cf. Hinman, 1973, 1975): v_c, commanded speed; v, real speed; U_A, motor voltage. (a) Detailed block diagram. (b) Transfer function of the closed loop.

- The necessary limitation of acceleration and jerk magnitudes is ensured by means of a low-pass filter, which changes the commanded speed v_c into a modified signal $\tilde{v}_c$ containing no sudden speed changes.
- The proportional and integral (PI) controller ensures a sufficiently large static accuracy, i.e., small deviations between the real vehicle speed v and the filtered commanded speed $\tilde{v}_c$.
- The compensation filter put into the feedback loop ensures a sufficiently large distance from the stability limit and reduces the sensitivity of the control systems dynamics against the above-mentioned parameter change. It can be shown (cf. Hinman 1975) that a proper choice of the filter parameters leads to the transfer function of the closed control loop, which does not depend significantly on changes in parameters and other disturbances (cf. Figure 16.4):

$$G_v(p) = \frac{v(p)}{\tilde{v}_c(p)} \approx \frac{\omega_0^2}{p^2 + 2\omega_0 p + \omega_0^2} \qquad (16.1)$$

16.3.2 AUTOMATED POSITION CONTROL

The position control system has to ensure that the vehicle automatically follows commanded positions s_0 prescribed by the corresponding stationary control computer. One basic principle for solving this task is illustrated in Figure 16.5. The whole control system consists of an inner speed control loop in accordance with Figure 16.4, an additional outer speed control loop, and an outer position control loop (cf. Hinman 1973, 1975). If suitable parameters are used in these outer loops, then the band width of the inner speed control loop is kept higher, i.e, about $\omega_0 \approx 7$ radians per second, compared with that of the outer ones at $\omega_0 \approx 0.3$ radians per second. Therefore the dynamics of the inner loop can be neglected, i.e., the transfer function $G_v(p)$ given by eq. (16.1) can be approximated by

$$G_v(p) \approx 1 \text{ for } |\omega| \leqslant 0.3 \text{ radians/second}$$

This leads to the simple transfer function

$$G(p) = \frac{S(p)}{S_p(p)} = \frac{v(p)}{v_0(p)} \approx \frac{pK_v + K_S}{p^2 + p[K_v + 1/T_F] + K_s} \qquad (16.2)$$

for the closed position control loop depending on the low-pass filter time constant T_F and the controller parameters K_v and K_S alone. A suitable choice of these parameters guarantees a sufficiently fast and accurate operation of the combined position and speed control sytem. For further details concerning the control system shown in Figure 16.5, as well as regarding control systems developed in connection with other driving units, e.g. linear motors, the reader is referred to Brown 1973, 1974, Demag 1971, Fling and Olson 1972, Hinman 1973, 1975, and Ishii et al. 1974.

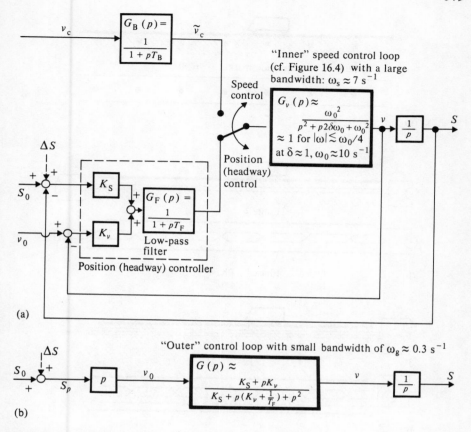

FIGURE 16.5 Position (headway) control system (cf. Hinman 1975): S_0, commanded position.

The implementation of any position control system requires special means for position measurement. The following two principles are used (cf. Figure 16.6):

1. *Use of discrete position markers.* The vehicle passes certain position markers located within or at the guideway (cf. Figure 16.6(a)) in the form of

- permanent or electromagnets
- inductive loops
- other electronic position transmitters

The onboard measurement system counts the markers or decodes the transmitted positions. Between two markers the position can be determined sufficiently accurately by counting the number of wheel rotations. If this method is also applied in

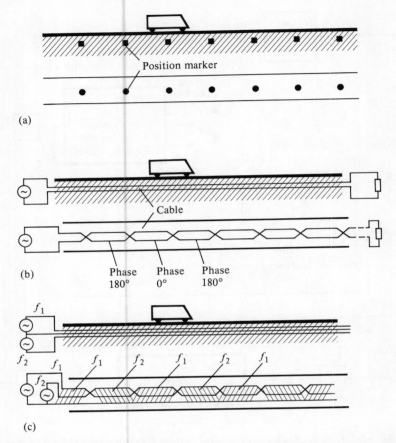

FIGURE 16.6 Methods of position measurement (cf. Burrow and Thomas 1976): (a) using fixed position markers; (b) using cables supplied with one frequency; (c) using cables supplied with two frequencies.

emergency braking situations, then the distances between two adjacent markers must be kept small in order to enable a sufficiently high accuracy. Here, the application of cheap permanent magnets, which do not require frequent maintenance is appropriate.

2. *Using electric cables embedded in the guideway surface.* If the data exchange between the stationary control computer and the onboard system is established by means of an electric cable embedded in the guideway, then this cable can be used to identify vehicle positions. Two wires are used, which cross each other at constant intervals. At these crossings, electric phase changes between 0° and 180° occur, which can be detected by the onboard system. Counting these phase changes and interpolating on the basis of the number of wheel rotations, gives position

measurements that are sufficiently accurate (cf. Figure 16.6(b)). As is illustrated in Figure 16.6(c), a similar effect can be achieved using cables supplied with different frequencies.

For details the reader is referred to Burrow and Thomas (1976).

16.4 AUTOMATIC LONGITUDINAL VEHICLE GUIDANCE: HEADWAY REGULATION

The distance regulating system has to fulfill two objectives (cf. Figure 16.1, level (3) longitudinal):

- minimize the danger of a rear-end collision (safety problem)
- maximize the vehicle throughput (capacity problem)

Two basic principles were developed for implementing distance regulation systems (cf. Figures 16.1 and 16.2): (1) asynchronous, and (2) synchronous control methods.

16.4.1 ASYNCHRONOUS HEADWAY CONTROL METHODS

This control concept is often called the vehicle-follower control principle (Anderson and Powner 1970, Brown 1972, 1973, 1974, Bender et al. 1971, Chu, 1974, Chiu et al. 1976, Candill and Garrard 1976, Garrard et al. 1972, Hesse 1972, Starr and Horowitz 1972, Thomas 1974). Its implementation requires the availability of information on the real state of the moving vehicle string, i.e., on the distances between vehicles, the vehicle speeds, etc., to derive driving instructions for the individual cars. Three different modifications can be distinguished:

    a. *Centralized control.* Here the string of vehicles is considered as one process. A central controller, i.e., a computer, has to be provided with the measured values of the distances and speeds. It determines the optimal driving instructions and transmits them to the individual vehicles. This principle, first studied by Levine and Athans (1966) leads to the theoretical advantage that the operation of the whole vehicle system can be optimized. However, it is not feasible for PRT systems because of the large number of feedback-loops required. Therefore, a partially decentralized principle was proposed.

    b. *Partially decentralized control principle.* This principle according to Figure 16.7(b), assigns one controller to groups of three vehicles. Here, the driving regime of a vehicle is determined by means of information on the moving states of the vehicle in front of it as well as of that behind it. However this principle also did not reach practical feasibility with respect to AGT systems. Therefore, a restriction is made in the following to decentralized control principles.

    c. *Decentralized control principles.* Here the driving regime of a car depends only on the distance to and the speed of the vehicle in front of it. Four modifications of this principle have to be taken into consideration:

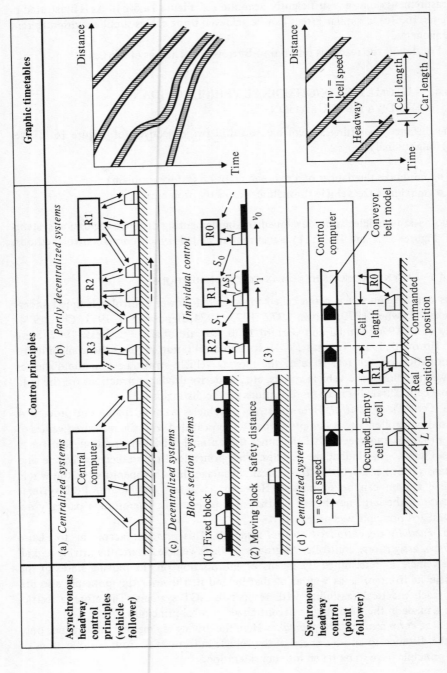

FIGURE 16.7 Survey of headway control principles.

i. The block section system used in railway signalling technology is characterized by dividing the guideway in block sections of a certain length. A vehicle is permitted to enter a block section if the vehicle in front of it has cleared this section. This method is characterized by the advantage that it can be implemented by means of well-proven technology. On the other hand, the block section length limits the number of vehicles that can pass a lane per time unit, i.e., the lane capacity. The use of very short block sections requires expensive installations. Therefore, the block section principle has been applied only in GRT systems operating at relatively long vehicle headways of more than 8–12 seconds (cf. Hinman 1975 and Figure 15.2).

ii. The moving block section principle modifies the above mentioned system in the sense that a vehicle is not allowed to enter a constant safety region moving together with the leading car, namely, in the form of a "red tail" (cf. Figure 16.7(c), part (2)). This principle, which is based on operation at constant spatial distances, allows the lane capacity to be increased. However, essential shorter headways in the range of a few seconds can only be achieved by using the principles that are normal in car driving on freeways, but that until now, have not been accepted by public transit authorities.

iii. Operation at speed-dependent distances. this control principle aims to ensure collision-free operation under the following conditions. If the leading car carries out a so-called "brick-wall stop", i.e., it stops at a certain moment of time with a stopping distance equal to zero, then the following vehicle should be able to stop collision-free. This requires, as is well known, the distance between the two vehicles to be proportional to the square of the speed of the following car. Such a driving regime has some similarities to that recommended by highway codes to car drivers. But even this principle, being unconventional for existing public transit systems like railways, does not overcome the headway barrier indicated in Figure 15.2, i.e., for achieving headways less than one second. This requires the following.

iv. Operation at difference-speed-dependent distances: here it is assumed that the leading car will stop within a stopping distance larger than zero, even in emergency situations, i.e., with a limited emergency deceleration. If the following car uses the same deceleration after a very short delay then it is — at least theoretically — possible to stop collision-free.

The implementation of the operating principles (iii) and (iv), which are of vital importance for enabling PRT vehicle operation at short headways (cf. Figure 15.2), requires the development of highly sophisticated distance regulation systems characterized by short response times and high accuracy.

The block diagram shown in Figure 16.5 may be considered a suitable control system if v_0 and S_0 represent the measured values of the speed and position, respectively, of the leading vehicle. The term ΔS denotes an additional safety distance. But the use of the position control system in accordance with Figure 16.5 as a distance regulation system, requires the consideration of a further design criterion, which concerns the interaction of a whole string of vehicles, i.e., the so-called queue stability problem well-known from highway accidents. A small thought experiment illustrates the nature of that problem and the way to solve it. Assume that the speed of the leading car is changing periodically according to

$$v_0(t) = A_0 \sin \omega t \qquad (16.3)$$

Then the speed

$$v(t) = A_1 \sin(\omega t + \varphi_1) \qquad (16.4)$$

of the following car will also carry out oscillating speed changes. The magnitude A_1 of these oscillations depends on the characteristics of the distance regulation system shown in Figure 16.5, i.e., on the transfer function $G(p)$ given by eq. (16.2) and the corresponding frequency response magnitude. For $p = j\omega$,

$$|G(j\omega)| = \frac{|V(j\omega)|}{|V_0(j\omega)|} = \frac{A_1}{A_0} \qquad (16.5)$$

The oscillation magnitude then takes the form

$$A_1 = |G(j\omega)|A_0 = \left| \frac{K_S + j\omega K_v}{K_S + j\omega[K_v + (1/T_F)] + (j\omega)^2} \right| A_0 \qquad (16.6)$$

(cf. eq. (16.2) and Figure 16.5). If there exists one frequency ω for which $|G(j\omega)|$ is larger than 1, then for this frequency the speed oscillation of the first vehicle given by A_0 will be amplified by the second car; the oscillation amplitude $A_1 > A_0$ will be amplified again by the third vehicle of the string leading to an oscillation magnitude $A_2 > A_1 > A_0$, and so forth. One can imagine that these amplified oscillations can lead to collisions a certain distance from the head of the vehicle queue. How can this queue instability be avoided? The design criterion takes a very simple form: it has to be ensured, by a proper selection of the control parameters K_S and K_v and the low-pass filter time constant T_F (cf. Figure 16.5), that the magnitude $|G(j\omega)|$ of the frequency response $G(j\omega)$ of the closed distance control loop remains smaller than or at most equal to 1:

$$|G(j\omega)| \leq 1 \text{ for } 0 \leq \omega < \infty \qquad (16.7)$$

Equation (16.6) illustrates that under these conditions speed oscillations of the leading car cannot be amplified by the following vehicles; they will be kept constant (for $|G(j\omega)| = 1$) or absorbed (for $|G(j\omega)| < 1$) (cf. Hinman 1975).

The following question must now be addressed: What engineering means are needed to implement a high performance asynchronous distance regulation system? Two different principles may be distinguished:

1. *Direct communication between two vehicles.* Here a direct measurement of the distance between two vehicles is required (cf. Figure 16.8(a), and Demag 1971). It is assumed that, for PRT systems characterized by short headways, the leading

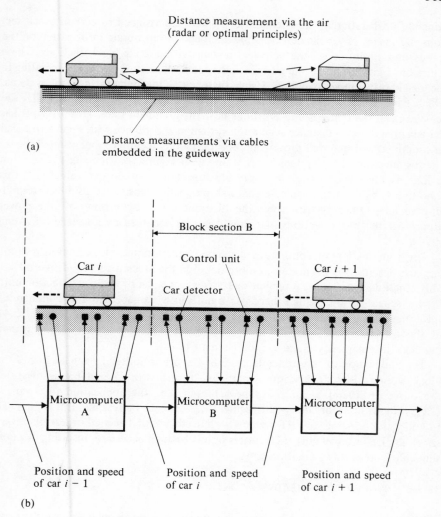

FIGURE 16.8 Asynchronous headway control methods: (a) using onboard systems; (b) using stationary systems (computers).

vehicle can transmit a signal to the following vehicle via the air or a cable embedded in the guideway, permitting the onboard control system of the following car to identify both the distance to and speed of the vehicle in front of it. To ensure a sufficiently high reliability the onboard microcomputer control systems will be doubled. Systems of this type are considered suitable for operation at headways of one second or less (cf. Demag 1971, and Section 19.2).

2. *Communication between two vehicles via stationary microcomputers.* This

principle uses stationary microcomputers, which are assigned to certain block sections. By means of vehicle detectors the microcomputers obtain information on the speeds and positions of the vehicles moving within their block sections. These values are transmitted via the adjacent microcomputers to the vehicles operating in the neighboring block sections. Using these data, the onboard control systems, which have, for example, structures as shown in Figure 16.5, will ensure that the vehicles keep their speeds and positions near the nominal values. This principle has the advantage that a distance regulation system in accordance with Figure 16.5 can be combined with the well-proven block-section safety technology mentioned above. The distance regulating system in according with Figure 16.5 remains in operation as long as there is at least one empty block section between two vehicles. If this condition is violated then an emergency stop signal is transmitted to the corresponding vehicle. Other systems solve this blocking effect by turning off the power supply for the block sections that should not be occupied by a vehicle (cf. Allen 1974).

The combination of continuous distance regulation and block-section technology just described is illustrated in more detail in the block diagram of Figure 16.9. This block diagram shows a further safety device, i.e., an over-speed detection unit. If the speed of the vehicle becomes significantly larger than the commanded speed because of a failure of the speed controller (in accordance with Figure 16.4), of the driving motor, or for other reasons, then the over-speed safety system switches the emergency brakes on (cf. Hinman 1975).

The application of the principle described above is limited by the lengths of the block sections. For block-section lengths of 20 meters or so, minimal headways of the order of several seconds can be achieved, i.e., the application to advanced GRT systems in accordance with Figure 15.2 seems to be feasible. If it is intended to apply the principle to PRT systems with headways of 1 second or less, then very short block-section lengths of 10 meters or less become necessary, resulting in a considerable increase of installation costs.

16.4.2 SYNCHRONOUS HEADWAY CONTROL METHODS

The problems of queue stability described in the previous paragraph, as well as certain questions of traffic flow control (cf. Section 16.5), motivated the introduction of a second basic distance-regulation principle, which is known under the names point follower control, moving target control and moving cell principle (Bender and Fenton 1972, Boyd and Lukas 1972, Burke and Ormsby 1973, Dietrich et al. 1975, Fenton et al 1974, Garrard and Kornhauser 1973, Hinman 1975, Kornhauser et al. 1974, Morag 1974, Rumsey and Powner 1973, Wilkie 1970, Whitney and Tomizuka 1972).

The basic idea of this control principle is to use a certain reference system for determining the nominal vehicle positions and speeds (cf. Figure 16.7(d)). These nominal values are generated by a model in the form of a computer program or a

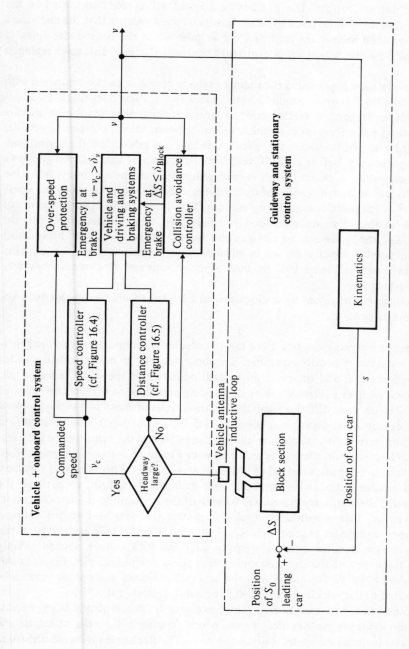

FIGURE 16.9 Principle of a headway control system with fail-safe features.

special electronic device in such a way that collision-free operation of the whole vehicle string is feasible. The position and speed values are transmitted to the onboard control systems (cf. Figure 16.5), which have to ensure that the real values of positions and speeds are kept as close as possible to the commanded ones. If these tasks can be solved by the onboard systems, the safe distance keeping is ensured.

The model used for defining the required vehicle positions can be compared with a "conveyor belt" moving parallel to the guideway (cf. Figure 16.7(d)). This conveyor belt model carries certain points or targets at equal distances, which indicate the positions prescribed for the vehicle heads. Between these points or targets lies a so-called cell, which reserves the space needed for one vehicle and the corresponding safety distance. Before an AGT vehicle starts its trip, it has to be assigned to a special cell, point, or target, which it must follow for the whole journey. When the speed changes so do the lengths of the cells, i.e., they are proportional to the nominal speed of the individual guideway sections. This illustrates an essential feature of the synchronous distance regulation principle: the time distances between the vehicles, i.e., the headways, are always equal and the travel time needed for reaching a certain point is exactly known in advance. This is illustrated by the graphical schedules shown in Figure 16.7 for both asynchronous and synchronous headway control principles.

Two different ways can be distinguished for implementing the synchronous distance control concept.

1. *Direct computer control.* Here the so-called cells, points or targets are generated in a stationary control computer. The computer calls up the individual vehicles in a cyclic manner and transmits the normal position and speed values via certain data links, such as guideway cables and vehicle antennas, to the onboard control system. The onboard distance regulation system can be designed in a similar manner to that described for asynchronous control (cf. Figure 16.10(a)). The assumed data connection between the stationary control computer and the vehicle allows the distance and speed controllers in accordance with Figure 16.5 to be implemented in the stationary computers instead of in the onboard system. This leads to a very simple and reliable onboard system, but to an increased data flow rate between the vehicles and the stationary computer. Because of the availability of microcomputers it is expected that the distance regulation system will also be installed onboard using microcomputers as controllers (cf. Ishii et al. 1974). The applicability of this variant of synchronous control systems to GRT and PRT systems depends mainly on the frequency of the connections between computer and vehicles. For the short headways of the PRT-type systems the nominal positions and speeds have to be transmitted to the vehicles every 0.1–0.5 seconds (cf. Ishii et al. 1974).

2. *Hardwired reference system.* The basic idea of this principle is to generate along the guideway an electrical signal, which represents the cells or targets and which can be detected by the onboard system. The electronic system illustrated in

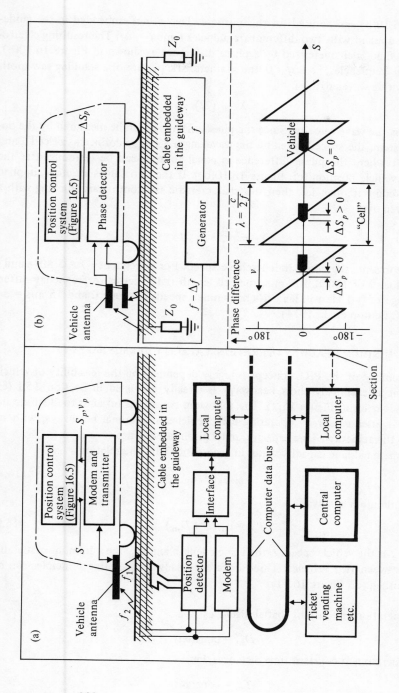

FIGURE 16.10 Synchronous headway control methods: (a) direct digital control (DDC); (b) electronic reference system.

Figure 16.10(b) can be used for this purpose. Two cables embedded in the guideway are supplied with two different frequencies f and $f - \Delta f$. The resulting electromagnetic field is characterized by a phase difference as shown in Figure 16.10(b). For equal frequencies, i.e., $\Delta f = 0$, the characteristic picture of a standing saw-tooth occurs, with a wavelength of

$$\lambda = c/(2f)$$

where c is the wave velocity. Since the phase difference can be measured by the onboard system, the space limited by the wavelength λ can be defined as a "cell", and the point where the phase difference is equal to zero can be introduced as the nominal vehicle position (cf. Figure 16.10(b)). If the two cables are now supplied with different frequencies, then the characteristic saw-tooth starts moving with a speed given by

$$v_c = \frac{1}{2}\frac{\Delta f}{f} c = \Delta f \lambda$$

which represents the commanded vehicle speed. For a frequency $f = 5$ MHz and a wave velocity of 150×10^6 m/s, the cell length is 15 metres. A frequency difference range of 0-1 Hz will lead to commanded speeds between 0 and 15 m/s = 54 km/h (cf. Fenton et al., 1974).

16.4.3 MINIMUM HEADWAY OPERATION: CAPACITY VERSUS SAFETY

The feasibility of the PRT concept is highly dependent on the feasibility of vehicle operation at extremely short headways as already discussed in Section 15.1 (cf. headway barrier in Figure 15.2). The following consideration aims to illustrate the possible consequences of operating PRT vehicles at minimal headways, with respect to the achievable lane capacity under given safety rules.

The maximal lane capacity varies in accordance with the relation

$$C_{max} = 1/T_m \tag{16.8}$$

on the minimum headway

$$T_m = v_0/(L + D_m) \tag{16.9}$$

where v_0 is the vehicle speed, L the vehicle length and D_m the minimum spatial distance between two vehicles. Three different headway operating principles can be distinguished (cf. Figure 16.11):

- operation at constant spatial distances, i.e.,

$$D_m = \text{constant}$$

- operation at constant time distances, i.e.,

$$T_m = \text{constant}$$

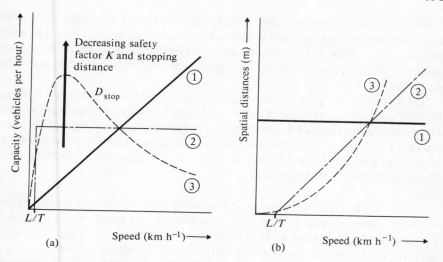

FIGURE 16.11 Relations between lane capacity and the various headway operating principles: 1, constant spatial distances; 2, constant headways (time distances); 3, constant safety factor.

- operation at a constant safety factor defined by

$$K = D_m/D_{stop} = \text{constant} \tag{16.10}$$

where D_{stop} is the minimal stopping distance needed for a collision-free stop.

All three headway operating principles are feasible using the asynchronous distance regulation concept described in paragraph 16.4.1. The synchronous principle, in accordance with paragraph 16.4.2 is, by definition, characterized by constant headway operation, leading to a constant line capacity and spatial distances D_m that increase proportionally with the nominal speed v_0 as shown in Figure 16.11 (curves 2). The first headway operating principle, i.e., D_m = constant, leads to unrealistic conditions if D_m is not chosen large enough to avoid a collision, even at the maximum speed. This is obviously the case in the so-called block-section technology sketched in paragraph 16.4.1 (cf. Figure 16.7(c)).

Of special interest is the third headway operating principle in accordance with equation (16.10), which leads to a maximum of the lane capacity curve within a certain speed range. This maximum depends significantly on the safety factor K, i.e., on the safety rules that can be accepted. Three conditions have to be considered:

$$K \begin{cases} > 1, \text{ i.e., } D_m > D_{stop} \text{ (collision-free stop)} \\ = 1, \text{ i.e., } D_m = D_{stop} \text{ (touching the frontcar with speed } v_R = 0) \\ < 1, \text{ i.e., } D_m < D_{stop} \text{ (collision with speed } v_R > 0) \end{cases} \tag{16.11}$$

Some PRT systems consider a safety factor of $K < 1$ to be acceptable if the speed v_R at the moment of collision is sufficiently small, e.g., $v_R < 10\text{-}15$ km/h (cf. Demag, 1971, Hesse 1972). This would be an extremely rare situation, and the passenger should be protected by auto-type passive safety systems like seat belts, shock absorbers, etc.

The minimum stopping distance D_{stop} represents the second parameter influencing the maximum of the capacity curve 3 in Figure 16.11(a). The minimum value of D_{stop} that can be assumed depends on two distinct factors: the capabilities of the distance regulation systems described above; and the acceptability of certain safety rules. In this respect three different cases have to be analyzed (cf. MacGean 1972, Hinman and Pitts 1974, Miller 1974).

1. *The brick-wall stop criterion.* This criterion assumes that the leading vehicle will stop immediately in an emergency situation, i.e., its speed will be reduced to zero without any delay and within a vanishing stopping distance. In such a situation the stopping distance of the following vehicle has to be computed from

$$D_{stop} = v_0 \tau + v_0 b/2r + v_0^2/2b \tag{16.12}$$

which contains the following three characteristic parts:

- the distance $v_0 \tau$ that the vehicle will travel during the reaction time delay τ of the emergency braking system, i.e., till the brakes start operating
- the distance $v_0 b/2r$ that will be travelled during the time interval needed to increase the braking deceleration from zero to the nominal value b considering a limited jerk r
- the breaking distance $v_0^2/2b$ needed to reduce the vehicle speed from the initial value v_0 to zero

Equations (16.12), (16.10), and (16.8), give the maximal lane capacity

$$\widetilde{C}_{max} = \frac{3600N}{K[\tau + (b/2r) + (v_0/2b)] + L/v_0} \tag{16.13}$$

expressed as the number of persons per hour per guideway, if N denotes the vehicle capacity in number of persons per vehicle.

2. *Emergency stop of the leading vehicle with limited deceleration.* Here it is assumed that a vehicle forced to stop in an emergency situation will use a limited deceleration value b_0. The stopping distance needed for a collision-free stop of the following car then takes the form

$$D_{stop} = v_0 \tau + v_0 \frac{b}{2r} + v_0^2 \left(\frac{1}{2b} - \frac{1}{2b_0}\right) \tag{16.14}$$

and instead of eq. (16.13),

$$\widetilde{C}_{max} = \frac{3600N}{K[\tau + (b/2r) + (v_0/2b) - (v_0/2b_0)] + L/v_0} \tag{16.15}$$

3. *Deceleration controlled stop.* If one assumes that the distance control system of the following car has the capability to identify the deceleration b_0 of the leading vehicle and to choose the same deceleration $b = b_0$ to stop its own car, then eqs. (16.14) and (16.15) become

$$D_{stop} = v_0 \tau + v_0 b/2r \qquad (16.16)$$

and

$$\tilde{C}_{max} = \frac{3600N}{K[\tau + (b/2r)] + L/v_0} \qquad (16.17)$$

The practical importance of these three criteria and conditions with respect to the feasibility of the PRT concept are illustrated by the numerical example given in Table 16.1. It can be observed that for the example considered the safety factor $K = 1$ leads to headways larger than 1 second. Thus, a breakthrough to headways smaller than 1 second will very likely require the acceptance of vehicle operation at safety factors $K < 1$, i.e., the occurrence of collisions at low speeds must be accepted in emergency situations.

In conclusion one may state that overcoming the headway barrier shown in Figure 15.2 requires two kinds of measures:

- the introduction of very advanced headway regulation systems
- the introduction of new safety rules that so far did not find acceptance for public transit systems; these new safety rules would have to tolerate rear-end collisions in emergency situations, and include the protection of passengers by passive systems like shock absorbers, seat belts, etc.

In view of this situation it seems to be an open question whether the headway barrier illustrated in Figure 15.2 can really be broken, i.e., whether the desired features of the PRT concept can be reached in real passenger service operation (Bernstein and Schmitt 1972, Hinman 1975, Hinman and Pitts 1974, Lobsniger 1974, MacKinnon 1975, MacGean 1972, 1974, Miller 1974, Rahimi et al. 1971, Stepner et al. 1972, Whitney and Tomizuka 1972).

16.5 TRAFFIC FLOW CONTROL

The control methods described in the previous section are needed to enable the automated guidance of the individual vehicles along the guideway and through branching points. But, what kind of control principles has one to take into account to control the operation of a system of vehicles at conflicting points, etc.? This question, which refers to level (2) of the control tasks hierarchy shown in Figure 16.1, is considered in the following.

The existing control concepts can be divided into three groups:

TABLE 16.1 Guideway Capacities and Minimum Headways for Different Safety Factors Obtained for a PRT System[a]

Safety criterion	Guideway capacities and minimum headways for safety factor $K = 1$			Safety factors K for minimum headway $T_m = 1$ s
	Maximum guideway capacity		Minimum headway T_m (s)	
	C_{max} (10^3 cars/h)	$\tilde{C}_{max}$ (10^3 persons/h)		
(1) Brick-wall stop (cf. eqs. (16.12), (16.13))	1.2	4.9	2.9	0.3
(2) Leading car stopped with limited deceleration b_o $\begin{cases} b_o = 1g \\ b_o = 0.6g \end{cases}$	1.6 / 2.1	6.5 / 8.3	2.2 / 1.7	0.4 / 0.5
(3) Controlled deceleration stop (cf. eqs. (16.16), (16.17))	2.4	9.9	1.5	0.6

[a] Vehicle capacity N, 4 persons; speed v_o, 50 km/h; emergency deceleration b, 0.5 g = 4.9 m/s^2; jerk r, 2.5 m/s^3; vehicle length L, 3 m; delay of emergency brake $\tau = 0.3$ s.

- synchronous (centralized) control
- asynchronous (decentralized) control
- quasi-synchronous (decentralized) control

(cf. Figures 16.1, 16.2, and Bender and Fenton 1972, Brinner 1973, Breeding 1970, Brown 1973, 1974, Boyd and Lukas 1972, Chu 1974, Eliassi-Rad and Fenton 1972, Hinman 1973, 1975, Kiselewich et al. 1976, Kornhauser and McEvaddy 1975, Munson et al. 1972, Stefanek and Wilkie 1973, Sarachik and Chu 1975, York 1974.)

16.5.1 SYNCHRONOUS TRAFFIC FLOW CONTROL

This control concept is based on the application of a synchronous headway control system, i.e., of the so-called moving-cell, moving-target, or point-follower control principle described in paragraph 16.4.2 (cf. Figures 16.7 and 16.12).

In using this headway control principle the occurrence of collisions at intersections and merging points can be avoided by means of the following (cf. Figure 16.12(a)):

- The whole network is divided into cells with a length proportional to the required speed.
- The speed "zero" would, therefore, correspond to a vanishing cell length. For this reason, stations lie in the form of "off-line stations" on by-passes, i.e., essentially outside the network.
- The division of the network in to cells has to be planned in such a manner that at merging points two cells unite into one cell (cf. Figure 16.12). If it can be guaranteed that at least one of these two cells is not occupied by a vehicle then collision-free merging of the two vehicle strings is feasible.

This control philosophy can be compared with a traffic system in which imaginary, i.e., really non existent, model vehicles or containers representing the "cells" are permanently moving on fixed routes according to fixed time tables. For example, a vehicle which is to be moved from station A to station B (cf. Figure 16.12(d)) has first to be assigned to an empty container, i.e., an empty cell. Let us assume cell N is chosen for this purpose. This cell is scheduled to pass the station exit ramp at time $t = t_0$. Such an assignment is admissable only if cell N', which is scheduled to merge with cell N in front of station B at time $t = t_1$, remains empty (cf. Figure 16.12(d)).

It becomes obvious that two different tasks need to be solved for the implementation of such a control concept:

1. *Automated cell reservation.* Before a vehicle starts its journey a driving route and a corresponding timetable have to be planned by selecting a suitable cell number. This "seat-reservation" procedure has to consider cell changes, e.g., the change

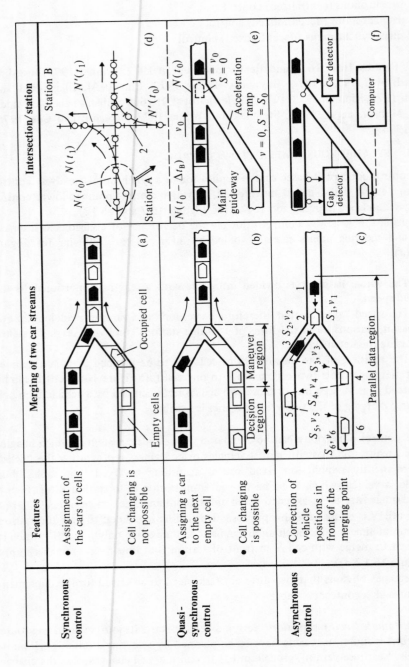

FIGURE 16.12 Survey of traffic flow control principles.

of the vehicle considered above (cf. Figure 16.12(d)) from cell N to cell N' at time $t = t_1$. Such cell reservation tasks can be solved only by a central computer, which is informed of the status of all cells "travelling" through the different parts of the network. This feature characterizes the synchronous control concept as a centralized control strategy. The solution of the cell reservation task is connected with the selection of optimal routes. This question has been assigned to level 1 of the control hierarchy (cf. Figure 16.1) and is, therefore, discussed later in Section 16.6.

2. *Automated merging control.* This second task can also be explained in terms of the above-mentioned example (cf. Figure 16.12(a), (b)). Let us assume that a vehicle waiting in station A is ready for departure and has been assigned to cell N, which is scheduled to pass the station exit ramp at time $t = t_0$. Then, the following merging control problem occurs: the vehicle has to be accelerated along the exit ramp in such a way that it meets with the assigned cell N with the prescribed cell speed, at time $t = t_0$. Thus, the initial state of the vehicle at time $t = t_0 - \Delta t_B$, which is characterized by its intitial position and vanishing values of speed and acceleration, has to be changed to the final state at time $t = t_0$, characterized by the final position, the speed $v = v_0$, and an acceleration of zero.

This state transition problem can be handled as an optimal control problem using well developed methods of modern control theory (cf. Rumsey and Powner 1973, for details). As a result, nominal time functions of the commanded position and the corresponding speed are obtained. The values of these time functions have to be transmitted from the stationary computer to the position/speed control system onboard the vehicle illustrated in Figure 16.5. Another possibility is to memorize the nominal time functions in the onboard control system. In this case the stationary computer has to transmit only a departure signal at the right time, i.e., at $t = t_0 - \Delta t_B$.

What assessment can be made of the synchronous control concept described here? One basic advantage is that two vehicle strings can merge collision-free and without any additional delay. Moreover, the trip-time between certain origin and destination stations is fixed and known beforehand. These advantages are paid for by the need to solve the above-mentioned cell reservation problem within a very short time. The solution of this problem requires the availability of a powerful real-time computing system. Moreover, certain additional waiting times at the stations may occur until an empty cell becomes available. This waiting time obviously depends on the existing travel demand, i.e., on the number of cells occupied by vehicles.

An unsolved problem connected with the introduction of the synchronous control concept is vehicle operation management in an emergency situation, i.e., in case of a stopped vehicle. It is obvious that a stopped vehicle is forced to leave its cell thus blocking the way for the following cells occupied by other vehicles. To avoid a breakdown of the vehicle operation in the whole network, a completely new plan for the cell routes and the cell time tables has to be set up within seconds. It is argued that this task cannot be solved. For this reason only a modified variant of the synchronous control principle, i.e., the so-called quasisynchronous control concept is considered feasible.

16.5.2 QUASISYNCHRONOUS TRAFFIC FLOW CONTROL

This concept no longer uses strict assignment of a certain vehicle to a specific cell, which the vehicle is not allowed to leave during the whole trip. A vehicle that is ready for departure in station A of Figure 16.12(e), for example, will be assigned by the station computer to the next empty cell without checking all the possibilities of conflicts that may occur in meeting with other occupied cells. Therefore, the following control principle is applied to avoid collisions during the vehicle merging process: a stationary control computer checks the state of the individual cells moving within a certain "decision region" covering certain parts of the two guideways in front of the merging point (cf. Figure 16.12(b)). If the computer discovers that the cells that will flow into each other at the merging point are occupied by vehicles, then the following two tasks have to be solved.

1. *The reassignment problem.* The computer has to determine new cells for those vehicles that are moving within those pairs of cells that will meet at the merging point. The new cells are chosen in such a way that

- collisions cannot occur
- the cell changing process is feasible with a minimum of additional vehicle acceleration and deceleration maneuvers (cf. Chu 1974)

2. *The state transition control problem.* This involves guidance of the vehicles from their old cells to the new ones. This position changing process has to be carried out while the vehicles are operating within the so-called "maneuver region", which covers certain parts of the two guideways lying between the "decision region" and the merging point (cf. Figure 16.12(b)). At the end of the maneuver region the vehicles must have reached their new cells.

Figure 16.13 illustrates the cell changing process as an optimal state transition control problem, which can be solved by means of advanced control theoretic methods (cf. Chu 1974, Stefanek and Wilkie 1973). As a result, nominal time functions are obtained of the commanded positions and speeds, which have to be transmitted to the onboard position/speed control system according to Figure 16.5. Another approach is to implement the position transition control algorithm onboard the vehicle and to solve only the assignment problem by means of the stationary computer. In this case the stationary computer has only to transmit the modified cell numbers to the corresponding vehicles. This requires the use of a

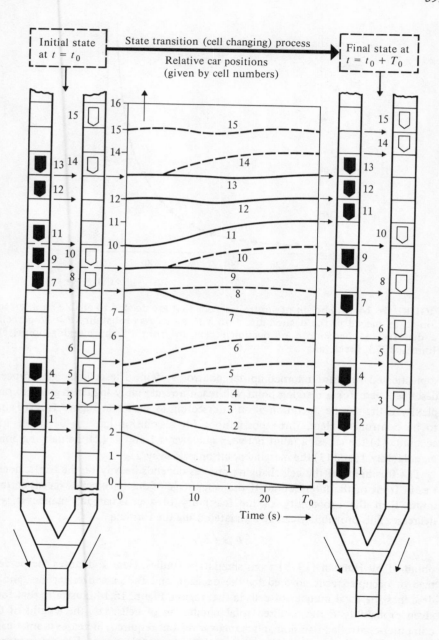

FIGURE 16.13 The cell changing process within the maneuver region (cf. Chu 1974, for details concerning control theoretical foundations).

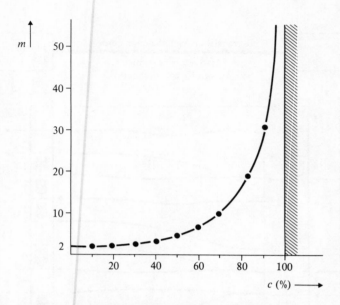

FIGURE 16.14 Number m of controlled cells that are needed to enable an intersection to be passed in the desired direction with a success probability $P \geqslant 96$ percent (c denotes the percentage of occupied cells in front of the merging point; cf. Hinman 1975, for details).

sophisticated onboard microcomputer control system. The cell-changing process described here for a merging point will reach a remarkably higher degree of complexity if the vehicle operation at an intersection as shown in Figure 16.12(d) has to be controlled. Here, three possibilities for a collision have to be taken into account: (1) the crossing point between guideways 1 and 2; (2) the merging point on guideway 1; and (3) the merging point on guideway 2.

The feasibility of the cell-changing process depends heavily on the traffic densities in front of the intersection. If the percentage of occupied cells exceeds certain limits then the probability of the feasibility of a cell-changing maneuver will decrease. This probability can be estimated using the formula

$$P \geqslant 1 - c^m$$

obtained by Hinman (1975) from simulation studies. Here, c denotes the percentage of occupied cells covered by the decision and the maneuver regions, and m describes the total number of cells in that region. Figure 16.14 illustrates that for a given probability P the required total number m of cells, i.e., the lengths of the computer-controlled decision and maneuver regions required, increases dramatically if the traffic density c exceeds values of the order of 80 percent. For a success probability P of 96 percent, i.e., where it is expected that no more than 4 percent of the vehicles can pass the intersection in the desired direction, the decision and maneuver regions would have to cover at least $m = 5$ cells if the percentage c of occupied cells reaches the value $c = 50$ percent. For $c = 90$ percent, i.e., where only every tenth cell is empty, the stationary computer has to check and control the vehicle positions in decision and maneuver regions covering at least 31 cells. This would lead to very long decision and maneuver regions of several hundreds of meters, which are not realizable. To avoid this problem the traffic density c has to be kept below these critical values. This represents a task of control level (1) (cf. Figure 16.1), i.e., of the route guidance system (cf. Section 16.6).

16.5.3 ASYNCHRONOUS TRAFFIC FLOW CONTROL

The asynchronous control concept does not use the synchronous headway control principle discussed in paragraph 16.4.2, but rather the asynchronous headway regulation methods described in paragraph 16.4.1. Two different control principles have to be taken into account.

1. *Traffic-light type control.* Here a procedure is applied that is similar to traffic-light control methods used for controlling the traffic flow through single intersections. For a certain period of time one guideway has "green", i.e., the vehicles approaching the merging point via that guideway are allowed to pass the merging point while the vehicles moving along the other guideway have to be stopped. Such an operating principle seems to be feasible for AGT systems characterized by relatively long headways as, for example, a certain class of GRT systems (cf. Figure 15.2). For GRT and PRT systems operating at very short headways of a few seconds or fractions of a second, the following disadvantages of the traffic-light type control method excludes its implementation:

- congestion similar to that occurring in ordinary street traffic cannot be avoided
- the basic advantage of GRT and PRT vehicles, being individually controllable, cannot be used

For these reasons the following principle is in general preferred:

2. *"Mirror-inverted" or "zip-fastener-type" control.* As described for the quasi-synchronous control concept, certain decision and maneuver regions are used for ordering the vehicles, in such a manner that they will pass the merging point at safe headways. For this reason the corresponding control computer deals with the vehicles operating on the two guideways as if they were moving on a single guideway, i.e., the computer transmits commands for correcting the vehicle positions and speeds that will ensure an occupied place on one guideway corresponds to an empty place on the other guideway (cf. Figure 16.12(c)). This obviously leads to a merging process, which has certain similarities to closing a "zip-fastener" or to reflecting an occupied place on one lane to an empty one on the other (Demag 1971).

Powerful methods have been developed to solve this control task, using again the position/speed control system in accordance with Figure 16.5 (cf. Brown 1974, Hinman 1975). As for the quasisynchronous control concept, a relation exists between the length needed for the decision and maneuver regions and the traffic densities occurring in front of the merging point or intersection. Results obtained from simulation studies by Hinman (1975) lead to the conclusion that a decision and maneuver region with a length of about 60 meters is needed if vehicle operation at headways of 4 seconds and speeds of about 54 km/h is required. The speed reduction occurring within the decision and maneuver region is on the order of 5 percent if a speed of about 55 km/h occurs in the guideway section behind the merging point, and the traffic volume reaches 90 percent of the maximal capacity (cf. Brown 1974, for details).

One basic advantage of the asynchronous control principle is that on-line stations are permitted though off-line stations are, in general, preferred in order to avoid significant reductions in line capacity. In the case of off-line stations the same merging problem occurs, as for both synchronous and quasisynchronous control (cf. Figure 16.12(f)), if a vehicle leaves the station and enters the main line. For this purpose, the station computer has first to identify an acceptable gap within the vehicle stream moving towards the station exit ramp. Then, it has to accelerate the waiting vehicle in such a way that it will meet with that gap at the prescribed speed. One observes that this merging control problem is from a methodological point of view very similar to that occurring at freeway on-ramps (cf. Chapter 5).

16.6 ROUTE GUIDANCE AND EMPTY-CAR DISPOSITION

The highest level of the control hierarchy according to Figure 16.1 concerns tasks that are of importance for the operation of the whole traffic system. Two problems are of special interest:

- Planning and control of driving routes, i.e., assigning a given traffic demand to the different network parts (cf. Boyd and Lukas 1972, Cunningham 1975, Demag 1971, Hesse 1972, Hinman 1975, Ford et al. 1972, Garrard and Kornhauser 1973, Roesler et al. 1974, Rubin 1975).
- Empty-car disposition, i.e., assigning the available vehicles to the different stations (cf. Hinman 1975, Waddell et al. 1974).

16.6.1 THE ROUTE GUIDANCE SYSTEM

The automated route guidance system has to solve two different tasks:

- determine the optimal routes connecting given origins and destinations, i.e., obtain the control programs for guiding the vehicles through the network (optimal route determination)

- automated guidance of the vehicles along the determined routes (automated route control)

The second of these tasks is considered first:

1. *Automated route control.* Consider the situation illustrated in the network shown in Figure 16.15(a). A vehicle is to be guided from the station of origin A to the target station B along route 1. How can the vehicle find its way automatically? To answer this question, two control principles have to be considered: centralized, and decentralized route control.

In the case of a centralized control system the central computer has to memorize the route connecting A and B as well as all other routes connecting all possible origins and destinations with each other. If an asynchronous longitudinal vehicle guidance system is used (cf. Section 16.4.1) then the central computer has to read the numbers of the individual vehicles approaching a branching point (cf. Figure 16.2, vehicle position 3). By means of the memorized routes the computer identifies the appropriate driving direction and transmits a corresponding command to the lateral vehicle guidance system (cf. point A shown in Figure 16.2). Such a procedure is not needed for the synchronous longitudinal vehicle guidance system discussed in paragraph 16.4.2, and the corresponding synchronous traffic flow principle described in paragraph 16.5.1. Here the assignment of the vehicle to a certain cell will already determine the vehicle guidance along a certain route. For the quasi-synchronous traffic flow control concept this statement holds only partly, since the vehicles can be forced to change their cells as illustrated in paragraph 16.5.2, and in Figure 16.13. Thus, it can happen that a vehicle has to change its cell in order to maintain the route prescribed for reaching the target station.

The centralized route control concept just described is suitable for application in connection with the centralized synchronous traffic flow control principle (cf. paragraph 16.5.1 and Figure 16.12(a)). In case of quasisynchronous, and especially asynchronous flow control (cf. paragraphs 16.5.2 and 16.5.3, and Figure 16.12(c), (b)), however, centralized route control has the following disadvantages:

- the central control computer must have the potential to memorize and process a large amount of data
- a very large data flow rate occurs between the central computer and the local units
- a breakdown of the central computer causes a breakdown of the whole route control system

In view of these difficulties, a decentralized control principle is preferred, in general, in which two distinct possibilities have to be taken into consideration:

- the onboard control system has stored the whole driving route
- the onboard system has stored only the address of the target station, while

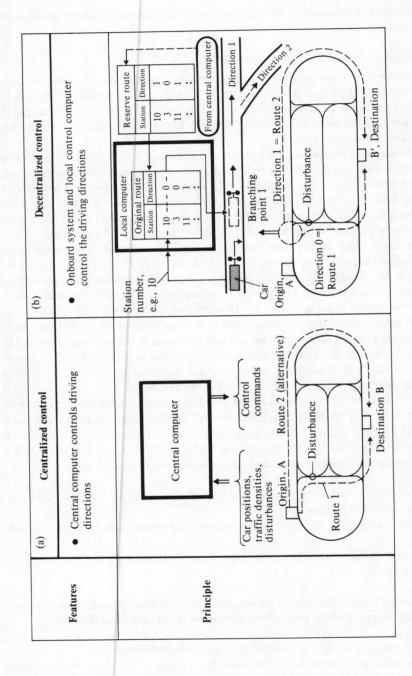

FIGURE 16.15 Principles of route control.

the driving directions that the vehicle has to follow at branching points are memorized in stationary computers located at those diversion points

In the case of the first principle the vehicle "knows" which driving directions it has to choose at branching points. The implementation of this principle would require the installation of an onboard microcomputer with a sufficiently large memory capacity. The second variant, on the other hand, requires a microcomputer with only a small memory capacity or else no onboard computer at all for storing the target station number. This number has to be transmitted to the onboard control system at the station of origin A by the ticket reading device or by the corresponding station computer. If the vehicle approaches a diversion point, this target station number will be transmitted to the stationary computer located there. As illustrated in Figure 16.15(b), this computer selects the direction that the vehicle has to follow to reach its target from a stored driving direction table, and transmits a corresponding command to the lateral vehicle guidance systems (cf. Figure 16.2, point B). This procedure has the advantage that in an emergency situation, e.g., in case of a blocked section of the original route 1, the stationary computer can obtain an alternative route 2 using a stored reserve driving direction table (cf. Figure 16.15(b)). Moreover, this table can be exchanged easily for another one by the central computer or by an adjacent computer. It is argued (cf. Demag 1971) that this route control concept, being applicable in connection with both quasisynchronous and asynchronous traffic flow principles, is the most promising one from the perspective of practical feasibility.

2. *Optimal route determination.* The selection of optimal driving routes involves two main tasks:

• minimizing travel times, thus representing the main objective of the individual PRT system users
• maximizing the overall network capacity by making equable use of the available network parts, thus representing a major objective of the public transit company or the travellers as a whole

Therefore, a method for determining optimal routes should consider both criteria. To what extent this can be done depends on the information available on the existing traffic flow conditions. Three different situations and their respective methods have to be taken into account (cf. Figure 16.16).

1. *Route planning.* This involves determining optimal driving routes using longterm mean values of the traffic demand occurring between the different origins and destinations. A task of this kind has to be solved in planning the routes and timetables of the moving cells in the case of the synchronous vehicle guidance and traffic-flow control system (cf. 16.5.1). Here, the problem of obtaining an optimal route for a specific vehicle and the corresponding pair of stations is reduced to the selection of

	Planning	Selection	Real-time computing
Driving route determination	• Planning driving route by means of long-term mean values of original-destination trip demands *Open-loop route control*	• Time- or traffic-responsive selection of an optimal route from a set of memorized (precomputed) routes *Adaptive open-loop control*	• Traffic-responsive computation of optimal routes *Closed-loop route control*
Empty car disposition	• Assigning a mean number of empty cars to each station using long-term mean values of traffic demand	• Time- or-traffic-responsive selection of an optimal empty car distribution plan	• Traffic-responsive distribution of empty cars considering the existing and a predicted traffic demand

FIGURE 16.16 Survey of methods for determining routes and empty car disposition.

a suitable cell (cf. Boyd and Lukas 1972). Routes that have been determined using long-term mean values are characterized by a limited adaptability to changing traffic conditions. They can become useless in emergency situations, e.g., in the form of blocked guideway sections. For this reason the following procedure has to be considered.

2. *Time- or traffic-responsive route selection.* Here it is assumed that a set of alternative routes is precomputed and memorized in the central computer or in decentralized ones. The time-dependent route selection takes into account the variations in the origin–destination relations that occur during the course of the day. Traffic-responsive route selection, on the other hand, provides a certain feedback leading to an adaptive control system. For this purpose, the corresponding computers have to be provided with the actual values of traffic volumes occurring in the different parts of the network.

3. *Real-time computing of optimal routes.* This concerns the solution of a dynamic routing problem using measurement data on:

- the travel times in the different network links

- the corresponding traffic densities.

Moreover, the changing tendencies of these variables have to be identified, which in turn requires the solution of a short-time prediction problem.

It is expected that the solution of this real-time optimizing problem by means of a central computer would lead to unacceptably high requirements concerning computing speed, memory-size and data transmission rates. Therefore, the development of decentralized routing algorithms, which can be implemented by local computers, represents an actual problem of fundamental research (cf. Cunningham 1975, Rubin 1975).

16.6.2 EMPTY-CAR DISPOSITION

The aims of the empty-car disposition scheme are

- to provide the stations with as many empty cars as necessary
- avoid unproductive empty-car trips

As for the routing problem, a distinction may be made between the following three categories of methods:

1. Planning of empty-car distribution, which is based on long-time mean values of the traffic volumes and travel times for the different origin–destination relations. As a result one gets

- an average number of empty cars per station
- certain rules prescribing the manner in which an empty-car shortage at a station should be remedied

2. Time- or traffic-responsive selection of empty-car distribution schemes, in which a set of empty-car distribution plans is precomputed and a suitable one is selected, depending either on the real traffic situation or on the time of day.

3. Real-time empty-car distribution strategies, which permit an optimal empty-car distribution plan to be determined in a real-time operation mode using information on:

- the waiting times of the passengers at the different origin stations
- the empty-car surpluses and shortages at the stations
- the number of cars that have already been assigned to passengers but that are still empty
- the real travel times between stations

The computer control system has first to deal with a prediction problem, which concerns the following questions:

- Which station will experience an empty-car shortage during the next few minutes and how large will it be?
- Where will empty-car surpluses occur during the same time interval?

From the results of this prediction task, certain surpluses have to be assigned to the corresponding shortages in an optimal manner. Here, the need may arise for the introduction of additional vehicles into the network from the depot.

To solve this dynamic assignment problem, it seems useful to employ a spatial decomposition of the following form:

- The whole network is divided into sub-networks, each containing only a relatively small number of stations. It is the task of the local computers to handle the empty-car distribution problem for these parts.
- If an empty-car shortage cannot be removed by local computers on the sub-network level, then an empty-car exchange between the subnetworks has to be induced by the central computer.

The feasibility of this concept could be adequately tested by means of simulation studies. Nevertheless, it should be stated that the development of robust and powerful real-time methods of empty-car disposition is still a subject of fundamental research (cf. Ford et al. 1972, Hinman 1975, Waddell et al 1974, for details).

REFERENCES

Allen, T. E. 1974. The LECTRAVIA: an active-guideway personal rapid transit system. In J. E. Anderson and S. Romig, eds., Personal Rapid Transit II, University of Minnesota, Minneapolis, Minnesota, USA, pp. 501–516 (published by the University).

Anderson, J. H. and E. T. Powner. 1970. Optimal digital computer control of cascaded vehicles in high-speed transportation systems in the presence of measurement noise and stochastic input disturbances, Transportation Research, 4: 185–195.

Bender, J. G., R. E. Fenton, and K. W. Olson. 1971. An experimental study of vehicle automatic longitudinal control. IEEE Transactions on Vehicular Technology, VT-20 (4): 114–123.

Bender, J. G. and R. E. Fenton. 1972. Synchronous control for automated ground transport—some practical limitations. Western Electronics Show and Convention, Los Angeles, Vol. 16.

Bernstein, H. and A. Schmitt. 1972. Emergency strategies for safe close-headway operation of PRT vehicles. In J. E. Anderson et al., eds., Personal Rapid Transit, University of Minnesota, Minneapolis, Minnesota, USA, pp. 351–360 (published by the University).

Boyd, R. K. and M. P. Lukas. 1972. How to run an automated transportation system. IEEE Transactions on Systems, Man, and Cybernetics, SMC-2 (3): 331.

Breeding, K. J. 1970. Minor to major stream merging within an automated highway system. Transportation Research, 4: 301–324.

Brinner, R. E. 1973. Continuous synchronous longitudinal guidance of automated highway vehicles. PhD Dissertation, The Ohio State University, Ohio, USA.

Brown, S. J. 1972. Characteristics of a linear regulation control law for vehicles in an automated transit system. Report No. UMTA-MD-060011-72-1 from the, Applied Physics Laboratory, The Johns Hopkins University, Silver Spring, Maryland, USA. Contract No. CP 009/TRP 020, Washington, DC.

Brown, S. J. 1973. Design of car-follower type control system with finite bandwidth plants.

Seventh Annual Princeton Conference on Information Science and Systems, Princeton, New Jersey, USA, pp. 57–62.

Brown, S. J. 1973. Merge control in automated transit system networks. Intersociety Conference on Transportation, Denver, Colorado, USA.

Brown, S. J. 1974. Adaptive merging under car-follower control. Report No. UMTA-MD-06-0018-74-4 from the, Applied Physics Laboratory, The Johns Hopkins University, Silver Spring, Maryland, USA.

Brown, S. J. 1974. Design considerations for vehicle state control by the point-follower method. In J. E. Anderson and S. Romig, eds., Personal Rapid Transit II, University of Minnesota, Minneapolis, Minnesota, USA, pp. 381–389 (published by the University).

Burke, H. B. and J. L. Ormsby. 1973. Synchrotrac – an off-board control system for automated mass transit. Proceedings of the IEEE, 61 (5): 644–646.

Burrow, L. and T. Thomas. 1976. Measurement and communications in automated transport. Report from the Urban Transport Research Group, Warwick University, Coventry, UK.

Candill, R. J. and W. L. Garrard. 1976. Vehicle-follower, longitudinal control for automated transit vehicles. In Control in Transportation Systems. Proceedings of the IFAC/IFIP/IFORS Third International Symposium, Columbus, Ohio, pp. 195–209 (published by the International Federation of Automatic Control, distributed by Instrument Society of America, Pittsburgh, Pennsylvania, 1976).

Chiu, H. Y., G. B. Stupp, and S. J. Brown. 1976. Vehicle-follower control with variable-gains for short-headway automated transit systems. Proceedings of the Fourth Annual Intersociety Conference on Transportation, Los Angeles, California, USA.

Chu, K.-C. 1974. Optimal decentralized regulation for a string of coupled systems. IEEE Transactions on Automatic Control, AC-19 (3): 243–246.

Chu, K.-C. 1974. Decentralized control of high-speed vehicular strings. Transportation Science, November.

Cunningham, E. P. 1975. A dynamical model for vehicle routing in a two-way PRT network. Transportation Research, 9 (6): 323–328.

Demag. 1971. Systems analysis of personal rapid transit system Cabintaxi (CAT). Report by DEMAG Foerdertechnik, Wetter (Ruhr), FRG (in German).

Dietrich, E. et al. 1975. Dual-mode bus systems. A study by order of the Bundeministerium für Forschung und Technologie, Bonn, Friedrichshafen, Munich, Cologne, FRG (in German).

Eliassi-Rad, T. and R. E. Fenton. 1972. On computer control in the merging of automated vehicles. Paper presented at The Technical Conference Islands of Application, US Conference and Exhibition, Harumi Pier, Tokyo, Japan.

Eliassi-Rad, T. and R. E. Fenton. 1972. On the longitudinal control of ramp vehicles, Proceedings of the IEEE Vehicular Technology 23rd Annual Conference. New York: IEEE, pp. 102–106.

Fenton, R. E. 1970. Automatic vehicle guidance and control – A state-of-the-art survey. IEEE Transactions on Vehicular Technology, VT-19 (1): 153–161.

Fenton, R. E. et al. 1974. Synchronous longitudinal control for automated ground-transport theory and experiment. In Traffic Control and Transportation Systems, Proceedings of the Second IFAC/IFIP/IFORS World Symposium, Monte Carlo, Monaco, Amsterdam: North-Holland Publishing Co., pp. 285–296.

Fling, R. B. and C. L. Olson. 1972. An integrated concept for propulsion, braking, control, and switching of vehicles operating at close headways. In J. E. Anderson, et al. eds., Personal Rapid Transit, University of Minnesota, Minneapolis, Minnesota, USA, pp. 361–382 (published by the University).

Ford, B. M., W. J. Roesler, and M. C. Waddell. 1972. Vehicle management for PRT systems. In J. E. Anderson, et al., eds., Personal Rapid Transit, University of Minnesota, Minneapolis, Minnesota, USA, pp. 411–434 (published by the University).

Garrard, W. L., G. R. Hand, and R. Raemer. 1972. Suboptimal feedback control of a string of vehicles moving in a single guideway. Transportation Research, 6.

Garrard, W. L. and A. L. Kornhauser. 1973. Optimal control of automated transit vehicles. Transportation Research, 7: 125–144.

Garrard, W. L. and A. L. Kornhauser. 1973. Use of state observers in the optimal feedback control of automated transit vehicles. Transactions of the ASME, Journal of Dynamic Systems, Measurement, and Control, 95 (2): 220–227.

Hesse, R. 1972. Problems of full automation of the Cabintaxi operation and of the man/automatics interaction, Nahverkehrspraxis, 20 (4): 154–155 (in German).

Hinman, E. J. 1973. Command and control studies for personal rapid transit. Program Status 1973, Applied Physics Laboratory, The Johns Hopkins University, Silver Spring, Maryland, USA.

Hinman, E. J. and G. L. Pitts. 1974. Practical safety considerations for short-headway automated transit systems. In J. E. Anderson and S. Romig, eds., Personal Rapid Transit II, University of Minnesota, Minneapolis, Minnesota, USA, pp. 375–380 (published by the University).

Hinman, E. J. 1975. Command and control studies for personal rapid transit. Program Status 1974, Applied Physics Laboratory, The Johns Hopkins University, Silver Spring, Maryland, USA.

Ishii, T., K. Kashida, K. Kinoshita, and H. Takaoka. 1974. The control system of CVS using the two-target-tracking scheme. In J. E. Anderson and S. Romig eds., Personal Rapid Transit II, University of Minnesota, Minneapolis, Minnesota, USA, pp. 325–334 (published by the University).

Kästner, E. 1971. Possibilities for inserting electronically guided ground vehicles. Wissenschaftliche Zeitschrift der Hochschule für Verkehrswesen "Friedrich List" Dresden, GDR, 18 (4): 961–971 (in German).

Kiselewich, S. I., Y. M. Tong, and A. S. Morse. 1976. Vehicle management policies for automated transportation systems. In Control in Transportation Systems, Proceedings of the IFAC/IFIP/IFORS Third International Symposium, Columbus, Ohio, pp. 223–227.

Kornhauser, A. L., P. M. Lion, P. I. McEvaddy, and W. L. Garrard. 1974. Optimal sampled-data control of PRT vehicles. In J. E. Anderson and S. Romig, eds., Personal Rapid Transit II, University of Minnesota, Minneapolis, Minnesota, USA, pp. 359–365 (published by the University).

Kornhauser, A. L. and P. McEvaddy. 1975. A quantitative analysis of synchronous versus quasi-synchronous network operations of automated transit systems. Transportation Research, 9: 241–248.

Levine, W. S. and M. Athans. 1966. On the optimal error regulation in a string of moving vehicles. IEEE Transactions on Automatic Control, AC-11 (3).

Lobsniger, D. I. 1974. An analysis of minimum safe headway for no collisions. In J. E. Anderson and S. Romig, eds., Personal Rapid Transit II, University of Minnesota, Minneapolis, Minnesota, USA, pp. 391–398 (published by the University).

MacGean, T. J. 1974. Headway limitations for short-term people mover programs. In J. E. Anderson and S. Romig, eds., Personal Rapid Transit II, University of Minnesota, Minneapolis, Minnesota, USA, pp. 349–354 (published by the University).

MacGean, T. J. 1972. Some performance factors relevant to the evaluation of people movers. In J. E. Anderson et al., eds., Personal Rapid Transit, University of Minnesota, Minneapolis, Minnesota, USA, pp. 295–311 (published by the University).

MacKinnon, D. D. 1974. Technology development for advanced personal rapid transit. In J. E. Anderson and S. Romig, eds., Personal Rapid Transit II, University of Minnesota, Minneapolis, Minnesota, USA, pp. 57–64 (published by the University).

MacKinnon, D. D. 1975. Longitudinal control techniques for automated guideway transit. Paper presented at IEEE IAS Tenth Annual Meeting, Hyatt Regency Atlanta, New York, pp. 232–237.

Miller, C. A. 1974. Comment on 'Headway limitations for short-term people mover programs'. In J. E. Anderson and S. Romig eds., Personal Rapid Transit II, University of Minnesota, Minneapolis, Minnesota, USA, pp. 355–357 (published by the University).

Morag. D. 1974. Operating Policies for personal rapid transit. Report No. RDD-8-74-2 from US Department of Transportation, Office of the Secretary, Washington.

Müller, S. 1975. New simulation results and their influence on network planning and operation modes under synchronous operation. In Nahverkehrsforschung '75, Status Seminar II, Bundesministerium für Forschung und Technologie, Bonn-Bad Godesberg, FRG (in German).

Munson, A. V., H. Bernstein, J. R. Buyan, K. J. Liopiros, and T. E. Travis. 1972. Quasi-synchronous control of high-capacity PRT networks. In J. E. Anderson et al., eds., Personal

Rapid Transit, University of Minnesota, Minneapolis, Minnesota, USA, pp. 325-350 (published by the University).

Oshima, Y., K. Kikuchi, M. Kimura, and S. Matsumoto. 1965. Control system for automatic automobile driving. Proceedings of the IFAC Symposium on Systems Engineering for Control Systems Design, Tokyo, Japan, pp. 347-357.

Rahimi, A., L. P. Hajdu, and H. L. Macomber. 1971. Safety analysis for automated transportation systems. Paper presented at the Joint Automatic Control Conference '71, St Louis, Missouri, USA.

Roesler, W. J., M. B. Williams, B. M. Ford, and M. C. Waddell. 1974. Comparisons of synchronous and quasi-synchronous PRT vehicle management and some alternative routing algorithms. In J. E. Anderson and S. Romig eds., Personal Rapid Transit II, University of Minnesota, Minneapolis, Minnesota, USA, pp. 425-438 (published by the University).

Rubin, F. 1975. Routing algorithms for urban rapid transit. Transportation Research, 9 (4): 215-225.

Rumsey, A. F. and E. T. Powner. 1973. Digital-computer control of vehicles in an automated transportation system. Proceedings of the IEE (GB), 120 (10): 1267-1272.

Sarachik, P. E. and K. -C. Chu. 1975. Real-time merging of high-speed vehicular strings. Transportation Science, 9 (2): 122-138.

Starr, S. H. and B. M. Horowitz. 1972. A control theoretic approach to the analysis and design of vehicular convoys. Transportation Research, 6: 143-155.

Stefanek, R. C. and D. F. Wilkie. 1973. Control aspects of a dual-mode transportation system. IEEE Transactions on Vehicular Technology, VT-22 (1): 7-13.

Stepner, D. E., L. P. Hajdu, and A. Rahimi. 1972. Safety considerations for personal rapid transit. In J. E. Anderson et al., eds., Personal Rapid Transit II, University of Minnesota, Minneapolis, Minnesota, USA, pp. 457-472 (published by the University).

Strobel, H. and E. Kästner. 1970. On the automatic course control of ground vehicles. Wissenschaftliche Zeitschrift der Hochschule für Verkehrswesen "Friedrich List" Dresden, GDR, 17 (3): 613-622 (in German).

Thomas, T. H. 1974. Control techniques for PRT. Railway Gazette International, 130 (1): 15-17.

Waddell, M. C., M. B. Williams, and B. M. Ford. 1974. Redistribution of empty vehicles in a personal rapid transit system. In Traffic Control and Transportation Systems. Proceedings of the Second IFAC/IFIP/IFORS World Symposium, Monte Carlo, Monaco, Amsterdam: North-Holland Publishing Co. pp. 533-564.

Wilkie, D. F. 1970. A moving cell control scheme for automated transportation systems. Transportation Science, 4 (4): 347-363.

Whitney, D. E. and M. Tomizuka. 1972. Normal and emergency control of a string of vehicles by fixed reference sampled-data control. In J. E. Anderson et al, eds., Personal Rapid Transit II, University of Minnesota, Minneapolis, Minnesota, USA, pp. 384-404 (published by the University); and IEEE Transactions on Vehicular technology, VT-21 (4); 128-138.

York, H. L. 1974. The simulation of a PRT system operating under quasi-synchronous control. In J. E. Anderson and S. Romig, eds., Personal Rapid Transit II, University of Minnesota, Minneapolis, Minnesota, USA, pp. 439-447 (published by the University).

Zimdahl, W. 1971. Control problems of electronically guided motor cars, Dissertation, Technische Hochschule Darmstadt, FRG (in German).

17 International SLT Systems Experiences

The survey of control concepts and methods presented in Chapter 16 illustrated that a fairly well developed methodology for designing computerized control systems for new modes of urban transport has been created during the last five to eight years. But what is the status of the practical application of this methodology? Can this control methodology make a contribution to the creation of the new urban transportation systems, which are expected to open the door to a new age of urban transport as has been discussed in the Introduction and Chapter 3 (cf. Section 3.2)?

These questions are analyzed in this chapter with respect to SLT systems, while Chapters 18 and 19 summarize the corresponding international experience gained so far with GRT and PRT systems.

SLT systems, characterized by the features given in Section 15.1, and illustrated in Figure 15.2, have reached the status of passenger service in more than ten cases during the last eight years (cf. Table 17.1). They have been installed in (1) expanded airports, (2) recreation parks and expositions, (3) shopping centers, (4) a hospital, and (5) a suburban area. The currently most important area of application of SLT systems may be at expanded airports, where they can serve as the main ground transportation system carrying airline passengers between landside and air terminals, parking lots, and the hotel. This application area is considered first.

17.1 AIRPORT GROUND TRANSPORTATION SYSTEMS

In the following the experience gained in three characteristic airport installations is sketched briefly, i.e.,

- the shuttle system at Tampa Airport
- the loop system at Houston Airport
- the combined shuttle-loop system at Seattle-Tacoma Airport

TABLE 17.1 Survey of Selected SLT Systems in Operation

Installation site	SLT installation site	Year of introduction	Number of loops (L) and shuttles (S)	Main tasks	References
Airports	Houston, Texas, USA	1969	1 L	Transporting airline passengers between landside and air terminals (at Houston airport, moreover, hotel and parking lots)	OTA 1975, Mason 1971, Brux 1974
	Tampa, Florida, USA	1971	8 S		
	Seattle–Tacoma, Washington, USA	1973	2 L 1 S		
Recreation parks, expositions	Hershey, Pennsylvania, USA (Amusement Park)	1969	1 L	Transporting visitors through recreation and exposition areas	OTA 1975
	Valencia, California, USA (Magic Mountain)	1971	1 L		
	Sacramento, California, USA (California Exposition)	1975	1 L		
	Yatsu, Chiba Prefecture, Japan, System VONA (Amusement Park)	1973	1 L		Oku 1973, NTS 1974
Shopping centers	Pearl Ridge, Honolulu, Hawaii, USA	1975	1 S	Linking two shopping centers	OTA 1975
	Fairlane Town Center, USA	1976	1 S	Linking a hotel and a shopping center	
Hospitals	Ziegenhain at Kassel, FRG (System Cabinlift)	1976	1 S	Transporting persons and goods between two remote buildings	DHF 1975
Urban cities	Lille, France (System VAL)	1975	1 double tracked line (= 1 L)	Linking a remote university campus with the city center and railway stations	Ralite and Gabillard 1973, Frybourg 1974, Hughes 1974, RGI 1974

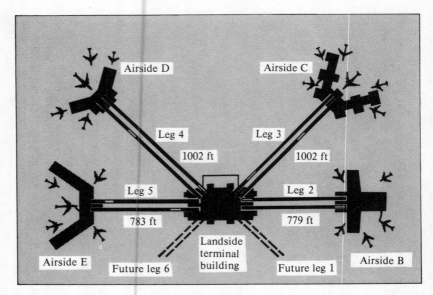

FIGURE 17.1 The automatic shuttle system at the Tampa Airport (cf. OTA 1975).

17.1.1 THE SHUTTLE SYSTEM AT TAMPA AIRPORT

In 1971 a new international airport was put into service in Tampa (Florida, USA), and was provided with an automated shuttle system for transporting passengers between the landside terminal building and four airside terminals located at distances of about 300 meters (cf. Figure 17.1 and photograph in Figure 15.2). Each airside terminal is connected with the landside terminal building via two elevated guideways. On each of the guideways one driverless vehicle with a capacity of 100 passengers operates at a maximum speed of 50 km/h with a travel time of 40 seconds. Each vehicle waits 30 seconds at the corresponding station and then starts its trip. With 25 forward and return trips per hour the line capacity reaches a value on the order of 2500 passengers per hour in each direction at a mean waiting time of 1.25 minutes. The cost of the whole system is estimated to be $8.25 million, 51 percent of which was used for providing the vehicles and the electrical and electronic systems. Only six employees are needed to operate the eight vehicles 24 hours a day. During the first four years of operation about four million passengers were transported; the current figure is about 37,000 per day (cf. OTA 1975).

For the year 1973 the availability of the system reached the value 99.96 percent for one guideway. If an error occurs, the corresponding vehicle is stopped automatically while the vehicle on the parallel guideway remains operating. This happened on average every 20 hours; the mean time for repair is on the order of seven minutes. Only very seldom — not more than once a week — are both vehicles forced to stop because of failures. In such cases the passengers are asked to leave the vehicles and to walk on special sidewalks. During the first four to five years of operation the vehicles covered a distance of 1.5 million kilometers without any serious accident. The number of passenger accidents is considerably lower than for systems such as elevators and escalators, which have similar usages (cf. Mason 1971, OTA 1975, for more details).

17.1.2 THE LOOP SYSTEM AT HOUSTON AIRPORT

The intercontinental airport of Houston (Texas, USA) opened in 1969, and has been provided with an automated loop transit system for transporting the passengers between the different terminals, the hotel, and parking lots. The vehicles operate in a tunnel 1 km long on guideways lying to the left and the right of the pedestrian subway. The whole loop is 2.06 km long and contains eight on-line stations; one at the hotel, two on the terminals and five at the parking lots. Three small vehicles are coupled to a train with a total passenger capacity of 18 seats and 18 standing places. These trains operate driverless at headways of about three minutes, resulting in a systems capacity of 720 persons per hour per lane (cf. Figure 15.2). The capacity can be increased to 2160 persons per hour per lane, if the feasible enlargement to 18 trains and the possible reduction of headways to one minute were used. The trains operate at a maximum speed of 12.8 km/h, which is further reduced when passing sharp turns. The introduction of the system was beset with certain initial reliability problems, which however were solved during the first years of operation. The six trains cover a distance of about 54,000 km per year without any serious accidents (cf. OTA 1975, for more details).

17.1.3 THE SLT SYSTEM AT SEATTLE-TACOMA AIRPORT

In mid-1973 a combined shuttle and loop transit system was put into operation at the international airport of Seattle-Tacoma (Washington, USA). It consists of two loops 1.1 and 1.2 km long and a shuttle line 300 m long. These guideways connect the landside terminal with two new remote airside terminals. Four vehicles operate on the two loops and one on the shuttle line. The vehicles provide space for 102 passengers but for only 12 seats. The lane capacity reaches a value of 1,800 persons per hour for the shuttle line and 4,800 persons per hour for each loop. The maximum speed is about 43 km/h, resulting in a mean travel time of about 1.8 minutes for the shuttle and 3.3 minutes for the loop. The system is used by six million passengers per year and no accident has been reported so far. The mean time between failures (MTBF) is estimated to lie on the order of one week and the mean time to repair is as low as six minutes (OTA 1975, Brux 1974, for more details).

17.2 SLT SYSTEMS IN URBAN CITIES

At present, there is only one automated guideway system with SLT features that has been put into passenger service in a real urban environment. This is the French system VAL (Villeneuve d'Asq à Lille), which has been installed in the city of Lille and was opened for passenger revenue service in 1975 (Ralite and Gabillard 1973, Frybourg 1974, Hughes 1974, RGI 1974). The VAL system provides a completely automated passenger service between university buildings located in Lille-East and a station of the French state railway SNCF eight kilometers distant. The system is operated 20 hours a day and at headways of one minute during rush-hour traffic and four minutes during the remaining time. The 8 km long double-tracked guideway with 8 stations and 30 vehicles is estimated to have cost 120 million Francs.

Two other systems have been installed for special tasks within cities: a shuttle system for connecting to shopping centers in Pearl Ridge (Honolulu, Hawaii) in 1975; and a shuttle system in the Fairland Town Center of Dearborn (Michigan), which serves for passenger transport between the Hyatt-Regency Hotel and a shopping center (cf. OTA 1975).

17.3   FURTHER SLT INSTALLATIONS

In addition to the SLT systems mentioned above, the following systems are in operation or under construction (cf. Table 17.1):

- Six SLT systems have been developed by the USA company Universal Mobility Inc., for passenger transport within expanded exhibition areas as, for example, EXPO '67 in Montreal, and the "California Exposition and State Fare" (1975) (cf. OTA 1975).
- Since 1973, an automated loop-transit system named VONA (vehicle of new age) has been operating in the Yatsu Amusement Park in the Chiba Prefecture (NTS 1974, Oku 1963). The underlying system concept has features of a GRT system.
- In the FRG the shuttle system Cabinenlift was developed and put into operation in a hospital in December 1975. This system is a simplified version of the PRT system Cabinentaxi that is described in detail in Section 19.2 (cf. DHF 1975).

In conclusion it can be stated that SLT systems could be successfully applied to the following categories of transportation tasks:

- passenger transport within expanded airports and recreation areas (Mason 1971, MPC 1967, NTS 1974, Oku 1973, OTA 1975)
- connecting the separate buildings of universities and hospitals with each other or with distant railway stations (Hughes 1974, RGI 1974, DHF 1975)
- passenger transport between two centers of activity like hotels and congress centers, parking lots and shopping centers (OTA 1975)

To date the potential of SLT systems for solving real urban transport problems has only partly been tested. The French VAL system may be considered the first

SLT system that demonstrates SLT concepts can be implemented, not only under the specific conditions of an airport, a recreation area, etc., but also fulfilling real urban transportation tasks in cooperation with other conventional public transit systems.

REFERENCES

Brux, G. 1974. The Seattle-Tacoma airport in the USA, with an automated guideway system. Der Stadtverkehr, 19 (4): 136-137 (in German).
DHF. 1975. The cabinlift used as a linking elevator in a hospital. Deutsche Hebe- und Fördertechnik, No. 6 (in German).
Frybourg, M. 1974. The new systems of urban transport. Revue Générale des Chemins de Fer, 93 (3): 127-154 (in French).
Hughes, M. 1974. Automated rapid transit in Lille. Modern Railways, 31 (305): 72-73.
Mason, R. 1971. Automated transit system reduces walking in expanding airport. Westinghouse Engineer, January: 8-14.
MPC. 1967. Report on the testing and evaluation of the transit expressway, MPC Corporation, Pittsburgh, Pennsylvania, USA.
NTS. 1974. NTS under development in Japan. Look Japan, 19 (217): 10-13.
Oku, T. 1973. Development of urban transportation systems in Japan. Japanese Railway Engineering, 14 (3-4): 4-7.
OTA. 1975. Automated guideway transit — an assessment of PRT and other new systems. United States Congress, Office of Technology Assessment, Washington, DC (published by US Government Printing Office).
Ralite, J. G. and R. Gabillard. 1973. The new automated subway of Villeneuve d'Ascq-Lille (the VAL). Rail International, March: 497-514 (in French).
RGI. 1974. VAL may be the world's first fully automated public transport. Railway Gazette International, 130 (1): 27-29.

18 International GRT Systems Experiences

GRT systems, characterized by the features described in Section 15.1 (cf. Figure 15.2), are in the process of development in the USA, Japan, the FRG, France, and the UK. A relatively large number of GRT systems has reached the state of prototype construction and large-scale experiments (cf. Aston et al. 1972, Barsony 1977, Corbin 1973, Dobler and Rahn 1975, DOT 1974, Elias 1974, Frederich and Müller 1975, 1976, Hamada 1975, Hillmer and Uyanik 1976, LTV 1974, MacKinnon 1974, Marten et al. 1977, Müller and Frederich 1974, Nickel 1976, Oku 1973, Russell 1974, Wienecke 1975). However, only three systems have been opened for passenger service:

- the Airtrans system at Dallas–Fort Worth airport, USA
- the Morgantown system, USA
- the KRT (Kabe Rapid Transit) system in Japan

The Airtrans and Morgantown systems were the first GRT installations put into passenger service. They are analyzed in detail in the following.

18.1 AIRTRANS AT DALLAS–FORT WORTH AIRPORT*

18.1.1 CASE HISTORY

The Dallas–Fort Worth airport is located midway between the cities of Dallas and Fort Worth, Texas. The boundaries of the airport encompass over 18,000 acres, stretching approximately nine miles from north to south, and eight miles from east to west (Figure 18.1).

The passenger terminal complex at the airport consists of 13 semicircular buildings; the aircraft parking positions are along the exterior perimeters of the buildings, while the passenger access roadways and parking facilities occupy the interior areas of the semicircles. Five of the thirteen terminals in the airport master plan have been completed and are currently in operation (Figure 18.1).

*Based on a case description specially prepared by D. M. Elliott, Director of Engineering, Dallas–Fort Worth Airport, Texas, USA.

FIGURE 18.1 Aerial view of the Dallas-Fort Worth airport and the airport transportation system (Airtrans).

The need for an efficient airport transportation system was recognized early in the planning for the airport, to enable the huge widely separated terminals to function together as an integrated facility. Feasibility studies indicated that an automated system would be financially superior to conventional methods (buses and trucks); thus the Dallas/Fort Worth Airport Board of Directors elected to install an automated transportation system to carry passengers and cargo between the terminals and other airport facilities.

Since no such system was then in existence, the Airport Board set about sponsoring its development. Conceptual proposals were requested from the transportation industry in mid-1968. From these proposals, two companies (Varo Incorporated, and the Dashaveyor Company) were selected to assist the Airport Board in the development of Airtrans (an acronym for *Air*port *tran*sportation *system*).

In their technical studies, both Varo and Dashaveyor concluded that a multi-purpose automated system was entirely feasible at the Airport. Also, both firms identified several alternative system configurations for deployment at the Airport, representing varying levels of sophistication and service, and, of course, different costs. Based on these studies, the Airport Board proceeded to define, in detail, the type of Airtrans service that would be required at the Airport. This system definition task began in early 1970.

Because of its complex nature, Airtrans proved to be difficult to analyze by manual methods. Thus an early task in the system definition was to develop a computer simulation of the Airtrans operation, for use as an analytical tool. The first step was the actual design and programming of the model, which was done in IBM GPSS 360 language. A second step was the development of ridership data. For this, a projected Airline Flight Schedule previously developed for the Airport was used to develop an origin–destination matrix of transferring passengers for each five minute increment of the Airport's "design day" (the average day of the peak month in the year 1975). After developing the transfer passenger demand matrix, analyses were conducted to identify and assign other potential transit riders, including Airport terminal employees, who park at remote parking lots and ride Airtrans to and from work. To obtain the best possible projection of employee ridership, questionnaires were distributed to the airlines, and to 26 major airports in the United States, requesting detailed employment forecasts. From these questionnaires, the projected demand for employee ridership was developed, and added to the previously developed passenger demand, to complete the total Airtrans ridership data. The computer simulation was then used extensively to evaluate alternative routing schemes, vehicle operating characteristics, and other passenger and employee system variables. Finally, the automated movement of cargo was evaluated in the Airtrans computer simulation, simultaneously with the passenger and employee system operation. The results of this combined simulation substantiated the feasibility of Airtrans as a total airport transportation system.

The culmination of the Airtrans system definition was the preparation of comprehensive performance specifications. These specifications prescribed the various system requirements for Airtrans, such as capacities of people and cargo to be transported, allowable trip times, and permissible train spacing deviations. Also, the specifications included detailed subsystem criteria, such as the size of the vehicles and their operating characteristics, the guideway configuration, system safety standards, power distribution provisions, and system reliability and system maintenance requirements. The performance specifications were the basis for the detailed design and construction of Airtrans.

Competitive cost and design proposals for Airtrans, based on the performance specifications, were received from Varo and Dashaveyor in February of 1971. Unfortunately, both manufacturer's cost proposals exceeded the available funds. As a result, additional competitive cost and design proposals were solicited, and in July of 1971, the Airport Board awarded a contract for Airtrans to LTV Aerospace Corporation. The amount of the contract was $34 million; this covered the design, construction, and testing of the system. A three-year maintenance contract was included at an additional cost of $4.9 million.

18.1.2 IMPLEMENTATION AND OPERATION

In the following a brief survey is presented of the essential features of Airtrans with regard to (1) system requirements, (2) system design, (3) system components, i.e., guideway and vehicles, (4) automation and operation.

18.1.2.1 System requirements. The major requirements for Airtrans contained in the performance specifications may be summarized as follows:

- transport 9000 people/hour, with separate trains for passengers and employees
- transport 6000 bags/hour
- transport 70,000 pounds of mail/hour
- transport containerized trash and supplies
- 20 minute maximum trip time (including waiting time) between terminals, with a 10 minute average trip time
- 30 minute maximum trip time (including waiting time) to remote parking lots, with a 20 minute average trip time
- 30 minute (maximum) delivery of baggage and airmail
- maximum 30 second deviation on train headways
- 30 minute (average) restoration time in case of system breakdown
- 500 hour mean time between failure (MTBF) for each vehicle
- 50 hour mean time between failure (MTBF) for non-vehicle systems
- one-year warranty on all components
- 30 year design life (20 years on vehicles)
- expandable to meet future airport needs

18.1.2.2 System Design. The first stage Airtrans includes provisions for moving passengers, employees, interline baggage, mail, supplies, and trash. A diagram of the Airtrans guideway network is shown in Figure 18.2. Major features of the guideway include two main lines running north and south through the passenger terminal complex, with loop guideways circling through each terminal building and the remote parking lots. At intervals, guideways cross the Airport highway system to provide for turnaround. Circulation on the guideway network is oneway, in a counterclockwise direction. The total guideway system is 67,697 ft (13 miles) long, extending over a straight-line distance of 3.2 miles. Airtrans vehicles travel over this guideway on a series of dedicated routes by switching at predetermined locations in the guideway. There are five routes for passengers, two for employees, one for baggage, and two for mail (Figure 18.3). There are also four routes for supply and trash movements. These routes connect the four Airport terminal buildings, two remote parking lots, the Airport Hotel, and the Transportation Center.

18.1.2.3 System components.
Guideway. The Airtrans guideway consists of a reinforced concrete running surface 8 inches thick, with parapet walls 5 inches thick, and 24 inches high. 80 percent of the Airtrans guideway is on grade; the remaining 20 percent is elevated on precast, prestressed beam and column bridges, as shown in Figure 18.4. To harmonize with other Airport facilities, the guideway is constructed entirely of light-brown, sandblasted concrete. The Airtrans vehicles are supported by the horizontal running surface, while the parapet walls provide vehicle guidance, power rail support and switch mounting. Also, the parapet walls form a visual screen for the vehicle wheels and underchassis, enhancing the appearance. Grade changes on the guideway vary from -8 to $+4$ percent.

Power Distribution. Electric power is distributed to the Airtrans vehicles from 14 substations, strategically located throughout the Airport. In the event of a failure of any substation, the adjacent substations will assume the load and permit uninterrupted operation. Three-phase, 489 V a.c. electric power is supplied to the vehicles through three copper-clad steel rails. The three power rails are mounted between a safety ground rail, below, and a signal rail, above. All five rails are mounted on the side of the concrete parapet wall on plastic insulators, and the power rails are recessed, to prevent inadvertent contact.

Vehicles. The Airtrans vehicle fleet consists of 51 passenger vehicles and 17 utility (cargo) vehicles (Figure 18.4). The passenger vehicles operate either singly, or as two-car trains. the utility vehicles operate only as single vehicles. The Airtrans passenger vehicles are 21 feet long, 7 feet wide, 10 feet high, and weigh 14,000 pounds empty. Passenger access and egress is from the side through automatic, bi-parting doors. In addition, there are also emergency exit doors on each end of the vehicles. The bodies of the Airtrans vehicles are fiberglass, with a laminated

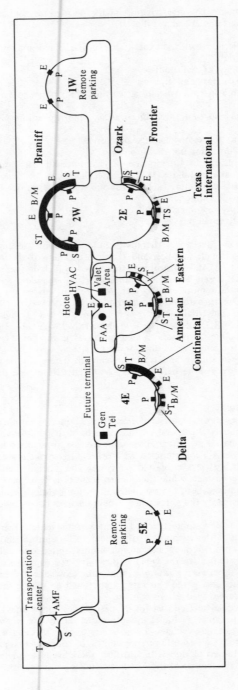

FIGURE 18.2 The guideway configuration of the Airtrans system: P, passenger station; E, employee station; B/M, baggage and mail station; T, trash station; S, supply station.

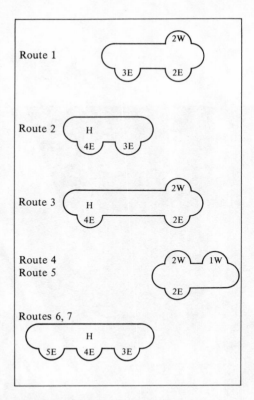

FIGURE 18.3 Passenger routes (routes 5 and 7 are employee routes).

acrylic exterior surface. The acrylic contains an integral color for appearance, and is a durable, low-maintenance material. Inside, the passenger vehicles feature upholstered seats for 16 people, and standing room for up to 24 more, giving a capacity of 40 passengers. For passenger convenience and safety, the vehicles are equipped with two-way radio communication equipment, and an automatic announcement system to give advance notice of station arrivals. When used to transport Airport employees, the vehicles are turned end-to-end in the guideway, to properly present the entrance doors to the employee stations, which are on the opposite side of the guideway from the passenger stations.

LTV's design approach for Airtrans was to make maximum use of off-the-shelf components where possible, to minimize cost and procurement lead time. To this end, the vehicles are supported by foam-filled truck tires mounted on standard truck axles, and feature a commercial air suspension system, modified to be self-leveling. The vehicle brakes are pneumatically activated, drum type, truck units. On each corner of the vehicles are power and signal collector assemblies, featuring copper/graphite brushes to collect the 480 V a.c. power for propulsion, and 48 V d.c. signals for vehicle control. There is also a ground brush, for electrical grounding of the vehicle and vehicle presence detection. The a.c. power is rectified onboard each vehicle, and fed to a 60 horsepower d.c. motor, which drives the vehicle through a commercial truck differential. The maximum speed of the vehicles is 17 mph. A.c. power is also used to operate the vehicles' heating, air conditioning, and air supply systems.

Onboard, the Airtrans vehicles carry sophisticated electronic controls, which prescribe their speed, acceleration, braking, station stopping, and route through the guideway. A unique feature of the vehicles is their all-wheel steering, which provides precise rear-wheel tracking, to minimize the required width of the guideway.

FIGURE 18.4 Airtrans utility (cargo) vehicle transporting containers (cf. Airtrans passenger vehicle in middle photograph in Figure 15.2).

The Airtrans utility vehicles are similar to the passenger vehicles in many respects. Wheels, axles, suspension, propulsion, steering, and controls are identical on both vehicles. Also, the utility vehicles use the same mainline guideway as the passenger vehicles. Instead of passenger bodies, however, the utility vehicles are equipped with three container positions for transporting containers of baggage, mail, trash, and supplies between the various Airport facilities. These container positions consist of powered, chain-driven conveyor modules that support the containers in transit, and automatically load and unload them at the cargo stations.

The Airtrans containers are made of fiberglass-covered plywood, mounted on an aluminum structural framework. Each container can hold about 150 cubic feet of cargo; for example 60 full size suitcases, or 3000 pounds of trash or supplies. The baggage/mail and supply containers feature spring-loaded, roll-up doors for ease of access. Also, aircraft "LD-3" type containers can be transported on Airtrans without special adapters.

Stations. There are a total of 52 independent stations in the Airtrans system, located on different sidings along the guideway (see Figure 18.2). The passenger stations feature a glass-enclosed waiting platform, with automatic bi-parting doors that open with the vehicle doors. Access to the stations is gained by depositing a quarter (25¢) in the entrance turn-stiles. Inside, a static sign and map explains how to use Airtrans. As different vehicles arrive, their destinations are automatically displayed on lighted signs above the boarding doors.

Each of the four Airport terminals is also equipped with a baggage and mail station, and at least two trash and supply stations. At the baggage and mail stations, containers are filled by airline employees. Then, when the proper utility vehicle arrives, the filled containers are automatically moved to an elevator, lowered, and loaded aboard the vehicle. The reverse operation occurs at the receiving station. In this way, interconnecting baggage and airmail can be transported among the various terminals, and between the terminals and the on-site postal facility. Similarly, Airtrans carries concession supplies (food, magazines, gifts, etc.) from a remote warehouse to the terminals, and can remove trash from the terminals to an incinerator facility for disposal.

18.1.2.4 Automation and Operation.

Airtrans control system. Airtrans is a fully automatic transportation system — there are no drivers or attendants on any of the vehicles. The control system that accomplishes this is a marriage of two technologies: traditional, proven, train control equipment assures vehicle (and passenger) safety, while modern digital computers monitor and supervise the system's operation.

The Airtrans control system is composed of three subsystems: automatic vehicle protection (AVP), automatic vehicle operation (AVO), and central control (CC). The functions of these subsystems, which represent different levels of a control hierarchy,* can be summarized as follows:

Level 1   Central control (CC)

- System status monitoring
- Supervisory controls
   speed commands
   switch positioning
   route changes
   bunch control
- Station monitoring
- Power distribution monitoring and control
- Voice, video, data communications

---

*This hierarchy of control systems may be considered as a special implementation of the general task hierarchy illustrated by Figure 16.1 (cf. Section 16.1).

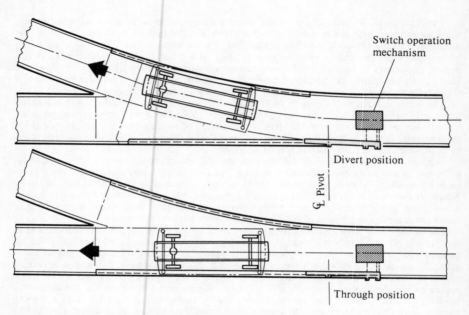

FIGURE 18.5 Typical guideway switch operation (cf. Figure 16.2, column 1).

Level 2   Automatic vehicle operation (AVO)

- Route control
- Position stopping
- Door controls
- Speed controls

Level 3   Automatic vehicle protection (AVP)

- Assures safe train spacing
- Safe switching
- Speed limits
- Vehicle safety systems

The AVP subsystem performs parts of the lateral and longitudinal vehicle guidance tasks described in Sections 16.2–16.4 (cf. Figures 16.2, 16.7). To assure reliable and safe switching, LTV developed a new switch that utilizes proven railroad technology, together with the all-wheel steering capability of the Airtrans vehicles. Basically, the switch consists of fixed and moveable entrapment blades, attached to the side of the guideway (Figure 18.5). To accomplish vehicle switching, the moveable entrapment blade is actuated by a railroad-type automatic switch machine, and moved into a position where it physically "captures" a small wheel attached to the vehicle steering mechanism, causing the vehicle to steer in the desired direction. This design provides positive, safe switching, as the vehicles are mechanically entrapped throughout the switch area. Also, the position of each switch is electrically interlocked with the Airtrans automatic control system to prevent any unsafe vehicle movements. There are a total of 33 diverge switches and 38 merge switches in the Airtrans system.

In the AVP, conventional railroad block control equipment governs normal system activities; it assures safe vehicle separation and switching, enforces maximum speed limits, and prevents unsafe operating conditions (such as vehicle movement with open doors). This time-tested, fail-safe equipment provides maximum confidence that Airtrans will conform to existing standards of public transit safety.

Vehicle operating safety is obtained through a five block control system. The guideway is divided into 708 blocks by insulators in the signal rail. A normal block is 90 feet long, except at the station sidings, where a closer vehicle spacing is desired; here the blocks are 45 feet long. At least one full block separates the vehicles under all conditions. At high speed, five blocks separate the vehicles. At 25 ft/s the minimum vehicle headway (time between trains passing a given point) is 18 seconds. The average system operating headway is approximately 30 seconds (cf. Figure 15.2).

Both railroad-type train control equipment and modern digital computers are used in the AVO. For example, vehicle speeds are prescribed by conventional cab signalling equipment that is interactive with the AVP. However, a computer onboard each vehicle (designed especially for Airtrans) designates the route for the vehicle to follow, directs it to the proper stations, and initiates recorded onboard announcements. Also, this same computer constantly monitors the vehicle's operating equipment, analyses and diagnoses malfunctions, and transmits this information to five larger wayside computers. These wayside computers accumulate operating data from all the vehicles, and transmit it to a large central computer, which formats messages, and displays information for the benefit of operating personnel.

The nerve center of Airtrans is the central control room, where all aspects of the system's operation are constantly monitored by Airport personnel (cf. Figure 18.6). Here, the location and status of all vehicles are continuously displayed on a lighted schematic of the guideway. Using the console, the central control operators can impose numerous supervisory overrides on the system, such as reducing the speed of a vehicle, changing individual switch positions, changing vehicle routes, and bypassing stations. Further, the central control operators monitor and control the power distribution system. They also have radio communication with all vehicles, and can watch activities at all stations by closed circuit television. Finally without requiring operator assistance, the central control computer performs top-level system management functions, such as preventing vehicles on a common route from "bunching".

Maintenance. The Airtrans maintenance area is located in the Transportation Center and contains the Airtrans maintenance building, the departure test track, and the vehicle "ready" track. The maintenance building includes ten bays for servicing vehicles, together with laboratories for maintenance of electronic equipment, and the control room for the departure test. Vehicles are removed from the guideway with aircraft-type tow tractors and towed to the service bays for maintenance after which they are towed back to the departure for reinsertion into the guideway. For retrieval of stalled vehicles, seven tow tractors are equipped with special steering wheels and presence detection brushes, and stationed at special sidings throughout the guideway.

The departure test track automatically checks vehicle equipment and performance prior to their entering the guideway for operation under automatic control. After completing a departure test, a vehicle may be stored in the "ready" track, with all power on, until it is dispatched into the system by central control.

18.1.3 OPERATIONAL EXPERIENCES

The Airtrans system was designed, constructed, tested, and deployed for revenue service in only 30 months. Construction of the system was accomplished by the prime contractor, LTV Aerospace Corporation, and four prime subcontractors responsible for: guideway, electrification, train controls, and cargo equipment. Contract administration was performed by the Dallas–Fort Worth Airport Staff, with consultant assistance for field inspection and safety.

Airtrans opened for revenue service on January 13, 1974, concurrently with the opening of the Dallas–Fort Worth Airport. For the first few weeks, only interline passenger service was provided, for 15 hours each day. As operating experience and reliability increased, service was extended to 24 hours a day, seven days a week.

Airmail service was initiated in March of 1974, and continued until October 1974. A total of over 30,000 tons of mail was carried during this period. In July of 1974, the employee and interline baggage/mail systems were tested in actual service for a short time. During these tests, it was established that an improved level of service, over and above that originally specified by the Airport, would be required. Thus, these two services were not continued.

FIGURE 18.6 Airtrans control center.

In late 1974, the entire Airtrans system was mobilized for integrated testing. Passenger service was operated "wet" (passengers transported) while the other services were operated "dry"; that is, the systems operated exactly as designed, but no cargo was actually transported. These tests showed that Airtrans had achieved in excess of 90 percent of specified performance. The deficiencies noted at that time appeared to be a direct result of inadequate system reliability; under the contract, full specified reliability was to be developed over the three-year contract maintenance period, during which the contractor was to perform the maintenance work.

Unfortunately, disagreements between the Airport Board and the system contractor about the system's performance and reliability prompted the contractor to withdraw its maintenance forces in March of 1975, forcing a complete system shutdown for ten days. Service was resumed under a six-month, interim maintenance agreement.

Again, in September of 1975, unresolved contractual differences between the parties resulted in another system shutdown, and the filing of litigation. After extensive negotiation, this litigation was settled out-of-court in December of 1975. Under the terms of the settlement agreement, LTV received an additional $7 million (total payment of $41 million) and the Airport assumed the complete responsibility for operating and maintaining Airtrans.

The system was re-opened for passenger service (under Airport operation) in January, 1976. Shortly thereafter, supply service was initiated. In March, 1976, after significant design and operational changes, employee service was activated.

Airtrans is now carrying all transferring passengers and remote-parked passengers at the Dallas–Fort Worth Airport. Also, it is carrying all terminal employees and flight crews to and from work each day. Finally, the system carries all concession supplies to the terminal buildings from a remote warehouse. These services operate 24 hours a day, seven days a week.

Since the Board assumed the total operation and maintenance responsibility for Airtrans, costs and operating experience have been carefully documented, in the belief that this information is crucial to individuals and agencies contemplating the installation of a system similar to Airtrans. In this regard, there are a total of 55 Airtrans vehicles in use; these vehicles operate throughout the entire guideway system. Since the initiation of employee service in April of 1976, the vehicle fleet has accumulated an average of 330,000 miles per month, or approximately 6,000 miles per vehicle per month.

In terms of ridership, the Airtrans passenger and employee system carries an average of 515,000 riders per month, or approximately 18,000 riders per day. On the cargo side, approximately 9,000 pounds of supplies are delivered to the terminals each day.

The Airtrans staff consists of 110 employees. These employees are assigned as follows:

- Management and administration   3
- Administrative services (purchasing)   9
- Operations   11
- Engineering   3
- Maintenance   84

This total of 110 compares with approximately 150 personnel performing the same tasks in mid-1975.

LTV has recently compiled and released its total costs for construction of the Airtrans system. These amount to a total of $53.5 million, as compared with contract payments of $41 million.

The operating and maintenance costs for Airtrans in 1977 were as follows:

- Operating and maintenance costs per mile of vehicle operation   80¢
- Operating and maintenance cost per passenger carried   50¢

The above figures include the costs of the backup buses and passenger service agents that are used in conjunction with Airtrans.

There are some significant comparisons that can be made with the above numbers. For example, a recent study by LTV identified operating and maintenance costs per vehicle mile for various rapid transit and bus systems throughout the nation. These ranged from a high of

$2.90 per vehicle mile for the New York City Bus System, to a low of $1.08 per vehicle mile for the Chicago Rapid Transit System. The Dallas Transit System, in 1975, experienced a cost per vehicle mile of $1.30. When compared with the above figures, the Airtrans operating and maintenance cost of 80¢ per vehicle mile is very good. Also, it is probable that this cost will be further reduced in the future.

The matter of reliability and product improvement has received considerable attention since the airport began operating Airtrans. Originally, the contract called for a three-year reliability improvement program, to be performed by the system contractor concurrently with the maintenance task. This reliability improvement program was never performed, because of the various contract difficulties.

Accordingly, the Board has given the matter of product improvement great emphasis and priority, since only by improving the reliability of the Airtrans equipment can the operating and maintenance costs be further reduced. Some examples of changes designed to improve the life and reliability of the Airtrans system are:

- a fully redundant, backup computer at Central Control
- redesigned ground brushes
- improved vehicle control equipment, for improved precision at station stops
- recalibrated block signal levels
- numerous improved components and parts

Also, in some areas where equipment has experienced high malfunction rates, and ways to improve the reliability of the equipment have not been immediately known, changes have been made to improve the availability of the equipment, without necessarily improving the reliability.

In the maintenance area, operating cost data have revealed that considerable expense is involved in replacing high wear items, such as vehicle guidewheels, which are constantly in contact with the parapet walls, the brushes that run on the rails, the tires, the power collector arms, the pneumatic compressors, and various brake components.

From an operational standpoint, the Airtrans system has not been without problems; however, no one who has experienced the implementation of an advanced system like Airtrans would have expected it to be. In the early months, equipment malfunctions and computer software "bugs" prevented the system from providing a suitable level of service, with the result that many passengers using the airport were delayed and inconvenienced. Backup systems (trucks and buses) were necessary to transport passengers and the various cargos when Airtrans outages occurred. However, LTV Aerospace Corporation thoroughly investigated all malfunctions and failures, and initiated extensive modifications to both hardware and software to improve system reliability. These efforts resulted in tremendously improved system performance; during mid-1975, the system operated for six consecutive months, 24 hours a day (1.6 million vehicle miles) without a service interruption. Also, the system's performance since its reopening under full Airport control has been excellent. In two and a half years of operation, Airtrans accumulated a total of 7 million vehicle miles, and carried over 6 million passengers. It is especially important to stress the high level of safety that has accompanied this performance in that period. Airtrans has never had an unsafe train movement, nor have there been any major passenger accidents onboard the system.

The public acceptance of Airtrans has been very good, especially in recent months, as improved reliability has resulted in excellent performance and service. Both passengers and employees have expressed great satisfaction with the high quality, reliable service that the system now provides.

18.2 THE MORGANTOWN GROUP RAPID TRANSIT (GRT) SYSTEM*

18.2.1 CASE HISTORY

The Morgantown Group Rapid Transit System (MGRT) is a revolutionary automated public transportation system built as a research, development, and demonstration (RDD) project by the Urban Mass Transportation Administration (UMTA) of the US Department of Transportation.

*Based on a case description specially prepared by Steven A. Barsony, Director, Office of AGT Applications, UMTA, USA Department of Transportation, Washington, DC, USA.

While many factors and interests entered into the decision to design and build an automated transit system in Morgantown, West Virginia, the principal ones were: (1) the significant transportation problem existing in Morgantown due to students travelling between classes at the three separate campuses of West Virginia University, and (2) the UMTA RDD program objectives of developing, testing, and evaluating an advanced automated transit system in an urban environment where the results would be nationally relevant and meaningful.

Morgantown, a city of 30,000 population, lies in a valley between the Monongahela River on one side and hilly, mountainous terrain on the other. It is the home of the West Virginia University (WVU) with a student enrollment and faculty of about 23,000. All WVU buildings were originally concentrated in an area adjacent to the central business district (CBD) of Morgantown. As the University grew, classroom buildings and athletic facilities were added about a mile from the original, main downtown campus. Subsequently, additional classroom buildings, dormitories and a regional medical center were built at the new campus about 3.2 miles from the main campus.

There are only two thoroughfares on which traffic can travel between the three campus locations: University Avenue and Beechurst Avenue (both are single lanes). The northern segment of road beginning from Beechurst Avenue at Eighth Street is two lanes and is called Monongehela Boulevard. The congestion that occurs on these roads when classes change brings all traffic, including University operated buses, almost to a standstill. Travel times between campuses have ranged up to 20 minutes in good weather and up to an hour in rain, snow, and ice. Because of terrain topography, widening the existing roads or adding a new road were judged either too expensive or not feasible.

As relief for these congested conditions, the University turned to a new concept of transportation called automated guideway transit. In 1969, WVU applied for and received an UMTA grant to conduct a feasibility study of new mass transportation concepts to determine if any could be used to solve the Morgantown transportation problems and, if, consistent with UMTA RD program goals, it would qualify them for an UMTA assisted procurement. Additional UMTA funds were provided for the study of specific alternative systems. WVU concluded that the most promising concept involved small, electrically powered, computer-dispatched, and computer-controlled vehicles running on an exclusive roadway between stations located at each of the WVU campuses and the city downtown. The University proposed a group rapid transit (GRT) system consisting of six stations, 3.6 miles of double lane guideways, and 100 vehicles.

From the outset is was clear that the Federal Government could not very well delegate the creation of a major prototype system for national application to an organization whose sole objective was to meet the local transportation requirements. The only way to conduct the project consistent with the UMTA objective appeared to be to conduct it under direct UMTA control. Close cooperation with WVU would have to be maintained to provide consideration to local requirements.

Since we do not have available the depth of staffing required for managing and integrating several subsystem contracts, we decided to use one system contractor. We selected the Jet Propulsion Laboratory (JPL) of the California Institute of Technology as the system manager. When the cost of a six station, 100 vehicle system became evident from the first JPL cost estimates, the scope of the project was reduced for budgetary reasons to a three station, five vehicles, and 2.1 miles double lane guideway. UMTA considered this to be adequate for the purposes of the RD program and for the verification of the technology proposed for this GRT system.

Owing to the inability of UMTA to reach a contractually acceptable arrangement with the JPL, they were replaced by the Boeing Aerospace Company in August 1971. Since then, Boeing has been responsible for overall system management integration, including design and development of the three major system components (vehicles, structures and power distribution, and control and communications subsystems) (cf. Figure 18.7).

The guideway and station constructions were completed by late Spring of 1973, and full scale testing on the entire 2.1 mile system followed. These tests provided conclusive evidence that the Morgantown GRT technology was sound and warranted further commitment by the Government to obtain data on operating parameters, maintenance requirements, operating and maintenance costs, and public acceptance under actual operating conditions. To obtain these kind of data it was necessary to make the three station configuration operational by providing operational software, additional vehicles and fare collection and destination selection equip-

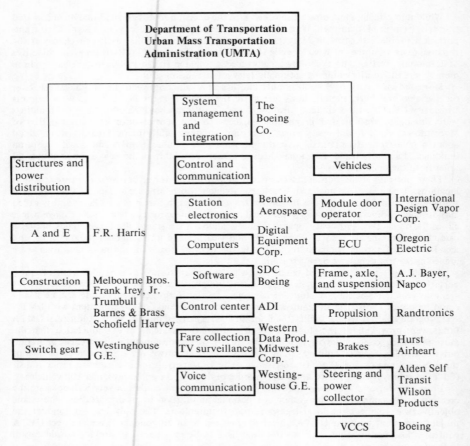

FIGURE 18.7 Morgantown program organization and system suppliers.

ment as well as the additional equipment for the central control facility and station areas (TV monitor, public address system, etc.).

When these additional equipments were installed and fully tested, it provided UMTA with sufficient data to declare the system safe for passenger service operation. The three station system, with the vehicle fleet size increased to 45, entered a one-year period of operational evaluation in passenger carrying revenue service on October 3, 1975. In that year more than 2.0 million passengers were carried and more than 1,000,000 miles accumulated on the vehicle fleet without any injuries to passengers or maintenance personnel. When the University is in session, the system normally operates over 13 hours on weekdays and 5½ hours on weekends. The ratio of actual hours operated to the hours scheduled, called system availability, has typically been higher than 95 percent for the year. At the end of the one-year evaluation, it was concluded that the system was operating successfully and had met its specification requirements.

18.2.2 IMPLEMENTATION AND OPERATION

The MGRT is a bold step forward both in the development of automated transportation technology and in the implementation of a full-scale system in an urban area. Summarized below are the principal distinguishing subsystem characteristics of the MGRT:

- Vehicles: capacity for 21 passengers, electrically propelled, onboard switching and steering, humidity and temperature controlled environment, four-wheel steering, four-wheel disc-type brakes, rubber tires for low noise.
- Stations and guideway: off-line stations for direct origin-to-destination trip without intermediate stops. Full-time remote television surveillance for safety and passenger security. Guideway heating system for reliable traction in cold weather or snow.
- Control and communications: fully automatic, computer-controlled merging and de-merging of vehicles, 15 second vehicle headways, scheduled and demand-responsive vehicle dispatching, redundant (backup) computers with automatic switch-over, software collision avoidance system with hardware collision avoidance system backup, system of inductive loops (FSK) and sensors in reed relays in guideway provide speed and spacing control of vehicles and inputs to computer (software) control system.

These three categories of subsystems are described in more detail in the following

18.2.2.1 Vehicle subsystems. The Morgantown vehicle has ten major subsystems: passenger module, environmental control unit, chassis, hydraulics, pneumatics, electrical power, propulsion, steering, braking, and vehicle control and communication systems (cf. Figure 18.8).

Commands are transmitted to the vehicle from buried communication loops and are received by the onboard vehicle control and communication system (VCCS). The commands operate the vehicle motor, brakes, steering, and doors. Three-phase, 575 V a.c. electrical power is received from the power rail, rectified, and controlled for the operation of the 70 horsepower, d.c. motor. The electrical power also operates the lights, the air conditioner, the hydraulic and pneumatic pumps, the control system, and charges the batteries. The pneumatic system provides an automatic vehicle leveling control and extends the power collector arms to contact the guideway power rail. The redundant four-wheel disc brakes are hydraulically operated in response to input commands and are actuated automatically under emergency conditions. Independent parking brakes operate when the hydraulic pressure is below a safe level. Guide wheels control the steering of the vehicle via the hydraulic, four-wheel, power-steering subsystem (cf. Figures 18.8 and 16.2, column 1). Normal door operation is electrical in response to input commands from the control and communication system (CCS). Functional details of the vehicle subsystems are given in Table 18.1.

18.2.2.2 Stations, Guideway, and Power Supply Subsystems.

Guideway. The guideway structure is double, approximately 65 percent elevated, the remainder being at ground level. The running surface is concrete containing distribution piping for guideway heating to allow all weather operation. Inductive communication loops, also contained in the running surface, enable messages to be transmitted and received between the vehicle and the control and communications equipment. Steering and electrical power rails are mounted vertically along the side of the guideway. Emergency sidewalks, handrails and guideway lighting are provided for passenger safety if egress is required.

A total of 27,776 linear feet of guideway network is installed with grades up to 10 percent. Curves that are superelevated as well as spiraled offer comfortable ride characteristics. Thirty-foot radius curves are used in station areas, resulting in compact station design. Guideway speeds up to 30 mph enable passengers to depart from downtown (Walnut Street Station) and arrive non-stop 6.5 minutes later at the Engineering Station a distance of 2.1 miles, any time of day or night.

Passenger stations. The station facilities provide access to the system, directing passengers to and from the vehicle loading area. The facilities also house control and communications equipment required for controlling vehicle operations within the station area.

All stations have two levels, the entry or concourse level and the loading platform level. This eliminates interference of vehicle and passenger movement. Each platform has one loading position and two or three unloading positions.

Maintenance facility. The maintenance facility provides for operation, maintenance, test, cleaning, and storage of the vehicles. The facility is comprised of a maintenance building and associated guideway. The building houses a maintenance control room, maintenance shops, a central control room and the communications equipment and personnel necessary to operate

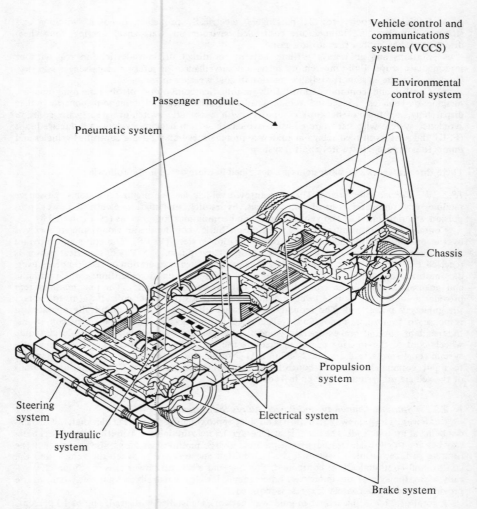

FIGURE 18.8 Morgantown vehicle subsystem.

and maintain the system. The associated maintenance guideway contains a test loop for post maintenance check and a vehicle wash area.

Guideway heating system. The guideway has pipes embedded within the concrete where hot water/glycol solution is circulated to melt ice or snow. The guideway is serviced by three boiler plants. Each boiler plant is different in the number and capacity of the boilers, pumps, and expansion tanks it contains but the function of each plant is identical.

The guideway heating system is under the control of the system operator and he must turn each boiler plant "ON" or "OFF" from his console at central control. Once turned on the boiler plant operation is automatic and normal operation will be indicated to central control unless a malfunction occurs.

Power distribution details. The propulsion substations receive power from the 23 kV distribution cables and deliver 575 V, three-phase power to the power rails. Substation spacing pre-

TABLE 18.1 Characteristics of the Morgantown Vehicle

Physical characteristics
Length	15 ft 6 in (4.72 m)
Height	8 ft 9 in (2.67 m)
Width	6 ft 8 in (2.03 m)
Weight	8,750 lb empty (3,969 kg)
Wheel Base	127 in (3.23 m)
Tread width	62 in (1.57 m)
Accommodation	21 passengers

Performance characteristics
Control	Automatic remote
Propulsion	70 hp electric motor
Velocity	44 ft/s (30 mph = 48.3 kmph) maximum
Suspension	Air bag – automatic leveling
Tires	Dual chamber (1.5 in = 3.8 cm deflation)
Steering	Side sensing (1.2 s transfer)
Brakes	Redundant dual-piston caliper
Conveniences	Environmentally controlled, quiet, comfortable, safe
Turning	30 ft (9.14 m) radius

vents the overall guideway voltage variations from exceeding ±7 percent (exclusive of the power company regulation, which is +0 and −5 percent). A 1000 kVA, three-phase transformer provides the power for the rails.

The housekeeping substations are located within the passenger stations and maintenance center, and provide power for lighting, heating, cooling, and operation of noncritical displays and the uninterrruptable power supplies (UPS). The housekeeping substations also provide for operation of pumps and boiler controls for the guideway heating system.

A UPS is capable of supplying power to critical loads for 15 minutes in case of loss of primary power. The critical loads include the computers, the processors, and the critical communication circuits. The UPS is composed of batteries, switching gear, and the equipment necessary to detect primary power interruptions. Standby power generators at each passenger station and at the maintenance facility are able to start automatically, with a manual start override, and will assume some of the loads of the housekeeping power within one minute of power loss. The station platform and guideway emergency lighting, the Radio Frequency (RF) Voice Communication system, the PA system, the TV system, and the passenger assistance telephone are powered by the standby power generator.

Propulsion power distribution. Power rails along the guideway distribute the three-phase, 575 V a.c. 60 Hz power to the vehicles. The rails are compatible with the maximum total current demand of the expected vehicles between propulsion substations.

The power rails are securely anchored to the guideway. Rail joints allow thermal expansion of the rails. Electrical continuity is maintained across the expansion joints to avoid arcing of the collector brushes.

The guideway power rails are connected to the propulsion power substation transformer secondaries through remotely controlled circuit breakers operated from the control center. Independent circuit breakers are provided so that the main guideway on either side of a passenger station can be operated independently; we refer to this as power isolation. It permits station guideways and the maintenance facility guideway to be isolated from the main guideway power for maintenance and fault correction. The 575 V bus at each propulsion power substation is connected to the transformer secondary by a circuit breaker equipped with overcurrent trips, undervoltage trips, and reverse power sensing to protect the propulsion power system from internal transformer faults fed from the other propulsion power substation transformers via the guideway power rails.

There are seven guideway segments that are automatically controlled by the central computer to power down the guideway. This is done if a vehicle is stopped on the guideway or a vehicle door is opened. Automatic power down is provided to protect passengers on the guideway. Power up can only be accomplished manually, and only when the software senses that there are no doors open.

The 575 V a.c. power is distributed by copper bus bars set in and attached to the plastic carrier. A smooth copper surface is provided to reduce brush wear and arcing by attaching the copper bus bars by means of welded-on studs everywhere on the guideway except at the leading edge of curves. Here through-bolts are added for an extra margin of safety for protection against stud breakage.

Vehicle power is picked up from the power rail by the power collector that rides on the power rails as the vehicle travels along the guideway.

18.2.2.3 Automation and Operation.

Operation. From the passenger's point of view, the use of the system is very easy. At single platform stations — Walnut Street and Engineering — the passenger enters the station at the street level and proceeds to the platform level. At Beechurst Station, which has two platforms, the passenger reads the Platform Assignment Display at the entry to the concourse to determine the proper platform to obtain service to his desired destination. The Platform Assignment Display is controlled by the central control (system) operator.

Use of a coded magnetic code at the Fare Collection Unit is required for passage through the entrance gate. A one-way fare card dispenser is available at stations. A valid fare card operates both the Distribution Selection Unit and the entrance gate.

After the passenger has inserted his card in the Fare Collection Unit, he pushes the button for his desired destination. A legend then lights to acknowledge the selection. The passenger proceeds through the gate to the vehicle loading area. The Fare Collection and the Destination Selection Unit is reset when the passenger proceeds through the entrance gate (cf. Figure 18.9).

During the scheduled mode the passenger boards the next vehicle scheduled to his destination. During the demand mode the station computer begins a sequence of searches. First, the computer looks for an empty vehicle in the station loading position. Second, the computer looks for an empty vehicle in the station and directs it to the loading position. Otherwise, the computer finds the nearest available vehicle and directs it to the station where the passenger is waiting.

Passenger Destination Request dispatch of vehicles is presented pictorially in Figure 18.9. When the vehicle with the desired destination is ready for boarding, a display above the vehicle is illuminated and the vehicle doors open for boarding. After allowing 15 seconds for passenger boarding, the vehicle doors close. The dispatch of the vehicle is presented pictorially in Figure 18.10. A dispatch time is determined by the central computer such that the vehicle, following the nominal acceleration/position profile for that station, will arrive at the main guideway synchronously with the position of the computer assigned, vacant, moving space.* The station computer time clock is synchronized with centrol control computer time clock and then, at the pre-determined dispatch time, the stop tone is removed. The vehicle then accelerates to 8 feet per second. In the zone immediately before the acceleration ramp, the communication channel to the vehicle for receiving a switching command is opened and a switching command transmitted to steer right (cf. point A in Figure 16.2).

After switching right the vehicle transmits positive confirmation of switching to the station computer. Failure to verify switching on time triggers a command from the computer to the vehicle for emergency braking. On the acceleration ramp, speed tone loops cause the vehicle to accelerate at 2 feet/sec^2 until the specified guideway speed is reached. If the vehicle speed and position (reflected by actuation of a presence detector buried in the running surface) are within tolerances, the vehicle proceeds to fill its computer-assigned slot position moving along the main guideway. After passing the merge point, the vehicle is commanded to steer left and proceeds non-stop to the destination. Along the main guideway, the civil speed varies: 22, 33, or 44 feet per second depending on the guideway configuration. When the vehicle enters a section with a different civil speed, a speed change is commanded by a frequency change in the speed

---

*This operation mode results from the use of the synchronous longitudinal vehicle guidance system described in Section 16.14 (cf. Figures 16.7 and 16.12).

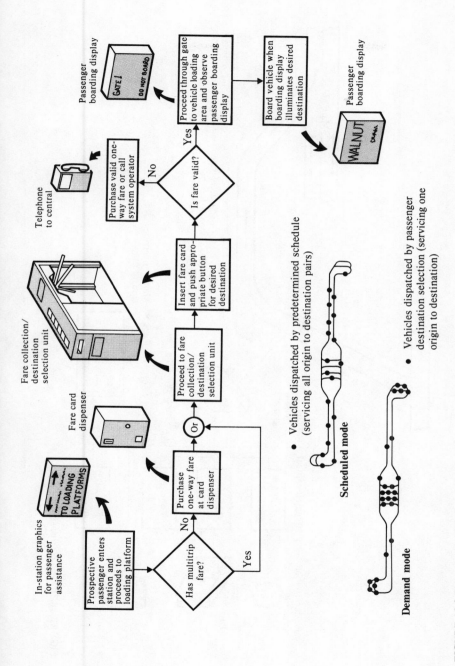

FIGURE 18.9 Passenger destination request.

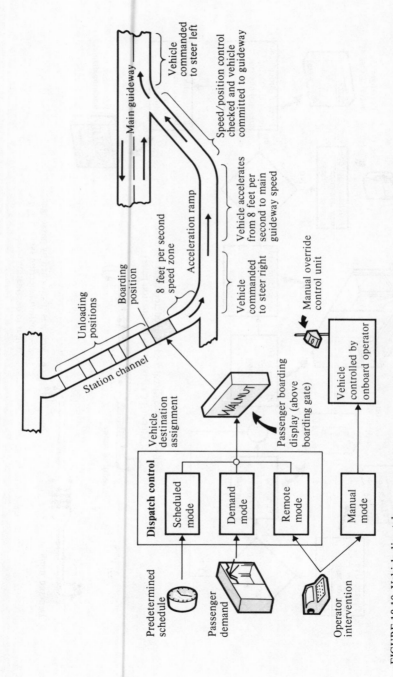

FIGURE 18.10 Vehicle dispatch.

tone loop in the guideway. The frequency change is detected by the antenna underneath the vehicle, which transmits the information to the vehicle control and communication system (VCCS) and a smooth, controlled transition to the new speed is effected at 2 feet/sec^2.

A sequence similar to that for acceleration from the station and merging onto the guideway, but reversed, causes the vehicle to be identified for demerging to its destination, and decelerated to a stop at the unloading gate. There, the doors open for the passengers to deboard. Passengers then leave the station through an exit turnstile gate. Stations are equipped with closed circuit television cameras for safety and passenger security monitoring by the central control operator. A telephone to the central control operator is also provided on each station platform.

Control and communication systems (CCS) (Figure 18.11). The Morgantown GRT System is operated automatically by the CCS, which implements the general control tasks hierarchy shown in Figure 16.1 by means of the following subsystems:

- central control and communication system (CCCS)
- station and guideway control and communication systems (SCCS and GCCS)
- vehicle control and communication system (VCCS)

The primary purpose of the CCS is to provide automatic control, communications, and monitoring of the movement of vehicles along the guideway. The CCS controls vehicle movements on the main guideway, within each station area, at guideway and station interchanges, and at the maintenance facility. All communications, commands, station signals, and the management thereof are the responsibility of the CCS. The CCS provides dynamic graphics and other communications for passenger assistance.

The CCS consists of dual central supervisory computers, dual station control computers, and the communication links between central control and each station. The central computer carries out the automatic system management functions, receiving destination service requests from the stations and transmitting commands to the stations. Duplex communications with the stations are through asynchronous 2400 baud data lines. The interface between computers is through standard modems at both central control and the stations. The station computer receives inputs from the destination selection units and provides passenger instructions via the passenger advisory displays. The station computer manages vehicle movements and receives status information via the data handling unit. Speed commands, station stop commands, steering switch signals, and calibration signals are received by the vehicle through the inductive communication loops buried in the guideway.

a. Central control and communications characteristics: The central control equipment includes the central computers, peripherals, control console/displays, and communications equipment. The system operators, located at central control, monitor and exercise direct control over the system during conditions of initialization, failure, or shutdown. At all other times, the central computer provides control and supervision of vehicles in the station, on the guideway and at the maintenance facility. The system operators merely monitor the operation. All commands are routed from the central control console through the central control computer to the remote computers located at each facility. The operators can call on certain software routines by typing the required message on a control console keyboard.

Software routines allow the operator to restart the system, run vehicles at reduced performance levels, assign vehicles to various locations, and perform other system control and override actions. Performance level modification involves running the vehicles at speeds lower than normal for use during abnormal or emergency conditions.

In the scheduled mode of operation, the central computer manages vehicles by assigning destinations and dispatch times to each vehicle in the system. The passenger enters the station and boards a vehicle assigned to his destination. In the demand mode, the central computer allocates vehicles only if the number of vehicles within the station is inadequate to handle passenger demands. Dispatch times are assigned by the central computer in both the schedule and demand modes to ensure that no conflicts arise at guideway merge points between vehicles enroute to their destinations.

The central console equipment permits the operators to monitor and control the transit system. The consoles include display and control equipment, as well as communications equipment. The central control room also includes a mimic display, which permits the operators to

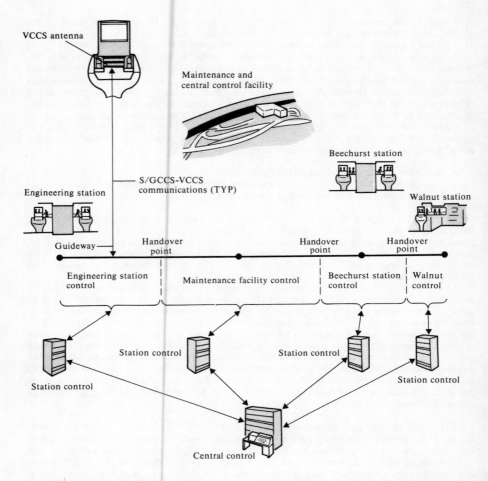

FIGURE 18.11 Control and communications system.

monitor the progress of each vehicle operating in the system, and closed circuit TV monitors for system security and passenger safety.

b. Station control and communications characteristics: The SCCS controls vehicles and station operations in response to central supervisory commands. Communication of control signals to the vehicle is accomplished through inductive communication loops embedded in the guideway. Communication is in the form of coded FSK messages and fixed frequency control tones. The station computer controls vehicle switching, stopping, and door operations in the station. The station computer also operates the station dynamic boarding displays and responds to inputs from the passenger activated destination selection units. The computer in the maintenance facility performs the same types of functions as the station computer and also controls the test track and maintenance ready storage positions.

Each station has a collision avoidance system (CAS), which acts to prevent vehicle collisions in case the primary CCCS, SCCS, and VCCS controls fail. The principal elements of CAS consist of redundant sensors that detect vehicle entry into a control block, inductive communication loops that transmit a safetone to the vehicle in the block, and redundant control electronics (and software) that determine correct occupancy of the block. As a vehicle progresses along the guideway, the CAS control electronics removes safetone from the block immediately behind. If a trailing vehicle violates the "OFF" block, it stops on emergency brakes. In each leg of a guideway merge area, one safetone is normally off. This safetone is turned on allowing a vehicle to proceed when vehicle priority at the merge is established by the CAS control electronics. At each switch point on the guideway, one safetone is normally off. This safetone is turned on allowing the vehicle to proceed when verification of proper switching action has been received.

c. Guideway control and communication characteristics (Figure 18.12): The GCCS consists of the equipment installed on the guideway. This equipment includes digital data cables, tone signal cables, passive presence detectors, and the cable and hardware required to connect the GCCS equipment to the SCCS equipment. All active electronics, which drive the cabling, are located in station and maintenance facility SCCS equipment rooms. Station-generated commands are inductively coupled to the vehicle from the loops buried in the guideway surface. The function of these guideway mounted control loops is as follows:

- Station stop loops: the station stop tone transmitter generates a signal to decelerate and stop the vehicle ±6 inches from the center of the station platform unloading/loading gates. The vehicle enters the stop loop at 4 feet per second and is decelerated to a precise stop as brakes are applied.

- Switching tone loops: the switching tone transmitter generates a signal to command the vehicles to "steer left" or "steer right". The vehicle is sent a switch command at every guideway juncture (merge and demerge). The vehicle must verify that switching has been accomplished or it is brought to an immediate stop.

- Calibration loops: the calibration tone generator transmits a signal to the vehicle to provide a measured distance reference. This nonvital signal is used by the VCCS as a reference for calibrating the vehicle's odometer. The vehicle measures distance travelled and calibrates the odometer, removing any error accumulated since the last loop.

- FSK and speed tone loops: the FSK transceiver unit transmits performance level, brake commands, door commands and identification request to the vehicles operating in the system. These commands are transmitted over one set of loops. A second set of loops is used for vehicle identification, door responses and fault status.

d. Vehicle control and communications characteristics (Figures 18.8 and 18.12): The VCCS is that portion of the automatic control system that is carried onboard the vehicle. The VCCS controls vehicle movements and operations from commands generated by the SCCS; it also identifies and transmits vehicle status to the SCCS. The data link between the VCCS and the station (SCCS) is an inductive communications link via the GCCS over which vital signals are transmitted by tones, and nonvital signals are transmitted by digital messages. The VCCS consists of antennas, communications unit, data handling unit, control unit, and support unit, which perform the following functions:

- antenna: two antenna assemblies provide the VCCS two-way communication with the CCS through buried loops in the guideway. There is one dual antenna assembly for receiving and one antenna for transmitting low frequency electromagnetic signals. The antennas are mechanically fixed to the vehicle and electrically linked to the VCCS.

- communications unit: the communications unit receives low frequency signals from the receiving antenna. These signals are conditioned and transferred to the data handling unit. The communications unit also receives signals from the data handling unit, conditions and transmits them through the transmitting antenna to the guideway.

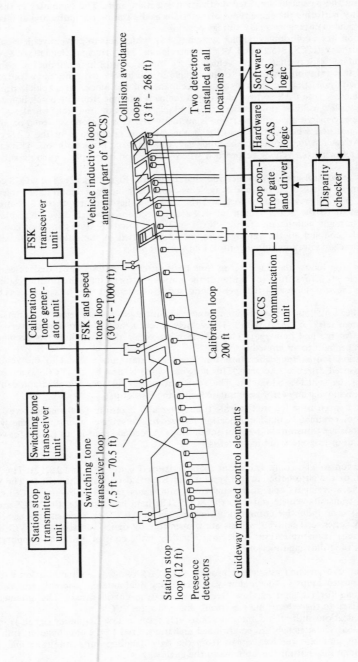

FIGURE 18.12 Station/guideway CCS elements (FSK = frequency shifted keying).

- data handling unit: the data handling unit (DHU) receives conditioned logic signals from the communication unit. The DHU decodes the signals and produces logical instruction and response sequences unique to the input. This unit will initiate logic commands and messages when vehicle conditions change.
- control unit: the control unit reacts to signals from the vehicle and the DHU to control the brakes, steering, doors and propulsion.
- support unit: the support unit provides synchronization of logic signals between units, power conditioners, test circuit isolation and interface signals receivers and transmitters.

18.2.3 CONCLUSION

Urban planners, in their effort to understand the practical impact of their decisions, have carried out intensive studies of the urban growth process and only now are beginning to confirm what has long been suspected — that the quality of urban life and form is deeply influenced by the quality of its transportation. Consequently, transportation was used as a means of shifting populations for encouraging or discouraging the growth of market centers as well as for controlling population density.

At the present time, the demand for urban transportation is predominantly served by the private automobile. But as the cost of ownership (capital investment) and the operating (fuel) and maintenance costs increase, and as the environmental and land use constraints (parking, auto-free zone) became more restricted the automobile will become too expensive for simple passenger trips. The public will not be using their cars for such purposes as to go to work or to go shopping, etc. A new mode of mass transportation will be required to accommodate the needs of the urban population efficiently. This system will have to provide access and service to many potential origins and destinations in a metropolitan area.

Systems such as the Morgantown GRT system provide the needed assurance that these types of people movers can perform well in a relatively restricted area. To obtain further assurance that these technologies can be deployed effectively in the harsher downtown environments, UMTA announced in April 1976 a demonstration program. The intent of this program was to show whether relatively simple, fully automated systems can provide a reliable and economical solution to local circulation problems in the congested downtown areas. In December 1976, four cities — Cleveland, Houston, Los Angeles and St. Paul — were chosen by UMTA as demonstration sites. The major goals of these demonstrations are

- to test the operating cost savings that automated transit might provide
- to assess the economic impact of improved downtown circulation systems on the central city
- to test the feasibility of surface or elevated people movers both as a feeder distributor and as a potential substitute for certain functions now performed by more expensive fixed guideway systems, such as subways

It is argued that these demonstrations will conclusively prove that automated systems can perform a large part of the urban transportation functions in a way that is cheaper and pollution free, and without adding to the congestion of the urban traffic.

18.3 FURTHER GRT SYSTEMS

Table 18.2 illustrates that more than 10 GRT systems have been designed, fabricated, and tested in the USA, Japan, the FRG, and the UK, during the last 5-10 years.

TABLE 18.2 Characteristics of Selected GRT Systems

System		Developer	Development stage	Vehicle capacity		Speed (km/h)	Minimum headway (s)	References
				Seated	Total			
Airtrans	USA	LTV Aerospace Corporation, Dallas, Texas	Passenger service since January 1974	16	40	50	18	Corbin 1973, LTV 1974
Morgantown		Boeing Aerospace Co. Seattle, Washington	Passenger service since October 1975	8	21	48	15	Aston et al. 1972, Bendix 1972, Chamberlain and Kleine 1972, Buchner et al 1976, Elias 1974
Dashaveyor		Bendix Corporation, Ann Arbor, Michigan	Test track (demonstrated during TRANSPO '72)	12	32	48	15	MacKinnon 1974, DOT 1974, OTA 1975
Ford		Ford Motor Company, Dearborn, Michigan		12	24	48	8	
Monocab		Rohr Industries Inc., Chula Vista, California		6	6	48	8	

TTI		Otis Elevator Company, Denver, Colorado		6 or 10	6 or 10	48	12	
H-Bahn	FRG	Siemens, AG, Duewag	Test track	8	16	35	8	Frederich and Müller 1975, 1976, Birnfeld 1975, Marten et al. 1977, Hillmer and Uyanik 1976
Kompakt-Bahn		Krupp Industrie- und Stahlbau	Design proto- type car	24	48	70	60	Dobler and Rahn 1975, Wienecke 1975
Transurban		Krauss-Maffei AG, Standard Elektrik Lorenz AG	Test track	12	18–20	48 (72)	10 (15)	Dobler and Eltzschig 1975, Pöschel 1975, Schindler 1975
KRT	Japan	Kobe Steel, Nisho Iwai Trading	EXPO '75; passenger service since 1980 in Kobe	8	23	48	15	Hamada 1975, Nickel 1976 Buchner 1976, Oku 1973
NTC		Niigata Tekko Sumitomo Electric	Test track	20	50	60	20	Oku 1973
Minitram	UK	Hawker Siddeley Dynamics Limited, Hertfordshire	Design	6	12	54	40–10	Groves and Baker 1974, Russell 1974

18.3.1 FURTHER USA DEVELOPMENTS

In the preparation of the International Transportation Exposition, TRANSPO '72, held in Washington, DC, from May 26 to June 4, 1972, the US government supported the development of the following GRT systems (cf. Table 18.2 and DOT 1974):

- The Bendix-Dashaveyor system, characterized by rubber tired vehicles with a capacity of 32 persons, a headway of 15 seconds at 48 km/h, and an asynchronous headway regulation system, using fixed block sections (cf. Figure 16.7(c)(1)).
- The Ford GRT system, characterized by rubber tired vehicles with a capacity of 24 persons, a headway of 8 seconds at 48 km/h, and a synchronous headway, i.e., a point-follower control system (cf. Figure 16.7(d)).
- The Rohr-Monocab system, which uses a suspended monorail, vehicles with a capacity of 6 persons, and an asynchronous (moving block) headway control system (cf. Figure 16.7(c)(2)), permitting operation at headways of 8 seconds at a speed of 48 km/h.
- The TTI (Transportation Technology Incorporated) system of Otis Elevator Company, characterized by air-cushion-suspended and linear-motor-driven vehicles with a capacity of 6 or 10 vehicles (two configurations), a headway of 12 seconds at 48 km/h and a fixed-block headway protection system (cf. DOT 1974, for more details).

18.3.2 JAPANESE DEVELOPMENTS

About five GRT systems are under development by Japanese Industries (cf. Table 18.2). An advanced status has been reached by the so-called Kobe Rapid Transit (KRT) system, which was designed and fabricated by Kobe Steel Ltd, in cooperation with the Nisho Iwai Trading Company using licensing agreements with the Boeing Co., with respect to the Morgantown vehicle. This system has been demonstrated to the public during the International Ocean Exposition, held in Okinawa during July, 1975. A guideway of length 2.8 kilometers, with three stations was constructed. Exposition visitors were transported by 16 driverless vehicles (cf. Buchner et al. 1976, Oku 1973).

18.3.3 FRG DEVELOPMENTS

Three major development efforts should be mentioned:

- The H-Bahn (Haenge-Bahn), designed with the track above the vehicle was developed by Siemens and the DUEWAG-Waggonfabrik Uerdingen AG, with financial assistance from the Federal Ministry for Research and Technology. System design and component tests began in 1973. A 180 meter full-scale guideway section was erected at the DUEWAG plant in Duesseldorf. The first tests were carried out

successfully with a prototype car in October, 1974 (cf. Frederich and Müller 1975, 1976, Marten et al. 1977).
- The Kompaktbahn, representing a medium capacity AGT system has been designed by Krupp Industrie- und Stahlbau; work began in July, 1974 (cf. Dobler and Rahn 1975, Wienecke 1975).
- The Transurbahn GRT system has been under development by Krauss-Maffei AG, since 1970. In contrast to most of the other GRT systems, Transurbahn uses the magnetic levitation principle. The experiences obtained in prototype testing since 1973 made it clear that the introduction of magnetic levitated GRT vehicles involves unexpected engineering problems (additional weight of the electromagnetic systems resulting in more costly guideways). From their experiences it may be concluded that the magnetic levitation principle is more suitable for high-speed trains using large vehicles (cf. Dobler and Eltzschig 1976, Pöschel 1975).

REFERENCES

Aston, W. W., et al. 1972. Personal rapid transit, computerized, in Morgantown, West Virginia. Computers and Automation, June: 11–17.
Barsony, S. A. 1977. Purpose and plans for the downtown people-mover project. Paper presented at the International Congress '77 of The Society of Automotive Engineers, Detroit, Michigan, USA.
Bendix. 1972. Morgantown PRT control system. Report from Bendix Corporation, Ann Arbor, Michigan, USA.
Birnfeld. 1975. The H-Bahn system: Development status of the operating control system. In Nahverkehrsforschung '75, Status Seminar II, Bundesministerium für Forschung und Technologie, Bonn-Bad Godesberg, FRG (in German).
Buchner, B. C., M. J. Christianson, and W. D. Osmer. 1976. A tale of two systems – Operating experience with automated transportation systems at Morgantown, West Virginia, and Expo '75, Okinawa, Japan. Proceedings of the Fourth Annual Intersociety Conference On Transportation, Los Angeles, California, USA.
Chamberlain, R. G. and H. Kleine. 1972. A simulation model of the Morgantown personal rapid transit system design. In J. B. Anderson et al., eds., Personal Rapid Transit, University of Minnesota, Minneapolis, Minnesota, USA, pp. 405–410 (published by the University).
Corbin. A. 1973. AIRTRANS: Intra-airport transportation system. Society of Automotive Engineers, Air-Transportation Meeting, Miami, Florida, USA.
Dobler and Eltzschig. 1975. The transurbahn system: the SEL operation control system SEL-TRAC, development status, and further extension. In Nahverkehrsforschung '75, Status Seminar II, Bundesministerium für Forschung und Technologie, Bonn-Bad Godesberg, FRG (in German).
Dobler and Rahn. 1975. The Kompaktbahn system – the SEL operation-control system SEL-TRAC, concept and technological status. In Nahverkehrsforschung '75, Status Seminar II, Bundesministerium für Forschung und Technologie, Bonn-Bad Godesberg, FRG, pp. 75–81 (in German).
DOT. 1974. High performance personal rapid transit system development – statement of the work. Report from the US Department of Transportation, Urban Mass Transportation Administration, Office of Research and Development, Washington, DC, USA.
DOT. 1974. Urban deployability studies for high performance personal rapid transit – statement-of-the-work. Report from the US Department of Transportation, Urban Mass Transportation Administration, Office of Research and Development, Washington.
DOT. 1974. United States Department of Transportation automated urban transportation

system developments. Report from the US Department of Transportation, Urban Mass Transportation Administration, Office of Research and Development, Washington, DC.

Elias, S. E. G. 1974. The West Virginia University – Morgantown personal rapid transit system. In J. E. Anderson and S. Romig, eds., Personal Rapid Transit II, University of Minnesota, Minneapolis, Minnesota, USA, pp. 15–33 (published by the University).

Frederich, F. and S. Müller. 1976. The H-Bahn: technological status and results of operation on the test section in Duesseldorf. Verkehr und Technik, 29 (2): 43–46 (in German).

Frederich, F. and S. Müller. 1975. The H-Bahn: technological status of operational tests on the experimental installation in Duesseldorf. In Nahverkehrsforschung '75, Status Seminar II, Bundesministerium für Forschung und Technology, Bonn-Bad Godesberg, FRG, pp. 31–40 (in German).

Groves, H. W. and R. C. Baker. 1974. Minitram in Britain: automatic urban transportation. In Traffic Control and Transportation Systems, Proceedings of the Second IFAC/IFIP/IFORS World Symposium, Monte Carlo, Monaco. Amsterdam: North Holland Publishing Co., pp. 529–540.

Hamada, H. 1975. New KRT (Kobe Rapid Transit) transportation system for Okinawa Marine Expo. Denshin Gijuton (Electronic Technology), 16 (10): 1–8 (in Japanese).

Hillmer, A. and A. Uyanik. 1976. The H-Bahn: development of the guideway elements. Verkehr und Technik, 29 (10): 367–370, 372–373.

LTV. 1974. Airtrans automatic transportation system technical orientation. Report from LTV Aerospace Corporation, Texas, USA.

MacKinnon, D. D. 1974. United States Department of Transportation personal rapid transit system developments. In Traffic Control and Transportation Systems, Proceedings of the Second IFAC/IFIP/IFORS World's Symposium, Monte Carlo, Monaco. Amsterdam: North Holland Publishing Co., pp. 578–586.

Marten, F., S. Müller, and F. Frederich. 1977. The H-Bahn used as an automated guideway transit system for line traffic. Verkehr und Technik, 30 (1): 27–30; 30 (3): 105–107, 110–111 (in German).

Müller, S. and F. Frederich. 1974. The H-Bahn: one solution for an urban public traffic means. Verkehr und Technik, (1): 8–12 (in German).

Nickel, B. 1976. A new horizontal elevator: the KRT system in Japan. Nahverkehrspraxis, 24: 418–420 (in German).

Oku, T. 1973. Development of urban transportation systems in Japan. Japanese Railway Engineering, 14 (3–4): 4–7.

OTA. 1975. Automated guideway transit – an assessment of PRT and other new systems. United States Congress Office of Technology Assessment, Washington (published by the US Government Printing Office).

Pöschel. 1975. The transurbahn system: results of the technological development of the driving control. In Nahverkehrsforschung '75, Status Seminar II, Bundesministerium für Forschung und Technologie, Bonn-Bad Godesberg, FRG (in German).

Russell, W. 1974. The Minitram concept: prospects for development. In J. A. Anderson and S. Romig, eds., Personal Rapid Transit II, University of Minnesota, Minneapolis, Minnesota, USA, pp. 231–240 (published by the University).

Schindler. 1975. The transurbahn system: considerations on its economy. In Nahverkehrsforschung '75, Status Seminar II, Bundesministerium für Forschung und Technologie, Bonn-Bad Godesberg, FRG, pp. 101–104 (in German).

Wienecke. 1975. The Kompaktbahn system: A presentation of the system concept. In Nahverkehrsforschung '75, Status Seminar II, Bundesministerium für Forschung und Technologie, Bonn-Bad Godesberg, FRG, pp. 63–67 (in German).

19 International PRT Systems Experiences

There is no PRT system that has been put into passenger service, up to the present time. But, three systems have reached the development stage of full-scale demonstration projects:

- the Japanese CVS (computer-controlled vehicle systems)
- the Cabinentaxi (CAT) system in the FRG
- the ARAMIS system in France

The first two are presented in detail by means of specially prepared case descriptions.

19.1 COMPUTER-CONTROLLED VEHICLE SYSTEM (CVS): PERSONAL RAPID TRANSIT IN JAPAN*

19.1.1 CASE HISTORY

CVS, which stands for computer-controlled vehicle system, is a pure personal rapid transit system. In contrast to most other new urban transportation systems, it does not group passengers. An individual can utilize a vehicle just as he presently uses a taxi. As a necessary consequence, it aims at a network traffic system rather than a linear one.

CVS guideways consist of a low speed network called a path and a high speed network called superway. The path network consists of a minimum grid of 100 meters. The off-line stations called stops are placed along the path network and, if necessary, can be placed in buildings. Paths can cross each other on the same level, providing a high density network.

The driving speed is 40 km/h on the path and 60 km/h on the superway. A portion of the network, usually a unit grid of the superway and the path network inside it, is called a module. The superway network is connected with the path network via module gates, which are located

---

*Based on a case description specially prepared by Takemochi Ishii, Masakazu Iguchi, and Masaki Kochi, The University of Tokyo, Tokyo, Japan.

on the links of the superway. Since CVS allows for lane changing, the superways can be of the multi-lane type.

The CVS vehicle is rubber tired and is driven by an electric motor. There are two types of vehicles, one for passenger transportation and the other for freight transportation. Both have several variations to meet special requirements. The standard passenger vehicle is a four-seater. It is also possible to connect vehicles together with bellows-covered joints, so as to carry 20-30 passengers per train in order to cope with commuting rush hour demand.

The minimum time headway of the present system is 1.0 second and hence the basic capacity of a lane reaches 3,600 vehicles per hour. In real operation, the practical capacity of a path lane is estimated at about one-third of this figure, due to mergings, branching, and grade crossings.

The branching system is mechanical and is controlled on the vehicle, not on the guideway. There is no moving part on the guideway. This branching system was chosen to enable vehicles to be operated with short time headways.

The first step to CVS development began in 1968. From March to September 1970, The World Expo was held in Osaka. The "traffic game" was demonstrated in the Automobile Industry Pavilion, in which several dozen electric vehicles were operated individually under computer control and ran on a checkerboard-like guideway network with intersections every five meters.

With this background, research into its practical applications began with the support of the Ministry of International Trade and Industry. The basic concept of CVS was formulated in July of that year. In the autumn of 1970, work on the basic design of the system began. Miniature models of vehicles and a guideway were constructed and a total system with 1,000 vehicles was simulated using a large scale computer.

Based on this research, a reduced scale experiment, in which 60 miniature cars reduced to one-twentieth scale were operated under computer control on a scale model of the Ginza area of Tokyo, was made public at the 18th Tokyo Motor Show in November, 1971.

The conclusion of this experimental project is the Higashimurayama experiment. The experimental facilities are located in Higashimurayama City, about 30 kilometers to the west of Tokyo (cf. Figure 19.1). The facilities are as follows: the total length of the guideway is 4.8 kilometers of which two kilometers, constructed as an outer loop, is called superway for high speed running. On the northern side, there are two traffic lanes running parallel to each other, where high speed lane changing tests are conducted. A diamond shaped part in the center of the guideway is called path. It contains a sample of the grid in the low speed network. One side of the grid is 100 meters long, which is the minimum length of the path grid in the CVS specifications. Two stations called stops are provided, each of which has two berths, one for passengers and the other for freight containers, equipped with an automatic loading and unloading device. Two grade crossings are set on the path network.

This Higashimurayama project was begun in 1971 and in the autumn of 1973 the full length of the guideway was completed and the full scale experiment started with about sixty vehicles.

Meantime, CVS had the opportunity for practical application in the Ocean Expo held in Okinawa from July 19, 1975 to January 18, 1976. A system consisting of a guideway 1.6 kilometers long with a grade crossing, 5 stops and 16 vehicles, each of which can seat 6 passengers, was provided. This system transported 800,000 passengers in the 183 days of its operation. The technological level of this Okinawa CVS was naturally not as high as that under experiment in Higashimurayama, but even so, many valuable results were obtained from this field-test demonstration experiment of CVS, especially with regard to transporting the general public.

19.1.2 IMPLEMENTATION AND OPERATION

In the following a brief summary of the guideway and vehicle design principles, and a more detailed description of the methodology used for operation, control, and guidance of the system are presented:

19.1.2.1 Vehicle and guideway charateristics. CVS has been designed in such a way that it can be operated with

- passenger cars with a seating capacity for 4 persons (no standing is allowed; cf. parameters summarized in Table 19.1)

FIGURE 19.1 CVS Higashimurayama experimental site, which is 800 m long and 160 m wide. The white line in the photograph shows a CVS guideway 4.8 km long. The outer oval-shaped guideway is called superway and the center diamond-shaped part is called path.

- cargo vehicles transporting containers (cf. Figure 19.2(a))

Moreover, the use of special dual-mode vehicles (cf. Section 15.3 and Figure 15.4) is possible. Figure 19.2(b) shows such a dual-mode car as it was used for the Okinawa CVS. The design principles used for elevated guideways, at grade intersections, and off-line stations are illustrated in Figure 19.3 for the Higashimurayama experimental system. A passenger car stops at a platform located on a by-pass and can be entered by a passenger, if both the vehicle and platform doors are opened. An automatic handling machine is installed near the platform for loading and unloading the container vehicles carrying cargo (cf. equipment at the right-hand side of the passenger platform shown in Figure 19.3).

19.1.2.2 Automation and Operation.

Function sharing. The general control tasks hierarchy in accordance with Figure 16.1 is implemented in the CVS system at the following levels:

- system management control
- macroscopic control of the entire traffic flow
- microscopic control of the vehicle operations

These operations are performed by the multilevel computer hierarchy system in which computers are assigned to control the different levels of operation.

One path link and the stop on it are under the microscopic control of a small-scale "quantum computer". Each of the quantum computers therefore should control only a limited number of vehicles moving in its control section. A huge network of CVS, where a large number of vehicles is to be controlled simultaneously on a real time basis, thus becomes possible by dividing the whole network into subsections.

TABLE 19.1 CVS Vehicle Characteristics

• *Dimensions*	
Length	3,000 mm
Width	1,600 mm
Height	1,850 mm
Wheel base	1,850 mm
Tread	1,350 mm
• *Total Weight*	1,200 kg
• *Capacity*	
Seating	4
Standing	0
• *Power supply*	200 V a.c. single phase
• *Driving motor*	12 kW d.c. motor, rear wheel drive
• *Driving performance*	
Maximum vehicle speed	80 km/h
Operational maximum speed	60 km/h
Maximum acceleration	0.2 g
Operational maximum deceleration	0.2 g
Emergency brake deceleration	2.0 g
• *Guidance*	Mechanical guidance with a center groove
Steering	Front wheel steering with an Ackermann linkage
Minimum turning radius	5 m
• *Body support*	Pneumatic rubber tires
	Spring and oil shock absorbers
• *Passenger service equipment*	Air conditioning
	Public telephone
	Radio
	Communication device to control center

Each vehicle on a path link communicates frequently with the quantum computer by polling and is controlled so that it follows an imaginary point, called target, which moves along the guideway. In the CVS system, this method is called the "moving target system".*

The upper level computer called the "module computer" informs each quantum computer under it which moving target each vehicle should follow and in which direction they should proceed at a branching point. The module computer is a medium scale computer provided for each unit grid of the superway and is in charge of the macroscopic control. For each superway link, one "superway computer" is provided to control the vehicles on the superway. At the highest level, a large-scale computer system called the "city computer" is provided and is in charge of the overall management such as vehicle maintenance, records of operation, fare calculation, and analysis of the traffic demand patterns.

*Moving target system**. The moving target system is characterized by the mutually independent behaviour of two successive vehicles, which is quite different from the car-following characteristics observed in automobile traffic flow. The targets are moved along the guideway with the velocity patterns predetermined depending on the geometry of the guideway. The spacing of targets is so determined that two successive vehicles should not collide even in an

*See synchronous headway control methods described in paragraph 16.4.2 (Figure 16.7(d)).

(a)

(b)

FIGURE 19.2 CVS vehicles: (a) CVS wagon at the Higashimurayama test site carrying a container on its flat floor; (b) a dual-mode car of Okinawa CVS completely computer controlled on exclusive guideways.

FIGURE 19.3 A CVS off-line station is called stop and is composed of a passenger berth and a container berth; a personal car has stopped at the passenger berth. The photograph also shows an automatic handling machine (far right-hand side) and an at-grade crossing (top).

emergency stop. Each target has its own identification number and at intersections or merging areas two targets that cross or merge with each other have the same number (Figure 19.4). Thus, all vehicles can be operated safely without stopping even at the intersections and merging areas. This method is considered suitable for computer control because the functions of scheduling, routing, and control of vehicles are separated from each other. In addition, this method is not subjected to instability in the control of vehicle spacing and enables stable operation of even long platoons of vehicles (cf. paragraph 16.4.2).

In order for the moving target system to be possible, however, there must be assurance that the vehicles are controlled within a limited position error.

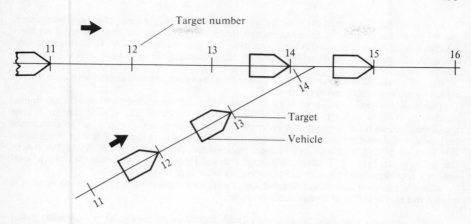

FIGURE 19.4 The moving target system (cf. Figures 16.7(d) and 16.12(a), (b)).

Vehicle-computer communication. A quantum computer or a superway computer can send commands to only one of the vehicles in its control section at a time through the data transmission system (inductive radio). Therefore each vehicle is called up or polled periodically by the computer. The position of each vehicle is identified by an absolute address when it answers the computer polling. The resolution of vehicle position is 0.5-2.0 meters.

Various control systems proposed were tested and analyzed with computer simulation. Figure 19.5 shows a block diagram of the resultant control system (cf. Figure 16.5).

Each vehicle has two identification codes. When a computer polls a vehicle a common frequency is used to specify the vehicle by one of those codes, i.e., each vehicle does not have its own frequency. The first code is permanent and is usually of long code length because very many vehicles are to be operated in a system. The second code, on the other hand, is a temporary code, which is assigned to a vehicle by the quantum computer and is only effective while the vehicle is operated in the control area of the quantum computer. A vehicle, therefore, changes its second code as it proceeds from the control area of one quantum computer to that of the next. Except in a case of initialization, the second code is used for vehicle operation because of the greater data transmission efficiency.

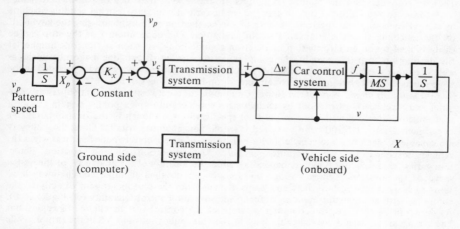

FIGURE 19.5 Position/speed control system for CVS cars (cf. Figures 16.4, 16.5).

In communications with the vehicles, the following seven types of polling are performed, depending on the content of information and the priority: (1) urgent stopping of all vehicles, (2) checking transmission activities, (3) receiving data on vehicle position (4) transmitting control data, (5) initial setting of polling number, (6) changing polling number, and (7) transmitting service data.

By one of the above seven types of polling, the computer communicates with each of the vehicles once in a certain period of time.

The polling period is determined on the basis of the minimum headway and the maximum number of vehicles under the control of the quantum computer. The second type of polling mentioned above is to suppress the onboard buil in function, which activates the emergency brake when communication with the computer is suspended for a certain length of time, hence this type of polling is carried out at least once within that time interval. As to the fourth through seventh types of polling, transmission is conducted by a complementary repeat-and-compare technique, with the aim of improving the reliability of data.

Route assignment and target designation. Target designation by the module computers is carried out taking into consideration the following:

a. It is obviously important that vacant vehicles should be assigned as soon as possible, but it is also important to relocate the vacant vehicles depending on the demand distribution prediction based on the historical data.

b. Since there are a large number of alternative routes in the CVS network for a single origin–destination pair, the actual routes of many vehicles should be determined taking into consideration their mutual interference. The shortest path, therefore, is not always selected.

c. There are several stochastic factors such as time for boarding and alighting of passengers, trip cancelling en route and vehicle failures.

d. Where at-grade intersections and merging areas are densely located, a considerable number of targets should be kept vacant.

e. With regard to (c) and (d) above, it may become inevitable that vehicles have to change the targets from one to the next (cf. quasisynchronous flow control according to Figures 16.12b and 16.13). This target changing, however, should be as rare as possible.

The role of the highest level computer (city computer) in route determination is to make rough route distributions of traffic, to module over the whole CVS network based on the total traffic demand distribution. The module computers then take care of the target assignment and route selection for each vehicle and of the traffic flow thus allocated to the module. The determined route and target of each vehicle are then transmitted to the quantum computer. The city computer and the module computers should be kept informed of passenger demand.

In order to use CVS, the user inserts the credit card or coins into a ticket vending machine located at each stop and the code number of the destination stop by depressing the keyboard on the machine, which is connected to the computer. The code number of the stop can be easily looked up on an information board or in a guide book. As soon as the code number of the destination stop is keyed in, the computer system begins dispatching a vacant vehicle, as well as issuing the ticket simultaneously. This kind of computer activity can be regarded as the same job as seat reservation systems of air lines or on-line banking systems.

Communication and control hardware. Communication between a quantum/superway computer and a vehicle is carried out by an inductive wireless full duplex polling system at a data transmission rate of 4,800 baud in the final specifications, although in the Higashimuriyama experiment a rate of 1,200 baud was used. Twisted pair antennas installed along the guideway are capable of detecting the vehicle position in terms of an absolutely coded address when the vehicle responds to the computer polling. The resolution of the position detection varies from 0.5 to 2.0 meters depending upon the location of the guideway, or the design speed of the location. Vehicle passage sensors are also installed at important guideway locations such as crossings, branches, and stops. Signals from the sensors are sent directly to the computer with an interrupt mode in order to improve the control reliability (cf. Figure 19.6). As shown in Figure 19.6, the computer network of the control system will be of a typical bus line arrangement using a coaxial cable loop in the real applications of CVS in the future, while the traditional star arrangement was adopted in the Higashimuriyama experiment. The bus line, with a data transmission rate of more than 2M baud, not only distributes information among

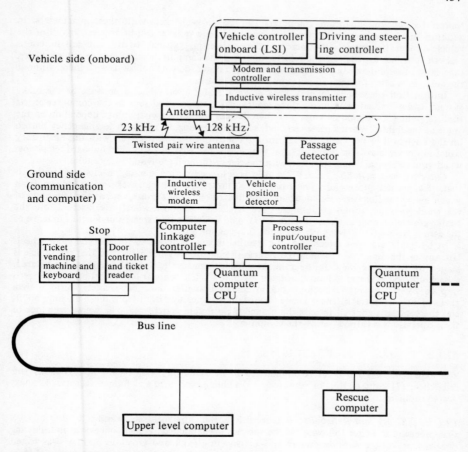

FIGURE 19.6 Schematic of the communication system.

the quantum computers but also enables those computers to exchange data manually, which is impossible in star arrangement systems.

It is also shown in Figure 19.6 that one rescue computer for several quantum/superway computers is provided as a back-up function for unexpected failure of the quantum computers.

On-vehicle electronic devices in the Higashimurayama experiment and Okinawa Expo CVS have two main functions; first to memorize the temporary code number of the vehicle for reference when the vehicle is polled by the computer, and second to count up the time elapsed after last communication of polling type 2. These functions are of a relatively simple level because at the design stage it was considered that the environmental effect or the severity factor of the onboard electronic devices would be about ten times larger than those installed on the ground. It was found, however, as a result of the experiments that the failure rates of onboard data processing and communication were almost zero and the MTBF was so long that its measurement was difficult. The recent advance in technology on LSI (large scale integration) solid state circuits will further increase the reliability of the onboard electronic devices.

It would be better, therefore, to design onboard electronic devices so that they have the function of microcomputers in order to increase the reliability of control. With the extended function of onboard microcomputers, parameters of the guideway network such as distance, curvature, etc., are memorized on the vehicles and the instructions on speed, switching, etc., transmitted by the quantum computers can be cross-checked with these parameters. In addition, since the instructions from a computer to the vehicles can be received by all of the vehicles under its control, the increased data processing capacity of the onboard functions makes it possible for each of the vehicles to utilize the computer instructions to other vehicles, especially those immediately in front or behind, in order to avoid collisions with them, without the delay of waiting for the computer instructions to do so.

Information on vehicle failures, such as abnormally low hydraulic pressure of the brake system, is determined in terms of three grades of seriousness and sent to the computer at the time of polling. The module computer then decides the proper instructions depending on the grade of the failure. When a passenger finds something wrong, he can push an emergency button on the keyboard in the vehicle to inform the control center, which deals with this emergency signal in the same manner as for physical vehicle failures. He can also use an onboard telephone, which provides vocal communication with the center through radio and leaky cables.

Guidance and switching. A CVS vehicle is equipped with a steering mechanism similar to that of a normal automobile. The front wheels are steered with an Ackermann linkage, from which a steering linkage is extended forward like an elephant's trunk, and at the front end of the trunk two horizontal guide wheels are attached and those wheels are inserted into the center groove of the guideway (Figure 19.7). The vehicle is thus steered under the guidance of the center groove (cf. Figures 19.2(a), and 16.2, column 2).

A hydraulic actuator attached to the steering linkage is activated to push the guide wheels laterally in the direction in which the vehicle should proceed at a branch, so that the vehicle is guided to the groove for that direction. At the same time as the hydraulic actuator is activated, a back-up device for switching is rotated so that a small cylinder-like wheel on the device is trapped by the guideway back-up rail in the desired direction. The vehicle status is then checked by an on-track sensor located a certain distance upstream from the branching point, and the vehicle is allowed to pass the branch if no failure is detected. The whole sequence of these operations is carried out by the computer.

19.1.3 EVALUATION OF TEST RUNS

The experience gained so far is summarized in the following with regard to: (1) safety and reliability, (2) conservation of resources, (3) public acceptance and human factors, (4) unexpected problems.

19.1.3.1 Safety and Reliability. It is most important for both safety and reliability of CVS that precision in target following of the vehicles is guaranteed. Vehicle behaviour in following decelerating or accelerating targets and in changing from one target to another was tested experimentally in detail and very satisfactory results were obtained. Even in such special cases as eight percent up- and down-grade sections, where there are borders between quantum and superway computers, deviations of the actual vehicle positions from the targets were found to be within the permissible ranges (1.5 m lead and 2.5 m lag at a speed of 35–40 km/h), which are quite satisfactorily controllable.

In lane changing tests on the superway it was proved that a vehicle with switching failure could stop safely with the emergency brake.

Intersection tests at grades were also successful. Two fleets of five vehicles each with three seconds time headway at a speed of 30 km/h crossed each other and in the case of failures the vehicles stopped safely with the emergency brakes.

Close headway running tests were the most highlighted and severe experiments in the course of research and development of CVS. A fleet of three vehicles could run with a time headway of 0.95 seconds at 63 km/h for the first time in January, 1976. A headway of 0.95 seconds in the test still includes a considerable safety margin and a data transmission rate of 1,200 baud can also be improved. With a 4,800 baud transmission rate, which is the future CVS standard, the headway can be shortened to 0.84 second with the same safety margin.

Onboard devices for detecting vehicle spacing and to be used as an independent system were manufactured on a trial basis and have been tested. Rubber roller power collectors were also tested for decreasing noise level. The most frequent cause of suspension of the total system was human errors of the computer operators.

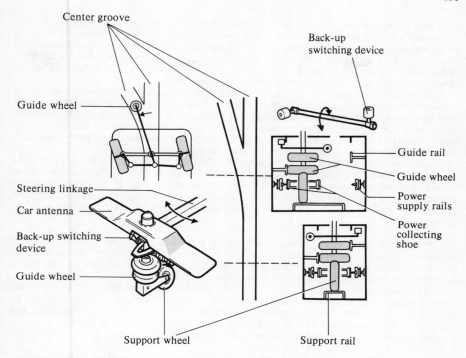

FIGURE 19.7 The guidance and switching mechanism.

As to computer failures or destruction, in addition to computer self-checking such as parity checks, the computers mutually monitor through the bus line of the computer network whether their functions are normal or not. In Okinawa CVS, two computers of the same type were used, one for vehicle control on the main line and another for automatic vehicle testing in the maintenance shop; the latter was so designed that it could function as a back-up computer in the case of failure of the former.

Although a usual way to increase system reliability is to have dual or triple sytems of computers and other components, a fine-meshed guideway network of CVS provides many alternative detour routes even in the case of failure of one of the quantum computers, and this structural redundancy can maintain the total system performance.

The highest level computer, the city computer, is designed in a dual or triple structure to make failure probability extremely low. Even in the case of the city computer failures, however, the module computers can control the trips in their modules based on mutual communication through the bus line, though optimization of traffic flow has to be abandoned. A similar relationship holds true between the module and the quantum computers when the module computers fail. The vehicle safety problem is concerned with vehicle operation at stations and on the main line.

Boarding on and off at stop. A door is provided on the platform and is opened or closed automatically together with the door of the vehicle stopped behind. This design prevents passengers' intentional or accidental stepping on the guideway and also makes it possible to aircondition the platform.

Running on the main line. Collision avoidance is currently fully dependent on moving target following control by computers. For future specifications, however, a dual collision avoidance function will be provided with onboard devices, which measure the clearance to the vehicle in front.

A safety code on the minimum headway has not yet been established. An emergency brake mechanism with a deceleration rate of 2g has been developed in order to prepare for the more severe regulation that vehicle spacing should be longer than the emergency brake distance. This emergency brake is applied only when a collision cannot be avoided without it. In future design, this 2g brake will be called "life brake" and a new brake system of 0.3–0.5g will be provided as an emergency braking system in the conventional sense.

Braking at a deceleration of 2g is fulfilled with brake shoes clamping the steel flange of the guideway. As to protection of passengers at the time of 2g deceleration, a series of dummy tests have been conducted and it was recognized that passengers are safe enough if they are seated facing backward. However, further study is needed for protecting children standing on the floor. Those vehicles that are in service for the sick or the handicapped should be operated with long enough headways so that 2g braking does not become necessary.

In the case of vehicle troubles on the main line, the vehicle is directed to and stopped at the nearest stop. When this vehicle is not able to run by itself, it should be pushed by another vehicle or towed by a maintenance vehicle to get to the nearest stop. If the vehicle cannot be moved at all, the passengers are instructed by the control center through the onboard telephone to leave the vehicle and walk to the nearest stop. The guideway is provided with a sidewalk for this purpose.

19.1.3.2 Conservation of Resources. A normal motor vehicle in city traffic, which is subject to repeated stop and go operation due to many traffic signals, converts its kinematic energy into heat energy by its friction brake every time it is braked. Energy spent in this way is approximately half the energy needed to propel the vehicle.

In the case of CVS, there is no need to stop at intermediate stops because CVS is a purely personal transit and therefore non-stop operation at a fairly constant speed from origin to destination stop is possible. Due to this feature the energy efficiency of CVS is much higher than that of normal automobiles, although the level of service does not differ very much.

CVS vehicles are, in addition, equipped with a driving system of high power efficiency as well as with a power regenerating brake system.

It is a basic design philosophy of CVS vehicles to make them as small and light as possible, with the constraint that comfort of passengers is maintained. The target of the total weight of a vehicle is 1,000 kg although in the Higashimurayama experiment it reached 1,200 kg, because ready-made components were put together instead of assembling newly designed ones.

Lighter vehicles not only bring about conservation of materials and energy for vehicles but also make it possible to build light and slender guideways, which helps to solve the aesthetic problem as well as to save the guideway costs.

19.1.3.3 Public Acceptance and Human Factors. In connection with the CVS demonstration show at the 1971 Tokyo Motor Show, a public acceptance survey was made through the questionaire method. In the results, 87 percent of the responders answered that they looked forward to the realization of CVS, 72 percent considered CVS to be in keeping with existing cities, 78 percent said that elevated guideways would be acceptable if they were adequately designed, and 68 percent answered that they would use CVS if the fare rate were similar to or less than that of taxis.

A further systematic human engineering survey was made in the Higashimurayama experiment. As to desirable spacing of CVS stops, 10 percent of the interviewed preferred 100 meters, 34 percent accepted 200 meters and 47 percent tolerated 300 meters; the minimum spacing in the present CVS specifications is 100 meters.

Since CVS is a purely personal system, it is possible to provide specially designed vehicles for exclusive use by the handicapped and to dispatch them on call. In the Higashimurayama experiment, a vehicle for a wheel chair was manufactured and earned a good reputation in the demonstration.

With regard to automated operation of CVS, even though self-operated elevators have already become quite common, many people still have a feeling of unrest about automatic vehicle operation. It is expected that the uneasiness of passengers will gradually decrease as the safety of the CVS operation is proved and as passengers become accustomed to using the CVS. It will be necessary, even so, to give proper information to passengers on opening or closing the doors, starting, stopping, unexpected standstill, and other changes of vehicle state. Important information should be given both visually and vocally.

In order to cope with vehicle failures and in-car crimes, an emergency push button and an onboard telephone connected directly to the control center are essential. With respect to in-car crime, prevention is easier in CVS than in conventional groups transits, as in CVS a passenger is not forced to share a ride with someone else.

Mischief is a big problem for the driverless vehicles of CVS. In particular, children are usually curious enough to touch all the equipment in the vehicle. Some passengers may soil the vehicle interior. Fire protection, to cope with the dangers of smoking, as well as cleaning of the vehicle interior should be considered carefully.

Passengers are seated facing backward in the CVS vehicles. It is sometimes believed that passengers get sick more easily when seated facing backward than when seated facing forward. The results of the test showed that seating direction does not make any significant difference when lateral acceleration rate is not high. The maximum lateral acceleration rate of $0.1g$ in the present design specifications of CVS is low enough for this purpose. In addition, measurements of heart rates and galvanic skin reactions of passengers reveals that psychological uneasiness is smaller but physiological tension is larger in the case of forward facing seats.

19.1.3.4 Unexpected Problems. The most significant problem recognised throughout the experiment was communication-dead sections, which are gaps between two adjacent ground antennas connected to different data channels. These dead sections will arise in the future in real systems, in which the whole network has to be divided into a number of subsections in order to be controlled by many quantum computers. Vehicles cannot communicate with the computer in the dead sections and the emergency brake may possibly be activated automatically when speed is low because of too long suspension of polling. The significace of this problem was underestimated in the design stage and an unexpectedly long time had to be spent in coping with this problem.

19.2   CABINENTAXI — A NEW CONCEPT OF URBAN TRANSPORT DEVELOPED IN THE FRG*

19.2.1   CASE HISTORY

The development of a new urban transport system called Cabinentaxi (cabintaxi) began in 1970 and led to the formation of a joint venture in 1972 between Demag Fördertechnik of Wetter (Ruhr) and Messerschmitt-Bölkow-Blohm GmbH of Ottobrunn. The research and development programmes of these two industrial companies are such that their products and know-how complement each other. The main proportion of Demag's share of the tasks were mechanical engineering in nature, whereas MBB were responsible for the automatic feed-back control system. The cabintaxi joint venture has been backed by the Federal Ministry of Research and Technology for 80 percent of the development costs.

After testing the most important components of the system in laboratory tests in 1971 and 1972, testing facilities for practical trials were built on a 9000 m² plot of land in Hagen, over several phases of construction, beginning in 1973 (cf. Figure 19.8). The most important route components of the guideway, such as straight sections, clothoids, minimum and standard radii and different designs of supports such as mushroom, cantilever T, gantry and pylon, were built of steel and reinforced concrete.

A major part of the research and development phase of the overall project was completed in 1976. By the end of 1976, a guideway section with a 15 percent gradient was added to the trial circuit and improved prototypes of small-capacity cabs and twelve-seater large-capacity cabs were subjected to endurance tests. Furthermore, preventive maintenance with check-out techniques has been tested, and special service vehicles were put into operation in order to

---

*Based on a case description specially prepared by Klaus Becker, Head of Division, Central Department for New Product Development, Demag Fördertechnik, Wetter (Ruhr), FRG.

FIGURE 19.8 Aerial photograph of the cabintaxi testing grounds at Hagen, FRG.

study and demonstrate simple means of rescuing and recovering stranded passengers and vehicles.

19.2.2 IMPLEMENTATION AND OPERATION

In the following a brief description of the main components and subsystems, i.e., of the vehicles, the guideway system, and the automation system, is presented.

19.2.2.1 The Main System Components. The broad range of possible research and development directions, which have to be taken into account in dealing with the creation of AGT systems, is illustrated in Figure 19.9.

The proposed system features for the cabintaxi system are indicated by heavy lines. They can be summarized as follows:

Vehicle sub-system (cf. Figures 19.9, 19,10, and 15.2)

- small and large capacity cabs with seating accommodation and sufficient space for luggage, designed as top-running or underrunning vehicles
- electric drive unit with quiet wheel travel units
- high acceleration and braking rates, ability to climb gradients and operate in all types of weather because of no-contact power transmission by linear induction motor (LIM)

Guideway subsystem (cf. Figure 19.9 and 19.11)

- two separate traffic levels for overrunning and underrunning tracks along a track girder mounted on pillars, laid at ground level or underground
- track girders with gradients of up to 10 percent and suspension at points or linear suspension
- on- or off-line stations

Automation subsystem (cf. Figure 19.9 and 19.12)

- automation system subdivided into three hierarchical operating levels: vehicle, station, and network
- autonomous headway measuring system in every vehicle and a network computer for supervisory tasks
- automatic system for guiding vehicles to set destinations via fixed track switches, and constant supply of vehicles at stations

The overall concept of system development covers both small capacity cabs, with seating accommodation for three, and large capacity cabs for 12 seated persons. Both types of cabs can be operated to provide a demand-responsive service.

The size of the vehicle depends on the application in question and the type of service. Some possible areas of application for the cabintaxi are:

- overall service for an average city
- service to cover areas of satellite towns and their connections to underground and commuter railway systems
- a service for subregions and centers, e.g., large area pedestrian zones in cities with parks and shopping centers

The application of large cabs can be advantageous in a timetable-controlled operation. Therefore the term cab-way is frequently used in the case of large-capacity cabs travelling according to a schedule, whereas the term cabintaxi is employed for the mainly demand-responsive transport.

The cab-way variant of the cabintaxi system can be used to implement an SLT system with features illustrated by Figure 15.2. Such an SLT system was developed and constructed in a large hospital for connecting the main clinic with a remote convalescence building (cf. Figure

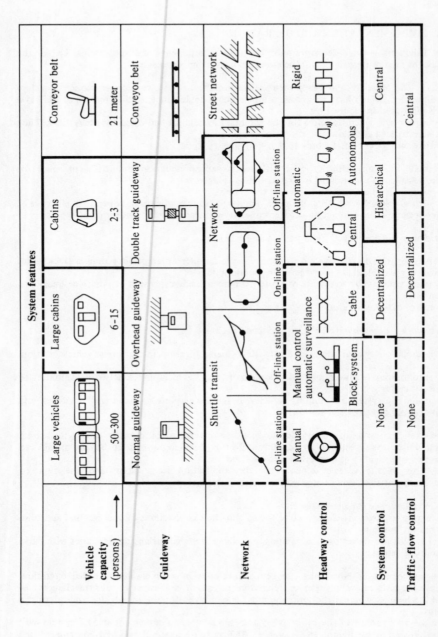

FIGURE 19.9 Morphological review of urban transport systems: ——, prototype development; ———, full range of research and development.

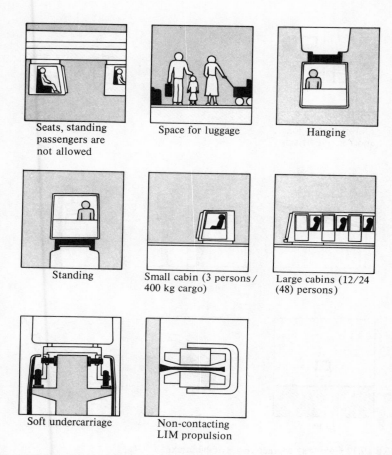

FIGURE 19.10 Essential vehicle characteristics of cabintaxi.

19.13 and Section 17.3). At the beginning of 1976, this link-up lift or cabinlift was put into operation for transporting persons and goods automatically within the hospital complex.

The design principles of the guideway network and the stations are illustrated in Figures 19.14, 19.15 and 15.2, which show the experimental system at Hagen. One basic feature is the capability to operate the vehicles overrunning and underrunning on a single track girder (cf. Figure 19.14).

19.2.2.2 Automation and Operation. Each cabin is equipped with a storage unit to register selected station addresses, an automatic headway measuring device, a drive control unit, and a mechanical switching device to divert it at track switching points. The stations (cf. Figures 19.14, 19.15) are equipped with travel destination purchasing machines, travel destination reading machines, and a station-controlled unit. Each track switch has its own control unit. The whole system is coordinated through a central network computer. The passenger selects an address on the travel destination purchasing machine; this address is inductively recorded on a band, which is part of the ticket. The ticket is then fed into the travel destination reading machine, which transfers the selected address to the storage unit in the appropriate cabin. After the cabin is set in motion by the passenger it travels automatically to the selected destination.

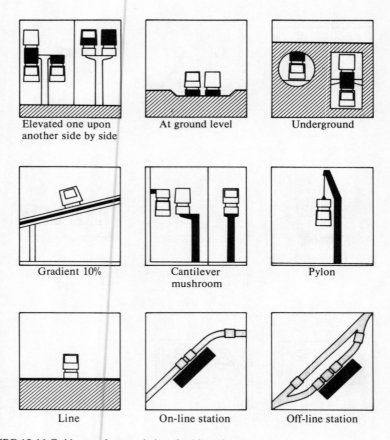

FIGURE 19.11 Guideway characteristics of cabintaxi.

The acceleration and speed are controlled by the drive control unit in conjunction with the headway measuring device, which determines the travel of the cabin for particular situations. When the cabin approaches a track switching point the address that is stored in it is inductively read at an appropriate distance away from the point, and is fed into the track switch control unit. In this control unit a list of all station addresses that can be reached from that point is stored in conjunction with the switching patterns necessary to reach the stations. The address is read by the point control unit and the cabin is then directed to its appropriate route. An exception to this operation occurs when the central network computer intervenes in order to reduce the travel time through a change of route. Normally the shortest distance is selected for the travel route but because of heavy traffic or a breakdown it may be desirable to change the route in order to cut the travel time. In the event that the central network computer breaks down a standby route list comes into operation in order to direct the cabin to its desired destination.

Each cab is accelerated by two double comb linear motors at a rate of 2.5 m/s² to a maximum speed of 36 km/h. The deceleration rate of the service brake is 2.5 m/s². For emergency braking, a maximum rate of 5.1 m/s² is planned, which will be provided by a combination of the linear brake and the hydraulic wheel brakes. The switch wheels are required for selecting the correct direction when passing over rigid track switches.* A photo of the rigid track switch

*A switching principle as illustrated in column 3 of Figure 16.2 is used.

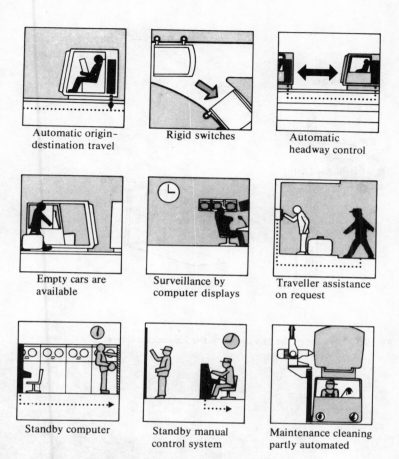

FIGURE 19.12 Automation and operation of cabintaxi.

on the overrunning track is shown in Figure 19.15. It can be clearly seen how the travel trajectories cut through each other.

The drive control unit influences the speed of the cabin through the power supply to the linear motors, which is controlled by the phase cutting technique (rectifier, thyristor). A headway measuring device is built into each vehicle and consists of a transmitter and receiver, the transmitter emitting a signal that is passed along a conductor mounted on the guideway. When the cabin approaches a second vehicle the transmitted signal is received, the distance registered and the required speed compensation is automatically performed.* Due to the full automation of the system, the manual effort required to operate it is naturally cut to a minimum and is limited to a control operating team and maintenance personnel.

19.2.3 SYSTEM EVALUATION

In the course of laboratory tests and trials with the test vehicles on the trial circuit, which were carried out as a means of "maturing" the engineering, several project studies for the cabintaxi

*This corresponds to the asynchronous headway control principle described in paragraph 16.4.1 (cf. Figure 16.7(c)(3)).

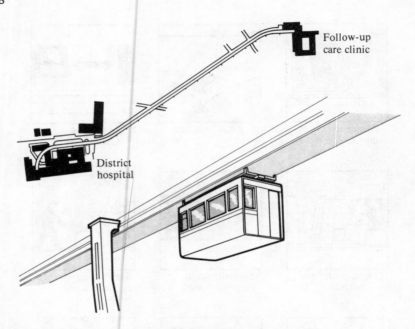

FIGURE 19.13 Drawing of the link-up lift to the Ziegenhain Hospital.

FIGURE 19.14 First construction phase of the trial circuit with station 01 and small-capacity cabs.

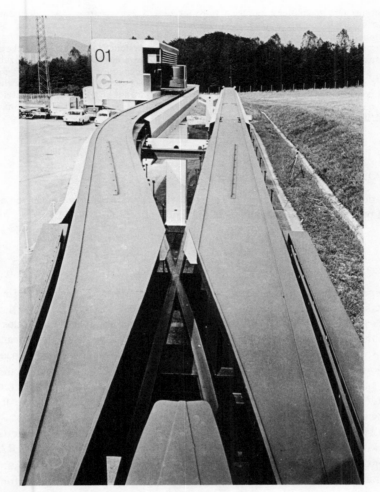

FIGURE 19.15 Rigid track switch on the overrunning track.

for transport companies and city planning offices were prepared. Among these are the studies for the cities of Freiburg, Hagen, Bremen, Marl, and Hamburg.

With the two latest feasibility studies, a major area of application for the new track technology in cities is being covered.

The economic importance of the cabintaxi development project for the future of public urban transport is corroborated by the results of the "Cabintaxi cost–benefit analysis". In this study for the year 1990, which was carried out for the Federal Ministry of Research and Technology, the independent economic consultants WIBERA AG of Düsseldorf came to the conclusion that the cabintaxi may be a city's sole means of public transport and is superior to a bus system in running costs and overall efficiency. These results confirm the development target of the cabintaxi research and development project.

In 1976 the chief remaining technological problems were studied in full scale on the trial circuit in Hagen, involving large-capacity vehicles and maintenance systems. The whole range of this new technology has been fully tested and seven years of research and development work have led to promising results. Therefore planning activities for installing the cabintaxi system in agglomeration centers were started in 1976.

Before the introduction of the new urban transport system, however, further clarification work must be carried out which, as was already shown in a market analysis, affects different interest groups, i.e., passengers, city dwellers, and transport companies (cf. Figure 3.2). In this connection there are still groups of problems that cannot be mastered by the development companies alone. The system's success as a means of public transport will be determined on the basis of a reference installation. This will require active cooperation from federal, regional, municipal, and transport authorities.

19.3 OTHER PRT PROJECTS

In addition to CVS and cabintaxi, the following three projects are or have been the subject of extensive research and development work (cf. Table 19.2).

19.3.1 THE FRENCH PRT SYSTEM ARAMIS

In comparison with other existing PRT concepts, ARAMIS is characterized by a remarkable speciality (cf. Engins Matra 1973, Kaiser 1974, OTA 1975): several vehicles with a capacity of four passengers can be coupled together as trains. However, this is not done by mechanical devices, but electronically, by means of a highly sophisticated distance regulation system. In this way, very short headways of the order of 0.2 seconds between vehicles (cf. table 19.2) and a remarkably high lane capacity of about 75,000 seats per hour per lane are expected to be achievable. The individual vehicles can join a train on the main guideway or leave it when entering a station.

The feasibility of this platooning technology was demonstrated on a test guideway 1 km long near Paris-Orly airpoirt. For the further development of ARAMIS a 30 month research and development program was funded by the Regie Autonome des Transports Parisiens (RATP), the regional transport authority for Paris. The expenses of this development phase are estimated at 40–50 million francs; RATP funded 70 percent of that sum, and the remaining 30 percent were raised by the system developer Engins Matra.

The target of the research and development program was to create an operational PRT system that could be implemented in a real urban environment at the beginning of the eighties. For this purpose the following case studies of possible ARAMIS installations were prepared:

- Paris (linking the commuter lines on the outskirts of Paris in an arc about 70 km in length)
- Nice (installing ARAMIS as the main public transportation system for that city with a population of 300,000)

TABLE 19.2 Characteristics of Selected PRT Systems

System	Country	Developer	Development stage	Vehicle capacity (seats)	Speed (km/h)	Minimum headway (s)	Maximum lane capacity (seats/h)	References
CVS	Japan	JSPM[a]	Full scale test Tokyo and Expo '75 Okinawa	4	60	1	~15,000	JSPMI 1972, 1975, Ishii et al. 1972, Ishii 1977, Iguchi 1977, MacKinnon 1975, OTA 1975
CAT (Cabintaxi)	FRG	Demag and MBB[b]	Full scale test Hagen in Westphalia	3	36	0.5 to 1	~15,000	Becker, 1972, 1974, 1975 Demag/MBB 1974, BMFT 1975, Haubitz 1975, Hesse 1975, NVP 1976, TÜ 1977, Tappert and Heinrich 1975, Runkel and Krauss 1975, Hübner and Herkenfeld 1976
ARAMIS	France	Engins Matra	Full scale test Paris Orly	4	50	0.2 (with platooning)	~75,000	Engins Matra 1973, Frybourg 1974, Kaiser 1974, MacKinnon 1975, OTA 1975
Aerospace	USA	Aerospace Corporation	Design	6	32	0.5	~28,800	Bernstein and Olson 1974, DOT 1973, OTA 1975, MacKinnon 1975
CABTRACK	UK	Royal Aircraft Establishment	Conceptual	4	36	0.6	~24,000	Gibbs and Leedham 1974, RAE 1969, OTA 1975, MacKinnon 1975

[a] Japanese Society for the Promotion of Machine Industry (cf. Section 19.1).
[b] cf. Section 19.2.

19.3.2 THE AEROSPACE-PRT SYSTEM IN THE USA

The Aerospace-PRT system, designed by The Aerospace Cooperation in El Sugundo (California) uses vehicles with a capacity of six persons, which are operated at headways as low as 0.5 seconds, and speeds of 32 km/h (OTA 1975, MacKinnon 1975). A case study for possible installation of the system was carried out for Los Angeles, resulting in a proposal for a guideway network 1,020 km long, with 1,084 stations and 64,000 vehicles. As yet, no large-scale demonstration project comparable with those installed for CVS and the Cabintaxi system has been built.

19.3.3 THE CABTRACK SYSTEM IN THE UK

During the years 1967-1971 the Royal Aircraft Establishment developed a PRT concept called CABTRACK. Case studies prepared for the installation of CABTRACK in the Westend district of London came to the conclusion that the elevated guideways would cause an unacceptable visual intrusion. For this and other reasons the development of CABTRACK was abandoned in 1971, and the available research and development capacities were concentrated in the creation of a GRT system called MINITRAM (cf. Table 18.3).

REFERENCES

Becker, K. 1972. An interdependency: man in automated transit. Foerdern und Heben, 22 (4): 159-163 (in German).
Becker, K. 1974. On the economy and problems of the cabintaxi system. Nahverkehrspraxis, 21 (9): (in German).
Becker, K. 1975. On the influence of driving speed and guideway network on the traffic volume and costs structure of a new PRT system. DEMAG Fördertechnik, Wetter (in German).
Bernstein, H. and C. L. Olson. 1974. High-capacity personal rapid transit. Paper prepared for Los Angeles Council of Engineers and Scientists, Transpo LA: The future is now, The Aerospace Corporation, El Segundo, California.
BMFT. 1975. Benefit-cost analysis for the cabintaxi, Bundesministerium für Forschung und Technologie, Bonn-Bad Godesberg, FRG (in German).
Demag/MBB. 1974. The cabintaxi transit system: a system description, Report from DEMAG, MBB, FRG (in German).
DOT. 1973. High-capacity personal rapid transit systems developments. US Department of Transportation, New Systems Division, Office of Research and Development, Urban Mass Transportation Administration, Washington, DC.
Engins Matra. 1973. "ARAMIS", a new generation of urban transportation, Engins Matra, Avenue Luis, Brequet, Velizy, France.
Frybourg, M. 1974. New Urban Transport Systems. Revue Générale des Chemins de Fer, 93 (3), 127-154 (in French).
Gibbs, E. W. and H. C. Leedham. 1974. Cabtrack: communication and control instrumentation for a one-fifth scale test track. Royal Aircraft Establishment Technical Report 73169, UK.
Haubitz, G. 1975. Development Status of Cabintaxi. Eisenbahningenieur, 26 (7): 256-257 (in German).
Hesse, R. 1975. The cabintaxi system: its operation and first results of dynamic simulation. In Nahverkehrsforschung '75, Status Seminar II, Bundesministerium für Forschung und Technologie, Bonn-Bad Godesberg, FRG (in German).

Hübner, E. and M. Herkenfeld. 1976. Cost–Benefit Analysis of the Cabintaxi. Verkehr und Technik, 29 (7): 266–268 (in German).
Iguchi, M. 1977. New technology in personal rapid transit, Journal of the Japan Society of Mechanical Engineers, 80 (698): 28–33 (in Japanese).
Ishii, T., M. Iguchi, T. Nekahara, Y. Kohsaka, and Y. Doi. 1972. Computer-controlled minicar system in Expo '70: an experiment in a new personal urban transportation system. IEEE Transactions on Vehicular Technology, VT-21(3): 77–91.
Ishii, T., 1977. Assessment of new urban transportation systems from the energy and resource saving viewpoint. Journal of the Society of Automotive Engineers, Japan, 31 (2): 73–76 (in Japanese).
JSPMI. 1972. A new urban traffic system: CVS (computer-controlled vehicle system), Japan Society for the Promotion of Machine Industry, Tokyo, Japan.
JSPMI. 1975. A new urban transit system: CVS (computer-controlled vehicle system). Japan Society for the Promotion of Machine Industry, Tokyo, Japan.
Kaiser, R. G. 1974. New technology in urban transportation in France. In J. E. Anderson and S. Romig, eds., Personal Rapid Transit II. University of Minnesota, Minneapolis, Minnesota, USA, pp. 47–55 (published by the University).
MacKinnon, D. D. 1975. Longitudinal control policies for automated guideway transit. Proceedings of the Conference on Information Sciences and Systems, Johns Hopkins University, Silver Spring, Maryland, USA.
MacKinnon, D. D. 1975. High-capacity personal rapid transit developments. IEEE Transactions on Vehicular Technology, VT-24.
NVP. 1976. Fourth construction phase for testing cabintaxi. Nahverkehrspraxis, 24 (11): 461 (in German).
OTA. 1975. Automated guideway transit – an assessment of PRT and other new systems. United States Congress, Office of Technology Assessment, Washington, DC (Published by US Government Printing Office).
RAE. 1969. Assessment of auto-taxi urban transportation, Parts 1 and 2. Technical Report 68287 from the Royal Aircraft Establishment, UK.
Runkel and P. Krauss. 1975. The cabintaxi system: passenger information and dispatch system. In Nahverkehrsforschung '75, Status Seminar II, Bundesministerium für Forschung und Technologie, Bonn-Bad Godesberg, FRG (in German).
Tappert, H. and K. Heinrich. 1975. Users' opinions on PRT systems. Nahverkehrspraxis, No. 2 (in German).
TÜ. 1977. Cabintaxi: an individual traffic means, TÜ, 18 (3): 93–94 (in German).

20 Findings and Summary

Chapters 15-19 analyzed the contribution of advanced computer control technology to the creation of totally new modes of urban transportation by presenting a survey of

- basic systems concepts (Chapter 15)
- concepts and methods of control and automation (Chapter 16)
- international experience gained so far in real passenger service or full-scale test runs (Chapters 17-19)

Three basic systems concepts were discussed in Chapter 15:

- automated guideway transit (AGT)
- combined AGT and dial-a-ride systems
- the dual-mode concept

It was shown that the AGT concept has received the greatest attention and reached the highest level with respect to research, development, and practical implementation.

The control concepts and methods developed for automated vehicle guidance (cf. Sections 16.2-16.4), traffic-flow control (cf. Section 16.5), and route guidance (cf. Section 16.6) are highly sophisticated and have been extensively studied — theoretically, in simulation runs, and in full-scale tests.

Most of the AGT systems suppliers are engaged in research and development work related to space flight problems. Therefore, it is not surprising that their know-how concerning advanced space flight control and guidance technologies has been used for the development of AGT control systems.

The status reached in the development and implementation of AGT systems is different for the three categories, SLT, GRT, and PRT. From the experience gained so far, the following conclusions may be drawn focussing on technical, social, and economic issues.

20.1 SLT SYSTEMS

20.1.1 TECHNICAL ISSUES (AUTOMATION AND OPERATION)

It was reported that SLT systems that have been in operation for several years (cf. Table 17.1) have provided highly satisfactory service. The SLT systems installed in the USA have transported more than 200 million passengers with only one serious accident. The initial problems, which occurred in connection with the fully automated vehicle operation, especially with respect to overall reliability of the system, have been solved.

20.1.2 SOCIAL ISSUES

Problems with public acceptability of the systems were not observed. However, to date no SLT system has been installed within a city center. The construction of huge elevated SLT guideways above downtown streets would very likely cause public opposition. For this reason the future role of SLT systems, as a special component of the whole urban public transport system, cannot be judged reliably at present.

20.1.3 ECONOMIC ISSUES

For the operation of an SLT system 24 hours a day and 7 days per week, a labor force of about one employee per vehicle is required (OTA 1975). The solution of the same transport task using buses would require 3-5 employees per vehicle. The guideway costs are comparable with those for exclusive elevated bus lanes.

Summarizing, the following conclusions can be drawn:

- the main technological problems connected with completely automated operation of SLT systems are solved
- the acceptance of SLT systems by the public is favorable as far as installations at airports, recreation parks, etc., are concerned
- the relation between benefits and costs is advantageous for special applications

However the following question has to be answered and requires further studies or perhaps the creation of a full-scale demonstration project: What could be and should be the role of an automated SLT system in an urban city, i.e., as a special mode of public transport?

20.2 GRT SYSTEMS

20.2.1 TECHNICAL ISSUES (AUTOMATION AND OPERATION)

The two GRT systems discussed in Chapter 19 that have reached the status of real passenger service, i.e., Airtrans and the Morgantown GRT, were faced with serious reliability problems at the beginning of their operation.

However, it is interesting to note that the reliability problems were not caused by the control computers, and that the safety of the systems was never a problem, i.e., no accident occurred.

A well-proven control technology is now available for GRT systems of the Airtrans and Morgantown GRT type. But, further research and development work is needed regarding the development of

- headway regulation systems to ensure reliable and safe vehicle operation at minimal headways in the range 3–10 seconds
- computer software needed for monitoring and controlling a large vehicle fleet

20.2.2 SOCIAL AND TRANSPORT ISSUES

The existing GRT installations have demonstrated their potential to solve transport tasks such as exist in medium-sized and small cities with populations of about 30,000–100,000, as well as in the downtown areas of larger cities.

The operational experience gained so far has illustrated, however, that the use of an automated GRT system requires more active cooperation between the passenger and the corresponding subsystems than is necessary in using conventional public transport means. This involves, for example, operating automatic ticket selling machines, and gates, entering a station without any personnel, and riding a driverless vehicle. Therefore, the installation of a clearly arranged and easily understandable passenger information and guidance system at stations and within cars is of basic importance.

20.2.3 ECONOMIC ISSUES

The development and installation of GRT systems has proved to be much more costly than originally expected: for Airtrans $35 million were planned and about $53 million were finally needed (cf. Section 18.1). The expenses of the Morgantown GRT, which were estimated to be $18 million in 1970, and which were corrected to $37 million in 1977, finally amounted to $64 million (cf. OTA 1975). It is obvious that a large part of the additional funds resulted from unexpected technological and other problems, which quite naturally occur if completely new technologies have to be developed and implemented. Moreover, it is expected that the costs of the most expensive subsystem, i.e., of the guideway, can be reduced significantly by introducing rational construction technologies such as are used today in road building.

Summarizing, the status of automated GRT systems may be characterized by the following observations:

- the main technological problems, especially the computer control problems, may be considered solved for vehicle operation at minimal headways larger than 8-15 seconds
- for favorable public acceptance of GRT systems the installation of a sophisticated computer assisted passenger information and guidance system is indispensable
- it has been demonstrated that GRT systems can meet the requirements of an urban public transport system with respect to both passenger and goods transport

Nevertheless, the installation of GRT systems in a real urban environment remains a pioneering effort, which will require the solution of numerous detailed problems. One main question concerns the identification of the optimal areas for application of GRT systems within cities.

20.3 PRT SYSTEMS

20.3.1 TECHNICAL ISSUES (AUTOMATION AND OPERATION)

The feasibility of the PRT concept has been demonstrated by means of full-scale experiments carried out in Japan, the FRG, and France (cf. Chapter 19). This concerns mainly the complicated control tasks that are connected with automatic vehicle operation at minimal headways of one second or less. Therefore, it may be justified to state that the overcoming of the headway barrier given in Figure 15.2 will in principle be possible. Whether this statement holds true, not only under the relatively "clean" conditions of a full-scale experiment, but also under the rough operational conditions of a real passenger service in an urban city, has to be proven by special demonstration projects. The installation of such demonstration projects was planned for the seventies. However, this target has not yet been reached.

The main problems requiring clarification are:

- development and testing of the software for the control system, which consists of a large number of spatially distributed control computers that have to monitor and control the operation of hundreds of PRT vehicles in a dense network
- ensuring safe operation of dense vehicle strings moving at minimal headways less than 1 second and at speeds of about 30-60 km/h
- ensuring sufficiently high reliability and availability of the whole PRT system

20.3.2 SOCIAL ISSUES

As discussed in Section 15.1, the PRT concept is aimed at reducing automobile use within cities. It is obvious that only the installation of a PRT system within a city can provide the answer as to whether this goal is really achievable. Moreover, it must be considered an open question whether elevated PRT guideways installed above narrow city streets will receive the necessary public acceptance.

20.3.3 ECONOMIC ISSUES

The economic characteristics of PRT are so unclear that a meaningful analysis is difficult at present (OTA 1975). Proponents of the PRT concept claim that a large number of components of the same type, such as vehicles, stations, guideway elements, control systems, will lead to excellent possibilities for standardization and series production, resulting in significant decreases in expenses. The most expensive subsystem is — as for all AGT systems — the guideway network. But in this respect PRT systems seem to be much more economical than SLT and GRT systems. This is illustrated by Figure 20.1, which shows, for the three AGT categories as well as for elevated auto roads and urban railways, the relation between the three parameters (cf. MacKinnon 1974):

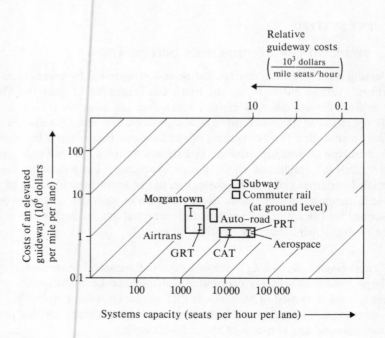

FIGURE 20.1 Comparison of guideway costs (CAT, cabintaxi).

- lane capacity in seats per hour per lane
- guideway costs per mile per lane
- guideway costs per seat and miles per hour

The given cost estimates illustrate that the relative PRT guideway expenses are about one-tenth as high as those of a subway and about half as high as those of an elevated auto road.

20.3.4 LONG-TERM IMPACTS

The final and basic question is: What will be the long-term impacts of a PRT system installation in an urban city, with regard to the solution of the main urban traffic problems summarized in Chapter 1, considering a time horizon of 10–20 years from now? It is clear that this question is even more complicated to answer at present than the one of economic efficiency. However, rough estimates can be obtained using special system analytic methods, i.e., the so-called scenario technique. Such a study has been carried out by Tsuchiya et al. (1976) for a typical Japanese city, which is characterized by the parameters summarized in Table 20.1. The authors compared the following three strategies (scenarios) for a forecasting time-range of 20 years:

TABLE 20.1 Characteristics of the City Analyzed by Tsuchiya et al. (1976) with Regard to the Long-term Effects of Different Development Strategies, Including the Installation of a PRT System

• Population	542,000
• Area	7,740 ha
• Area utilization	
housing	18.2 percent
industry	7.7 percent
roads	8.0 percent
agriculture	5.3 percent
forest	29.6 percent
other	29.1 percent
• Traffic demand	2,500 trips per day
• Increase of traffic demand	3 percent per year (resulting from population growth)
• Existing public transport systems:	14 bus lines 9 stations of regional railway lines

TABLE 20.2 Admissable Emission Rates for Carbon Monoxide, Hydrocarbons, and Nitrogen Oxides

Admissable emission rates (g/car/km)	1973	1975	1976 and later
Carbon monoxide CO	18.4	2.1	2.1
Hydrocarbons $H_n C_m$	2.94	0.25	0.25
Nitrogen oxides NO_x	2.18	1.20	0.6 or 0.85

Strategy I

- No changes regarding the existing public transport systems (14 bus lines and 9 urban railway stations).
- Introduction of stricter emission standards for motorcars according to Table 20.2.

Figure 20.2 illustrates that a significant improvement in air quality can be achieved by this strategy. But, the problem of energy consumption remains unchanged and worse conditions will develop regarding the general mobility and occupancy rates of the conventional public transport means: The mean travel time for all modes of urban transport is expected to increase by 50 percent, and the mean automobile travel speed will decrease from about 39 km/h to 29 km/h. The rush-hour occupancy rates of the urban railways and buses are predicted to increase by factors of 1.8–2.1 (cf. Figure 20.2).

Strategy II

In addition to the stricter emission standards just mentioned, the introduction of a PRT system, i.e., of the CVS described in Section 19.1, is assumed. The PRT network will be put into operation step by step, i.e., every four years a new part will be opened for passenger service. Special attention was paid in the simulation studies to integrating the CVS with the existing public transit systems. The expected results are illustrated by curve II in Figure 20.2. It can be observed that a further reduction of the air-pollution levels by 10–20 percent seems to be achievable. The heavy increase of the mean travel time and the corresponding decrease of the mean automobile travel speed can be avoided. This is so far a remarkable result, since an increase of the population has been assumed, which results in an increase of the traffic demand by 3 percent per year, i.e., the number of trips per day is assumed to be 1.8 times larger in the twentieth year than in the first year.

The occupancy rates of the existing public transport means are predicted to increase by only 10–20 percent, but a significant decrease of the overall energy consumption can be observed.

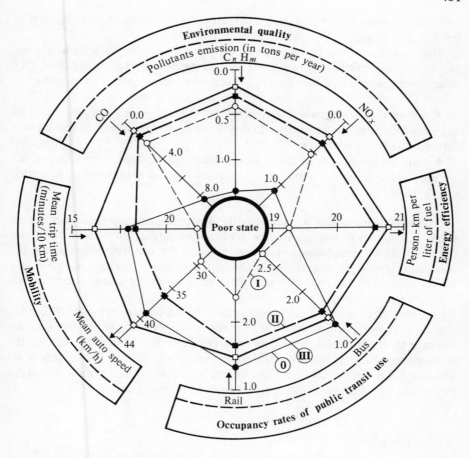

FIGURE 20.2 Status of the urban transport system: O, during the 1st year (initial state); I, II, III, during the 17th year (final state). I, II and III represent Strategies I, II and III, respectively.

Strategy III
In addition to strategy II, the following two measures are assumed:

• installation of computerized route guidance systems for automobile traffic, e.g., the comprehensive automobile control (CAC) system mentioned in Section 4.2
• continuous replacing of 20 percent of the gasoline-powered motorcars by electric cars during a time interval of 20 years

The resulting curve III in Figure 20.2 makes it clear that a further improvement of all parameters describing the environmental quality, the mobility, the occupancy rates of conventional public transit, and the energy efficiency seems to be possible (cf. Tsuchiya et al. 1976, for more details).

These simulation results, which predict the future role of PRT systems and the technical, social, and economic issues characterizing the present status of PRT development summarized above, seem to justify the following final conclusions:

- a long-term urban transport development strategy, which is mainly based on the introduction of PRT systems, will very likely be able to succeed in dealing with future urban traffic problems
- but, in spite of extraordinary progress in the technological development and testing of PRT systems, several economic, social, and other problems need further clarification

At present one cannot be sure that the automated transportation modes now being developed will really bring the needed breakthrough to better urban transportation. This is especially true if one considers the essential differences in the economic and social structures of different countries. Nevertheless, there is a strong motivation for proceeding with the development of new automated demand-responsive urban transportation systems.

Any fundamental change in transportation will need a certain period of time for experimentation: this was certainly true for railway systems, whose development began with the invention of the steam engine. It will doubtless be true for the development from the invention of the new "systems technology" — the large-scale integrated digital computers and the related automation techniques — to totally new urban transportation systems (cf. Figure 1 in the Introduction).

REFERENCES

Ashford, N. et al. 1976. Passenger behaviour studies for automatic transit systems. In Transportation Research Board, National Research Council, Transportation Research Record 559, Washington, DC, pp. 63–72.

MacKinnon, D. D. 1974. Technology development for advanced personal rapid transit. In J. E. Anderson and S. Romig, eds. Personal Rapid Transit II, University of Minnesota, Minneapolis, Minnesota, USA, pp. 57–64 (published by the University).

OTA. 1975. Automated guideway transit — An assessment of PRT and other new systems. United States Congress, Office of Technology Assessment, Washington, DC (published by the US Government Printing Office).

Tsuchiya, H., M. Abe, A. Uehara, M. Ishikawa, F. Harashima, and T. Itoh. 1976. A study of the effects of new transportation systems on urban transportation and environment by computer simulation. In Control in Transportation Systems, Proceedings of the IFAC/IFIP/IFORS Third International Symposium, Columbus, Ohio, USA, pp. 245–251.

Author Index

Adams, C.J., 355, 367
Adlerstein, S., 35, 47, 48
AEG, 280, 281, 352
Afanasyev, L.L., 13, 17
Alden, 355, 367
Alimanestianu, M., 355, 367
Allen, D.C., 258, 260, 281
Allen, T.E., 386, 408
Allsop, R.E., 86, 87, 95, 120
Almond, J., 140, 143, 147
Anderson, D.G., 9, 13, 17
Anderson, J.E., 355, 361, 366, 367
Anderson, J.H., 381, 408
Anderson, P.A., 362, 368
Anderson, P.E., 85, 120
Anke, K., 39, 45, 48
ARRB, 86, 120
Ashford, N., 482
Aston, W.W., 418, 444, 447
Astrakhan, V.I., 321
Athans, M., 101, 122, 123, 381, 410
Athol, P., 60, 63
Atkinson, W.J., 292, 296
Audinot, P., 343, 346
Autruffé, H., 343, 346

Baerwald, J.E., 13, 17, 58, 63
Bahke, E., 25
Baier, W., 60, 63
Bailey, S.J., 47, 48
Baker, J.L., 58, 63
Baker, R.C., 445, 448
Bång, K.L., 79, 85, 86, 88, 90, 91, 92, 120

Barker, J., 85, 120
Baron, P., 66, 126
Barsony, S.A., 418, 447
Barthel, S., 10, 18
Bartolo, R., 292, 296
Barwell, F.T., 35, 48
Bauer, H., 292, 296
Beatty, R.L., 12, 13, 17
Becker, H., 275, 282
Becker, K., 361, 367, 471, 472
Belenfant, G., 71, 120
Bellis, W.R., 86, 121
Bender, J.G., 366, 368, 381, 386, 395, 408
Bendix, 444, 447
Benjamin, P., 366, 367
Bennett, B.T., 65, 100, 122
Bennewitz, E., 78, 121
Berger, C.R., 72, 73, 74, 78, 121
Berg von Linde, O., 343, 345, 346
Bernstein, H., 393, 395, 408, 410, 471, 472
Besacier, G., 343, 344, 346
Birnfield, 445, 447
Blaise, R., 343, 346
Bleyl, R., 93, 94, 95, 121
Blumentritt, C.W., 193
Blunden, W.R., 87, 121
Bly, P.H., 311, 313
BMFT, 25, 26, 471, 472
Böttger, O., 366, 367
Bolle, G., 60, 63
Bone, A.J., 86, 121
Bonsall, J.A., 292, 296

Bopp, K., 263, 274, 281
Boura, J., 260, 281
Boyd, R.K., 386, 395, 402, 406, 408
Brand, D., 25, 26
Bredendieck, R., 292, 293, 296
Breeding, K.J., 395, 408
Breuer, M.W.K.A., 248, 281
Brinkman, A., 79, 84, 85, 86, 88, 121
Brinner, R.E., 395, 408
Brooks, W.D., 93, 94, 95, 121
Brown, S.J., 376, 378, 381, 395, 402, 408, 409
Bruggeman, J.M., 93, 95, 121
Brux, G., 413, 415, 417
Buchner, B.C., 444, 445, 446, 447
Buck, R., 251, 281
Burke, H.B., 386, 409
Burmeister, P., 292, 293, 297
Burrow, L., 260, 281, 380, 381, 409
Butrimenko, A., 75, 76, 121
Buyan, J.R., 395, 410
Byrd, L.G., 58, 64

Cabeza, C.M., 310, 313
Camp, S., 355, 368
Candill, R.J., 381, 409
Cannon, R.H., 2, 3, 4
Carroll, J., 25, 26
Carter, A.A., 15, 17
Carvell, J.D., 187, 193
Casciato, L., 79, 121
Cass, S., 79, 121
Cassy, M., 311, 312, 313
Chamberlain, R.G., 444, 447
Chang, M.F., 102, 121
Chiu, H.Y., 381, 409
Cholley, J., 343, 346
Chreswick, F.A., 13, 14, 17
Christianson, M.J., 444, 446, 447
Chu, K.-C., 381, 395, 398, 399, 409
Ciessow, G., 263, 274, 281
Citron, S.J., 118, 124
Claffey, P.J., 15, 17
Cleveland, D.E., 66, 121
Cleven, G.W., 66, 123
CMEA, 10, 17
Cobbe, B.M., 96, 139, 142, 145, 147, 148
Cohen, N.V., 311, 312, 314
Committee of Urban Traffic Control Systems, 162, 163, 167, 168, 170

Cook, A.R., 111, 121
Copper, J., 248, 281
Corbin, A., 418, 444, 447
Covault, D.C., 60, 64
Cunningham, E.P., 402, 407, 409
Curry, D.A., 9, 13, 17
Cushman, R.H., 42, 43, 48
Cutlar, S.E., 79, 85, 121

Dais, J.L., 355, 366, 367
Dare, C.E., 13, 17
Darroch, J.N., 86, 121
Day, J.B., 25, 26
Delpy, A., 336, 343
Demag, 19, 21, 26, 361, 368, 378, 384, 385, 392, 401, 402, 405, 409, 471, 472
Derrish, T., 60, 64
Devlin, S.S., 45, 49, 118, 124, 126
DHF, 413, 416, 417
DHUD, 25, 26
Dietrich, E., 366, 367, 368, 375, 386, 409
Dobler, 272, 273, 281, 418, 445, 447
Doi, Y., 471, 473
Donner, R.L., 99, 102, 121
Dörrscheidt, F., 35, 48
DOT, 25, 26, 418, 444, 446, 447, 471, 472
Drew, D. R., 35, 48, 86, 108, 111, 121, 124
Dudek, C.L., 193
Dunne, M.C., 86, 87, 121

ECE, 9, 10, 11, 15, 17, 18
Edie, L.C., 65, 100, 122
Elias, S.E.G., 418, 444, 448
Eliassi-Rad, T., 395, 409
Eltzschig, 272, 273, 281, 445, 447
Engins Matra, 470, 471, 472
ERA, 42, 48
Estournet, G., 277, 344, 346
Etschberger, K., 248, 292, 294, 296
Etschmaier, M.M., 35, 49
Eustis, G.F., 79, 85, 122
Everall, P.F., 58, 60, 64, 65, 66, 103
Everts, K., 66, 121

Färker, G., 47, 49
Faggin, F., 39, 43, 48, 49
Faulhaber, F., 248, 281
Favaut, R., 60, 64

Federal Highway Administration, 183, 193
Fenton, R. E., 366, 368, 375, 381, 386, 395, 408, 409
FHWA, 65, 106, 108, 109, 110, 121, 138
Fichter, D., 361, 368
Filion, A., 25, 26
Finnamore, A.J., 256, 311, 312, 313
Fischer, P., 10, 18
Fling, R.B., 376, 378, 409
Fong, B.C., 99, 102, 121
Footh, R.S., 65, 100, 101, 122
Ford, B.M., 402, 408, 409, 411
Ford Motor Co., 292, 297
Foth, J.R., 106, 122
Fowler, P., 111, 122
Frederich, F., 418, 445, 447, 448
Freidl, W., 281
French A., 14, 15, 18
French, R.J., 60, 64
Frey, H., 256, 311, 312, 313
Friedl, H., 60, 64, 70,
Frimstein, M.I., 176
Frimstein, M.N., 176
Frost, C.R., 41, 49
Frybourg, 471, 472
Frybourgh, M., 413, 416, 417
Fuehrer, H.H., 93, 96, 122
FuH, 355, 368

Gabbay, H., 93, 95, 122
Gabillard, R., 413, 416, 417
Garrard, W.L., 355, 366, 367, 381, 386, 402, 409, 410
Gartner, N.H., 93, 95, 122
Gazis, D.C., 7, 8, 18, 35, 49, 65, 79, 100, 101, 102, 111, 112, 115, 121, 122, 125
Geber, W., 343, 346
Geist, 313
Genser, R., 35, 49
Gershwin, S.B., 99, 122
Gervais, E.F., 111, 122
Ghahraman, D., 248, 281
Gibbs, E.W., 471, 472
Giraud, A., 256, 311, 312, 313
Goetz, W.R., 85, 120
Goodson, R.E., 118, 124
Goodwine, D.N., 102, 106, 124
Gould, A.V., 251, 281
Grafton, R.B., 87, 122
Green, F.B., 141, 147

Green, R.H., 105, 106, 107, 122
Griffe, P., 346
Groth, G., 60, 64, 70, 122
Grover, A.L., 105, 122
Groves, H.W., 445, 448
Guenther, K., 245, 292, 297
Gurevich, M., 176

Hagland, C., 343, 344, 345, 346
Hahlgnass, G., 114, 122
Hahn, H.-J., 260, 281
Hahn, J.F., 79, 85, 122
Hahn, L., 114, 122
Haikalis, G., 355, 368
Hajdu, L.P., 393, 411
Halton, D., 141, 147
Ham, R., 141, 147
Hamada, 418, 445, 448
Hamada, T., 65, 86, 87, 93, 94, 96, 123
Hamilton, C.W., 25, 26
Hamilton, W.F., 362, 368
Hand, G.R., 381, 409
Hansen, G.R., 251, 253, 281
Hanshin Expressway Corporation, 204, 205
Hanysz, E.A., 60, 64
Harrison, R.P., 343, 345, 346
Harvey, T.N., 86, 121
Hasegawa, T., 162, 170, 196, 205
Haslböck, 15, 16, 18
Haubitz, G., 471, 472
Heathington, K.W., 66, 122, 248, 282
de Heer, J.J., 343, 344, 345, 346
Heinrich, K., 471, 473
Henry, J.-J. 311, 313
Herkenfeld, M., 471, 473
Herman, R., 258, 281
Hess, E.A., 15, 17
Hesse, R., 355, 368, 381, 392, 402, 409, 410, 471, 472
Hewton, J.T., 79, 122, 140, 147
Hillier, J.A., 58, 64, 65, 93, 95, 122, 139, 142, 144, 147
Hillmer, A., 418, 445, 448
Hinman, E.J., 376, 377, 378, 379, 383, 384, 386, 392, 393, 395, 400, 402, 408, 410
Hirao, O., 14, 18, 31
Hirten, J.E., 15, 18
Hobbs, L.C., 41, 45, 49
Hodkins, E.A., 15, 17
Hoff, G.C., 66, 122

Hollingworth, D., 41, 45, 49
Holroyd, J., 7, 18, 65, 81, 83, 93, 96, 122, 142, 144, 145, 147
Hoppe, K., 258, 260, 281, 283
Horn, P., 260, 267, 268, 281, 282
Hornby, D.G., 147, 149
Horowitz, B.M., 381, 411
Horowitz, J.L., 13, 14, 18
Houpt, P.K., 101, 122, 123
Howson, L., 248, 282
HRB, 86, 123
Huddart, K.W., 142, 147
Hübner, E., 471, 473
Hughes, M., 413, 416, 417
Hupkes, G., 292, 297
Huttman, 15, 16, 18

IFAC, 35, 49
Iguchi, M., 471, 473
Iida, Y., 71, 123
Inada, S., 196, 205
Inose, H., 65, 86, 87, 93, 94, 96, 123
Inouye, H., 71, 125
IRF, 10, 11, 18
Isaksen, L., 98, 99, 101, 123
Ishii, T., 376, 378, 388, 410, 471, 473
Ives, A.P., 114, 116, 123

Jackson, P.M., 114, 116, 123
Jacoub, M., 277, 279, 344, 346
Jasper, L., 66, 78, 123
Jauquet, C., 343, 346
Jennings, S.W., 142, 147
JSPMI, 361, 368, 471, 473
Jürgen, R.K., 41, 45, 49
Junt, J.F., 85, 120

Kästner, E., 375, 410
Kahn, R., 106, 107, 123
Kaiser, R.G., 470, 471, 473
Kaltenecker, H., 39, 48
Kamada, S., 260, 267, 282
Kanen, A.C., 60, 64
Kaplun, G.F., 176
Kariva, S., 343, 346
Kashida, K., 376, 378, 388, 410
Kavriga, V.P., 343, 346
Kent, J., 275, 282
Khilazhev, E.B., 176
Khorovich, B.G., 176
Kieffer, I.A., 19, 25, 26
Kikuchi, K., 375, 411

Kimura, M., 375, 411
King, J.H., Jr., 332
Kinoshita, K., 376, 378, 388, 410
Kirby, R.F., 235, 240, 241
Kiselwich, S.J., 366, 368, 395, 410
Kleine, H., 444, 447
Knapp, C.H., 79, 102, 122
Knoll, E., 66, 123
Kockelkorn, E., 260, 266, 283
Köhler, R., 41, 49
Kohsaka, Y., 471, 473
Kometani, E., 196, 205
Kopp, H., 275, 282
Kornhauser, A.L., 355, 366, 367, 386, 395, 402, 409, 410
Kosemund, M., 267, 268, 282
Koshi, M., 13, 18
Kovatch, G., 355, 368
Krauss, P., 471, 473
Krayenbrink, C.J., 79, 85, 123
Krell, K., 58, 60, 64, 65, 97, 103, 104
Kubo, S., 260, 267, 282
Kull, W.F.E., 41, 49
Kuntze, H.B., 65, 66, 123

Lam, T., 258, 281
Lang, G. M., 60, 64
Laurance, N.L., 116, 117, 118, 123, 124
Leedhan, H.C., 471, 472
Lehmann, S., 260, 282
Lemaire, A., 343, 346
Le Pera, R., 103, 123
Leutzbach, W., 361, 368
Levine, S.T., 47, 49
Levine, W.S., 381, 410
Lex Systems, Inc., 292, 297
Lindberg, A.L., 66, 123
Linde, H., 263, 274, 281
Lion, P.M., 386, 410
Liopiros, K.J., 395, 410
Lisitsyn, V.M., 321
Little, J.D.C., 93, 94, 95, 122, 123
Lobsniger, D.I., 393, 410
Loder, J.L., 25, 26, 361, 366, 368
Looze, D.P., 101, 123
Lott, R.S., 140, 143, 147
LTV, 414, 444, 448
Lukas, M.P., 386, 395, 402, 406, 408
Lutman, P., 292, 297

MacGean, T.J., 392, 393, 410
MacKinnon, D.D., 355, 368, 373, 393, 410, 418, 444, 448, 471, 472, 473, 478, 482
Macomber, H.L., 393, 411
Magnien, C., 343, 346
Mahoney, M.A., 66, 123
Majou, J., 343, 344, 346
Maksimov, V.M., 321
Maleyev, V.V., 321
Mammano, F.J., 60, 64
Mampey, R., 260, 266, 282
Mangrulkar, S.M., 118, 124
Manlow, M.J., 248, 282
Marcus, A.H., 13, 14, 18
Marsh, B.W., 79, 123
Marten, F., 418, 445, 447, 448
Martin, B.V., 86, 93, 94, 95, 121, 123
Masher, D.P., 14, 18, 65, 111, 123
Mason, F.J., 248, 282
Mason, R., 413, 415, 416, 417
Massachusetts Institute of Technology, 292
Mathson, R., 260, 272, 273, 282
Matsumoto, S., 375, 411
Matsunobu, M., 366, 368
Maxwell, W.W., 343, 346
May, A.D., 60, 64, 65, 87, 94, 95, 123, 124
McCasland, W.R., 111, 124, 187, 188, 193
McCoy, J., 251, 252, 253, 282
McDermott, J.M., 111, 124
McEvaddy, P.I., 386, 395, 410
McGlynn, D.R., 42, 43, 44, 45, 49, 255
Meditch, J.S., 75, 124
Mertens, F.H., 258, 260, 282
Messer, G.J., 111, 124, 196
Meyer, H., 292, 293, 294, 297
Mies, A., 343, 344, 346
Miller, A.J., 86, 87, 88, 124
Miller, C.A., 392, 393, 410
Mishina, T., 162, 170
Mitchell, G., 139, 147
MITI, 22, 23, 26, 60, 61, 62, 63, 64, 78, 223
Morag, D., 386, 410
Morgan, J.T., 93, 94, 95, 123
Morris, R.W.J., 86, 121
Morse, A.S., 395, 410
Moyer, D.F., 118, 124
MPC, 416, 417

Müller, S., 418, 445, 447, 448
Münchrath, R., 41, 49
Mumford, J.R., 248, 282
Munson, A.V., 395, 410
Myojin, S., 197, 205

Nahi, N.E., 102, 124
Nakahara, T., 93, 124
Nance, D.K., 25, 26, 362, 368
Navin, F.P.D., 292, 296, 362, 368
Neininger, G., 114, 124
Nekahara, T., 471, 473
Nenzi, R., 103, 123
Neuberger, H., 9, 18
Neubert, H., 47, 49
New Developments in Ticket Issuing, 275, 282
Newell, G.F., 86, 87, 121, 122, 124
Newman, L., 111, 124
New Scientist, 311, 313
Nguyen, S., 72, 124
Nickel, B., 418, 445, 448
Nijmeyer, J., 79, 84, 85, 86, 88, 121
Nilsson, L.E., 79, 85, 86, 88, 90, 91, 92, 120
NTS, 413, 416, 417
NVP, 471, 473

OECD, 60, 64
Oetker, R., 39, 48
Okada, Y., 275, 277, 283
Oku, T., 413, 416, 417, 418, 445, 446, 448
Okura, I., 13, 18
Oliver, B.W., 248, 282
Olson, C.L., 376, 378, 409, 471, 472
Olson, K.W., 366, 368
Ontario Department of Transportation and Communication, 292, 297
Oom, R., 355, 368
Orlhac, D., 102, 124
Ormsby, J.L., 386, 409
Oshima, Y., 375, 411
Osmer, W.D., 446, 447
Oswald, R.S., 118, 124
OTA, 355, 357, 361, 368, 413, 414, 415, 416, 417, 444, 448, 470, 471, 472, 473, 475, 476, 478, 482
Owens, D., 145, 147
Oxley, P.R., 292, 297

Pählig, K., 114, 124
Papworth, G.R., 343, 346
Parker, H., 85, 125
Paulignan, J.F., 260, 266, 282
Pavlenko, G.P., 176
Payne, H.J., 98, 99, 101, 102, 106, 107, 123, 124
Peat Marwick Livingston & Co., 142, 147
Pechersky, M.P., 176
Peckmann, M., 239, 246, 248, 249, 292, 294, 297
van Peeteren, H., 79, 84, 85, 86, 88, 121
Penoyre, S., 366, 368
Perrin, J.-P., 256, 311, 312, 313
Pierce, J.R., 14, 18
Pignatoro, L.J., 111, 126
Pilsack, D., 70, 122
Pilsack, O., 60, 64
Pinnell, C., 111, 124
Pins, F., 343, 344, 346
Pitts, G.L., 392, 393, 410
Pitzinger, P., 65, 124
Plotkin, S.C., 366, 368
Pöschel, 445, 447, 448
Police Department of the Osaka Prefecture, 162, 168, 170
Potts, R.B., 86, 87, 121
Powner, E.T., 381, 386, 397, 408, 411
Prabhaker, R., 118, 124

Quonten, R., 343, 346

Rach, L., 65, 93, 95, 96, 124, 125
Rackoff, N.J., 114, 125
Radtke, Th., 114, 125
RAE, 471, 473
Raemer, R., 381, 409
Rahimi, A., 393, 411
Rahn, 418, 445, 447
Ralite, J.G., 413, 416, 417
Raus, I., 15, 17
Rebibo, K.K., 248, 282
Reyling, G., Jr., 47, 49
RGI, 413, 416, 417
de la Ricci, S., 93, 96, 126
Ritter, S., 251, 252, 253, 282
Robertson, D.J., 7, 18, 65, 81, 83, 93, 95, 96, 122, 125, 142, 145, 147
Robinson, D., 245, 282
Rockwell, T.H., 114, 125

Roesler, W.J., 402, 408, 409, 411
von Rohr, J., 275, 282
Romig, S.H., 355, 361, 366, 367
Roos, D., 292
Rosen, D.A., 60, 64, 109, 111, 126
Ross, D.W., 65, 88, 93, 96, 125
Ross, H.R., 25, 26
Rossberg, R.R., 78, 125
Roth, S.H., 251, 282
Rothery, R., 95, 122, 258, 281
Rottenburg, J., 275, 282
RRL, 60, 64, 93, 125
Rubin, F., 402, 407, 411
Rumsey, A.F., 386, 397, 411
Runkel, 471, 473
Russell, W., 418, 445, 448

Saidenberg, Ya.I., 176
Salwen, H., 251, 281
Sarachik, P.E., 395, 411
Sasaki, T., 71, 125, 197, 205
Sassmannshausen, G., 343, 344, 346
Sato, T., 162, 170
Savage, M.J., 260, 281, 343, 344, 345, 346
Sawaragi, T., 14, 18
Schenk, O., 260, 282, 336, 343
Schindler, 445, 448
Schlaefli, J.L., 65, 125
Schmitt, A., 393, 408
Schnabel, W., 65, 87, 93, 94, 125
Schwedrat, K., 292, 293, 297
Scott, E., 85, 125
Sealbury, T., 79, 85, 125
Shaw, L., 72, 73, 74, 78, 121
Shea, R., 311, 312, 314
Shefer, J., 114, 125
Sher, N.C., 362, 368
Shields, C.B., 25, 26
Shima, M., 43, 49
Shinkarev, N.I., 176
Siemens, 79, 85, 125
SIGOP, 93, 95, 125
Simpkins, B.D., 292, 296
Slevin, R., 235, 240, 292, 297
Smith, P.G., 366, 368
Snider, J.N., 114, 125
Snitter, A., 311, 313
Southall, A., 245, 282
Special Report on Automative Electronics 1973, 118, 125
Spies, G., 66, 121

Spring, A., 15, 18
Stablo, J., 343, 346
Stanford, M.R., 85, 125
Stapff, A., 343
Starr, S.H., 381, 411
Stefanek, R.C., 366, 368, 395, 398, 409, 411
von Stein, W., 79, 125
Steinfeld, H., 343, 346
Stepner, E.E., 393, 411
Stockfisch, C.R., 65, 125
Stockton, R., 193
Strobel, H., 35, 49, 65, 74, 75, 76, 112, 121, 125, 260, 266, 267, 268, 270, 282, 375, 411
Stupp, G.B., 381, 409
Sue, W., 86, 126
Sulzer, R.E., 65, 124
Suwe, K.-H., 336, 337, 338, 343
Symes, D.J., 251, 253, 283
Szeto, M.W., 102, 125

Tabak, D., 35, 49, 65, 101, 125
Takaoka, H., 376, 378, 388, 410
Takaoka, T., 260, 283
Takashi, I., 366, 368
Takehara, J., 275, 277, 283
Takemura, S., 343, 344, 346
Tanaka, A., 93, 124
Tanizawa, T., 292, 297
Tappert, H., 471, 473
Teener, M.D., 102, 106, 124
Temple, R.H., 45, 49, 118, 126
Theis, D.J., 41, 43, 45, 49
Thomas, K.B., 66, 121
Thomas, T.H., 260, 281, 380, 381, 411
Thompson, W.A., 99, 124
Tomizuka, M., 386, 393, 411
Tong, Y.M., 395, 410
Toyota, 7, 13, 14, 15, 18, 60, 64
Traffic Research Corporation, 142, 148
Travis, T.E., 395, 410
TRB, 366, 368
Trivedi, A.N., 102, 124
TRRL, 258, 283
True, J., 109, 111, 126
Tsuchiya, H., 479, 482
TÜ, 471, 473
Turner, E.D., 142, 147

Ullman, W., 65, 112, 114, 115, 116, 125, 126, 256, 283
Ullrich, J., 66, 123
Ullrich, S., 13, 18
Urabe, M., 260, 267, 269, 283
Usmanov, F.G., 176
Uyanik, A., 418, 445, 448

Valdes, A., 93, 96, 126
Vasilyev, A.P., 176
de Veer, H., 13, 18
Vesval, Y., 71, 120
Vienna U-Bahn, 344, 346
Vincent, R.A., 142, 143, 145, 148, 258, 260, 281, 283
Vlaanderen, A., 79, 85, 123
Vollenwyder, K., 260, 267, 283
Volpe, J.A., 15, 18, 25, 26

Waddell, M.C., 402, 408, 409, 411
Ward, J.D., 14, 17, 18
Washington Subway Automates, 343, 344, 346
Watanabe, J., 260, 267, 269, 283, 343, 344, 346
Watkins, L.H., 13, 14, 18
Watson, B.K., 86, 126
Wattleworth, J.A., 111, 124
Wattleworth, J.S., 111, 126
Webster, F.V., 86, 87, 96, 126, 142, 148
Wehner, L., 333, 334, 338, 343
Weinberg, M.F., 88, 126
Weiss, C.D., 79, 86, 126
Weiss, G.H., 13, 18
Welsh, L.E., 99, 102, 121
Wetherbee, J.K., 25, 26
Wheat, M.H., 311, 312, 314
Wheele, D.W.E., 141, 148
White, B., 196
Whiting, P.G., 144, 148
Whitney, D.E., 386, 393, 411
Whitson, R.H., 187, 193
Wienand, K., 66, 78, 123
Wienecke, 418, 445, 447, 448
Wiener, R., 111, 126
Wilkie, D.F., 386, 395, 398, 411
Williams, D.A.B., 143, 145, 147, 148
Williams, M.B., 402, 408, 411
Williams, T.J., 39, 45, 49, 50
Wilson, N.H.M., 292
Windolph, J., 13, 14, 18

Winkler, A., 260, 267, 268, 270, 281, 283
Wittman, E., 292, 294, 297
Wocher, B., 114, 126
Wolf, H., 42, 50
Wolff, C.H., 147, 149
Woolcock, M., 140, 148
Worrall, R.D., 66, 122
Wüchner, E., 114, 126
Wunderlich, W., 260, 266, 283

Yagoda, H.N., 111, 126

Yamamoto, I., 260, 283
Yan, G., 86, 126
York, H.L., 395, 411
Yumoto, N., 93, 124

Zajkowski, M.M., 66, 126
Zames, G., 355, 368
Ziegler, E., 292
Ziegler, M., 66, 126
Zimdahl, W., 375, 411
Zöller, H.-I., 337, 338, 343

Subject Index

A *t* following a page number indicates a table. Page numbers in italic type indicate figures.

Acceleration
  in the optimum driving regime for trains, 269
Accidents
  frequency of as a function of traffic flow conditions, 12
Aerospace-PRT system
  PRT system in California, 472
AFC
  automatic fare collection 351, 265*t*
AGT systems
  automated guideway transit, 355, 362, *363*
  classification of, *358*
  combined with dial-a-ride, 357, 362, *363*
Airport ground transportation
  SLT systems installed as, 412
Airtrans
  aerial view of, *419*
  control center at, *428*
  control system of, 425
  GRT system installed at Dallas–Fort Worth Airport, Texas, 418
  GRT system, operational experiences with, 427
  guideway configuration of, and passenger routes, *422, 423*
  utility vehicle, *424*

ALI
  Autofahrer-Lenkungs und Informationssystem, 60
  use of for national freeway network of the FRG, 70
API
  automatic passenger information, 265*t*
ARAMIS
  PRT system in France, 470
Area traffic control systems
  technical, efficiency, economic and institutional issues, 223, 224, 225
ARS
  automated speed control and train protection, 319
Arterial road control
  at arterial urban streets, freeway sections, long tunnels, and bridges, 58
ASCOT
  adaptive signal control optimization technique, 96, 133, 225
ASU-PP system
  automated management and control system used in USSR for rapid rail transit control, 315
  structure of, *317*
ATC

automatic train control, 265t
ATO
   automatic train operation, 265t, 350
ATP
   automatic train protection, 265t, 271, 350
ATS
   automatic train supervision, 265t, 350
Automated control
   of urban transportation, hierarchy of tasks in, 369, *371*
Automated highway concept
   basic systems concept, *57*, 63
   in dual-mode systems, 366
Automatic stopping
   of trains at stations, *271*
Automatic vehicle guidance
   lateral, 373, *374*
   longitudinal: headway regulation, 381
   longitudinal: speed and position control, 375
Automatic vehicle monitoring
   (AVM) system in Dublin, 301
Automatic vehicle operation
   (AVO) in Airtrans GRT system, 426
Automatic vehicle protection
   (AVP) in Airtrans GRT system, 426
Automation subsystem
   in PRT system, Cabintaxi, 463
Automobile accident
   risk of being killed or injured in, 1
Automobiles
   air and noise pollution by, 14
   number of as a function of time, *11*
   risk of being killed by, 9
Avtovedeniye
   structure of train traffic control system, *318*

Bandwidth
   maximizing in traffic light control, *93*, 94
BART
   automated rapid rail transit system in San Francisco, 322
Bendix-Dashaveyor system
   GRT system in USA, 446
Bus monitoring
   case studies of, 298
   structure of the automatic system in Dublin, 302
   structure of the Hamburg system, 308
   survey of selected systems, *312*
   worldwide, 310
Bus priority system
   in Washington, 132
Buses/Trams
   computer-aided operation planning of, 236
   monitoring and control of, 249
   transit systems, technical, efficiency, and economic issues of, 348, 349

Cabintaxi
   automation and operation of, *467*
   automation subsystem in, 463
   essential vehicle characteristics of, *465*
   guideway characteristics of, *466*
   guideway subsystem in, 463
   PRT system in FRG, 461
   rigid track switch, *469*
   SLT system in the FRG, 416
   testing grounds for, *462*
   vehicle subsystem in, 463
CABTRACK
   PRT system in London, 472
CAC
   comprehensive automobile control, 60, 221
Carbon monoxide
   emission of by motor vehicles, 13
Cell changing process
   in quasisynchronous traffic control, *399*
CMOS technology
   for increasing computing speed and decreasing power dissipation, 43
COBSY
   computer-bus system in FRG, 293
Computer control
   centralized and distributed systems of, 45
Computer dispatching
   in dial-a-ride case studies, 290
Computerization
   broad application of control computers (mini- and low-cost), 41
   impacts of computer revolution, 39
   long-term strategies: total systems

innovations, 29
  of fare collection systems, 274, *276, 278, 280*
  of safeguarding systems for trains, 271
  of safeguarding systems with fail-safe features, 273
  of traffic control systems, contribution of to transport development strategies, *20*
  of vehicles, network and mode of operation, 30
  short-term strategies using operational innovations, 27
  various generations of and development of parameters over time, *40*
COMTRAC
  computer aided traffic control, 265*t*
Control algorithms
  for route control, 71
Control computers
  mini- and low-cost computers, 41
Control concepts
  applicability of, *36*
Controllers
  for single intersection control of urban streets, 91
Control systems
  various generations of, *46*
Control task hierarchy
  control criteria in, 34
  in transportation systems, 32
Control tasks
  hierarchy of, *33*
Control technology
  computer, of vehicles, network, and mode of operation, 30
Coordinated control
  of traffic-light controlled intersections in urban streets, 92, *93*
CTC
  centralized traffic control, 265*t*
CVS
  computer-controlled vehicle system (PRT) in Japan, 449
  experimental site, showing guideway, *451*
  guidance and switching mechanism of, *459*

human factors involved in, 460
moving target system used in, *455*
off-line birth, *454*
position/speed control system for, *455*
public acceptance of, 460
schematic of communication system in, *457*
vehicle characteristics of, 452*t*
wagon and dual-mode car on guideways, *453*

Demand-responsive para-transit
  basic modes of operation of, *244*
Detectors
  for single intersection control of urban streets, expenses of, 91
Dial-a-ride
  and computerized para-transit, 237
  basic control system structures of, 245, *247*
  basic operating principles of, 243
  case studies of, 284
  chronological survey of dial-a-ride systems with more than eight vehicles, 293*t*
  combined with AGT, 357, 362, *363*
  control and dispatching methods in, 248
  frequency distribution of area served per system, *295*
  frequency distribution of numbers of vehicles and inhabitants served per system, *295*
  route guidance in, 242
  tehnical, social, economic, and institutional issues of, 347, 348
Dispatching control
  of trains, 265
Dispatching methods
  and control in dial-a-ride systems, 248
  and control of buses/trams, 250
Driver displays
  onboard, route guidance by, *69*, 70
Dual-mode concept
  automated highways in, 366
  combination of conventional street traffic systems and AGT systems in, 357, 364
  survey of, *365*

use of electric cars in, 366
use of taxis and ordinary street vehicles in, 364
Dynamic optimization
and feedback control in ramp metering on freeways, 101

Electric cars
use of in the dual-mode concept, 366
Empty car disposition
and route planning, survey of methods for, *380*
in automated urban transportation, 407
Energy optimum operation
of trains, 267, *268, 270*
Engine control
by microcomputers, 116, *117*
EQUISAT
traffic responsive signal program, 96, 142

Fare collection
and passenger information, automatic, 274, *276, 278, 280*
Fatalities
frequency of, as a function of speed, 12
frequency of, as a function of traffic flow conditions, 12
relation of to increase in length of road network, 15
relationship of fatality rate to number of inhabitants and length of road network, *10*
risk of being killed as a result of an automobile accident, 9
Feedback control
and dynamic optimization in ramp metering on freeways, 101
Fixed-time control
at single intersections of urban streets, 85
Fixed-time plans
comparison of with traffic-responsive control methods, 96
traffic responsive modification of, 95
FLEXIPROG
vehicle-actuated flexible progressive system, for traffic light control, 96, 142

FORD GRT system
GRT system in USA, 446
Freeways
concept of incident detection on, *107*
control of traffic on by merging controls, 106
control of traffic on by ramp metering, 97
control of traffic on by speed and lane use control, 102, *103, 104*
control of traffic on by using reversible and reversed lanes, 104
merging control on, using open-loop and closed loop control of on-ramp vehicles, 108, 110
ramp metering, using static optimization and open-loop control, and dynamic optimization and feedback control, 101
traffic control systems on, technical, efficiency, economic, social, and institutional issues of, 226, 227, 228
traffic flow control on, 97
Fuel
consumption of as a function of acceleration and deceleration, 15
consumption of as a function of driving speed, 15
consumption of as a function of traffic congestion, 15
relationship of consumption of, to traffic flow conditions, traffic safety, and environmental quality, *13*
relation of consumption of, to travel speed, *16*

GOROD
computerized traffic control system in Alma-Ata, 174
GRT systems
capacity of, 360
characteristics of, 443*t*
group rapid transit, 357, 370
headways in, 360
off-line stations in, 360
passenger capacity of, 360
technical, social, transport, and economic issues in, 476

Guidance mechanism
   in CVS, a Japanese PRT system, *459*
Guidance network
   in PRT systems, 361
Guideways
   comparison of costs of, *478*
   configuration of in GRT system
      Airtrans, *424*
   switch operation in, *427*
Guideway subsystem
   in PRT system Cabintaxi, 463

H-Bahn
   GRT system in FRG, 446
Head-up displays
   headway control by means of, for
      vehicle control, 111
Headway barrier
   in PRT systems, 361
Headway control
   aided by microcomputers, 114, *115*
   asynchronous methods of, *385*
   details of control system, *379*
   for vehicle control, 111, *113, 115*
   survey of principles, 382
   synchronous methods of, *389*
   system with fail-safe features, *387*
   using head-up displays, 114, *115*
   using radar distance-warning system,
      114, *115*
Headway operation
   minimum: capacity versus safety for
      automatic longitudinal vehicle
      guidance, 390
Headway regulation
   asynchronous methods for automatic
      longitudinal vehicle guidance,
      381
   synchronous methods for automatic
      longitudinal vehicle guidance,
      381
Headways
   bus, dynamic instability of, *257*
   in GRT systems, 360
   minimum achievable with conven-
      tional block and long-loop
      track cable systems, *274*
Human factors
   involved in CVS, a Japanese PRT
      system, 460
Hydrocarbons
   emission of by motor vehicles, 13

Integrated Injection Logic ($I^2 L$)
   in fourth generation microprocessors,
      44
ISUMRUD
   computerized traffic control system
      in Moscow, 170

Junction control
   of traffic flow at intersections and
      freeway on-ramps, 58

Kompaktbahn
   AGT system in FRG, 447
KRT
   Kobe Rapid Transit, GRT system in
      Japan, 446

Land
   amount of reserved for parking and
      driving, 15
   ineffective use of, 15
Lane use control
   and speed, for controlling traffic on
      freeways, 102
Loop antenna
   routine guidance by onboard driver
      displays, 70
Loop system
   SLT, at Houston airport, Texas, 415

MAC
   multifunctional automobile com-
      munication system concept, 60
Many-to-many systems
   of route control, 75, *76*
Memory
   random access and read only, in
      combination with a micro-
      processor, 42
Merging control
   control of traffic on freeways by,
      106
   on freeways, using closed-loop
      control of on-ramp vehicles,
      110
   on freeways, using open-loop control
      of on-ramp vehicles, 108, *109*
MGRT
   control and communications system,
      440
   control system of, 439

Morgantown group rapid transit
system installed at Morgantown West Virginia, 431
passenger destination request, *437*
program organization in, *432*
stations, guideways, and power
supply system in, 433
vehicle characteristics of, 435*t*
vehicle dispatch in, *438*
vehicle subsystem in, *434*
Microcomputers
description and composition of, 42
for aiding headway control, 114, *115*
for controlling and surveilling
onboard vehicle tasks, *119*
for controlling engines, 116, *117*
in route guidance and control, 68
onboard, suitable for vehicle
monitoring systems, 255
onboard system for recording
operational data, *270*
use of in automatic ticket vending
machines for sales accounting,
276, 277, *278*, *280*
Microelectronics
impacts of on traffic control, and on
development of transportation
systems, 44
selected terms from, 43*t*
MINITRAM
GRT system in London, 472
Morgantown
GRT system installed between
university and Morgan in West
Virginia (see MGRT), 431
Moving target system
for CVS, a Japanese PRT system,
455

Network control
in urban street networks and
networks of urban freeways
and neighboring roads, 58
Network structure
in SLT systems, 359
Nitrogen oxides
emission of by motor vehicles, 14
NMOS technology
in second generation microprocessors, 42

Noise pollution
effect on by congestion and driving
speed, 14

Off-line stations
in GRT systems, 360
One-to-one systems
of route control, 72, *73*
On-line stations
in SLT systems, 359
Open-loop control
and static optimization in ramp
metering on freeways, 101

Para-transit systems
computerized, and dial-a-ride
concept, 237
demand-responsive, basic modes of
operation of, *244*
Passenger capacity
of GRT systems, 360
Passenger guidance
and service systems, technical,
economic, and attractiveness
issues, 351, 352
Passenger information
and fare collection, automatic, 274,
276, *278*, *280*
Passengers
guidance and information systems
for, 237
Passenger service system
computerized, for a large railway
station, *280*
PLIDENT
platoon identification method of
fully traffic responsive signal
plan generation, 96, 142
PMOS technology
in first generation microprocessors,
42
Pollution
air and noise, as a function of traffic
flow conditions, 13
relation of environmental quality to
traffic flow conditions, traffic
safety, and energy consumption, *13*
Position control
automatic, of AGT vehicles, 378
for cars in CVS, a Japanese PRT
system, *455*

Priority control
    for public transit, concepts of, *259*
    of public transit vehicles, 258
Private cars
    comparison of attractiveness of, over public transit, 21
PRT
    personal rapid transit, 357, 370
PRT systems
    costs, architectural aspects, operational advantages, passive safety measures, and individuality and crime in, 362
    guideway capacities and minimum headways for different safety factors, 394*t*
    guideway network in, 361
    headway barriers in, 361
    long-term impacts of, 479
    selected characteristics of, 471*t*
    technical, social, and economic issues, 477, 478
Public acceptance
    of CVS, Japanese PRT system, 460
Public transit
    advantages for the city from use of, 22
    effectiveness and attractiveness of, 16
    use of as a function of changing land-use patterns and use of private cars, 16, 17
Public transport
    automatic fare collection and passenger information, 274, *276, 278, 280*
    concepts of priority control for, *259*
    control concepts for, 37
    survey of basic urban transit control and surveillance concepts, *243*

Ramp metering
    for controlling traffic on freeways, 97
    for traffic control on Paris freeway corridors, 206
    traffic responsive, concept of, *98*
Random access memory
    in combination with a microprocessor, 42

Rapid rail systems
    computer controlled, 343
    computerized urban/suburban, survey of, 344
    technical, safety, and economic issues, 350, 351
Read only memory
    in combination with a microprocessor, 42
Resources
    ineffective consumption of, 14
RETAX
    computer-aided taxi-bus system in FRG, 294
Reversed lanes
    for controlling traffic on freeways, 104
RITA
    road information transmitted aurally, 60
Rohr-Monocab system
    GRT system in USA, 446
Route control
    control algorithms for, 71
    control criteria for, 71
    in automated urban transportation, principles of, *404*
    many-to-many systems, 75
    one-to-one systems, 72
Route guidance
    by changeable route signs, 66, *67, 68*
    by onboard driver displays, *69, 70*
    for traffic control on Paris freeway corridors, 206
    in automated urban transportation, 402
Route guidance displays
    onboard vehicles, 120
Route guidance systems
    technical, efficiency, economic, social, and institutional issues, 222, 223
Route planning
    and empty car disposition, survey of methods for, *408*
    and selection in automated urban transportation, 405, 406
RTOP
    real time optimization program of traffic signal control, 96, 225
RUFBUS
    'hail bus' system in FRG, 294

Safeguarding systems
    and signalling systems for trains, *263*
    computerized, for trains, 271
    computerized, with fail-safe features, *273*
Sales accounting
    in automatic ticket vending machines, 274, *276*, *278*, *280*
Schottky bipolar technology
    in third generation microprocessors, 43
SECAMA
    experimental automatic bus monitoring system in Paris, 311
Self-optimizing control
    at single intersections of urban streets, 87, *88, 89, 90*
Shuttle line
    in SLT systems, 359
Shuttle system
    SLT, at Tampa airport, Florida, 414
Signalling systems
    and safeguarding systems for trains, *263*
Signal plan generation
    fully traffic responsive, 96
SIGOP
    traffic signal optimization program, 95, 133, 142, 225
Single intersection control
    categories of for urban streets, 85, *86*
Single loop
    in SLT systems, 359
SLT
    shuttle loop transit, 357, 370
SLT systems
    combined shuttle and loop system at Seattle–Tacoma airport, Washington, 415
    in airport ground transportation systems, 412
    in urban cities, 415
    network structure in, 359, 370
    on-line stations in, 359
    shuttle line, 359
    single loop in, 359
    survey of systems in operation, 413*t*
    technical, social, and economic issues, 475
Speed control
    automatic, of AGT vehicles, 375
    for traffic control on Paris freeway corridors, 206
    in San Francisco BART system, *326*
    system for, basic structure of, *376, 377*
START
    computerized traffic control system in Moscow, 170
Static optimization
    and open-loop control in ramp metering on freeways, 99
Station stopping
    in San Francisco BART system, *327*
    precise, in San Francisco case study, 323
Stopping
    automatic, of trains at stations, *271*
Switching mechanism
    in CVS, a Japanese PRT system, *459*
Systems capacity
    of GRT systems, 360
System concepts
    man–machine–environment system, 53, *54*

Taxis
    and ordinary street vehicles, use of in dual-mode concept, 364
Traffic congestion
    effect of on accident frequency, 12
    effect of on air and noise pollution, 13
    effect of on traffic throughput and travel speed, 7
    effect of on travel time, 9
    time losses due to, 7
Traffic control
    by better temporal and spatial distribution of demand, 214
    by computerization of vehicles, network, and mode of operation, 30
    hierarchically structured system for, 35
    long-term strategies for, using new technology options, 29
    objectives and resulting functions of traffic control, 55
    of buses/trams, 236
    on freeways using ramp metering, 97
    short-term strategies using operational innovations, 27

simplified picture of, *28*
Traffic control centers
  Osaka, 169
  Tokyo, 148
Traffic control systems
  area traffic control, and freeway traffic surveillance and control, 58, 59
  computerized, GOROD in Alma-Ata, 174
  computerized, ISUMRUD and START in Moscow, 170
  computerized signal, basic concepts of, *59*
  coordination between, and urban management systems, *62*
Traffic corridor control
  in Dallas, 183
  in Paris, 205
Traffic density
  relation of to traffic volume and traffic-flow speed, 7
Traffic detectors
  inductive loop, 58
Traffic flow
  relation of to free-flow speed and jam concentrations, *8*
  relation of to traffic safety, environmental quality, and energy consumption, *12*
Traffic flow control
  asynchronous, in automated urban transportation, 401
  concept of distributed freeway inflow control, *99*
  in tunnels, *100*
  in urban streets, 78
  of automobiles, concepts and methods of, *67*
  on freeways, 97
  principles of, *396*
  quasisynchronous in automated urban transportation, 398
  synchronous in automated urban transportation, 395
Traffic-flow speed
  relation of to traffic volume and traffic density, 7
Traffic light control
  comparison of various control modes from field tests, 92*t*
  concepts of, for urban streets, 80
  systems cost for a centralized computer control system for 100 intersections, *83*
  systems structures of, for urban streets, 81, *82*
  using bandwidth maximization, *93, 94*
Traffic-responsive control
  comparison of with fixed-time plans, 96
Traffic safety
  risk of being killed as a result of an automobile accident, 9
Traffic signal control
  optimization of using TRANSYT, SIGOP, UTCS-1, and SIGRID, 95
  traffic lights and changeable route signs, speed indicators, 58
  various systems, 56
Traffic space
  as a function of time, *11*
Traffic supervision
  of trains, 236
Traffic throughput
  reduction in due to congestion, 7
Traffic volume
  relation of to traffic-flow speed and traffic density, 7
Trains
  automatic stopping of at stations, *271*
  basic safeguarding and signalling systems for, *263*
  centralized supervisory control of, 262
  computer-aided operation of, 266
  computer-aided operation planning of, 235
  control of in the automatic Munich system, 338, *339*
  detection and speed control of in San Francisco case study, 323
  dispatching control of, 265
  energy optimum operation of, 267, *268, 270*
  identification of in San Francisco BART system, 323, *328*
  operation and traffic control of, 260
  precise stopping of at stations, 269
  speed control of, 269
  survey of control and surveillance

tasks for, *261*
Transport
  and urban development, general concepts of, 19
Transport control
  concepts of, 32
Transport development
  objectives of, 19
  options for, 22
Transportation
  constraints on supply and demand, 24, 25
  demand for, and control of, 22, 23
  supply and control of, 22, 24
  supply and demand policies, long and short term 22, 23
Transportation systems
  control task hierarchy in, 32
  decision levels in management and control of, 32
  dominating system features of, 31
  new, completely automated, 38
Transurbahn
  GRT system in FRG, 447
TRANSYT
  for optimizing computing signal offsets, 95, 133, 142, 223, 225
  traffic network study tool, 95, 96, 142
Travel speed
  reduction in due to congestion, 7
  relation of to fuel consumption, *16*
Trip destination storage
  onboard vehicles, 120
TTI system
  GRT system in USA, 446

UART
  universal asynchronous receiver transmitter (LSI chip), suitable for onboard vehicle monitoring systems, *255*
Urban streets
  categories of single intersection control in, 85
  concepts of traffic light control in, 80
  traffic flow control in, 78
  traffic light control systems structures for, 81
Urban transportation

  hierarchy of tasks in automated control of, *369*, *371*
  new modes of, 359*t*
  new modes of, survey of basic systems concepts for, *356*
Urban transport systems
  morphological review of, *464*
UTCS-1
  urban traffic control system, 95, 128, 225
  experiences using, 136
Utility vehicle
  in GRT system Airtrans, *424*

VAL
  SLT system in Lille, France, 416
Vehicle-actuated control
  at single intersections of urban streets, 87
Vehicle characteristics
  of Cabintaxi, a PRT system in FRG, *465*
  of CVS, a Japanese PRT system, 452*t*
Vehicle control
  by headways, using head-up displays, 114
  technical, efficiency, economic, social, and institutional issues of, 229, 230
  using headway control, 111
Vehicle guidance
  automatic, lateral, 373, *374*
  automatic, longitudinal: headway regulation, 381
  automatic, longitudinal: speed and position control, 375
Vehicle location
  automatic, various methods of, *252*
Vehicle monitoring
  and dispatching control, automatic, basic scheme of, 250
  automatic, in case studies, 301
  automatic, of buses/trams, 250
  based on centralized or distributed computing systems, *254*
Vehicle subsystem
  in PRT system Cabintaxi, 463
VONA
  SLT system in Chiba, Japan, 416